Biology of Insect Eggs

IN THREE VOLUMES

Volume III

Biology of Insect Eggs

IN THREE VOLUMES

H. E. HINTON, F.R.S.

Late Professor and Head of the Department of Zoology, University of Bristol

Volume III

PERGAMON PRESS

OXFORD · NEW YORK · TORONTO · SYDNEY · PARIS · FRANKFURT

U.K.	Pergamon Press Ltd., Headington Hill Hall, Oxford OX3 0BW, England
U.S.A.	Pergamon Press Inc., Maxwell House, Fairview Park, Elmsford, New York 10523, U.S.A.
CANADA	Pergamon of Canada, Suite 104, 150 Consumers Road, Willowdale, Ontario M2J 1P9, Canada
AUSTRALIA	Pergamon Press (Aust.) Pty. Ltd., P.O. Box 544, Potts Point, N.S.W. 2011, Australia
FRANCE	Pergamon Press SARL, 24 rue des Ecoles, 75240 Paris, Cedex 05, France
FEDERAL REPUBLIC OF GERMANY	Pergamon Press GmbH, 6242 Kronberg-Taunus, Hammerweg 6, Federal Republic of Germany

Copyright © 1981 Pergamon Press Ltd.

All Rights Reserved. No part of this publication may be reproduced, stored in a retrieval system or transmitted in any form or by any means: electronic, electrostatic, magnetic tape, mechanical, photocopying, recording or otherwise, without permission in writing from the publishers

First edition 1981

British Library Cataloguing in Publication Data

Hinton, Howard Everest
Biology of insect eggs.
1. Insects – Eggs
I. Title
595.7'03'34   QL495.5   77-30390

ISBN 0-08-021539-4

Printed in Great Britain by A. Wheaton & Co. Ltd, Exeter

Contents

Volume III

BIBLIOGRAPHY	779
SPECIES INDEX	1001
AUTHOR INDEX	1053
SUBJECT INDEX	1103
PROFESSOR HINTON'S PUBLICATIONS, 1930–1977	1113

Volume I

ACKNOWLEDGEMENTS   xxiv

1. Introduction   1

Size of eggs	1
The kinds of metamorphosis	5
Apolysis and ecdysis	7
Arrangement of orders of insects	9

2. Number of Eggs   11

Oösorption	29
Effects of temperature	31
Fecundity and weight of pupae or pharate adults	35
Fecundity and weight of adult	36
Fecundity and larval density	38
Fecundity and adult density	41
Fecundity and longevity	42
Fecundity and mating	43
(1) *Stimulation of oviposition*	43
(2) *Fecundity and frequency of mating*	47
(3) *Duration of copulation*	49

3. Oviposition   51

Diversity of oviposition sites	51
Parasites that oviposit on their hosts while the latter are flying	56
Non-parasitic insects that lay on other insects or animals	57

Seasonal variation in selection of oviposition sites	58
Intraspecific variation in choice of oviposition site	59
Effects of temperature and humidity	60
Temperature	60
Humidity	61
Oviposition responses	63
Group oviposition and aggregation pheromones	63
Aggregation pheromones of the scolytid beetle Ips	65
Aggregation pheromones of the scolytid beetle Dendroctonus *and other insects*	66
Oviposition response to secondary plant substances	71
Oviposition affected by previous attack on plant	73
Attraction of parasite to odour of the food plant of its host	73
Attraction of parasite to sex pheromone of host	74
Effect of salts on selection of oviposition site	75
Oviposition stimulated by vibrations of host	75
Oviposition stimulated by sound of host	76
Oviposition and perception of quantity of food	76
Height preferences for oviposition	76
Oviposition and stem diameter	77
Oviposition and curvature of surface	77
Oviposition and size of wood vessels	77
Oviposition and surface texture	78
Oviposition and colour	78
Oviposition and light intensity	79
Oviposition and time of day	80
Circadian oviposition rhythms	81
Changes in response to stimuli associated with oviposition	82
Addendum	82
Resistance of plants to oviposition	86
Resistance to eggs of parasites	89
Sex of egg dependent on type of host chosen	94

4. Respiratory Systems   95

Shell with an air-containing meshwork	96
Gas filling of the chorionic respiratory system	97
Pressure of gases in chorionic respiratory system	98
"Primitive" intrachorionic meshworks	98
Specialized kinds of intrachorionic meshworks	100
Stalked aeropyles	102
Horns adapted only for atmospheric respiration	105
Compressible or shrinking physical gill	106
Plastron respiration	108
The plastron and the environment	111
The kinds of chorionic plastrons	112
Group I	113
Group II	115
Group III	125
Independent evolution of plastron-bearing horns	128
Hemiptera–Homoptera	130

Hemiptera–Heteroptera	130
Hymenoptera	132
Diptera	132
Utilization by larva of chorionic respiratory system	142
Subsidiary functions of air in plastron meshworks	144
Protective coloration by plastron	146
Plastron respiration of egg cocoons	147

5. Respiratory Efficiency of Egg Plastrons — 149

Resistance to wetting by hydrostatic pressures	149
Resistance to loss of waterproofing by surface active substances	150
Rate of oxygen uptake and area of plastron	151
Drop in pressure along the plastron	159

6. Hydropyles and Water Relations — 163

Hydropyles	164
Serosal hydropyles	166
Serosal cuticle hydropyles	169
Chorionic hydropyles	170
Absorption of liquid water	171
Resistance to desiccation and effects of relative humidity	176
Lipid layers	180
Metabolic water	182

7. Colleterial Glands — 183

Dictyoptera and Orthoptera	184
Hydrophiloidea	186
Chrysomelidae–Cassidinae	188
Trichoptera	192
Lepidoptera	196

8. Oöthecal and Shell Proteins — 201

Oöthecal proteins	201
Endocrine control of oöthecal secretions	204
Shell proteins	205
Helicoidal microfibrils in the chorion and serosal cuticle	209
Addenda	209
Hardening of oötheca	209
Shell proteins	210
Fine structure of shell	210

9. Enemies of Eggs — 211

Predators	211
Cannibalism	224
Myrmecophiles and termitophiles	226
Annotations and additions to the list of egg predators	226
Acari	226
Orthoptera	228

Dermaptera	228
Psocoptera	228
Thysanoptera	228
Hemiptera	228
Miridae	228
Nabidae	228
Anthocoridae	228
Reduviidae	229
Lygaeidae	229
Berytidae	229
Pentatomidae	229
Neuroptera	229
Coleoptera	230
Carabidae	230
Histeridae	230
Staphylinidae	230
Malachiidae	231
Melyridae (s. stricto)	231
Dermestidae	231
Rhizophagidae	231
Coccinellidae	231
Tenebrionidae	232
Meloidae	232
Hymenoptera	233
Formicidae	234
Lepidoptera	235
Cosmopterygidae	235
Blastobasidae	235
Heliodinidae	235
Tortricidae	235
Pyralidae	236
Noctuidae	236
Arctiidae	236
Diptera	236
Cecidomyiidae	236
Dolichopodidae	236
Bombyliidae	236
Syrphidae	237
Drosophilidae	237
Chloropidae	237
Chamaemyiidae	237
Sciomyzidae	237
Mollusca	237
Vertebrates	237
Parasites	238

10. Defensive Devices   240

Natural deception	240
Resemblance to an egg from which a parasite has emerged	241

Egg resembles background when spumaline remains sticky	242
Egg transparent and so resembles substrate	242
Resemblance to plant tendrils	242
Resemblance to a leaf gall	243
Resemblance to plant seeds	243
Disruptive coloration	247
Colour changes during development	249
Poisonous eggs	250
Defensive fluid of maternal origin on mosquito eggs	251
Poisonous larval setae used to protect eggs of next generation	256
Lymantriidae	256
Thaumetopoeidae	257
Notodontidae	257
Hemileucidae (= subfamily of Saturniidae auct.)	258
Mechanical protection from non-poisonous setae or scales of females	258
Protection by a palisade of scales around egg-mass	258
Serosal cuticle affords protection against predators	260
Extra-embryonic fluid seals egg punctures with a tanned membrane	261
Protective devices against high temperatures	261
Distortion of the surface film	265

11. Parental Care — 269

Mass provisioning	275
Scarabaeidae	275
Curculionidae	278
Scolytidae	279
Classification of subsocial insects	283
Cockroaches	284
Other insects	285
Orthoptera	286
Dermaptera	287
Hemiptera	291
Membracidae	291
Defence against predators	291
Switch in behaviour when female begins to brood	291
Amount of disturbance tolerated by a brooding female	292
Intraspecific brood care	292
Coccidae	293
Aradidae	294
Cydnidae	294
Pentatomidae	294
Phloeidae	296
Tingidae	296
Reduviidae	296
Coleoptera	297
Carabidae	297
Spercheidae	297
Hydrophilidae	298
Silphidae	298

Staphylinidae	298
Scarabaeidae	301
Passalidae	303
Heteroceridae	303
Chrysomelidae	303
Scolytidae (= Ipidae)	304
Platypodidae	308
Hymenoptera	310

12. Techniques — 312

Methods of obtaining and handling eggs	312
Extraction from soil or food material	313
Implanting techniques	313
Sampling	313
Use of dogs to detect egg-masses by their pheromones	314
Automatic devices for recording distribution and oviposition	314
Egg traps in the field	314
Cold storage	314
Surface sterilization	314
Marking methods	315
Dyes	315
Isotopes	316

PLATES — 317

Volume II

13. Ephemeroptera — 475

Parthenogenesis	475
Oviposition	476
Structure of the chorion	477
Micropyles	477
Egg-bursters	483
Quiescence and diapause	485

14. Odonata — 486

Oviposition	486
Structure of the chorion	491
Micropyle	493
Spumaline	493
Hatching	494

15. Dictyoptera, Isoptera and Zoraptera — 495

Dictyoptera — 495
SUBORDER BLATTARIA — 495
  Reproduction — 495
  Oötheca — 495
  Structure of the chorion — 499
SUBORDER MANTODEA — 499
  Oöthecae — 499
  Formation of the oötheca — 504
  Structure of the chorion — 506
  Hatching — 506
Isoptera — 507
Zoraptera — 507

16. Plecoptera and Embioptera — 508

Plecoptera — 508
  Oviposition — 508
  Structure of the shell — 508
  Micropyles — 510
Embioptera — 512
  Oviposition — 512
  Structure of the chorion — 512

17. Orthoptera — 513

Gryllidae — 513
Tetrigidae — 514
Tettigoniidae — 515
  Oviposition — 515
  Structure of the chorion — 515
  Taxonomy — 516
Acrididae and related families — 518
  Oviposition — 518
  Structure of the chorion — 520
  Hatching — 522
  Taxonomy — 522
    Key to North American grasshopper eggs — 523
    Key to egg-pods of grasshoppers of southern Ghana — 526

18. Cheleutoptera — 533

Oviposition — 533
Incubation period — 533
Colleterial glands — 534
Structure of shell — 534
  Structure of unspecialized part of shell — 539
  Operculum — 539
  Micropylar plate — 542

19. Psocoptera — 546

Oviposition — 546
Structure of the chorion — 547
Micropyles — 547
Egg-bursters — 548

20. Phthiraptera — 549

Oviposition — 549
Structure of the chorion — 551
Chorionic hydropyle — 553
Micropyles — 554
Egg-bursters — 555
Spumaline — 555
Effect of temperature on the eggs — 556
Effect of rain on the eggs — 561

21. Thysanoptera — 563

Oviposition — 563
Structure of the chorion — 563
Micropyle — 564

22. Hemiptera — 565

HOMOPTERA — 565
COLEORRHYNCHA — 566
AUCHENORRHYNCHA — 566
Cicadoidea — 566
   Cicadidae — 566
   Membracidae — 566
   Jassidae (= Cicadellidae) — 566
Fulgoroidea — 566
   Delphacidae — 566
STERNORRHYNCHA — 567
Psylloidea — 567
   Psyllidae (= Chermidae) — 567
Aleyrodoidea — 569
Aphidoidea — 569
Coccoidea — 570
HETEROPTERA — 570
Leptopodoidea — 570
   Saldidae — 570
     Oviposition — 570
     Structure of the chorion — 571
     Micropyle — 572
   Leptopodidae — 573
   Omaniidae — 574
Amphibicorisae — 574
   Hebridae — 574
   Mesoveliidae — 575
   Hydrometridae — 577

CONTENTS

Oviposition	577
Structure of the chorion	577
Micropyle	579
Veliidae	579
Oviposition	579
Structure of the chorion	579
Micropyles	579
Gerridae	580
Oviposition	580
Structure of the chorion	581
Micropyles	581
Pentatomomorpha	582
Aradidae	582
Idiostolidae	582
Thaumastellidae	582
Piesmatidae	582
Malcidae	583
Berytinidae (= Neididae)	583
Lygaeidae	583
Stenocephalidae	584
Pyrrhocoridae	585
Largidae	585
Colobathristidae	585
Coreidae	585
Key to the eggs of the subfamilies of Coreidae	587
Alydidae	588
Rhopalidae	589
Cydnidae	589
Key to the eggs of the families of Pentatomoidea	590
Acanthosomatidae	591
Urostylidae	591
Scutelleridae	591
Key to the eggs of European Scutelleridae	591
Dinidoridae	593
Tessaratomidae	593
Pentatomidae	594
Key to the eggs of North American Pentatomidae	595
Key to the eggs of European Pentatomidae	596
Plataspidae (= Coptosomidae)	607
Aphylidae	607
Cimicomorpha	608
Nabidae	608
Key to the eggs of some British species of Nabidae	609
Micropyles	609
Velocipedidae	609
Pachynomidae	610
Anthocoridae	610
Microphysidae	610
Plokiophilidae	610
Tingidae (+ Vianaididae)	611

Miridae (= Capsidae)	611
Oviposition	611
Structure of the chorion	612
Micropyles	613
Thaumastocoridae	613
Reduviidae	614
Oviposition	614
Structure of the chorion	614
Micropyles	617
Hydrocorisae	617
Corixidae	617
Oviposition	617
Structure of the chorion	617
Micropyles	618
Nepidae	621
Oviposition	621
Respiratory horns	622
Structure of the chorion of the main body of the shell	630
Selective advantage of the aeropyles	632
Micropyles	633
Spumaline	633
Key to the eggs of the Nepidae	634
Belostomatidae	635
Oviposition	635
Structure of the chorion	636
Chorionic hydropyle	636
Micropyles	636
Naucoridae	636
Oviposition	636
Structure of the chorion	637
Micropyles	637
Notonectidae	637
Oviposition	637
Structure of the chorion	637
Micropyle	638
Pleidae	638
Oviposition	638
Structure of the chorion	638
Micropyle	638
Helotrephidae	638
Ochteridae	638
Gelastocoridae	639
Oviposition	639
Structure of the chorion	639
Micropyles	639
Dipsocoroidea	640
Dipsocoridae (= Cryptostemmatidae)	640
Schizopteridae	640
Enicocephaloidea	641
Enicocephalidae	641

23. Megaloptera and Neuroptera — 642

Megaloptera — 642
Neuroptera — 643
   Oviposition — 643
   Structure of the chorion — 643
   Composition of egg-stalk of Chrysopa — 644
   Micropyles — 645
   Egg-bursters — 646
   Protective devices — 647

24. Coleoptera — 649

SUBORDER ADEPHAGA — 649
   Carabidae — 649
   Cicindelidae — 650
   Dytiscidae — 650
   Haliplidae — 653
   Hygrobiidae (= Pelobiidae) — 654
   Gyrinidae — 654
SUBORDER MYXOPHAGA — 654
SUBORDER POLYPHAGA — 656
   Hydrophiloidea — 656
      Key to the eggs of Hydrophiloidea — 657
   Staphylinidae — 659
      Key to the eggs of Staphylininae — 659
   Coccinellidae — 675
   Meloidae — 676
   Chrysomelidae — 676
      Oviposition — 676
      Structure of the chorion — 681
      Egg-bursters — 681
   Bruchidae — 683
      Oviposition — 683
   Cerambycidae — 685
      Oviposition — 685
      Structure of the chorion — 687
      Egg-bursters — 687
Mass provisioning in the Coleoptera — 688
   Scarabaeidae — 688
   Curculionidae — 688
   Scolytidae — 688

25. Hymenoptera — 704

SUBORDER SYMPHYTA — 705
   Pamphilidae — 705
   Siricidae — 705
   Diprionidae — 705
   Pergidae — 705
   Tenthredinidae — 705

SUBORDER APOCRITA	706
Trigonalidae	706
Ichneumonidae	706
Braconidae	706
Cynipidae	707
Torymidae	707
Pteromalidae	707
Encyrtidae	707
Eulophidae	707
Trichogrammatidae	708
Mymaridae	708
Scelionidae	709
Dryinidae	709

26. Trichoptera   710

27. Lepidoptera   712

Oviposition	712
Structure of the chorion	714
Flat and upright eggs, shape, thickness of shell	716
Micropyles	718
Egg-bursters	719
Respiration of eggs when covered by water	719
Additional papers	720

28. Mecoptera   722

Oviposition	722
Structure of the chorion	723
Micropyles	723
Egg-bursters	723

29. Diptera   724

Tipulidae	724
Oviposition	724
Structure of the chorion	725
Culicidae	727
Descriptions	728
Palaearctic	729
Ethiopian	729
Indo-Australian	729
American	729
Racial differences in shell structure	729
Facultative structural polymorphism	730
Resistance to desiccation	732
Hatching stimulus	732
Bibionidae	733
Nymphomyiidae	734
Simuliidae	734

CONTENTS

Chironomidae	735
Ceratopogonidae	736
Rhagionidae (= *Leptidae*)	737
Tabanidae	738
Acroceridae (= *Cyrtidae*)	738
Nemestrinidae	739
Bombyliidae	739
Therevidae	739
Apioceridae	740
Asilidae	740
Empididae	740
Syrphidae	741
Tephritidae (= *Trypetidae*)	741
Psilidae	743
Sciomyzidae	743
Dryomyzidae	744
Braulidae	744
Coelopidae	745
Lauxaniidae (= *Sapromyzidae*)	745
Lonchaeidae	745
Pallopteridae	745
Aulacigastridae	745
Helomyzidae	746
Sphaeroceridae (= *Borboridae*)	746
Ephydridae	748
Gasterophilidae	749
Oestridae	750
Cuterebridae	751
Calliphoridae	751
Sarcophagidae	752
Rhinophoridae	752
Tachinidae	752
(1) *Macrotype*	754
(2) *Microtype*	755
(3) *Membranous*	755
(4) *Pedicellate*	756
Scatophagidae (= *Cordiluridae*)	756
Anthomyiidae	757
Muscidae	758
Additional papers	761

ADDENDUM: REPRODUCTIVE SYSTEMS AND MICROPYLAR APPARATUS   763

TAXONOMIC LIST OF SPECIES ILLUSTRATED IN PLATES OR LINE DRAWINGS IN VOLUMES I AND II   765

Bibliography

Professor Hinton died during the stages of completion of this book, leaving the compilation and checking of the Bibliography and indexes to me. Due to the obscure nature of some of the references it is possible that occasional errors may be found although every effort has been made to check their accuracy.

J. ABLETT

ABASA R. O. (1972) The Mediterranean fruit fly, *Ceratitis capitata* Wied.: laboratory investigations of its reproductive behaviour in *Coffea arabica* in Kenya. *E. Afr. agric. For. J.* **37**, 181–184.

ABBAS H. M. and ANWAR M. S. (1963) Breeding of *Corcyra cephalonica* Stn. for the mass production of *Trichogramma minutum* Riley. *Agric. Pakist.* **14**, 209–214.

ABBOTT C. E. (1938) The development and general biology of *Creophilus villosus* Grav. *Jl NY ent. Soc.* **46**, 49–52.

ABDEL-MALEK A. (1947) A study of the biology of *Chelonella sulcata*. Nees (Hymenoptera: Braconidae). *Ohio J. Sci.* **47**, 206–216.

ABDEL-MALEK A. (1948) Plant-hormones (auxins) as a factor in the hatching of *Aedes trivittatus* (Coquillett) eggs. *Ann. ent. Soc. Am.* **41**, 51–57.

ABDELRAHMAN I. (1974) Studies in ovipositional behaviour and control of sex in *Aphytis melinus* DeBach, a parasite of California red scale, *Aonidiella aurantii* (Mask.). *Aust. J. Zool.* **22**, 231–247.

ABDINBEKOVA A. A. and AKHMEDOV R. M. (1973) The diel rhythm of oviposition, hatching of the larvae and adult emergence of the cabbage moth (*Barathra brassicae*) in the laboratory and in the field. *Zool. Zh.* **52**, 143–146 (in Russian; English summary).

ABLES J. R. and SHEPARD M. (1976) Influence of temperature on oviposition by the parasites *Spalangia endius* and *Muscidifurax raptor*. *Envir. Ent.* **5**, 511–513.

ABRAHAM V. A. and KURIAN C. (1974) *Chelisoches moris* F. (Forficulidae: Dermaptera), a predator on eggs and early instar grubs of the red palm weevil *Rhynchophorus ferrugineus* F. (Curculionidae: Coleoptera). *Proc. 1st Natn. Symp. Plantation Crops* 1 (suppl.) 1973.

ABRAHAMSON L. P., CHU H.-M., and NORRIS D. M. (1967) Symbiontic interrelationships between microbes and ambrosia beetles: 2, The organs of microbial transport and perpetuation in *Trypodendron betulae* and *T. retusum* (Coleoptera: Scolytidae). *Ann. ent. Soc. Am.* **60**, 1107–1110.

ABUL-NASR S., EL-SHERIF S. I., and NAGUIB M. A. (1972) Oviposition behaviour of the cotton leaf-worm *Spodoptera littoralis* (Boisd.) in clover fields (Lepidoptera: Agrotidae). *Z. angew. Ent.* **70**, 310–314.

ACHTELIG M. and KRISTENSEN N. P. (1973) A re-examination of the relationships of the Raphidioptera (Insecta). *Z. zool. Syst. Evolutionsforschung* **11**, 268–274.

ADAIR E. W. (1912) Notes sur la ponte et l'éclosion de *Miomantis savignyi* (Sauss.). *Bull. Soc. ent. Égypte* **3**, 117–127.

ADAIR E. W. (1914) Notes préliminaires pour servir à l'étude des Mantidae. *Bull. Soc. ent. Égypte* **6**, 21–36 (1913).

ADAIR E. W. (1925) On parthenogenesis in *Miomantis savignyi* Sauss. *Bull. Soc. ent. Égypte* **8**, 104–148 (1924).

ADAM D. S. and WATSON T. F. (1971) Adult biology of *Exorista mella*. *Ann. ent. Soc. Am.* **64**, 146–149.

ADAMS C. H., CROSS W. H., and MITCHELL H. C. (1969) Biology of *Bracon mellitor* a parasite of the boll weevil. *J. econ. Ent.* **62**, 889–896.

ADAMS T. S. and NELSON D. R. (1968) Bioassay of crude extracts for the factor that prevents second matings in female *Musca domestica*. *Ann. ent. Soc. Am.* **61**, 112–116.

ADARSH H. S. and SOHI G. S. (1969) Effects of temperature and humidity on the development, survival, fecundity and longevity of *Callosobruchus analis* (F.). *J. Res. Punjab agric. Univ.* **6** (suppl.), 207–213.

ADASHKEVICH B. P. and KARELIN V. D. (1975) The rearing of syrphids. *Zashch. Rast.* **7**, 22–23 (in Russian).

ADIYODI K. G. (1968) Left colleterial proteins in the viviparous cockroach *Nauphoeta cinerea*. *J. Insect Physiol.* **14**, 309–316.

ADKISSON P. L. (1959) The effect of various humidity levels on hatchability of pink bollworm eggs. *J. Kans. ent. Soc.* **32**, 189–190.

ADKISSON P. L., BULL D. L., and ALLISON W. E. (1960) A comparison of certain artificial diets for laboratory cultures of the pink bollworm. *J. econ. Ent.* **53**, 791–793.

ADLAKHA V. and PILLAI M. K. K. (1975) Involvement of male accessory gland substance in the fertility of mosquitoes. *J. Insect. Physiol.* **21**, 1453–1455.

AESCHLIMANN J. P. (1974) A method for the extraction of *Sitona* (Col.: Curculionidae) eggs from soil and occurrence of a mymarid (Hym.: Chalcidoidea) in the Mediterranean region. *Entomophaga* **20**, 403–408.

AFIFY A. M. and FARGHALY H. T. (1971) Comparative laboratory studies on the effectiveness of *Labidura riparia* Pall. and *Coccinella undecimpunctata* Reiche, as predators of eggs and newly hatched larvae of *Spodoptera littoralis* (Boisd.). *Bull. Soc. ent. Égypte* **54**, 277–282.

AFZAL M. and GHANI M. A. (1953) Cotton Jassid in the Punjab. *Sci. Monogr. Pakistan Ass. Adv. Sci.* no. 2, viii + 102 pp.

AGAPOVA E. G. (1966) The rôle of food-plants in the differentiation of species of *Oscinella*. *Tr. Nauch-issled Inst. Zashch. Rast. UZB SSR* **26**, 185–192 (in Russian; English summary).

AGARWALA S. B. D (1952) A comparative study of the ovipositor in the Acrididae: I, II. *Indian J. Ent.* **13**, 147–181 (1951); **15**, 53–69.

AGEE H. R. (1969) Mating behaviour of bollworm moths. *Ann. ent. Soc. Am.* **62**, 1120–1122.

AHMAD R. (1970) Studies in West Pakistan on the biology of one Nitidulid species and two Coccinellid species (Coleoptera) that attack scale insects (Hom., Coccoidea). *Bull. ent. Res.* **60**, 5–16.

AHMAD R. and GHANI M. A. (1971) The biology of *Sticholotis marginalis* Kapur (Col., Coccinellidae). *Tech. Bull. Commonw. Inst. biol. Control* **14**, 91–95.

AHMAD T. (1936) The influence of ecological factors on the Mediterranean flour moth *Ephestia kuehniella* and its parasite *Nemeritis canescens*. *J. Anim. Ecol.* **5**, 67–93.

AHMED M. K. (1957) Life-history and feeding habits of *Paederus alfierii* Koch (Coleoptera: Staphylinidae). *Bull. Soc. ent. Égypte* **41**, 129–143.

AINSLIE C. N. (1910) The New Mexico range caterpillar. *Bull. US Dep. Agric.* **85**, 59–96.

AINSLIE G. G. (1922) Contributions to a knowledge of the Crambinae: II, *Crambus laqueatellus* Clemens. *Ann. ent. Soc. Am.* **15**, 125–136.

AITKEN T. H. G. (1948) Recovery of anopheline eggs from natural habitats, an aid to rapid survey work. *Ann. ent. Soc. Am.* **41**, 327–329.

AKEY D. H. and JONES J. C. (1968) Sexual responses of adult male *Aedes aegypti* using the forced-copulation technique. *Biol. Bull. Woods Hole* **135**, 445–453.

ALAYO D. (1974) Los Hemipteros acuaticos de Cuba. *Torreia* no. 36, 1–62.
ALBRECHT F. O., VERDIER M., and BLACKITH R. E. (1958) Détermination de la fertilité par l'effet de groupe chez le criquet migrateur (*Locusta migratoria migratoroiodes* R. et F.). *Bull. biol. Fr. Belg.* **92**, 349–427.
ALDEN C. H. and FARLINGER D. F. (1931) The artificial rearing and colonization of *Trichogramma minutum*. *J. econ. Ent.* **24**, 480–483.
ALDRICH J. M. (1912) The biology of some western species of the dipterous genus *Ephydra*. *Jl NY ent. Soc.* **20**, 77–99.
ALEXANDER A. J. (1961) A study of the biology and behaviour of the caterpillars, pupae and emerging butterflies of the subfamily Heliconiinae in Trinidad, West Indies. Part I. Some aspects of larval behaviour. *Zoologica NY* **46**, 1–24.
ALEXANDER C. P. (1920) The crane-flies of New York: Part II, Biology and phylogeny. *Mem. Cornell agric. Exp. Stn* **38**, 691–1133.
ALEXANDER R. D. (1961) Aggressiveness, territoriality, and sexual behaviour in field crickets (Orthoptera: Gryllidae). *Behaviour* **17**, 130–223.
ALFIERI A. (1921) Description de l'oothèque inconnu de l'*Heterogamia aegyptiaca*. *Bull. Soc. ent. Égypte* **13**, 38–39.
ALFORD D. V. and DOCKERTY A. (1974) The distribution of eggs of *Acleris comariana* (Lepidoptera: Tortricidae) on strawberry plants. *Pl. Path.* **23**, 156–159.
ALGER N. E. and UNDEEN A. H. (1970) The control of a Microsporidian, *Nosema* sp., in an Anopheline colony by an egg-rinsing technique. *J. invert. Path.* **15**, 321–327.
ALLEMAND R. (1976) Influence des modifications des conditions lumineuses sur les rhythmes circadiens de vitellogèse et d'ovulation chez *Drosophila melanogaster*. *J. Insect Physiol.* **22**, 1075–1080.
ALLEN D. C. (1976) Biology of the green-striped mapleworm *Dryocampa rubicunda* (Lepidoptera: Saturniidae) in the northeastern United States. *Ann. ent. Soc. Am.* **69**, 857–862.
ALPATOV W. W. and BACHVALOVA T. T. (1930) Some data on variation of eggs in some insects (*Anopheles maculipennis* and *Bombyx mori*). *Rev. zool. russe* **10**, 32–44 (in Russian; English summary).
ALTAHTAWY M. M., EL-SAWAF S. K., and SHALABY F. F. (1972) Studies on morphology and development of immature stages of *Microplitis ruforentris* Kokujev (Hym. Braconidae). *Z. angew. Ent.* **71**, 134–139.
ALTSON A. M. (1923) On the method of oviposition and the egg of *Lyctus brunneus* Steph. *J. Linn. Soc. (Zool.) Lond.* **35**, 217–227.
ALTWEGG P. (1971) Ein semisynthetisches Nährmedium und Ersatzsubstrat für die Oviposition zur von der Jahreszeit unabhängigen Zucht des grauen Lärchenwicklers *Zeiraphera diniana* (Gn.) (Lep., Tortricidae). *Z. angew. Ent.* **69**, 135–170.
AMANTE E. (1965) Observações bionômicas sôbre *Hypocala andremona* (Cram.) (Lepidoptera, Noctuidae) praga do caquizeiro. *Biológico* **31**, 97–101.
AMMAN G. D. (1968) Effects of temperature and humidity on development and hatching of eggs of *Adelges piceae*. *Ann. ent. Soc. Am.* **61**, 1606–1611.
AMMAN G. D. (1972) Some factors affecting oviposition behaviour of the mountain pine beetle. *Envir. Ent.* **1**, 691–695.
AMMAR E. D. and FARRAG S. M. (1974) Studies on the behaviour and biology of the earwig *Labidura riparia* Pallas (Derm., Labiduridae). *Z. angew. Ent.* **75**, 189–196.
AMOS T. G. (1968) Some laboratory observation on the rates of development, mortality and oviposition of *Dermestes frischii* (Kug.) (Col., Dermestidae). *J. stored Prod. Res.* **4**, 103–117.
ANANTHASUBRAMANIAN K. S. and ANANTHAKRISHNAN T. N. (1959) The structure of the ootheca and egg laying habits of *Corydia petiveriana* L. (Blattidae). *Indian J. Ent.* **21**, 57–64.
ANCONA L. H. (1933) El ahuatle de Texcoco. *An. Inst. Biol. Univ. Mex.* **4**, 51–69.

ANDERSEN K. T. (1934) Biologie des Kornkäfers (*Calandra granaria* L.). *Nach. SchädlBekämpf. Florsheim.* (A) **9**, 105–131.

ANDERSEN N. M. (1973) Seasonal polymorphism and developmental changes in organs of flight and reproduction in bivoltine pondskaters (Hem., Ferridae). *Ent. Scand.* **4**, 1–20.

ANDERSEN N. M. and POLHEMUS J. T. (1976) Water-striders (Hemiptera: Gerridae, Vellidae, etc.). In *Marine Insects* (L. Cheng, ed.), North-Holland.

ANDERSON D. S. (1960) The respiratory system of the egg-shell of *Calliphora erythrocephala*. *J. Insect Physiol.* **5**, 120–228.

ANDERSON D. S. (1965) Observations on female accessory glands of some Acridoidea, with particular reference to *Pyrgomorpha dispar* I. Bolivar. *Entomologist's mon. Mag.* **101**, 16–17.

ANDERSON D. T. (1972a) The development of hemimetabolous insects. In *Development Systems: Insects* (S. J. Counce and C. H. W. Waddington, eds.) **1**, 95–163.

ANDERSON D. T. (1972b) The development of holometabolous insects. In *Development Systems: Insects* (S. J. Counce and C. H. W. Waddington, eds.) **1**, 165–242.

ANDERSON J. (1884) The urticating properties of the hairs of some Lepidoptera. *Entomologist* **17**, 275–276.

ANDERSON J. F. and HORSFALL W. R. (1963) Thermal stress and anomalous development of mosquitoes (Diptera: Culicidae). I. Effect of constant temperature on dimorphism of adults of *Aedes stimulans*. *J. exp. Zool.* **154**, 67–107.

ANDERSON J. F. and HORSFALL W. R. (1965a) Thermal stress and anomalous development of mosquitoes (Diptera: Culicidae). V. Effect of temperature on embryogeny of *Aedes stimulans*. *J. exp. Zool.* **158**, 211–222.

ANDERSON J. F. and HORSFALL W. R. (1965b) Dimorphic responses of transplanted gonadal anlagen of mosquitoes. *Proc. 12th Int. Congr. Ent.* 154–155 (1964).

ANDERSON J. F. and KAYA H. K. (1973) Influence of elm spanworm oviposition sites on parasitism by *Ooencyrtus clisiocampae* and *Telenomus alsophilae*. *Envir. Ent.* **2**, 705–711.

ANDERSON J. R. and TEMPELIS C. H. (1970) Precipitin test identification of blood meals of *Stomoxys calcitrans* (L.) caught on California poultry ranches, and observations of digestion rates of bovine and citrated human blood. *J. med. Ent.* **7**, 223–229.

ANDERSON R. C. and PASCHKE J. D. (1968) The biology and ecology of *Anaphes flavipes* (Hymenoptera: Mymaridae), an exotic egg parasite of the cereal leaf beetle. *Ann. ent. Soc. Am.* **61**, 1–5.

ANDREADIS T. G. and HALL D. W. (1976) *Neoaplectana carpocapsae*: encapsulation in *Aedes aegypti* and changes in host hemocytes and hemolymph proteins. *Expl. Parasit.* **39**, 252–261.

ANDRES A. (1914) L'oothèque de l'*Eremiaphila khamsin*. *Bull. Soc. ent. Égypte* **6**, 72–74.

ANDRES L. A. and ANGALET G. W. (1963) Notes on the ecology and host specificity of *Microlarinus lareynii* and *M. lypriformis* (Coleoptera: Curculionidae) and the biological control of puncture vine, *Tribulus terrestris*. *J. econ. Ent.* **56**, 333–340.

ANDREWARTHA H. G. (1939) The small plague grasshopper (*Austroicetes cruciata* Sauss.). Notes on the present position in South Australia and recommendations for control measures. *J. Dep. Agric. S. Aust.* **43**, 99–107.

ANGUS R. B. (1973) The habits, life histories and immature stages of *Helophorus* F. (Coleoptera: Hydrophilidae). *Trans. R. ent. Soc. Lond.* **125**, 1–26.

ANKERSMIT G. W. (1955) Over het verband tussen de aantasting door de koolzaadgalmug, *Dasyneura brassicae* Winn. (Diptera, Itonididae) en de koolzaadsnuitkever, *Ceuthorrhynchus assimilis* Payk. (Coleoptera, Curculionidae). *Tijdschr. PlZiekt.* **61**, 93–97.

ANKERSMIT G. W., POL B.-C. VAN DER, and WATER J. K. (1976) Temperature and mortality in the eggs of *Adoxophyes orana* (Lepidoptera, Tortricidae). *Neth. J. Pl. Path.* **82**, 173–180.

ANNILA E. (1969) Influence of temperature upon the development and voltinism of *Ips typographus* L. (Coleoptera, Scolytidae). *Ann. zool. fenn.* **6**, 161–207.

ANON. (1916a) *Report on the Great Invasion of Locusts in Egypt in 1915 and the Measures Adopted to deal with It*, Cairo.

ANON. (1916b) *Lyda hypotrophica*, a hymenopterous pest of *Epicea*, etc. *Z. angew. Ent.* **3**, 75–96.

ANON. (1950) The climbing cutworm moth in South Australia. *J. Agric. S. Aust.* (Nov.) 1–6.

ANON. (1970) Rearing stick insects. *Leafl. Amat. Ent.* **30**, 1–20.

ANSARI M. H. (1966) A note on the female genitalia of *Aschistonyx baranii* Grover (Diptera: Cecidomyiidae) with observation of egg-laying. *Proc. natn. Acad. Sci. India* **36** (B), 345–349.

ANTONIOU A. and HUNTER-JONES P. (1956) The life-history of *Eyprepocnemis capitata* Miller (Orth., Acrididae) in the laboratory. *Entomologist's mon. Mag.* **92**, 364–368.

ANWAR M., ASHRAF M., and ARIF M. D. (1973) Some aspects of mating and oviposition behaviour of the spotted bollworm of cotton, *Earias vittella* (F.). *Pakistan J. Sci. Ind. Res.* **16**, 28–31.

APPERT J. (1956) La bruche des arachides. *Bull. agron. Sect. tech. Agric. trop.* **13**, 181–190.

APPLEBAUM S. W. (1964) Physiological aspects of host specificity in the Bruchidae. I. General considerations of developmental compatibility. *J. Insect Physiol.* **10**, 783–788.

APPLEBAUM S. W., GESTETNER B., and BIRK Y. (1965) Physiological aspects of host specificity in the Bruchidae. IV. Developmental incompatibility of soybeans for *Callosobruchus*. *J. Insect Physiol.* **11**, 611–616.

ARBOGAST R. T., CARTHON M., and ROBERTS J. R. JR (1971) Developmental stages of *Xylocoris flavipes* (Hemiptera: Anthocoridae), a predator of stored-product insects. *Ann. ent. Soc. Am.* **64**, 1131–1134.

ARIAS GIRALDA A. and NIETO CALDERON J. (1973) Puesta y avivamiento de la "hoplocampa del peral" (*H. brevis* Kl.) en 1972 y 1973 en las vegas del Guadiana (Badajoz). *Bol. Inf. Plagas* **111**, 33–35.

ARKHIPOV G. E. (1965) *Trichogramma* in the control of the pea moth. *Trudȳ vses. Inst. Zashch. Rast.* **24**, 248–249 (in Russian; English summary).

ARMSTRONG E. A. (1950) The nature and functions of displacement activities. *Symp. Soc. exp. Biol.* **4**, 361–384.

ARMSTRONG J. S. (1958) The breeding habits of the Corduliidae (Odonata) in the Taupo district of New Zealand. *Trans. R. Soc. NZ* **85**, 275–282.

ARORA G. L. and PAJNI H. R. (1957) Some observations on the biology and oviposition of *Bruchus analis* F. (Bruchidae: Coleoptera). *Res. Bull. Panjab Univ. Sci.* **128**, 453–470.

ARORA G. L. and SINGH I. (1957) Manodean oothecae. *Res. Bull. Panjab Univ. Sci.* **105**, 261–267.

ARRAND J. C. and MCMAHON H. (1974) *Plagiognathus medicagus* (Hemiptera: Miridae) (Het.): descriptions of egg and 5 nymphal instars. *Can. Ent.* **106**, 433–435.

ARROW G. J. (1915) Notes on the coleopterous family Dermestidae and descriptions of some new forms in the British Museum. *Ann. Mag. nat. Hist.* (8) **15**, 425–451.

ARRU G. M. (1962) *Agrilus suvorovi populneus* Schaefer (Coleoptera: Buprestidae) dannoso ai pioppi nell' Italia settentrionale. *Boll. Zool. agr. Bachic.* **4**, 159–286.

ARTHUR A. P. (1961) The cleptoparasitic habits and the immature stages of *Eurytoma pini* Bugbee (Hymenoptera: Chalcidae), a parasite of the European pine shoot moth, *Rhyacionia buoliana* (Schiff.) (Lepidoptera: Olethreutidae). *Can. Ent.* **93**, 655–660.

ARTHUR A. P. (1962) Influence of host tree on abundance of *Itoplectis conquisitor* (Say) (Hymenoptera: Ichneumonidae), a polyphagous parasite of the European pine shoot moth, *Rhyacionia buoliana* (Schiff.) (Lepidoptera: Olethreutidae). *Can. Ent.* **94**, 337–347.

ARTHUR A. P. (1966) Associative learning in *Itoplectis conquisitor* (Say) (Hymenoptera: Ichneumonidae). *Can. Ent.* **98**, 213–223.

ARTHUR A. P. (1967) Influence of position and size of host shelter on host-searching by *Itoplectis conquisitor* (Hymenoptera: Ichneumonidae). *Can. Ent.* **99**, 877–886.

ARTHUR A. P., HEGDEKAR B. M., and BATSCH W. W. (1972) A chemically defined synthetic medium that induces oviposition in the parasite *Itoplectis conquisitor* (Hymenoptera: Ichneumonidae). *Can. Ent.* **104**, 1251–1258.

ARTHUR A. P., HEGDEKAR B. M., and ROLLINS L. (1969) Component of the host haemolymph that induces oviposition in a parasitic insect. *Nature, Lond.* **223**, 966–967.

ARTHUR A. P., STAINER J. E. R., and TURNBULL A. L. (1964) The interaction between *Orgilus obscurator* (Nees) (Hymenoptera: Braconidae) and *Temelucha interruptor* (Grav.) (Hymenoptera: Ichneumonidae), parasites of the pine shoot moth, *Rhyacionia buoliana* (Schiff.) (Lepidoptera: Olethreutidae). *Can. Ent.* **96**, 1030–1034.

ARTHUR A. P. and WYLIE H. G. (1959) Effects of host size on sex ratio, development time and size of *Pimpla turionellae* (L.) (Hymenoptera: Ichneumonidae). *Entomophaga* **4**, 297–301.

ASCHER K. R. S. (1957) Prevention of oviposition in the housefly through tarsal contact agents. *Science, Lancaster Pa* **125**, 938–939.

ASCHER K. R. S. (1959) "Di-(p-chlorophenyl) compounds" and oviposition in the housefly. *Riv. Parassit.* **20**, 143–144.

ASHBY D. G. and WRIGHT D. W. (1946) The immature stages of the carrot fly. *Trans. R. ent. Soc. Lond.* **97**, 355–379.

ASHBY J. W. (1974) A study of arthropod predation of *Pieris rapae* L. using serological and exclusion techniques. *J. appl. Ecol.* **11**, 419–425.

ASHBY K. R. (1961) The population dynamics of *Cryptolestes ferrugineus* (Stephens) (Col., Cucujidae) in flour and on Manitoba wheat. *Bull. ent. Res.* **52**, 363–379.

ASHER W. C. (1970) Olfactory response of *Dioryctria abietella* (Lepidoptera: Phycitidae) to slash pine cones. *Ann. ent. Soc. Am.* **63**, 474–476.

ASKEW R. R. (1967) Reactions of two species of Agromyzidae to parasitism by *Chrysocharis melaenis*. *J. econ. Ent.* **60**, 1453–1454.

ASKEW R. R. (1971) *Parasitic Insects*, Heinemann, London.

ASPÖCK H. and ASPÖCK U. (1971) Ordnung Raphidioptera (Kamelhalsfleigen). *Handbuch der Zoologie* **4** (2) 1–50.

ASSEM J. VAN DEN (1959) Notes on New Guinean species of *Tripteroides*, subgenus *Rachisoura* (Diptera, Culicidae), with descriptions of two new species. *Tijd. Ent.* **102**, 35–56.

ASSEM J. VAN DEN (1976) Queue here for mating: observations on the behaviour of unpaired *Melittobia* females in respect of a male of their own species. *Ent. Ber.* **36**, 74–78.

ASSEM J. VAN DEN and BRUIJN E. FEUTH-DE (1977) Second matings and their effect on the sex ratio of the offspring of *Nasonia vitripennis* (Hymenoptera: Pteromalidae). *Entomologia exp. appl.* **21**, 23–28.

ATKIN E. A. and BACOT A. W. (1917) The relation between the hatching of the eggs and the development of the larvae of *Stegomyia fasciata* (*Aedes calopus*) and the presence of bacteria and yeasts. *Parasitology* **9**, 482–536.

ATKINS E. D. T., FLOWER N. E., and KENCHINGTON W. (1966) Studies on the oöthecal protein of the tortoisebeetle, *Aspidomorpha*. *Jl R. microsc. Soc.* **86**, 123–135.

ATKINSON P. W., BROWN W. V., and GILBY A. R. (1973) Autoxidation of insect cuticular lipids: stabilization of alkyl dienes by 3,4-dihydric phenols. *Insect Biochem.* **3**, 103–112.

ATWAL A. S. (1955) Influence of temperature, photoperiod, and food on the speed of development, longevity, fecundity, and other qualities of the diamond-back moth *Plutella maculipennis* (Curtis) (Tineidae, Lepidoptera). *Aust. J. Zool.* **3**, 185–221.

ATWAL A. S. (1959a) Oviposition behaviour of *Diadromus (Thyraella) collaris* Gravenhorst (Ichneumonidae: Hym.), a parasite of cabbage diamond-back moth, *Plutella maculipennis* Curtis (Tineidae: Lep.). *Proc. natn. Inst. Sci. India* (B) **25**, 80–86.

ATWAL A. S (1959b) The oviposition behaviour of *Bagrada cruciferarum* Kirkaldy (Pentatomidae: Heteroptera) and the influence of temperature and humidity on the speed of development of eggs. *Proc. natn. Inst. Sci. India* (B) **25**, 65–67.

ATYEO W. T., WEEKMAN G. T., and LAWSON D. E. (1964) The identification of *Diabrotica* species by chorion sculpturing. *J. Kans. ent. Soc.* **37**, 9–11.

AUBERT J. F. (1961) L'expérience de la bourre de coton démontre que le volume de l'hôte intervient en tant que facteurs essentiels dans la détermination du sexe chez les Ichneumonides Pimplines (Hym.). *Bull. Soc. ent. Fr.* **66**, 89–93.

AUDEMARD H. (1968) L'élevage permanent de la mouche des semis *Phorbia platura* Meigen (*Hylemyia cilicrura* Rondani) (Diptera Muscidae). I. Pondoir artificiel. *Annls Épiphyt.* **18**, 551–555 (1967).

AUSTARA O. (1971) *Gonometa podocarpi* Aur. (Lepidoptera: Lasiocampidae), a defoliator of exotic softwoods in East Africa. The biology and life cycle at Muko, Kigezi District in Uganda. *E. Afr. agric. For. J.* **36**, 275–289.

AUSTARA O. and MIGUNDA J. (1971) *Orgyia mixta* Snell (Lepidoptera: Lymantriidae), a defoliator of exotic softwoods in Kenya. *E. Afr. agric. For. J.* **36**, 298–307.

AVIDOV Z., APPLEBAUM S. W., and BERLINGER M. J. (1965) Physiological aspects of host specificity in the Bruchidae. II. Ovipositional preference and behaviour of *Callosobruchus chinensis* L. *Entomologia exp. appl.* **8**, 96–106.

AVIDOV Z., BERLINGER M. J., and APPLEBAUM S. W. (1965) Physiological aspects of host specificity in the Bruchidae. III. Effect of curvature and surface area on oviposition of *Callosobruchus chinensis* L. *Anim. Behav.* **13**, 178–180.

AVIDOV Z. and GOTHILF S. (1960) Observations on the honeydew moth (*Cryptoblabes gnidiella* Milliere) in Israel. I. Biology, phenology and economic importance. *Ktavim* **10**, 109–124.

AVIDOV Z. and PODOLER H. (1968) Studies on the life history of *Metaphycus flavus* (How.) (Encyrtidae). *Israel J. Ent.* **3** (2) 1–15.

AVIDOV Z. and ROSEN D. (1961) Bionomics of the jasmine moth (*Glyphodes unionalis* Huebner) in the coastal plain of Israel. *Bull. Res. Coun. Israel* (B) **10**, 77–89.

AVIDOV Z. and ZAITZOV A. (1960) On the biology of the mango shield scale *Coccus mangiferae* (Green) in Israel. *Ktavim* **10**, 125–137.

AWADALLAH K. T. and SWAILEM S. M. (1971) On the bionomics of the sycamore fig psyllid *Pauropsylla trichaeta* Petty (Hem., Psyllidae) (Hom.). *Bull. Soc. ent. Égypte* **55**, 193–199.

AXTELL R. C. (1964) Phoretic relationship of some common manure-inhabiting Macrochelidae (Acarina: Mesostigmata) to the house fly. *Ann. ent. Soc. Am.* **57**, 584–587.

AXTELL R. C. (1968) Integrated house fly control: populations of fly larvae and predaceous mites, *Macrocheles muscaedomesticae*, in poultry manure after larvicide treatment. *J. econ. Ent.* **61**, 245–249.

AYATOLLAHI M. (1971) Importance of the study of Diptera and their role in biological control. *Ent. Phytopath. appl.* **31**, 20–28 (in Persian; English summary).

AYERTEY J. N. (1975) Egg laying by unmated females of *Sitotroga cerealella* (Lepidoptera: Gelechiidae). *J. stored Prod. Res.* **11**, 211–215.

AYYAPPA P. K., PERTI S. L., and WAL Y. C. (1964) A life-history study of the black carpet beetle *Attagenus alfierii* Pic. (Coleoptera: Dermestidae). *Indian J. Ent.* **26**, 275–280.

AYYAR T. V. R. (1920) Notes on the life-history of *Cantao ocellatus* Th. *Proc. ent. Meet. Pusa* **3**, 910–914.

AYYAR T. V. R. (1929) Notes on some Indian Lepidoptera with abnormal habits. *J. Bombay nat. Hist. Soc.* **33**, 668–675.

AZAB A. K., MEGAHED M. M., and EL-MIRSAWI D. H. (1971) On the biology of *Bemisia tabaci* (Genn.) (Hem., Hom.: Aleyrodidae). *Bull. Soc. ent. Égypte* **55**, 305–315.

AZAB A. K., TAWFIK M. F. S., and ABOUZEID N. A. (1973a) The biology of *Dermestes maculatus*

DeGreer (Coleoptera: Dermestidae). *Bull. Soc. ent. Égypte* **56**, 1–14.
AZAB A. K., TAWFIK M. F. S., and ABOUZEID N. A. (1973b) Factors affecting the rate of oviposition in *Dermestes maculatus* De Geer (Coleoptera: Dermestidae). *Bull. Soc. ent. Égypte* **56**, 49–59.
AZARYAN A. G. (1968) The diurnal rhythm of the emergence and oviposition of *Drosophila transversa* Fll. and *D. phalerata* Mg. (Diptera, Drosophilidae) under different photoperiodic conditions. *Ent. Obozr.* **47**, 809–814 (in Russian).
AZARYAN A. G. (1969) The circadian rhythm of oviposition and its connexion with photoperiodic reaction in different light–dark regimes in *Drosophila phalerata* Meig. *25th Ann. Armenian Acad. Sci. SSR*, 12–13 (in Russian).
AZARYAN G. KH., BABAYAN A. S., VASILYAN V. V., and MKRTUMYAN K. L. (1965) On the prospects of the irradiation method for the control of the malva moth (Lepidoptera, Gelechiidae). *Ent. Obozr.* **44**, 762–769 (in Russian; English summary).
AZIM A. (1963) The biology of *Microterys okitsuenis* Compere and its oviposition behaviour (Hymenoptera: Encyrtidae). *Mushi, Fukuoka* **37**, 65–78.
AZUMA K. and KITANO H. (1971) Experimental studies on the parasitism of *Apanteles glomeratus* Linné on the larvae of *Pieris melete* Ménétriès. *Kontyû Tokyo* **39**, 394–399 (in Japanese; English summary).

BACCETTI B. (1959a) Ricerche istochimiche sulla distribuzione di alcuni enzimi nelle ghiandole colleteriche degli Acridoidei (Orthoptera). *Boll. Zool.* **26**, 129–135.
BACCETTI B. (1959b) Reperti sull'accrescimento nucleare nelle ghiandole colleteriche degli Acridoidei (Osservazioni citologiche ed istofotometriche). *Boll. Zool.* **26**, 137–141.
BACCETTI B. (1960a) Indagini citologiche ed istofotometriche sull'accrescimento nucleare nelle ghiandole colleteriche di *Acrida bicolor mediterranea* Dirsh (Ins., Orth.). *Caryologia* **12**, 497–512.
BACCETTI B. (1960b) Le cocciniglie Italiane delle cupressacee. *Redia* **45**, 23–111.
BACCETTI B. (1961) Il problema della screzione della ooteca negli Ortotteriodei alla luce delle piu' recenti acquisizioni. *Atti Accad. naz. ital. Ent. Bologna Rc.* **8**, 112–136.
BACCETTI B. (1966a) Ultrastruttura e specializzazione cellulare in organi di insetti. 2. Un epitelio secretore specializzato: le due ghiandole colleteriche di *Periplaneta*. *Boll. Soc. ital. Biol. sper.* **42**, 796–798.
BACCETTI B. (1966b) Ultrastruttura e specializzazione cellulare in organi di insetti. 1. Un epitelio secretore non specializzato: le ghiandole pseudocolleteriche di *Acrida*. *Boll. Soc. ital. Biol. sper.* **42**, 794–796.
BACCETTI B. (1966c) Ultrastruttura e specializzazione cellulare in organi di insetti. 3. Un epitelio secretore fortementi specializzato: le cinque ghiandole colleteriche di *Mantis*. *Boll. Soc. ital. Biol. sper.* **42**, 798–800.
BACCETTI B. (1967) L'ultrastruttura delle ghiandole della ooteca in Ortotteri, Acridoidei, Blattoidei e Mantoidei. *Z. Zellforsch. mikrosk. Anat.* **77**, 64–79.
BACOT A. W. (1916) Report of the entomological investigation undertaken for the Commission for the year August 1914 to July 1915. *Rep. Yellow Fever Comm. (W. Afr.)* 119 pp.
BACOT A. W. (1917) The effect of the presence of bacteria or yeasts on the hatching of eggs of *Stegomyia fasciata* (the yellow fever mosquito). Summary. *Jl R. microsc. Soc.* **3**, 173–174.
BACOT A. W. (1918) A note on the period during which the eggs of *Stegomyia fasciata* (*Aedes calopus*) from Sierra Leone stock retain their vitality in a humid temperature. *Parasitology* **10**, 280–283.
BADAWY A. (1965) Some factors affecting the fecundity and longevity of *Trogoderma granarium granarium* Everts. (Coleoptera: Dermestidae). *Bull. Soc. ent. Égypte* **48**, 281–290 (1964).

BADCOCK R. M. (1953) Observation of oviposition under water of the aerial insect *Hydropsyche angustipennis* (Curtis) (Trichoptera). *Hydrobiologia* **5**, 222–225.

BADGLEY M. E. and FLESCHNER C. A. (1956) Biology of *Oligota oviformis* Casey (Coleoptera: Staphylinidae). *Ann. ent. Soc. Am.* **49**, 501–502.

BADONNEL A. (1933) Description de la ponte de *Psocus bipunctatus* L. *Bull. Soc. zool. Fr.* **58**, 69–71.

BAER W. (1920) Die Tachinen als Schmarotzer der schadlichen Insekten. Ihre Lebensweise, wirtschaftliche Bedeutung und systematische Kennzeichnung. *Z. angew. Ent.* **6**, 185–246; (1921) **7**, 97–163, 349–423.

BAGDAVADZE A. I. (1963) A study of the moth *Lithocolletis blancardella* F.—a pest of fruit trees in eastern Georgia. *Zool. Zh.* **42**, 1412–1413 (in Russian; English summary).

BAGGIOLINI M. and DUPERREX H. (1963) Observations sur la biologie et la nuisibilité de la sésie du groseillier et du cassis *Synanthedon tipuliformis* Clerck. (Lép. Aeg.). *Schweiz. landw. Forsch.* **2**, 13–32.

BAGNALL R. S. (1915) On a collection of Thysanoptera from the West Indies with descriptions of new genera and species. *J. Linn. Soc. Zool.* **32**, 495–507.

BAILEY J. C., MAXWELL F. G., and JENKINS J. N. (1972) Seasonal fluctuations in oviposition of the boll weevil in the laboratory. *J. Kans. ent. Soc.* **45**, 252–254.

BAISAS F. E. and HU S. M. K. (1936) *Anopheles hyrcanus* var. *sinensis* of the Philippines and certain parts of China, with some comments on *Anopheles hyrcanus* var. *nigerrimus* of the Philippines. *Mon. Bull. Bur. Hlth Philipp.* **16**, 205–242.

BAISAS F. E. and PAGAYON A. U. (1956) Notes on Philippine mosquitoes. XVII. The eggs and first instar larvae of some *Neomyzomyias*. *Philipp. J. Sci.* **85**, 215–227.

BAKER B. H. and TROSTLE G. C. (1973) Douglas fir beetle attraction and tree-group response. *J. econ. Ent.* **66**, 1002–1005.

BAKER C. R. R. (1969) Apparatus for studying nocturnal patterns of oviposition and larval eclosion and diel patterns of moth emergence. *Bull. ent. Res.* **58**, 553–557.

BAKER J. L. (1976) Determinants of host-selection for species of *Aphytis* (Hymenoptera: Aphelinidae), parasites for diaspine scales. *Hilgardia* **44**, 1–25.

BAKER J. R. and NEUNZIG H. H. (1968) The egg masses, eggs, and first-instar larvae of Eastern North American Corydalidae. *Ann. ent. Soc. Am.* **61**, 1181–1187.

BAKER W. A., BRADLEY W. G., and CLARK C. A. (1949) Biological control of the European corn borer in the United States. *Tech. Bull. US Dep. Agric.* **983**, 1–185.

BAKER W. A. and JONES L. G. (1934) Studies on *Exeristes roborator* (Fab.), a parasite of the European corn borer in the Lake Erie area. *Tech. Bull. US Dep. Agric.* **460**, 1–26.

BAKKE A. (1967) Pheromone in the bark beetle *Ips acuminatus* Gyll. *Z. angew. Ent.* **59**, 49–53.

BAKKE A. (1970) Evidence of a population aggregating pheromone in *Ips typographus* (Coleoptera: Scolytidae). *Contr. Boyce Thompson Inst.* **24**, 309–310.

BAKKE A. (1973) Bark beetle pheromones and their potential use in forestry. *Bull. Org. Européenne et Méd. Protect. Pl.* **9**, 5–15.

BAKKE A. (1975) Aggregation pheromone in the bark beetle *Ips duplicatus* (Sahlberg). *Norw. J. Ent.* **22**, 67–69.

BAKKENDORFF P. (1933) Biological investigations on some Danish hymenopterous egg parasites, especially in homopterous eggs, with taxonomic remarks and description of new species. *Ent. Medd.* **19**, 1–134.

BAKKER K., EIFSACKERS H. J. P., LENTEREN J. C. VAN, and MEELIS E. (1972) Some models describing the distribution of eggs of the parasite *Pseudeucoila bochei* (Hym., Cynip.) over its hosts, larvae of *Drosophila melanogaster*. *Oecologia* **10**, 29–57.

BALACHOWSKY A. (1928) Observations biologiques sur les parasites des coccides du Nord-Africain. *Annls Épiphyt.* **14**, 280–312.

BALACHOWSKY A. S. and CHARARAS C. (1964) Note complémentaire sur *Phloeosinus armatus* Reitter (Col. Scolytidae) nuisible au cyprès dans le bassin oriental de la Méditerranée. *Rev. Path. vég. Ent. agric. Fr.* **43**, 13–17.

BALAY G. (1964) Observations sur l'oviposition de *Simulium damnosum* Theobald et *Simulium adersi* Pomeroy (Diptera, Simuliidae) dans l'est de la Haute-Volta. *Bull. Soc. Path. exot.* **57**, 588–611.

BALDUF W. V. (1935) *The Bionomics of Entomophagous Coleoptera,* John Swift, New York.

BALDUF W. V. (1937) Bionomic notes on the common bagworm and its insect enemies. *Proc. ent. Soc. Wash.* **39**, 169–184.

BALDUF W. V. (1938) The rise of entomophagy among Lepidoptera. *Am. Nat.* **72**, 358–379.

BALDUF W. V. (1939) *The Bionomics of Entomophagous Insects,* Part II, New York.

BALFOUR-BROWNE F. (1909) The life-history of the agrionid dragonfly. *Proc. zool. Soc. Lond.* **1909**, 253–285.

BALFOUR-BROWNE F. (1910) On the life-history of *Hydrobius fuscipes* L. *Trans. R. Soc. Edinb.* **47**, 317–340.

BALFOUR-BROWNE F. (1922) The life-history of the water-beetle *Pelobius tardus* Herbst. *Proc. zool. Soc. Lond.* **1922**, 79–97.

BALL H. J. (1957) On the biology and egg-laying habits of the western corn rootworm. *J. econ. Ent.* **50**, 126–128.

BALL H. J. (1971) Laboratory observations on the daily oviposition cycle in the western corn rootworm. *J. econ. Ent.* **64**, 1319–1320.

BALLARD E. and HOLDAWAY F. G. (1926) The life-history of *Tectacoris lineola* F. and its connection with internal boll rots in Queensland. *Bull. ent. Res.* **16**, 329–346.

BALLARD E., MISTIKAWI A. M., and ZOHEIRY M. S. (1932) The desert locust *Schistocerca gregaria* Forsk. in Egypt. *Bull. Min. Agric. Egypt* **110**, 1–149.

BALOBESHKO V. S. (1968) The bionomics and ecology of the spruce Tortricid *Laspeyresia pactolana* Zell. (Lepidoptera, Tortricidae). *Ent. Obozr.* **47**, 31–40 (in Russian; English, *Ent. Rev.* **47**, 17–23).

BALOCH G. M., MOHYUDDIN A. I., and GHANI M. A. (1969) Biological control of *Cuscuta* spp. III. Phenology, biology and host-specificity of *Herpystis cuscutae* Bradley (Lep., Tortricidae). *Entomophaga* **14**, 119–128.

BALOGUN R. A. (1970) The life-history and habits of the larch bark beetle, *Ips cembrae* (Coleoptera: Scolytidae), in the north-east of Scotland. *Can. Ent.* **102**, 226–239.

BALOGUN R. A. (1974) A sex-specific ninhydrin-positive component detected in the accessory glands of adult male tsetse flies (Diptera, Glossinidae). *Nigerian J. Ent.* **1**, 13–16.

BALTER R. S. (1968a) Lice egg morphology as a guide to taxonomy. *Med. biol. Ill.* **18**, 94–95.

BALTER R. S. (1968b) The microphotography of avian lice eggs. *Med. biol. Ill.* **18**, 166–179.

BANERJEE A. C. and DECKER G. C. (1966a) Studies on sod webworms. 1. Emergence rhythm, mating, and oviposition behaviour under natural conditions. *J. econ. Ent.* **59**, 1237–1244.

BANERJEE A. C. and DECKER G. C. (1966b) Studies on sod web-worms. 2. Oviposition behaviour of *Crambus trisectus* under regulated light conditions in the laboratory. *J. econ. Ent.* **59**, 1245–1248.

BANHAM E. J. and CROOK L. J. (1966) Susceptibility of the confused flour beetle *Tribolium confusum* Duv. and the rust-red beetle *Tribolium castaneum* (Herbst) to gamma radiation. In *The Entomology of Radiation Disinfestation of Grain* (P. B. Cornwell, ed.), pp. 107–118, Oxford.

BANKS C. G. (1949) The absorption of water by the eggs of *Corixa punctata* Illig. (Hemiptera, Corixidae) under experimental conditions. *J. exp. Biol.* **26**, 131–136.

BANKS C. J. (1954) The searching behaviour of coccinellid larvae. *Br. J. anim. Behav.* **4**, 37–38.

BANKS C. J. (1955) An ecological study of Coccinellidae (Col.) associated with *Aphis fabae*

Scop. on *Vicia faba*. *Bull. ent. Res.* **46**, 561–587.
BANKS C. J. (1956a) Observations on the behaviour and mortality in Coccinellidae before dispersal from the egg shells. *Proc. R. ent. Soc. Lond.* (A) **31**, 56–60.
BANKS C. J. (1956b) The distribution of coccinellid egg batches and larvae in relation to numbers of *Aphis fabae* Scop. on *Vicia faba*. *Bull. ent. Res.* **47**, 47–56.
BANKS C. J. (1957) The behaviour of individual coccinellid larvae on plants. *Br. J. anim. Behav.* **5**, 12–24.
BARANOWSKI R. M. (1958) Notes on the biology of the royal palm bug *Xylastodoris luteolus* Barber (Hemiptera: Thaumastocoridae). *Ann. ent. Soc. Am.* **51**, 547–551.
BARBER G. W. (1926) A two-year study of the development of the European corn-borer in the New England area. *J. Agric. Res.* **32**, 1053–1068.
BARBER M. A. (1927) The food of culicine larvae. *Publ. Hlth Rep. Wash.* **43**, 11–17.
BARBER M. A. (1935) A method of detecting the eggs of *Anopheles* in breeding places and some of its applications. *Riv. Malar.* **14**, 146–149.
BARBIER R. and CHAUVIN G. (1972a) Origine et structure des enveloppes de l'oeuf et mise en place de la cuticule sérosale chez *Monopis rusticella* Clerk (Lépidoptère Tineidae). *C. r. hebd. Séanc. Acad. Sci. Paris* (D) **274**, 1079–1082.
BARBIER R. and CHAUVIN G. (1972b) Étude expérimentale de la perméabilité des enveloppes de l'oeuf et de la cuticule sérosale chez *Tinea pellionella* L. (Lépidoptère Tineidae). *C. r. hebd. Séanc. Acad. Sci. Paris* (D) **275**, 1003–1006.
BARBIER R. and CHAUVIN G. (1974a) Ultrastructure et rôle des aéropyles et des enveloppes de l'oeuf de *Galleria mellonella*. *J. Insect. Physiol.* **20**, 809–820.
BARBIER R. and CHAUVIN G. (1974b) The aquatic egg of *Nymphula nymphaeta* (Lepidoptera: Pyralidae): on the fine structure of the egg shell. *Cell. Tiss. Res.* **149**, 473–479.
BARDNER R. and KENTON J. (1957) Notes on the laboratory rearing and biology of the wheat bulb fly *Leptohylemyia coarctata* (Fall.). *Bull. ent. Res.* **48**, 821–831.
BARDNER R. and LOFTY J. R. (1971) The distribution of eggs, larvae and plants within crops attacked by wheat bulb fly *Leptohylemyia coarctata* (Fall.). *J. appl. Ecol.* **8**, 683–686.
BARE C. O. (1926) Life histories of some Kansas "backswimmers". *Ann. ent. Soc. Am.* **19**, 93–101.
BARE C. O. (1935) Some remarks concerning the egg parasite *Trichogramma minutum* Riley in Florida. *J. econ. Ent.* **28**, 803–815.
BARE C. O. (1942) Some natural enemies of stored-tobacco insects, with biological notes. *J. econ. Ent.* **35**, 185–189.
BARFIELD C. S., BOTTRELL D. G., and SMITH J. W. (1977) Influence of temperature on oviposition and adult longevity of *Bracon mellitor* reared on boll weevils. *Envir. Ent.* **6**, 133–137.
BARKER P. S. (1967) Bionomics of *Blattisocius keegani* (Fox) (Acarina: Ascidae), a predator on eggs of pests of stored grains. *Can. J. Zool.* **45**, 1093–1099.
BARLOW A. R. (1966) The relationship between resin pressure and Scolytid beetle activity. *Forest Rec. Lond.* **57**, 1–7.
BARLOW C. A. (1965) Stimulation of oviposition in the seed-corn maggot fly, *Hylemya cilicrura* (Rond.) (Diptera: Anthomyiidae). *Entomologia exp. appl.* **8**, 83–95.
BARNARD K. H. (1934) South African stoneflies (*Perlaria*) with descriptions of new species. *Ann. S. Afr. Mus.* **30**, 511–548.
BARNES H. F. (1924) Some observations on the mating habits and oviposition of the Limnobiidae (Diptera). *Entomologist's mon. Mag.* **60**, 71–73.
BARNES H. F. (1925) A short note on the viviparity of *Chrysocloa gloriosa* F. (Coleoptera, Chrysomelidae). *Entomologist's mon. Mag.* **61**, 243–245.
BARNES H. F. (1930) Gall midges (Cecidomyidae) as enemies of the Tingidae, Psyllidae,

Aleyrodidae and Coccidae. *Bull. ent. Res.* **21**, 319–327.

BARNES H. F. (1935) Some new coccid-eating gall midges (Cecidomyidae). *Bull. ent. Res.* **26**, 525–530.

BARNES H. F. and NIJVELDT W. (1954) A new gall midge, *Coccomyza leefmansi* sp.n., predaceous on the eggs of *Pulvinaria polygonata* in Indonesia. *Ent. Ber.* **15**, 91–93.

BARNES J. H. and GROVE A. J. (1916) The insects attacking stored wheat in the Punjab and the methods of combating them, including a chapter on the chemistry of respiration. *Mem. Dep. Agric. Pusa* **4** (6) 165–280.

BARNES J. K. (1976) Effect of temperature on development, survival, oviposition and diapause in laboratory populations of *Sepedon fuscipennis* (Diptera: Sciomyzidae). *Envir. Ent.* **5**, 1089–1098.

BARONIO P. (1971) Ricerche su un metodo di campionamento per rilevare la densità e la distribuzione delle uova di *Dysaphis plantaginea* Pass. (Hom. Aphididae) in un meleto. Nota preliminare. *Boll. Osservatorio Malattie Piante, Bologna* **2**, 1–13.

BARR A. R. and AL-AZAWI A. (1958) Notes on oviposition and the hatching of eggs of *Aedes* and *Psorophora* mosquitoes (Diptera: Culicidae). *Kans. Univ. Sci. Bull.* **39**, 263–273.

BARRERA A., RUANO R. G., and OROZCO F. (1971) Influencia de diversos factores en la puesta del *Tribolium castaneum* V. Influencia del corte de antenas. *An. Inst. nac. Invest. Agrarias, Gen.* **1**, 109–123.

BARROSA P. and PETERS T. M. (1970) Retardation of growth rate in *Aedes aegypti* (L.) larvae exposed to vital dyes. *J. med. Ent.* **7**, 693–696.

BARTER G. W. (1957) Studies of the bronze birch borer *Agrilus anxius* Gory in New Brunswick. *Can. Ent.* **89**, 12–36.

BARTER G. W. (1965) Survival and development of the bronze poplar borer *Agrilus liragus* Barter and Brown (Coleoptera: Buprestidae). *Can. Ent.* **97**, 1063–1068.

BARTH R. (1960) Sôbre a glândula do aviduto ímpar de *Odozana obscura* Schs. (Arctiidae, Lithosiinae). *Mem. Inst. Osw. Cruz, Rio de Jan.* **58**, 129–134.

BARTH R. and MUTH H. (1958) Estudos anatômicos e histológicos sôbre a subfamília Triatominae (Heteroptera, Reduviidae). VIII. parte: Observações sôbre a superfície dos ovos das espécies mais importantes. *Mem. Inst. Osw. Cruz, Rio de Jan.* **56**, 197–208.

BARTHOLOMAI C. W. (1954) Predatism of European corn borer eggs by arthropods. *J. econ. Ent.* **47**, 295–299.

BARTLETT B. R. and BALL J. C. (1964) The developmental biologies of two encyrtid parasites of *Coccus hesperidum* and their intrinsic competition. *Ann. ent. Soc. Am.* **57**, 496–503.

BARTLETT B. R. and FISHER T. W. (1950) Laboratory propagation of *Aphytis chrysomphali* for release to control California red scale. *J. econ. Ent.* **43**, 802–806.

BARTOLONI P. (1951) La *Phthorimaea operculella* Zeller (Lep. Gelechiidae) in Italia (note sulla morfologia, biologia e mezzi di lotta). *Redia* **36**, 301–379.

BARTON L. C. and STEHR F. W. (1970) Normal development of *Anaphes flavipes* in cereal leaf beetle eggs killed with X-radiation, and potential field use. *J. econ. Ent.* **63**, 128–130.

BAR-ZEEV M. (1967) Oviposition of *Aedes aegypti* L. on a dry surface and hygroreceptors. *Nature, Lond.* **213**, 737–738.

BASS J. A. and HAYS S. B. (1976) Predation by the mite *Tyrophagus putrescentiae* on eggs of the imported fire ant. *J. Ga ent. Soc.* **11**, 16.

BASU B. C., MENON P. B., and SEN GUPTA C. M. (1954) Field studies on the bionomics of *Tabanus* flies with a view to work out control measures. *Indian J. Ent.* **16**, 67–76.

BATES M. (1940) Oviposition experiments with anopheline mosquitoes. *Am. J. trop. Med.* **20**, 569–583.

BATES M. (1941) Field studies of the anopheline mosquitoes of Albania. *Proc. ent. Soc. Wash.* **43**, 37–58.

BATES M. (1949) *The Natural History of Mosquitoes*, Macmillan, New York.

BATES M. and HACKETT L. W. (1939) The distinguishing characteristics of the populations of *Anopheles maculipennis* found in southern Europe. *Proc. 6th Int. Congr. Ent.* **3**, 1555–1569 (1938).

BATISTE W. C. (1967) Biology of the trefoil seed Chalcid *Bruchophagus kolobovae* Fedoseeva (Hymenoptera; Eurytomidae). *Hilgardia* **38**, 427–469.

BATISTE W. C., BERLOWITZ A., OLSON W. H., DETAR J. E., and JOOS J. L. (1973) Codling moth: estimating time of first egg hatch in the field—a supplement to sex-attractant traps in integrated control. *Envir. Ent.* **2**, 387–391.

BATRA L. R. (1963) Ecology of ambrosia fungi and their dissemination by beetles. *Trans. Kans. Acad. Sci.* **66**, 213–236.

BAUER J. and VITE J. P. (1975) Host selection by *Trypodendron lineatum*. *Naturwissenschaften* **62**, 539.

BAUMANN H. (1974a) The isolation, partial characterization, and biosynthesis of the paragonial substances PS-1 and PS-2 of *Drosophila funebris*. *J. Insect Physiol.* **20**, 2181–2194.

BAUMANN H. (1974b) Biological effects of paragonial substances PS-1 and PS-2 in females of *Drosophila funebris*. *J. Insect Physiol.* **20**, 2347–2362.

BAY D. E., PITTS C. W., and WARD G. (1969) Influence of moisture content of bovine faeces on oviposition and development of the face fly. *J. econ. Ent.* **62**, 41–44.

BAZIRE-BÉNAZET M. (1957) Sur la formation de l'oeuf alimentaire chez *Atta sexdens rubropilosa*, Forel, 1908 (Hym. Formicidae). *C. r. hebd. Séanc. Acad. Sci. Paris* **244**, 1277–1280.

BAZIRE-BÉNAZET M. (1970) La pont des ouvrières d'*Atta laevigata* Fred. Smith 1858 (Hym. Form.). *C. r. hebd. Séanc. Acad. Sci. Paris* (D) **270**, 1614–1615.

BEAMENT J. W. L. (1946a) Waterproofing mechanism of an insect egg. *Nature, Lond.* **157**, 370.

BEAMENT J. W. L. (1946b) The formation and structure of the chorion of the egg in an Hemipteran *Rhodnius prolixus*. *Q. Jl microsc. Sci.* **87**, 393–439.

BEAMENT J. W. L. (1946c) The waterproofing process in eggs of *Rhodnius prolixus* Stål. *Proc. R. Soc. Lond.* (B) **133**, 407–418.

BEAMENT J. W. L. (1947) The formation and structure of the micropylar complex in the egg-shell of *Rhodnius prolixus* Stål (Hemiptera, Reduviidae). *J. exp. Biol.* **23**, 213–233.

BEAMENT J. W. L. (1948a) The penetration of the insect egg-shells. I. Penetration of the chorion of *Rhodnius prolixus* Stål. *Bull. ent. Res.* **39**, 359–383.

BEAMENT J. W. L. (1948b) The role of wax layers in the waterproofing of insect cuticle and egg-shell. *Discuss. Faraday Soc.* no. 3, 177–182.

BEAMENT J. W. L. (1949) The penetration of insect egg-shells. II. The properties and permeability of sub-choral membranes during development of *Rhodnius prolixus* Stål. *Bull. ent. Res.* **39**, 467–488.

BEAMENT J. W. L. (1961) The water relations of insect cuticle. *Biol. Rev.* **36**, 281–320.

BEAMENT J. W. L. (1964) The active transport and passive movement of water in insects. *Adv. Insect Physiol.* **2**, 67–129.

BEAMENT J. W. L. and LAL R. (1957) Penetration through the egg-shell of *Pieris brassicae* (L.). *Bull. ent. Res.* **48**, 109–125.

BEAMER R. H. (1930) Maternal instinct in a Membracid (*Platycotis vittata*) (Homop.). *Ent. News* **41**, 330–331.

BEAMS H. W. and KESSEL R. G. (1969) Synthesis and deposition of oocyte envelopes (vitelline membrane, chorion) and the uptake of yolk in the dragonfly (Odonata: Aeshnidae). *J. Cell Sci.* **4**, 241–264.

BEATTY A. F. and BEATTY G. H. (1970) Gregarious (?) oviposition of *Calopteryx amata* Hagen (Odonata). *Proc. Pa Acad. Sci.* **44**, 156–158.

BEATTY G. H. and BEATTY A. F. (1967) The unusual oviposition behaviour, habitat and the description of the larva of *Palaemnema* (Odonata: Zygoptera) of Mexico. *Am. Zool.* **7**, 787 (abst.).

BEAUCOURNU J.-C. (1968) L'oeuf d'*Hoplopleura captiosa* Johnson 1960 (Anoplura, Hoplopleuridae). *Annls Parasit. hum. comp.* **43**, 611–612.

BEAVER R. A. (1967) Notes on the biology of the bark beetles attacking elm in Wytham Wood, Berks. *Entomologist's mon. Mag.* **102**, 156–170 (1966).

BEAVERS J. B., DENMARK H. A., and SELHIME A. G. (1972) Predation by *Blattisocius keegani* on egg masses of *Diaprepes abbreviatus* in the laboratory. *J. econ. Ent.* **65**, 1483–1484.

BECK S. D. (1957) The European corn borer *Pyrausta nubilalis* (Hübn.) and its principal host plant. VI. Host plant resistance to larval establishment. *J. Insect Physiol.* **1**, 158–177.

BECK S. D. (1960) The European corn borer *Pyrausta nubilalis* (Hübn.) and its principal host plant. VII. Larval feeding behaviour and host plant resistance. *Ann. ent. Soc. Am.* **53**, 206–212.

BECK S. D. and STAUFFER J. F. (1950) An aseptic method for rearing corn borer larvae. *J. econ. Ent.* **43**, 4–6.

BECKEL W. E. (1953) Preparing mosquito eggs for embryological study. *Mosq. News* **13**, 235–237.

BECKEL W. E. (1955a) Oviposition site preference of *Aedes* mosquitoes (Culicidae) in the laboratory. *Mosq. News* **15**, 224–228.

BECKEL W. E. (1955b) A method of separating fertilised from non-fertilised mosquito eggs. *Can. Def. Res. Bd. North. Lab. DRNL Techn. Mem.* no. 6/55.

BECKEL W. E. (1958) Investigations of permeability, diapause, and hatching in the eggs of the mosquito *Aedes hexodontus* Dyar. *Can. J. Zool.* **36**, 541–554.

BECKER P. (1961) Observations on the life cycle and immature stages of *Culicoides circumscriptus* Kieff. (Diptera, Ceratopogonidae). *Proc. R. Soc. Edinb.* (B) **67**, 363–386.

BECKWITH R. C. (1970) Influence of host on larval survival and adult fecundity of *Choristoneura conflictana* (Lepidoptera: Tortricidae). *Can. Ent.* **102**, 1474–1480.

BEDARD W. D., TILDEN P. E., WOOD, D. L., SILVERSTEIN R. M., BROWNLEE R. G., and RODIN J.-O. (1969) Western pine beetle: field response to its sex pheromone and a synergistic host, terpene, myrcene. *Science, Wash.* **164**, 1284–1285.

BEDFORD G. O. (1970) The development of the egg of *Didymuria violescens* (Phasmatodea: Phasmatidae: Podacanthinae)—embryology and determination of the stage at which first diapause occurs. *Aust. J. Zool.* **18**, 155–169.

BEDFORD G. O. (1977) Description and development of the eggs of two stick insects (Phasmatodea: Phasmatidae) from New Britain. *J. Aust. Ent. Soc.* **15**, 389–393.

BEEBE W. (1947) Scale adaptation and utilization in *Aesiocopa patulana* Walker (Lepidoptera, Heterocera, Tortricidae). *Zoologica, NY* **32**, 147–152.

BEEDEN P. (1974) Bollworm oviposition on cotton in Malawi. *Cotton Grow. Rev.* **51**, 52–61.

BEEGLE C. C. and OATMAN E. R. (1976) Host discrimination by *Hyposoter exiguae* (Hymenoptera: Ichneumonidae). *Entomologia exp. appl.* **20**, 77–80.

BEESON C. F. C. (1931) The life-history and control of *Celosterna scabrator* F. (Col., Cerambycidae). *Indian For. Rec. (Ent. Ser.)* **16** (9) 1–16.

BEESON C. F. C. (1941) *The Ecology and Control of the Forest Insects of India and the Neighbouring Countries*, Dehra Dun.

BEESON C. F. C. and BHATIA B. M. (1939) On the biology of the Cerambycidae (Coleopt.). *Indian For. Rec.* **5**, 1–235.

BEESON C. F. C. and CHATTERJEE S. N. (1935) Biology of the Braconidae. *Indian For. Rec. (NS) Ent.* **1**, 105–138.

BEGG M. and SANG J. H. (1950) A method for collecting and sterilising large numbers of *Drosophila* eggs. *Science, Lancaster Pa* **112**, 11–12.

BEHRENZ W. (1952) Experimentelle und histologische Untersuchungen am weiblichen Genitalapparat von *Lymantria dispar* L. *Zool. Jb. (Anat.)* **72**, 147–215.

BEIER M. (1964) Ordnung: Mantodea Burmeister 1838. *Bronn's Kl. Ordn. Tierreichs* **5** (3, 2) 849–970.

BEILMANN A. P. (1937) Social behaviour in Homoptera. *Psyche, Camb.* **44**, 58–59.

BEINGOLEA G. O. (1958) Notes on a Chamaemyiid (Diptera) predator of the eggs of *Orthezia insignis* Douglas (Homoptera) in Peru. *Bull. Brooklyn ent. Soc.* **52**, 121.

BÉIQUE R. (1961) Étude sur la mouche-à-scie du peuplier, *Trichiocampus viminalis* (Fall.) (Hyménoptère: Tenthredinidae). *Can. Ent.* **93**, 1085–1097.

BELL K. O. JR and WHITCOMB W. H. (1962) Efficiency of egg predators of the bollworm. *Ark. Fm Res.* **11** (6) 9.

BELL K. O. and WHITCOMB W. H. (1964) Field studies on egg predators of the bollworm, *Heliothis zea* (Boddie). *Fla Ent.* **47**, 171–180.

BELL W. J. (1971) Starvation-induced oöcyte resorption and yolk protein salvage in *Periplaneta americana*. *J. Insect Physiol.* **17**, 1099–1111.

BELL W. J. and BOHM M. K. (1975) Oösorption in insects. *Biol. Rev.* **50**, 373–396.

BELL W. J., BURK T., and SAMS G. R. (1973) Cockroach aggregation pheromone: directional orientation. *Behav. Biol.* **9**, 251–255.

BELL W. J., PARSONS C., and MARTINKO E. A. (1972) Cockroach aggregation pheromones: analysis of aggregation tendency and species specificity (Orthoptera: Blattidae). *J. Kans. ent. Soc.* **45**, 414–421.

BELLAMY R. E. and REPASS R. P. (1950) Notes on the ova of *Anopheles georgianus* King. *J. natn. Malar. Soc.* **9**, 84–88.

BELTON P. (1967) The effect of illumination and pool brightness on oviposition by *Culex restuans* (Theo.) in the field. *Mosq. News* **27**, 66–68.

BELUR N. V. and HOLDAWAY F. G. (1962) Some insect predators of corn borer (*Ostrinia nubilalis* (Hb.)) eggs in Minnesota (abstract). *Proc. N. cent. Br. ent. Soc. Am.* **17**, 142–143.

BENASSY C. (1955) Remarques sur deux aphelinidés: *Aphytis mytilaspidis* Le Baron et *Aphytis proclia* Walker. *Annls Épiphyt.* **6**, 11–17.

BENEDEK P. (1967) On the *Eurydema* species in Hungary (Heteroptera, Pentatomidae). V. The natural enemies of Hungarian *Eurydema* species. *Állatt. Közl.* **54**, 29–34 (in Hungarian; English summary).

BENEDEK P. (1968a) On the *Eurydema* species in Hungary. VIII. The eggs of Eurydemae (Heteroptera, Pentatomidae). *Z. angew. Ent.* **61**, 113–118.

BENEDEK P. (1968b) *Kalmanius*, a new genus of the subfamily Nabinae (Heteroptera: Nabidae). *Acta zool. Acad. Sci. hung.* **14**, 295–300.

BENEDETTO L. A. (1970) Notes about the biology of *Jewettoperla munoai* Benedetto (Plecoptera, Gripopterygidae). *Limnologica* **7**, 383–389.

BENGTSSON S. (1913) Undersökningar öfrer äggen hos Ephemeriderna. *Ent. Tidskr.* **34**, 271–320.

BENNETT C. B. (1904) Earwigs (*Anisolabis maritima* Bonn.). *Psyche, Camb.* **11**, 47–53.

BENNETT F. D. (1969) Observations on the life-history and mode of parasitism of the Tachinid *Pseudochaeta syngamiae* Thompson. *Tech. Bull. Commonw. Inst. biol. Control* **12**, 37–41.

BENOIS A. and MARRO J. P. (1973) Action prédatrice des fourmis sur les oeufs de Bombylides. *Entomophaga* **18**, 321–331.

BEN SAAD A. A. and BISHOP G. W. (1969) Egg-laying by the alfalfa weevil in weeds. *J. econ. Ent.* **62**, 1226–1227.

BENSON J. F. (1973) Population dynamics of cabbage root fly in Canada and England. *J. appl. Ecol.* **10**, 437–446.

BENSON R. B. (1938) On the classification of sawflies (Hymenoptera: Symphyta). *Trans. R. ent. Soc. Lond.* **87**, 353–384.

BENSON R. B. (1963) Wear and damage of sawfly saws (Hymenoptera, Tenthredinidae). *Not. Ent.* **43**, 137–138.

BENSON W. W. and MORAIS H. C. DE (1975) Variação nas taxas de oviposição em *Triatoma infestans* controle intrínseco ou extrínseco? *Revta bras. Biol.* **35**, 325–329.

BENTLEY M. D., LEE H. P., MCDANIEL I. N., STIEHL B., and YATAGAI M. (1974) A mosquito egg counter by simple modification of a colorimeter. *J. econ. Ent.* **67**, 790–791.

BENTLEY M. D., MCDANIEL I. N., LEE, H. P., STIEHL B., and YATAGAI M. (1976) Studies of *Aedes triseriatus* oviposition attractants produced by larvae of *Aedes triseriatus* and *Aedes atropalpus* (Diptera: Culicidae). *J. med. Ent.* **13**, 112–115.

BENTUR J. S. and MATHAD S. B. (1975) Dual role of mating in egg production and survival in the cricket *Plebeiogryllus guttiventris* Walker. *Experientia* **31**, 539–540.

BENZ G. (1970) The influence of the presence of individuals of the opposite sex and some other stimuli on sexual activity, oogenesis and oviposition in five Lepidopterous species. *Colloques int. Centre natn. Rech. Sci.* no. **189**, 175–206.

BEQUAERT J. (1912) L'instinct maternel chez *Rhinocoris albospilosus* Sign. Hémiptère Réduviidae. *Rev. zool. afr.* **1**, 293–296.

BEQUAERT J. (1935) Presocial behaviour among the Hemiptera. *Bull. Brooklyn ent. Soc.* **30**, 177–191.

BÉRENGUIER P. (1907) Notes orthoptérologiques. II. Biologie de l'*Isophya pyrenaea* var. *nemausensis*. *Bull. Soc. Sci. nat. Nîmes* **35**, 1–13.

BERG C. O. (1960) Biology of snail-killing Sciomyzidae (Diptera) of North America and Europe. *Proc. 11th Int. Congr. Ent.* **1**, 197–202.

BERG C. O. (1973) Biological control of snail-borne diseases (with Sciomyzids): a review. *Expl Parasit.* **33**, 318–330.

BERG K. (1937) Contributions to the biology of *Corethra* Meigen (*Chaoborus* Lichtenstein). *Biol. Medd.* **13**, 1–101.

BERG K. (1938) Studies on the bottom animals of Ersom lake. *K. danske Vidensk. Selsk. Skv.* **8** (9) 1–255.

BERG K. (1941) Contributions to the biology of the aquatic moth *Acentropus niveus* (Oliv.). *Vidensk. Meddr. dansk naturh. Foren.* **105**, 59–139.

BERG M. A. VAN DEN (1972) Studies on the longevity of and number of eggs parasitised by *Mesocomys pulchriceps* Cam. (Hymenoptera: Eupelmidae), an egg parasite of Saturniidae (Lepidoptera). *Phytophylactica* **4**, 113–118.

BERGERARD J. (1962) Parthenogenesis in the Phasmidae. *Endeavour* **21**, 137–143.

BERI S. K. (1974) Biology of a leaf miner *Liriomyza brassicae* (Riley) (Diptera: Agromyzidae). *J. nat. Hist.* **8**, 143–151.

BERISFORD C. W. and FRANKLIN R. T. (1972) Tree host influence on some parasites of *Ips* spp. bark beetles (Coleoptera: Scolytidae) on four species of southern pines. *J. Ga ent. Soc.* **7**, 110–115.

BERKOWITZ S. (1945) Caterpillar dermatitis. *Bull. US Army med. Dep.* **4**, 464–467.

BERLEPSCH A. VON (1860) *Die Biene und die Bienenzucht*, Mülhausen, Heinrichshofensche Buchhandlung.

BERLESE A. (1913) Intorno alle metamorfosi degli insetti. *Redia* **9**, 121–138.

BERNARD F. (1936) La ponte et l'éclosion des *Anieles*. *Bull. Soc. ent. Fr.* **41**, 42–44.

BERNAYS E. A. (1970) Water uptake and exchange in the eggs of *Schistocerca gregaria* at a late stage in development. *J. Insect Physiol.* **16**, 825–832.

BERNAYS E. A. (1971) The vermiform larva of *Schistocerca gregaria* (Forskal): form and activity (Insecta, Orthoptera). *Z. Morph. Ökol. Tiere* **70**, 183–200.

BERNAYS E. A. (1972a) Hatching in *Schistocerca gregaria* (Forskal) (Orthoptera, Acrididae). *Acrida* **1**, 41–60.
BERNAYS E. A. (1972b) The muscles of newly hatched *Schistocerca gregaria* larvae and their possible functions in hatching, digging and ecdysial movements (Insecta: Acrididae). *J. Zool. Lond.* **166**, 141–158.
BERNHARD C. (1907) Über die vivipare Ephemeride *Chloën dipterum*. *Biol. Zbl.* **27**, 467–479.
BERRY S. J. (1968) The fine structure of the colleterial glands of *Hyalophora cecropia* (Lepidoptera). *J. Morph.* **125**, 259–279.
BERRYMAN A. A. (1966) Studies on the behaviour and development of *Enoclerus lecontei* (Wolcott), a predator of the western pine beetle. *Can. Ent.* **98**, 519–526.
BERRYMAN A. A. (1968) Estimation of oviposition by the fir engraver, *Scolytus ventralis* (Coleoptera: Scolytidae). *Ann. ent. Soc. Am.* **61**, 227–228.
BERRYMAN A. A. (1969) Responses of *Abies grandis* to attack by *Scolytus ventralis* (Coleoptera: Scolytidae). *Can. Ent.* **101**, 1033–1041.
BERRYMAN A. A. and ASHRAF M. (1970) Effects of *Abies grandis* resin on the attack behaviour and brood survival of *Scolytus ventralis* (Coleoptera: Scolytidae). *Can. Ent.* **102**, 1229–1236.
BERTRAND H. (1924) Éclosion de l'oeuf chez quelques Chrysomélides. *Bull. Soc. ent. Fr.* **1924**, 54–57.
BESS H. A. (1936) The biology of *Leschenaultia exul* Townsend, a tachinid parasite of *Malacosoma americana* Fabricius and *Malacosoma disstria* Hübner. *Ann. ent. Soc. Am.* **29**, 593–613.
BESS H. A. (1939) Investigations on the resistance of mealybugs (Homoptera) to parasitization by internal hymenopterous parasites, with special reference to phagocytosis. *Ann. ent. Soc. Am.* **32**, 189–226.
BESS H. A. (1961) Population ecology of the gypsy moth *Porthetria dispar* L. (Lepidoptera: Lymantridae). *Bull. Conn. agric. exp. Sta.* **646**, 1–43.
BETTEN C. (1934) The caddis flies or Trichoptera of New York State. *Bull. NY St. Mus.* **292**, 1–576.
BEVAN D. (1962) Pine shoot beetles. *For. Comm. Leaflet* **3**, 1–8.
BLVIS A. L. (1923) A lepidopterous parasite on a coccid. *S. Afr. J. nat. Hist.* **4**, 34–35.
BEYER A. H. (1922) A brief resumé of investigations made in 1913 on *Trogoderma inclusa* Lec. (a dermestid). *Kans. Univ. Sci. Bull.* **14** (15), 373–391.
BEZRUKOV J. G. (1922) A brief report on the work of the Omsk laboratory of the Siberian Entomological Bureau in 1919–1922. *Izv. sibirsk. ent. Byu.* **1**, 26–30 (in Russian).
BHANOT J. P. and KAPIL R. P. (1974) Effect of starvation on fecundity and hatchability of the eggs of *Acaropsis docta* (Berlese). *Indian J. Ent.* **35**, 67–69 (1973).
BHARADWAJ R. K. (1966) Observations on the bionomics of *Euborellia annulipes* (Dermaptera: Labiduridae). *Ann. ent. Soc. Am.* **59**, 441–450.
BHATIA K. R. and HARJAI S. C. (1968) Oviposition behaviour of the desert locust (*Schistocerca gregaria* Forsk.). *Indian J. Ent.* **30**, 109–112.
BHATIA M. L. (1939) Biology, morphology, and anatomy of aphidophagous syrphid larvae. *Parasitology* **31**, 78–129.
BHATIA M. L. and SHAFFI M. (1932) Life histories of some Indian Syrphidae. *Indian J. agric. Sci.* **2**, 543–570.
BHATIA S. K. and MAHTO Y. (1968) Notes on breeding of fruit flies *Dacus ciliatus* Loew and *D. cucurbitae* Coquillett in stem galls of *Coccinia indica* W. and A. *Indian J. Ent.* **30**, 244–245.
BHUANGPRAKONE S. and AREEKUL S. (1973) Biology and food habits of the snail-killing fly, *Sepedon plumbella* Weidemann (Sciomyzidae: Diptera). *Southeast Asian J. trop. Med. Public Health* **4**, 387–394.

BIBOLINI C. (1973) Contributo alla conoscenza dei crisomelidi italiani (Coleoptera-Chrysomelidae). II. Sulla biologia della *Cassida algirica* Lucas. *Frustula Ent.* **11** (3) 1–220.

BICK G. H. and BICK J. C. (1970) Oviposition in *Archilestes grandis* (Rambur) (Odonata: Lestidae). *Ent. News* **81**, 157–163.

BICKLEY W. E., HARRISON F. P., and DITMAN L. P. (1956) *Drosophila* as a pest of canning tomatoes. *J. econ. Ent.* **49**, 417–418.

BIER K. (1954) Über den Saisondimorphismus der Oogenesis von *Formica rufa rufo-pralensis* minor Grössw. und dessen Bedeutung für die Kasten-Determination. *Biol. Zbl.* **73**, 170–190.

BIERMANN G. and THALENHORST W. (1977) Zur Kenntnis des kleinen Buchdruckers *Ips amitinus* (Eichh.) (Col. Scolytidae). *Anz. Schädlingsk. Pflanzenschutz, Umweltschutz* **50**, 20–23.

BIGGER J. H. (1930) Notes on the life history of the clover-root curculio, *Sitona hispidula* Fab., in central Illinois. *J. econ. Ent.* **23**, 334–342.

BIGGER M. (1969) Partial resistance of arabica coffee to the coffee leaf miner *Leucoptera meyricki* Ghesq. (Lepidoptera, Lyonetiidae). *E. Afr. agric. For. J.* **34**, 441–445.

BILIOTTI E. (1956) Biologie de *Phryxe caudata* Rondani (Dipt.: Larvaevoridae), parasite de la chenille processionnaire du pin (*Thaumetopoea pityocampa* Schiff.). *Rev. Path. vég. Ent. agric. Fr.* **35**, 50–65.

BILIOTTI E. (1958) Les parasites et prédateurs de *Thaumetopoea pityocampa* Schiff (Lepidoptera). *Entomophaga* **3**, 23–34.

BILIOTTI E., DEMOLIN G., and DU MERLE P. (1966) Parasitisme de la processionnaire du pin par *Villa quinquefasciata* Wied. apud Meig. (Dipt.: Bombyliidae). Importance du comportement de ponte du parasite. *Annls Épiphyt.* **16**, 279–288 (1965).

BILLINGS R. F., GARA R. I., and HRUTFIORD B. F. (1976) Influence of ponderosa pine resin volatiles on the response of *Dendroctonus ponderosae* to synthetic *trans*-verbenol. *Envir. Ent.* **5**, 171–179.

BILLS G. T. (1973) Biological fly control in deep-pit poultry houses. *Br. Poultry Sci.* **14**, 209–212.

BINGHAM J. and LUPTON F. G. H. (1958) Breeding spring oats for resistance to frit-fly attack. *Ann. appl. Biol.* **46**, 493–497.

BIRCH L. C. (1944) The effect of temperature and dryness on the survival of the eggs of *Calandra oryzae* L. (small strain) and *Rhizopertha dominica* Fab. (Coleoptera). *Aust. J. exp. Biol. med. Sci.* **22**, 265–269.

BIRCH L. C. (1945) A contribution to the ecology of *Calandra oryzae* L. and *Rhizopertha dominica* Fab. (Coleoptera) in stored wheat. *Trans. R. Soc. S. Aust.* **69**, 140–149.

BIRCH L. C. and ANDREWARTHA H. G. (1942) The influence of moisture on the eggs of *Austroicetes cruciata* Saus. (Orthoptera) with reference to their ability to survive desiccation. *Aust. J. exp. Biol. med. Sci.* **20**, 1–8.

BIRCH L. C., PARK T., and FRANK M. B. (1951) The effect of intraspecies and interspecies competition on the fecundity of two species of flour beetle. *Evolution* **5**, 116–132.

BIRCH M. C. and WOOD D. L. (1975) Mutual inhibition of the attractant pheromone response by two species of *Ips* (Coleoptera: Scolytidae). *J. chem. Ecol.* **1**, 101–113.

BISHARA S. I. (1969) Factors involved in recognition of the oviposition sites of three species of *Sitophilus* (Coleoptera: Curculionidae). *Bull. Soc. ent. Égypte* **51**, 71–94.

BISHARA S. I., KOURA A., and EL-HALFAWY M. A. (1973) Oviposition preference of the granary and the rice weevils on the Egyptian rice varieties, and recommendations for grain protection. *Bull. Soc. ent. Égypte* **56**, 145–150.

BJEGOVIĆ P. (1957) Žitni (*Zabrus tenebrioides* Goeze) i njegova parazitska (*Viviania cinerea* Fall.). *Mem. Inst. Pl. Prot. Belgrade* no. 5, 1–104 (English summary).

BJEGOVIĆ P. (1968) Some biological characteristics of the damsel bug *Nabis feroides* Rm. (Hemiptera, Nabidae) and its role in the population dynamics of the cereal leaf beetle—*Lema melanopa* L. *Zašt. Bilja* **19**, 235–246 (in Yugoslav; English summary).

BLAGOVESHCHENSKIĬ D. I. (1955) Morphology of the ovum of bird lice (Mallophaga). *Trud. Zool. Inst. Akad. Nauk SSSR* **21**, 262–270 (in Russian).

BLAGOVESHCHENSKIĬ D. I. (1959) Fauna USSR: Mallophaga 1, no. 1. Mallophaga, pt. 1. Introduction. *Zool. Inst. Akad. Nauk SSSR* (NS) **72**, 1–202 (in Russian).

BLAKE G. M. (1961) Length of life, fecundity and the oviposition cycle in *Anthrenus verbasci* (L.) (Col., Dermestidae) as affected by adult diet. *Bull. ent. Res.* **52**, 459–472.

BLANCHARD E. E. (1937) Dipteros argentinos nuevos o poco conocidos. *Rev. Soc. ent. argent.* **9**, 35–58.

BLEWETT M. and FRAENKEL G. (1944) Intracellular symbiosis and vitamin requirements of two insects, *Lasioderma serricorne* and *Sitodrepa panicea*. *Proc. R. Soc.* (B) **132**, 212–221.

BLISS C. I. (1927) The oviposition rate of the grape leaf hoppers. *J. agric. Res.* **34**, 847–852.

BLISS M. JR. and KEARBY W. H. (1971) Observations on the oviposition sites (leaf scars on branches of Scots pine (*Pinus sylvestris*)) and laboratory development of the fundatrix and virginopara of the aphid *Eulachnus (Protolachnus) agilis*. *Ann. ent. Soc. Am.* **64**, 1407–1410.

BLUME R. R., MILLER J. A., ESCHLE J. L., MATTER J. J., and PICKENS M. O. (1972) Trapping Tabanids with modified Malaise traps baited with CO_2. *Mosq. News* **32**, 90–95.

BLUNCK H. (1913a) Kleine Beiträge zur Kenntnis des Geschlechtsleben und der Metamorphose der Dytisciden. 1. Teil. *Colymbetes fuscus* L. und *Agabus undulatus* Schrank. 2. Teil. *Acilius sulcatus* L. *Zool. Anz.* **41**, 534–546, 586–597.

BLUNCK H. (1913b) Das Geschlechtsleben des *Dytiscus marginalis* L. 2. Teil. Die Eiablage. *Z. wiss. Zool.* **104**, 157–179.

BLUNCK H. (1914) Die Entwicklung des *Dytiscus marginalis* L. vom Ei bis zur Imago. 1. Teil. Das Embryonalleben. *Z. wiss. Zool.* **111**, 76–151.

BLUNCK H. (1951) Zur Kenntnis der Hyperparasiten von *Pieris brassicae* L. 4. Beitrag: *Gelis* cf. *transfuga* Först. *Z. angew. Ent.* **33**, 217–267.

BOBB M. L. (1951) Life history of *Ochterus banksi* Barb. (Hemipt.: Ochteridae). *Bull. Brooklyn ent. Soc.* **46**, 92–100.

BOBB M. L. (1959) The biology of the lesser peach tree borer in Virginia. *J. econ. Ent.* **52**, 634–636.

BODENHEIMER F. S. and NERYA A. BEN (1937) One-year studies on the biology of the honeybee in Palestine. *Ann. appl. Biol.* **24**, 385–403.

BODENHEIMER F. S. and SHULOV A. (1951) Egg-development and diapause in the Moroccan locust (*Dociostaurus maroccanus* Thnb.). *Bull. Res. Coun. Israel* **1**, 59–75.

BODENSTEIN D. and SHAAYA E. (1968) The function of the accessory sex glands in *Periplaneta americana* (L.). I. A quantitative bioassay for the juvenile hormone. *Proc. natn. Acad. Sci. USA* **59**, 1223–1230.

BODENSTEIN D. and SPRAGUE I. B. (1959) The developmental capacities of the accessory sex glands in *Periplaneta americana*. *J. exp. Zool.* **142**, 177–202.

BODKIN G. E. (1917) Notes on the Coccidae of British Guiana. *Bull. ent. Res.* **8**, 103–109.

BOGOESCU C. (1933) Câtera consideraiuni asupra pontei la Ephemeroptere. *Notat. biol. Bucarest* **1**, 16–21.

BOGOESCU M. (1970) Contribuții la cunoașterea biologiei păduchelui rozător *Damalinia* (=*Bovicola*) *bovis* (Linné, 1758). I. Experiențe in vederea reproducerii in condiții de laborator a ciclului evolutiv pe cale partenogenetică. *Lucr. Ses. stiinț. Inst. agron. Nicolae Bălcescu* (C) **12** (1969) 491–500, Bucharest (summaries in English, French and Russian).

BOGUSH P. P. (1964) A predator of the Colorado beetle. *Zashch. Rast. Vredit. Bolez.* **42** (8) (in Russian).

BOHART G. E., STEPHEN W. P., and EPPLEY R. K. (1960) The biology of *Heterostylum robustum* (Diptera: Bombyliidae), a parasite of the alkali bee. *Ann. ent. Soc. Am.* **53**, 425–435.

BOHLEN E. (1967) Untersuchungen zum Verhalten der Möhrenfliege, *Psila rosae* Fab. (Dipt. Psilidae), im Eiablagefunktionskreis. *Z. angew. Ent.* **59**, 325–360.

BOLDORI L. (1933) Appunti biologici sul *Pterostrichus multipunctatus* Dej. *Studi trentini* **14**, 222–223.

BOLDT P. E. (1974) Effects of temperature and humidity on development and oviposition of *Sitotroga cerealella* (Lepidoptera: Gelechiidae). *J. Kans. ent. Soc.* **47**, 30–36.

BOLDȲREV M. I. and WILDE W. H. A. (1969) Food seeking and survival in predaceous coccinellid larvae. *Can. Ent.* **101**, 1218–1222.

BOLDȲREV M. I., WILDE W. H. A., and SMITH B. C. (1969) Predaceous coccinellid oviposition responses to *Juniperus* wood. *Can. Ent.* **101**, 1199–1206.

BOLES H. P. (1971) Ovipositional responses of the rice weevil, *Sitophilus oryzae* (L.), treated with synergized pyrethrins. *J. Kans. ent. Soc.* **44**, 70–75.

BOLES H. P. and ERNST R. L. (1976) Susceptibility of six wheat cultivars to oviposition by rice weevils reared on wheatcorn or sorghum. *J. econ. Ent.* **69**, 548–550.

BOLES H. P. and MAHANY P. G. A. (1967) A method of determining rate and time of egg deposition by insects. *J. econ. Ent.* **60**, 1763.

BOLING J. C. and PITRE H. N. (1971) Hostal plant preference for oviposition by *Trichoplusia ni* and efficiency of *Apanteles marginiventris* as a population regulator of *T. ni* in field-cage tests. *J. econ. Ent.* **64**, 411–412.

BOLLER E. F. (1965) Beitrag zur Kenntnis der Eiablage und Fertilität der Kirschenfliege *Rhagoletis cerasi*. *Mitt. schweiz. ent. Ges.* **38**, 194–202.

BOLLER E. F. (1968) An artificial oviposition device for the European cherry fruit fly, *Rhagoletis cerasi*. *J. econ. Ent.* **61**, 850–852.

BOLLER E. F., HAISCH A., and PROKOPY R. J. (1970) Ecological and behavioural studies preparing the application of the sterile insect release method (SIRM) against *Rhagoletis cerasi*. Int. Atom. Energy Agency Symp. Athens, 1970, *Sterility Principle for Insect Control or Eradication*.

BOMBOSCH S. (1962) Untersuchung über die Auslösung der Eiablage bei *Syrphus corollae* Fabr. (Dipt. Syrphidae). *Z. angew. Ent.* **50**, 81–88.

BOMBOSCH S. and VOLK S. (1968) Selection of the oviposition site by *Syrphus corollae* Fabr. In *Ecology of Phytophagous Insects* (I. Hodek, ed.), pp. 117–119 (1966), The Hague.

BOND E. J. and MONRO H. A. U. (1954) Rearing the Cadelle *Tenebroides mauritanicus* (L.) (Coleoptera: Ostomidae) as a test insect for insecticidal research. *Can. Ent.* **86**, 402–408.

BONDAR G. (1921) La larve de la noix des palmiers. *Broteria Serie zoológica* **19**, 125–135.

BONGERS W. (1970) Aspects of host–plant relationship of the Colorado beetle. *Meded. LandbHoogesch. Wageningen* **70** (10) 1–77.

BONNANFANT-JAÏS M. L. (1975) Morphologie de la glande lactée d'une glossine, *Glossina austeni* Newst, au cours du cycle de gestation. II. Aspects ultrastructuraux en période de repos et au cours des gestations successives. *J. microsc. Biol. Cell.* **24**, 295–314.

BONNEMAISON L. (1961a) A study of some factors affecting the fecundity and fertility of the cabbage armyworm (*Mamestra brassicae* L.) (Lep.). I. The effect of temperature. *Bull. Soc. ent. Fr.* **65**, 196–206 (1960) (in French).

BONNEMAISON L. (1961b) A study of some factors affecting the fecundity and fertility of the cabbage armyworm (*Mamestra brassicae* L.) (Lep.). II. The effect of light on the adults and on mating. *Bull. Soc. ent. Fr.* **66**, 62–70 (in French).

BONNEMAISON L. and JOURDHEUIL P. (1955) L'altise d'hiver du colza (*Psylliodes chrysocephala* L.). *Annls Épiphyt.* **5**, 345–524.

BOOTH D. C. and LANIER G. N. (1974) Evidence of an aggregating pheromone in *Pissodes approximatus* and *P. strobi*. *Ann. ent. Soc. Am.* **67**, 992–994.

Bordas L. (1908) Rôle physiologique des glandes arborescentes annexées à l'appareil générateur femelle des Blattes (*Periplaneta orientalis* L.). *C. r. hebd. Séanc. Acad. Sci. Paris* **147**, 1495–1497.

Bordas L. (1909) Recherches anatomiques, histologiques et physiologiques sur les organes appendiculaires de l'appareil reproducteur femelle des Blattes (*Periplaneta orientalis* L.). *Annls Sci. nat. (Zool.)* **9**, 71–121.

Borden J. H. (1968) Sex pheromone of *Trypodendron lineatum* (Coleoptera: Scolytidae): production, bioassay, and partial isolation. *Can. Ent.* **100**, 629–636.

Borden J. H. (1969) Observations on the life history and habits of *Alniphagus aspericollis* (Coleoptera: Scolytidae) in southwestern British Columbia. *Can. Ent.* **101**, 870–878.

Borden J. H. (1974a) Pheromone mask produced by male *Trypodendron lineatum* (Coleoptera: Scolytidae). *Can. J. Zool.* **52**, 533–536.

Borden J. H. (1974b) Aggregation pheromones in Scolytidae. In *Pheromones. Frontiers of Biology* (M. C. Birch, ed.) **32**, 135–160, London.

Borden J. H. and Fockler C. E. (1973) Emergence and orientation behaviour of brood *Trypodendron lineatum* (Coleoptera: Scolytidae). *J. ent. Soc. Br. Columb.* **70**, 34–38.

Borden J. H. and Slater C. E. (1969) Sex pheromone of *Trypodendron lineatum*: production in the female hindgut–Malpighian tubule region. *Ann. ent. Soc. Am.* **62**, 454–455.

Bordon J. (1971) Aspects de la physiologie de la reproduction et du développement chez *Oscinella pusilla* Meig. (Dipt.: Chloropidae). *Annls Zool. Écol. Anim.* **3**, 225–245.

Boreham P. F. L. and Snow W. F. (1973) Further information on the food sources of *Culex (Culex) decens* Theo. (Dipt. Culicidae). *Trans. R. Soc. trop. Med. Hyg.* **67**, 724–725.

Borg A. F. and Horsfall W. R. (1953) Eggs of floodwater mosquitoes. II. Hatching stimulus. *Ann. ent. Soc. Am.* **46**, 472–478.

Borkhausen M. B. (1770) *Naturgeschichte der europäischen Schmetterlinge von dem Verfasser des Nomenclator entomologicus* **3**, 140, Frankfurt.

Bornemissza G. F. (1966) Observations on the hunting and mating behaviour of two species of scorpion flies (Bittacidae: Mecoptera). *Aust. J. Zool.* **14**, 371–382.

Bornemissza G. F. (1968) Studies on the histerid beetle *Pachylister chinensis* in Fiji and its possible value in the control of buffalo-fly in Australia. *Aust. J. Zool.* **16**, 673–688.

Bos J. van den and Baltensweiler W. (1977) Oviposition efficiency of the grey larch bud moth (*Zeiraphera diniana*) on different substrates and its relevance to the population dynamics of the moth. *Entomologia exp. appl.* **21**, 88–97.

Bosch R. van den (1964) Encapsulation of the eggs of *Bathyplectes curculionis* (Thomson) (Hymenoptera: Ichneumonidae) in larvae of *Hypera brunneipennis* (Boheman) and *Hypera postica* (Gyllenhal) (Coleoptera: Curculionidae). *J. Insect Pathol.* **6**, 343–367.

Boscher J. (1975) Recherches préliminaires sur les substances émises par le poireau (*Allium porrum* L.) et stimulant la ponte de la teigne (*Acrolepia assectella* Zell.). *Annls Zool. Écol. anim.* **7**, 499–504.

Bose K. C. and Sinha P. B. (1965) Studies on the structure of eggs of an Indian stink bug *Halys dentatus* Fabricius. *Entomologist* **98**, 224–226.

Bose K. C. and Sinha P. B. (1968) Studies on the structure of the eggs of an Indian stink bug *Cyclopelta siccifolia* Westwood (Hemiptera, Heteroptera, Penthtomidae (*sic*)). *Entomologist* **101**, 221–223.

Boselli F. B. (1932) Istinti materni del *Sehirus sexmaculatus* Rbr. (Heteroptera: Cydnidae). *Boll. Lab. Zool. gen. agric. Portici* **26**, 1–8.

Boulard M. (1974) Comportement de ponte de *Cicadetta pygmea* Olivier et précisions sur la bionomie de cette cigale dans le Sud de la France (Hom. Tibicinidae). *Bull. Soc. ent. Fr.* **78**, 243–249 (1973).

Bouligand Y. (1965) Sur une architecture torsadée repandue dans de nombreuses cuticules

d'arthropodes. *C. r. hebd. Séanc. Acad. Sci. Paris* **261**, 3665–3668.
BOURGOGNE J. (1949) Un type nouveau d'appareil génital femelle chez les lépidoptères. *Annls Soc. ent. Fr.* **115**, 69–80 (1946).
BOURGOGNE J. (1950a) L'appareil génital femelle de quelques Hepialidae (Lépidoptères). *Bull. Soc. zool. Fr.* **74**, 284–291 (1949).
BOURGOGNE J. (1950b) Remarques sue l'appareil génital femelle des Psychidae (Lépidoptères) et sur l'importance de sa structure en systématique. *Bull. Soc. zool. Fr.* **75**, 104–109.
BOURNIER A. (1957) Un deuxième cas d'ovoviviparité chez les Thysanoptères *Caudothrips buffai* Karny (Tubulifère, Megathripidae). *C. r. hebd. Séanc. Acad. Sci. Paris* **244**, 506–508.
BOUVIER G. (1945) Note sur quelques oeufs d'ectoparasites se rencontrant sur les animaux domestiques (Anoplura et Mallophaga). *Schweiz. Arch. Tierheilk.* **87**, 273–276.
BOVEY P. (1936) Sur la ponte et la larve primaire d'*Oncodes pallipes* Latr. *Bull. Soc. vaud. Sci. nat.* **59**, 171–176.
BÖVING A. (1914) Notes on the larva of *Hydroscapha* and some other aquatic larvae from Arizona. *Proc. ent. Soc. Wash.* **26**, 169–174.
BØVING A. G. and HENRIKSEN K. L. (1938) The developmental stages of the Danish Hydrophilidae. *Vidensk. Meddr dansk naturh. Foren.* **102**, 27–162.
BOWDITCH F. C. (1884) *Hydrocharis obtusatus. J. Boston zool. Soc.* **3**, 1–7.
BOYCE A. M. (1934) Bionomics of the walnut husk fly, *Rhagoletis completa. Hilgardia* **8**, 363–379.
BOYCE J. M. (1946) The influence of fecundity and egg mortality on the population growth of *Tribolium confusum* Duval. *Ecology* **27**, 290–302.
BOYÉ R. (1932) La papillonite guyanaise. *Bull. Soc. Path. exot.* **25**, 1099–1107.
BOYER J. (1929) Chenilles et papillons venimeux. *Nature, Paris* 1929, 385–388.
BRACK-EGG A. (1973) Der Kopf *Rhynchites auratus* Scop. (Curculionidae)—Eine konstruktionsmorphologische Untersuchung mit einem Beitrag zum Brutfürsorgeverhalten. *Zool. Jb. (Anat.)* **91**, 500–545.
BRACKEN G. K. (1965) Effects of dietary components on fecundity of the parasitoid *Exeristes comstockii* (Cress.) (Hymenoptera: Ichneumonidae). *Can. Ent.* **97**, 1037–1041.
BRADER L. (1962) Observations on the life history of *Xyleborus morstatti* in the Ivory Coast. *Trop. Pl.-ziekten* **69**, 111–113.
BRADER L. (1964) Étude de la relation entre le scolyte des rameaux du cafeier *Xyleborus compactus* Eichh. (*X. morstatti* Hag.), et sa plante-hôte. *Meded. LandbHoogesch. Wageningen* **7**, 1–109.
BRADLEY G. H. and TRAVIS B. V. (1942) Soil sampling for studying distribution of mosquito eggs on salt marshes in Florida. *Proc. NJ Mosquito Extermination Ass.* **29**, 143–146.
BRAMMANIS L. (1963) Zum Vorkommen und zur Bekämpfung des kleinen Aspenbockes *Saperda populnea* L. in Schweden. *Z. angew. Ent.* **51**, 122–129.
BRANCH H. E. (1931) Identification of chironomid egg masses. II. *Trans. Kans. Acad. Sci.* **34**, 151–157.
BRAND J. M., BRACKE J. W., MARKOVETZ A. J., WOOD D. L. and BROWNE L. E. (1975) Production of verbenol pheromone by a bacterium isolated from bark beetles. *Nature, Lond.* **254**, 136–137.
BRANDT A. (1869) Beiträge zur Entwicklungsgeschichte der Libelluliden und Hemipteren, mit besonderer Berücksichtigung der Embronalhülle derselben. *Mem. Imp. Acad. Sci. St. Petersb.* **13**, 1–33.
BRANSON T. F. (1971) Resistance of spring wheat to the wheat stem maggot. *J. econ. Ent.* **64**, 941–945.
BRATT A. D., KNUTSON L. V., FOOTE B. A., and BERG C. O. (1969) Biology of *Pherbellia* (Diptera: Sciomyzidae). *Mem. Cornell Univ. agric. Exp. Stn* **404**, 1–246.

BRAUER F. (1869) Beschreibung der Verwandlungsgeschichte der *Mantispa styriaca* Poda; und Betrachtungen über die sogenannte Hypermentamorphose Fabre's. *Verh. zool. bot. Ges. Wien* **19**, 831–840.

BRAUER F. (1883) Erganzende Bermerkungen zu A. Handlirsch's Mittheilungen ueber *Hirmoneura obscura* Mg. *Wien. ent. Z.* **2**, 25–26.

BRAUER F. (1884) Zwei Parasiten des *Rhizotrogus solstitialis* auf der Ordnung der Dipteren. *Sber. Acad. Wiss. Wien* **88**, 865–875.

BRAUN A. F. (1927) A new species of *Holcocera* predaceous on mealybugs (Micro-lepidoptera). *Ent. News* **38**, 118.

BRAVERMAN Y., BOREHAM P. F. L., and GALUN R. (1971) The origin of blood meals of female *Culicoides pallidipennis* trapped in a sheepfold in Israel. *J. med. Ent.* **8**, 379–381.

BRAZZEL J. R. and MARTIN D. F. (1956) Resistance of cotton to pink bollworm damage. *Bull. Tex. agric. Exp. Stn* **843**, 20.

BRAZZEL J. R. and MARTIN D. F. (1957) Oviposition sites of the pink bollworm on the cotton plant. *J. econ. Ent.* **50**, 122–124.

BREADY J. K. and FRIEDMAN S. (1963) The nutritional requirements of termites in axenic cultures. 1. Sterilization of eggs of *Reticulitermes flavipes* and the requirements of first instar nymphs. *Ann. ent. Soc. Am.* **56**, 703–706.

BREDO H. J. (1936) Sommaire des observations faites au Congo Belge et projet des futures recherches sur les acridiens migrateurs. *Bull. agric. Congo belge* **27**, 298–302.

BREELAND S. G., PLETSCH D. J., and QUARTERMAN K. D. (1970) Laboratory studies on the suitability of mud as an oviposition substrate for *Anopheles albimanus* Wiedemann. *Mosq. News* **30**, 81–88.

BREESE M. H. (1948) Notes on the oviposition site and method of reproduction of the weevil *Strophosomus melanogrammus* (Förster) (Col.). *Proc. R. ent. Soc. Lond.* (A) **23**, 62–65.

BREESE M. H. (1961) The development and oviposition of *Oryzaephilus* spp. on unrefined sugars. *Bull. ent. Res.* **52**, 7–16.

BREESE M. H. (1963) Studies on the oviposition of *Rhyzopertha dominica* (F.) in rice and paddy. *Bull. ent. Res.* **53**, 621–637.

BRELAND O. P. (1941) Notes on the biology of *Stagmomantis carolina* (Joh.) (Orthoptera, Mantidae). *Bull. Brooklyn ent. Soc.* **36**, 170–177.

BRELAND O. P. (1949) The biology and the immature stages of *Aedes trivittatus* (Coquillett) (Diptera: Culicidae). *Ann. ent. Soc. Am.* **42**, 38–47.

BRELAND O. P. and DOBSON J. W. (1948) Specificity of Mantid oothecae (Orthoptera: Mantidae). *Ann. ent. Soc. Am.* **40**, 557–575 (1947).

BREMER H. (1950) Zur Rassenfrage bei *Pegomyia hyoscyami* Pz. *Z. PflKrankh.* **57**, 415–416.

BRENIÈRE J. (1965) Les trichogrammes parasites de *Proceras sacchariphagus* Boj., borer de la canne à sucre à Madagascar. Deuxième partie: étude biologique de *Trichogramma australicum* Gir. *Entomophaga* **10**, 99–117.

BRENNAN B. M. and CHENG T. C. (1975) Resistance of *Moniliformis dubius* to the defense reactions of the American cockroach *Periplaneta americana*. *J. invert. Path.* **26**, 65–73.

BRERETON J. LE GAY (1962) A laboratory study of population regulation in *Tribolium confusum*. *Ecology* **43**, 63–69.

BRESSLAU E. (1920) Eier und Eizahn der einbeimischen Stechmücken. V. Mitteilung der Beiträge zur Kenntnis der Lebensweise unserer Stechmücken. *Biol. Zbl.* **49**, 337–355.

BREWER F. D. and VINSON S. B. (1971) Chemicals affecting the encapsulation of foreign material in an insect. *J. invert. Path.* **18**, 287–289.

BRIAN M. V. (1953) Oviposition by the workers of the ant *Myrmica*. *Physiol. comp. oecol.* **3**, 25–36.

BRIAN M. V. and HIBBLE J. (1964) Studies of caste differentiation in *Myrmica rubra* L. 7. Caste bias, queen age and influence. *Insectes soc.* **11**, 223–238.

BRIAND L. J. (1931) Notes on *Chrysopa oculata*, etc. *Can. Ent.* **63**, 123–126.

BRIDGES E. T. and PASS B. C. (1970) Biology of *Draeculacephala mollipes* (Hem., Hom., Cicadellidae). *Ann. ent. Soc. Am.* **63**, 789–792.

BRIDWELL J. C. (1918) Notes on the Bruchidae and their parasites in the Hawaiian Islands. *Proc. Hawaii. ent. Soc.* **3**, 465–505.

BRIDWELL J. C. (1919) Some notes on Hawaiian and other Bethylidae (Hymenoptera) with descriptions of new species. *Proc. Hawaii. ent. Soc.* **4**, 21–38.

BRIEN P. (1923) Note sur *Phloea paradoxa* Burm. (1835). *Bull. Soc. ent. Belg.* **5**, 109–113.

BRIGGS E. M. (1905) The life history of case bearers. I. *Chlamys plicata*. *Cold Spr. Harb. Monogr.* **4**, 3–12.

BRIGHT D. E. JR and STARK R. W. (1973) The bark and ambrosia beetles of California (Coleoptera: Scolytidae and Platypodidae). *Bull. Calif. Insect Survey* **16**, 1–169.

BRINCK P. (1949) Studies on Swedish stoneflies. *Opusc. ent* (suppl.) **11**, 1–250.

BRINDLEY M. D. H. (1934) A note on the eggs and breeding habits of *Salda littoralis* L. *Proc. R. ent. Soc. Lond.* **9**, 10–11.

BRINKHURST R. O. (1959) Alary polymorphism in the Gerroidea (Hemiptera—Heteroptera). *J. Anim. Ecol.* **28**, 211–230.

BRISTOWE W. S. (1932) Insects and other invertebrates for human consumption in Siam. *Trans. ent. Soc. Lond.* **80**, 387–404.

BRITTON E. B. (1966) On the larva of *Sphaerius* and the systematic position of the Sphaeriidae (Coleoptera). *Aust. J. Zool.* **14**, 1193–1198.

BRITTON W. E. (1936) Clusters of flies mistaken for rust patches. *Bull. Conn. agric. Exp. Stn* **383**, 326–327.

BROADHEAD E. (1947) The life-history of *Embidopsocus enderleini* (Ribaga) (Corrodentia). *Entomologist's mon. Mag.* **83**, 200–203.

BROADHEAD E. (1958) The psocid fauna of larch trees in northern England—an ecological study of mixed species populations exploiting a common resource. *J. Anim. Ecol.* **27**, 217–263.

BROADHEAD E. (1961) The biology of *Psoquilla marginepunctata* (Hagen) (Corrodentia. Trogiidae). *Trans. Soc. Br. Ent.* **14**, 223–236.

BROADHEAD E. and HOBBY B. M. (1944) Studies on a species of *Liposcelis* (Corrodentia, Liposcelidae) occurring in stored products in Britain. Part I. *Entomologist's mon. Mag.* **80**, 45–59.

BROADHEAD E. and HOBBY B. M. (1945) The booklouse. *Discovery* **6**, 142–147.

BROADHEAD E. and WAPSHERE A. J. (1960) Notes on the eggs and nymphal instars of some psocid species. *Entomologist's mon. Mag.* **96**, 162–166.

BROADHEAD E. and WAPSHERE A. J. (1966) *Mesopsocus* populations on larch in England—the distribution and dynamics of two closely related coexisting species of Psocoptera sharing the same food resource. *Ecol. Monogr.* **36**, 327–388.

BROCHER F. (1911) Observations biologiques sur quelques insectes aquatiques. *Annls Biol. lacustre* **4**, 367–379.

BROCHER F. (1912a) Recherches sur la respiration des insectes aquatiques adultes. Les Haemonia *Annls Biol. lacustre* **5**, 5–26.

BROCHER F. (1912b) Recherches sur la respiration des insectes aquatiques adultes. Les Elmides. *Annls Biol. lacustre* **5**, 136–179.

BROCHER F. (1928) *Observations et réflexions d'un naturaliste dans sa campagne,* lre partie, Libraire Kundig, Geneva.

BROCK M. L., WIEGERT R. G., and BROCK T. D. (1969) Feeding by *Paracoenia* and *Ephydra* (Diptera: Ephydridae) on the microorganisms of hot springs. *Ecology* **50**, 192–200.

BRO LARSEN E. (1960) Birketaegen elasmucha. *Nat. Verden*, May, pp. 129–135.

BROMLEY S. W. (1928) A dragonfly ovipositing on a paved highway. *Bull. Brooklyn ent. Soc.* **23**, 69.
BROMLEY S. W. (1946) Family Asilidae. *Bull. Conn. geol. nat. Hist. Surv.* **69**, 2–51.
BRONGNIART C. (1881) Observations sur la manière dont les mantes construisent leur oothèques; sur la structure des oothèques; sur l'éclosion et la première mue des larves. *Annls Soc. ent. Fr.* **1** (6) 448–452.
BRONGNIART C. (1887) (No title) *Bull. Soc. ent. Fr.* **7** (6) lxxxiv–lxxxvii.
BRÖNNIMANN H. (1964) Rearing Anthocorids on an artificial medium. *Tech. Bull. Commonw. Inst. biol. Control* **4**, 147–150.
BROODRYK S. W. (1969) The biology of *Chelonus (Microchelonus) curvimaculatus* Cameron (Hymenoptera: Braconidae). *J. ent. Soc. S. Afr.* **32**, 169–189.
BROODRYK S. W. (1971a) The biology of *Diadegma stellenboschense* (Cameron) (Hymenoptera: Ichneumonidae), a parasitoid of potato tuber moth. *J. ent. Soc. S. Afr.* **34**, 413–423.
BROODRYK S. W. (1971b) Ecological investigations on the potato tuber moth, *Phthorimaea operculella* (Zeller) (Lepidoptera: Gelechiidae). *Phytophylactica* **3**, 73–84.
BROOKE M. M. and PROSKE H. O. (1946) Preciptin test for determining natural insect predators of immature mosquitoes. *J. Natn. Malar. Soc.* **5**, 45–56.
BROOKS F. E. (1926) Life history of the hickory spiral borer, *Agrilus arcuatus* Say. *J. agric. Res.* **33**, 331–338.
BROSSUT R., DUBOIS P., and RIGAUD J. (1973) Le grégarisme chez *Blaberus craniifer*: isolement et identification de la phéromone. *J. Insect Physiol.* **20**, 529–543.
BROVDY V. M. (1970) Viviparity in leaf beetle *Chrysochloa alpestris* Schumm. *Dopov. Akad. Nauk Ukr. URSR* (B) **1970**, 561–563 (in Russian; English summary).
BROWER L. P. (1961) Experimental analyses of egg cannibalism in the monarch and queen butterflies *Danaus plexippus* and *D. gilippus berenice*. *Physiol. Zoöl.* **34**, 287–296.
BROWN F. S. (1947) The distribution and vegetation of egg-laying sites of the desert locust (*Schistocerca gregaria* Forsk.) in Tripolitania in 1946. *Bull. Soc. Fouad I Ent. Cairo* **31**, 287–306.
BROWN H. D. (1972) The behaviour of newly hatched Coccinellid larvae (Coleoptera: Coccinellidae). *J. ent. Soc. S. Afr.* **35**, 149–157.
BROWN H. P. (1952) The life history of *Climacia areolaris* (Hagen), a neuropterous "parasite" of freshwater sponges. *Am. Midl. Nat.* **47**, 130–160.
BROWN J. G. (1966) Oviposition and egg incubation in three species of small green chafer (*Pyronota* Bois). *NZ Jl Sci.* **9**, 261–280.
BROWN K. W. (1973) Description of immature stages of *Philolithus densicollis* and *Stenomorpha puncticollis* with notes on their biology (Coleoptera, Tenebrionidae, Tentyriinae). *Postilla* **162**, 1–28.
BROWN R. W. (1957) Cockroach egg case from the Eocene of Wyoming. *J. Wash. Acad. Sci.* **47**, 340–342.
BROWN S. N. (1970) Distribution of *Menacanthus stramineus* in relation to chickens' surface temperatures. *J. Parasit.* **56**, 1205.
BROWNE F. G. (1959) Notes on two Malayan scolytid bark-beetles. *Malayan For.* **22**, 292–300.
BROWNE F. G. (1961) The biology of Malayan Scolytidae and Platypodidae. *Malayan For. Rec.* **22**, 1–255.
BROWNE F. G. (1963) Notes on the habits and distribution of some Ghanaian bark beetles and ambrosia beetles (Coleoptera: Scolytidae and Platypodidae). *Bull. ent. Res.* **54**, 229–266.
BROWNE F. G. (1971) *Austroplatypus*, a new genus of the Platypodidae (Coleoptera), infesting living eucalyptus trees in Australia. *Commonw. For. Rev.* **50** (1) 49–50.
BROWNE F. G. (1973) The African species of *Poecilips* Schaufuss (Coleoptera, Scolytidae). *Rev. Zool. Bot. afr.* **87**, 679–696.

Browne L. B. (1956) The effect of light on the fecundity of the Queensland fruit-fly, *Strumeta tryoni* (Frogg.). *Aust. J. Zool.* **4**, 125–145.

Browne L. B. (1957) The effect of light on the mating behaviour of the Queensland fruit-fly *Strumeta tryoni*. *Aust. J. Zool.* **5**, 145–158.

Browne L. B. (1958) The choice of communal oviposition sites by the Australian sheep blowfly *Lucilia cuprina*. *Aust. J. Zool.* **6**, 241–247.

Browne L. B. (1960) The role of olfaction in the stimulation of oviposition in the blowfly *Phormia regina*. *J. Insect Physiol.* **5**, 16–22.

Browne L. B. (1962) The relationship between oviposition in the blowfly *Lucilia cuprina* and the presence of water. *J. Insect Physiol.* **8**, 383–390.

Browne L. B. (1965) An analysis of the ovipositional responses of the blowfly *Lucilia cuprina* to ammonium carbonate and indole. *J. Insect Physiol.* **11**, 1131–1143.

Browne L. B., Bartell R. J., and Shorey H. H. (1969b) Pheromone mediated behaviour leading to group oviposition in the blowfly *Lucilia cuprina*. *J. Insect. Physiol.* **15**, 1003–1014.

Browne L. B., Soo Hoo C. F., Gerwen A. C. M. van, and Sherwell I. R. (1969a) Mating flight behaviour in three species of *Oncopera* moths (Lepidoptera: Hepialidae). *J. Aust. ent. Soc.* **8**, 168–172.

Browning F. R. (1947) Observations on the habits and oviposition of *Tettigonia viridissima* L. (Orth., Tettigoniidae). *Entomologist's mon. Mag.* **83**, 281–283.

Browning T. O. (1953) The influence of temperature and moisture on the uptake and loss of water in the eggs of *Gryllulus commodus* Walker (Orthoptera: Gryllidae). *J. exp. Biol.* **30**, 104–115.

Browning T. O. (1965) Observations on the absorption of water, diapause and embryogenesis in the eggs of the cricket *Teleogryllus commodus* Walker. *J. exp. Biol.* **43**, 433–439.

Browning T. O. (1967) Water and the eggs of insects. In *Insects and Physiology* (J. W. L. Beament and J. E. Treherne, eds.), pp. 315–328.

Browning T. O. (1969a) Permeability to water of the shell of the egg of *Locusta migratoria migratorioides*, with observations on the egg of *Teleogryllus commodus*. *J. exp. Biol.* **51**, 99–105.

Browning T. O. (1969b) The permeability of the shell of the egg of *Teleogryllus commodus* measured with the aid of tritiated water. *J. exp. Biol.* **51**, 397–405.

Browning T. O. (1972) The penetration of some non-polar molecules in solution through the egg-shells of *Locusta migratoria migratorioides* and *Teleogryllus commodus*. *J. exp. Biol.* **56**, 769–773.

Browning T. O. and Forrest W. W. (1960) The permeability of the shell of the egg of *Acheta commodus* Walker (Orthoptera, Gryllidae). *J. exp. Biol.* **37**, 213–217.

Brubaker R. W. (1968) Seasonal occurrence of *Voria ruralis*, a parasite of the cabbage looper, in Arizona, and its behaviour and development in laboratory culture. *J. econ. Ent.* **61**, 306–309.

Bruce-Chwatt L. J. and Service M. W. (1957) An aberrant form of *A. gambiae* Giles from Southern Nigeria (preliminary communication). *Nature, Lond.* **179**, 873.

Brückner W. R. (1935) Geschlechtsorgane und Eibildung des Neuropters *Chrysopa vulgaris* Schneid. *Jena Z. Naturw.* NF **69**, 469–506.

Bruel W. E. van den and Bollaerts D. (1960) Résistance des divers stades de développement de *Sitophilus granarius* L. et *S. oryzae* L. aux irradiations par les rayons γ (Co^{60}). *Bull. Inst. agron. Gembloux* (hors sér.) **2**, 883–905.

Bruner L. (1898) *The First Report of the Merchants' Locust Investigation Commission of Buenos Aires*, Buenos Aires.

Brunet P. C. J. (1951) The formation of the ootheca by *Periplaneta americana*. I. The microanatomy and histology of the posterior part of the abdomen. *Q. Jl microsc. Sci.* **92**, 113–127.

BRUNET P. C. J. (1952) The formation of the ootheca by *Periplaneta americana*. II. The structure and function of the left colleterial gland. *Q. Jl microsc. Sci.* **93**, 47–69.

BRUNET P. C. J. (1965) The metabolism of aromatic compounds. *Symp. Biochem. Soc.* **25**, 49–77.

BRUNET P. C. J. and KENT P. W. (1955) Observations on the mechanism of a tanning reaction in *Periplaneta* and *Blatta*. *Proc. R. Soc. Lond.* (B) **144**, 259–274.

BRUNSON M. H. (1938) Influence of Japanese beetle instar on the sex and population of the parasite *Tiphia popilliavora*. *J. agric. Res.* **57**, 379–386.

BRUST R. A. and COSTELLO R. A. (1969) Mosquitoes of Manitoba. 2. The effect of storage temperature and relative humidity on hatching of eggs of *Aedes vexans* and *Aedes abserratus* (Diptera: Culicidae). *Can. Ent.* **101**, 1285–1291.

BRUST R. A. and GIARDINO J. (1968) Illumination and photography of mosquito eggs. *Ann. ent. Soc. Am.* **61**, 1039–1041.

BRYANT D. G. and CLARK R. C. (1975) Spruce budworm egg-mass sampling in Newfoundland. *Bi-mon. Res. Notes* **31** (2) p. 12.

BRYANT E. H. and HALL A. E. (1975) The role of medium conditioning in the population dynamics of the housefly. *Res. Population Ecol.* **16**, 188–197.

BRYANTSEVA I. B. (1955) The characteristics of structure of the gonads in females of Acridoidea. *Sb. Rab. Inst. prikl. Zool. Fitopat.* **3**, 48–52.

BRYK F. (1918) Grundzüge der Sphragidologie. *Ark. Zool.* **11** (18), 1–38.

BRYK F. (1920) Bibliotheca sphragidologica. *Arch. Naturgesch.* (A) **85**, 102–183 (1919).

BUCHANAN W. D. (1960) Biology of the oak timberworm *Arrhenodes minutus*. *J. econ. Ent.* **53**, 510–513.

BUCHNER G. D., MARTIN J. H., and PETER A. M. (1925) Concerning the mode of transference of calcium from the shell of the hen's egg to the embryo during incubation. *Am. J. Physiol.* **72**, 253–255.

BUCK H. (1952) Untersuchungen und Beobachtungen über den Lebensablauf und das Verhalten des Trichterwicklers *Deporaus betulae* L. *Zool. Jb. (Physiol.)* **63**, 153–236.

BUCK H. (1953) Eine neue Art der Brutfürsorge bei einem Rüsselkäfer (*Polydrosus mollis* Ström.) (Vorläufige Mitteilung). *Jh. Ver. vaterl. Naturk. Württemb. Stuttgart* **108**, 66–70.

BUCKELL E. R. (1931) *Hemisarcoptes coccisugus* Lignieres, an enemy of the oyster-shell scale. *Proc. ent. Soc. Br. Columb.* **28**, 14–15.

BUÉI K. (1967) The relationship between the body weights of pupae of house fly and number of matured eggs in the ovaries. *Jap. J. sanit. Zool.* **18**, 18–20 (in Japanese; English summary).

BUENO J. R. DE LA TORRE (1902) Some preliminary notes on the early stages of *Notonecta*. *Proc. NY ent. Soc.* **10**, 250.

BUENO J. R. DE LA TORRE (1906) *Belostoma fluminea*, life history. *Can. Ent.* **38**, 189–197.

BUFFAM P. E. (1962) Observations on the effectiveness and biology of the European predator *Laricobius erichsonii* Rosen. (Coleoptera: Derodontidae) in Oregon and Washington. *Can. Ent.* **94**, 461–472.

BUGNION E. (1887) Recherches sur la ponte du *Phloeosinus thuyae*. *Rev. Ent.* **6**, 129–138.

BUGNION E. (1922) Note relative à l'*Ameles spallanziana*. Structure de l'oothèque, éclosion des jeunes larves. *Bull. Soc. zool. Fr.* **47**, 172–180.

BUGNION E. (1933) Le graphiptère égyptien: *Graphipterus serrator* Forskal. *Bull. Soc. ent. d'Égypte*, 1933, 28–63.

BULEZA V. V. (1971) Selectivity in the behaviour of females of certain egg-parasites (Hymenoptera, Scelionidae) when attacking their hosts. *Zool. Zh.* **50**, 1885–1888 (in Russian).

BULL R. M. (1973) A look at the biological control of the sugar cane leaf hopper. *Cane Grow. Q. Bull.* **37**, 40–42.

BULLINI L., COLUZZI M., and BIANCHI BULLINI A. P. (1976) Biochemical variants in the study of multiple insemination in *Culex pipiens* L. (Diptera, Culicidae). *Bull. ent. Res.* **65**, 683–685.

BULLOCK H. R., MANGUM C. L., and GUERRA A. A. (1969) Treatment of eggs of the pink bollworm, *Pectinophora gossypiella*, with formaldehyde to prevent infection with a cytoplasmic polyhedrosis virus. *J. Invert. Path.* **14**, 271–273.

BULLOCK R. (1960) Chorionic pattern of *Aedes* eggs by sump method. *Trans. Am. microsc. Soc.* **79**, 167–170.

BULYGINSKAYA M. A. (1965) On the prospects of control of some injurious Lepidoptera by the chemical sterilization method. *Ent. Obozr.* **44**, 738–749 (in Russian; English translation: *Ent. Rev.* **44**, 433–440).

BULYGINSKAYA M. A. and BRYANTSEVA I. B. (1969) The selective reaction of adults of the malva moth *Pectinophora malvella* Hb. (Lepidoptera) when ovipositing. *Ent. Obozr.* **48**, 57–60 (in Russian; English translation: *Ent. Rev.* **48**, 30–31).

BUNNETT E. J. (1947) The egg-raft and larva of *Culex*. *Proc. S. Lond. ent. nat. Hist. Soc.* **1946–7**, 183–184.

BURANDAY R. P. and RAROS R. S. (1975) Effects of cabbage-tomato intercropping on the incidence and oviposition of the diamond-back moth *Plutella xylostella* (L.). *Philipp. Ent.* **2** (5) 369–374.

BURDICK D. J. (1967) Oviposition behaviour and galls of *Andricus chrysolepidicola* (Ashmead) (Hymenoptera: Cynipidae). *Pan-Pacif. Ent.* **43**, 227–231.

BURGER T. L., BILLINGSLEY C. H., PHILIPS F. M., and HOSLER G. W. (1971) A process for deglutination of insect eggs. *J. econ. Ent.* **64**, 1253–1255.

BURGER T. L. and HOLMES M. C. (1972) A modified Buchner funnel for dispensing eggs of the cereal leaf beetle and the three-lined potato beetle for parasite production. *J. econ. Ent.* **65**, 1185–1186.

BURGES H. D. and CAMMELL M. E. (1964) Effect of temperature and humidity on *Trogoderma anthrenoides* (Sharp) (Coleoptera, Dermestidae) and comparisons with related species. *Bull. ent. Res.* **55**, 313–325.

BURGES H. D. and JARRETT P. (1976) Adult behaviour and oviposition of five noctuid and tortricid moth pests and their control in glasshouses. *Bull. ent. Res.* **65**, 501–510.

BURGESS A. F. (1911) *Calosoma sychophanta*: its life history, behaviour, and successful colonization in New England. *Bull. US Bur. Ent.* **101**, 1–94.

BURGESS A. F. and COLLINS C. W. (1915) The Calosoma beetle (*Calosoma sycophanta*) in New England. *Bull. US Dep. Agric.* **251**, 1–40.

BURK T. and BELL W. J. (1973) Cockroach aggregation pheromone: inhibition of locomotion (Orthoptera: Blattidae). *J. Kans. ent. Soc.* **46**, 36–41.

BURKE H. B. and MARTIN D. F. (1956) The biology of three Chrysopid predators of the cotton aphid. *J. econ. Ent.* **49**, 698–700.

BURKE H. E., HARTMAN R. D., and SNYDER T. E. (1923) The lead-cable borer or "short-circuit beetle" in California. *Bull. US Dep. Agric.* **1107**, 1–48.

BURNET B., CONNOLLY K., KEARNEY M., and COOK R. (1973) Effects of male paragonial gland secretion on sexual receptivity and courtship behaviour of female *Drosophila melanogaster*. *J. Insect Physiol.* **19**, 2421–2431.

BURNETT T. (1954) Influences of natural temperatures and controlled host densities on oviposition of an insect parasite. *Phys. Zool.* **27**, 239–248.

BURNETT T. (1956) Effects of temperatures on oviposition of various numbers of an insect parasite (Hymenoptera, Chalcididae, Tenthredinidae). *Ann. ent. Soc. Am.* **49**, 55–59.

BURNETT T. (1964) An Acarine predator–prey population infesting stored products. *Can. J. Zool.* **42**, 655–673.

BURNS A. N. and NEBOISS A. (1957) Two new species of Plecoptera from Victoria. *Mem. nat. Mus. Victoria* **21**, 92–100.

BURNS J. M. (1970) Duration of copulation in *Poanes hobomok* (Lepidoptera: Hesperiidae) and some broader speculations. *Psyche, Camb.* **77**, 127–130.

BUROV V. N. (1967) The role of population density in the population dynamics of the pea beetle *Bruchus pisorum* L. (Coleoptera, Bruchidae). *Zool. Zh.* **46**, 1357–1361 (in Russian; English summary).

BURTON C. J. (1920) Papulo-urticarial rashes caused by the hairlets of caterpillars of the moth *Euproctis edwardsi* Newm. *J. trop. Med. Hyg.* **23**, 148.

BURTON G. J. (1953) Some techniques for mounting mosquito eggs, larvae, pupae and adults on slides. *Mosq. News* **13**, 7–15.

BURTS E. C. and FISCHER W. R. (1967) Mating behaviour, egg production, and egg fertility in the pear psylla. *J. econ. Ent.* **60**, 1297–1300.

BUSCHING M. K. and TURPIN F. T. (1976) Oviposition preferences of black cutworm moths among various crop plants, weeds, and plant debris. *J. econ. Ent.* **69**, 587–590.

BUSCK A. (1913) New microlepidoptera from British Guiana. *Insecutor Inscit. menstr.* **1**, 88–92.

BUSHING R. W. and WOOD D. L. (1964) Rapid measurement of oleoresin exudation pressure in *Pinus ponderosa* Laws. *Can. Ent.* **96**, 510–513.

BUSHLAND R. C. (1934) A study of the sculpture of the chorion of the eggs of eighteen South Dakota grasshoppers (Acridiidae). Unpublished thesis, S. Dakota Coll. Agric. and Mech. Arts.

BUSSART J. E. (1937) The bionomics of *Chaetophleps setosa* Coquillet. *Ann. ent. Soc. Am.* **30**, 285–295.

BUTLER E. A. (1923) *A Biology of British Hemiptera–Heteroptera*, H. F. and G. Witherby Ltd., London.

BUTLER G. D. JR and MAY C. J. (1971) Laboratory studies of the searching capacity of larvae of *Chrysopa carnea* for eggs of *Heliothis* spp. *J. econ. Ent.* **64**, 1459–1461.

BUTLER L. (1966) Oviposition in the Chinese Mantid *Tenodera aridifolia sinensis* (Saussure) (Orthoptera: Mantidae). *J. Ga ent. Soc.* **1**, 5–7.

BUTOVITSCH V. VON (1939) Zur Kenntnis der Paarung, Eiablage und Ernährung der Cerambyciden. *Ent. Tidskr.* **60**, 206–258.

BUXTON J. H. and MADGE D. S. (1974) Artificial incubation of eggs of the common earwig *Forficula auricularia* (L.). *Entomologist's mon. Mag.* **110**, 55–57.

BUXTON P. A. (1930) Evaporation from the meal-worm (Tenebrio: Coleoptera) and atmospheric humidity. *Proc. R. Soc. Lond.* (B) **105**, 560–577.

BUXTON P. A. (1947) *The Louse. An Account of the Lice which Infest Man, their Medical Importance and Control*, 2nd edn., E. Arnold, London.

BUXTON P. A. and HOPKINS G. R. E. (1927) Researches in Polynesia and Melanesia. Parts I–IV. *Mem. Lond. Sch. Hyg. trop. Med.* **1**, 1–260.

BYERS G. W. (1961) The crane fly genus *Dolichopeza* in North America. *Kans. Univ. Sci. Bull.* **42**, 665–924.

BYERS G. W. (1963) The life history of *Panorpa nuptialis* (Mecoptera: Panorpidae). *Ann. ent. Soc. Am.* **56**, 142–149.

BYRNE H. D. (1969) The oviposition response of the alfalfa weevil *Hypera postica* (Gyllenhal). *Bull. Md agric. Exp. Stn* **160**, vi + 1–42.

BYRNE K. J., SWIGAR A. A., SILVERSTEIN R. M., BORDEN J. H., and STOKKINK E. (1974) Sulcatol: population aggregation pheromone in the scotytid beetle *Gnathotrichus sulcatus*. *J. Insect Physiol.* **20**, 1895–1900.

CABAL CONCHA A. (1956) Biologia y control del gorgojo del cafe: *Araecerus fasciculatus* de Geer, Fam: (Anthribiidae), en Barranquilla-Colombia. *Rev. Fac. nac. Agron.* **18**, 30–31.

CADAHIA A. (1965) Preferencias clonales del gorgojo perforador del chopo *Cryptorrhynchus lapathi* L. (Col. Curculionidae). *Boln Serv. Plagas for.* **8**, 115–125.

CAFFREY D. J. (1918) Notes on the poisonous urticating species of *Hemileuca oliviae* larvae. *J. econ. Ent.* **11**, 363–367.

CALABY J. H. (1970) Phthiraptera (lice): 376–386 pp, 5 figs, 1 tab. In (I. M. MacKerras, ed.), *The Insects of Australia. A textbook for Students and Research Workers*, CSIRO, Melbourne University Press, Victoria, Australia.

CALCOTE V. R. (1975) Pecan weevil: feeding and initial oviposition as related to nut development. *J. econ. Ent.* **68**, 4–6.

CALDERON M., DONAHAYE E., and NAVARRO S. (1967) The life cycle of the groundnut seed beetle *Caryedon serratus* (Ol.) in Israel. *Israel J. agric. Res.* **17**, 145–148.

CALDWELL R. L. and DINGLE H. (1967) The regulation of cyclic reproductive and feeding activity in the milkweed bug *Oncopeltus* by temperature and photoperiod. *Biol. Bull. Woods Hole* **133**, 510–525.

CALLAHAN P. S. (1957a) Oviposition response of the corn earworm to differences in surface texture. *J. Kans. ent. Soc.* **30**, 59–63.

CALLAHAN P. S. (1957b) Oviposition response of the imago of the corn earworm *Heliothis zea* (Boddie), to various wave lengths of light. *Ann. ent. Soc. Am.* **50**, 444–452.

CALLAHAN P. S. (1958) Serial morphology as a technique for determination of reproductive patterns in the corn-earworm *Heliothis zea* (Boddie). *Ann. ent. Soc. Am.* **51**, 413–428.

CALLAN E. McC. (1944a) A note on *Phanuropsis semiflaviventris* Girault (Hym., Scelionidae), an egg-parasite of cacao stink-bugs. *Proc. R. ent. Soc. Lond.* (A) **19**, 48–49.

CALLAN E. McC. (1944b) Cacao stink-bugs (Hem., Pentatomidae) in Trinidad, BWI. *Rev. Ent. Rio de J.* **15**, 321–324.

CALORI L. (1848) Sulla generazione vivipara della *Chloë diptera*. *Nuovi ann. Sci. nat. Bologna* (Z) **9**, 38–53.

CALVERT P. (1904) Oviposition by *Cordulegaster*. *Ent. News* **15**, 316.

CALVERT P. P. (1910a) A plant-dwelling Odonate larva. *Ent. News* **21**, 264.

CALVERT P. P. (1910b) Plant-dwelling Odonate larvae. *Ent. News* **21**, 365–366.

CAMERON A. E. (1913) On the life-history of *Lonchaea chorea* F. *Trans. ent. Soc. Lond.* **1913**, 314–322.

CAMERON A. E. (1918) Life-history of the leaf-eating crane-fly *Cylindrotoma splendens* Doane. *Ann. ent. Soc. Am.* **11**, 67–89.

CAMERON A. E. (1922) The morphology and biology of a Canadian cattle-infesting black-fly *Simulium simile* Mal. (Diptera, Simuliidae). *Bull. Can. Dep. Agric.* (NS) **5**, 1–26 (*Ent. Bull.* **20**, 1–26).

CAMERON A. E. (1926) The occurrence of *Cuterebra* (Diptera, Oesteridae) in Western Canada. *Parasitology* **18**, 430–435.

CAMERON J. W. and ANDERSON L. D. (1966) Husk tightness, earworm, egg numbers, and starchiness of kernels in relation to resistance of corn to the corn earworm. *J. econ. Ent.* **59**, 556–558.

CAMORS F. B. JR and PAYNE T. L. (1973) Sequence of arrival of entomophagous insects to trees infested with the southern pine beetle. *Envir. Ent.* **2**, 267–270.

CAMPAN M. (1968) Observations préliminaires sur le rhythme d'activité des femelles d'*Eristalis tenax* (diptères, syrphides) aux lieux de ponte influencé des facteurs météorologiques. *Rev. Comp. Anim.* **2**, 67–76.

CAMPAN M. (1973) Étude des variations saisonnières du rhythme du fréquentation des lieux de ponte chez les femelles d'*Eristalis tenax* (Diptères, Syrphides). *Bull. Soc. Hist. nat. Toulouse* **109**, 119–130.

CAMPAN M. (1974) Étude du comportement d'orientation des femelles de *Calliphora vomitoria* (diptères) vers l'odeur du lieu de ponte. Influence des sécrétions allates et premières données sur le rôle de l'ovaire. *Gen. comp. Endocr.* **22**, 177–183.

CAMPBELL F. L. (1929) The detection and estimation of insect chitin. *Ann. ent. Soc. Am.* **22**, 401–426.

CAMPBELL J. B. and HERMANUSSEN J. F. (1974) *Philonthus theveneti*: life history and predatory habits against stable flies, house flies, and face flies under laboratory conditions. *Envir. Ent.* **3**, 356–358.

CAMPBELL J. W. (1928) Notes on egg-laying and mating habits of *Myopsocus novae-zelandiae* Kolbe. *Bull. Brooklyn ent. Soc.* **23**, 124–128.

CAMPBELL K. G. (1964) Notes on the biology of *Tristaria grouvellei* Reitter (Lyctidae: Coleoptera). *J. ent. Soc. Aust. (NSW)* **1**, 28–31.

CAMPBELL R. W. (1973) Forecasting gypsy moth egg-mass density. *Res. Paper Northeast. For. Exp. Sta. For. Serv. US Dep. Agric.* NE-**268**, 1–19.

CAMPBELL W. V. and DUDLEY J. W. (1965) Differences among *Medicago* species in resistance to oviposition by the alfalfa weevil. *J. econ. Ent.* **58**, 245–248.

CAMPBELL W. V. and EMERY D. A. (1967) Some environmental factors affecting feeding, oviposition, and survival of the southern corn rootworm. *J. econ. Ent.* **60**, 1675–1678.

CAMPODONICO M. J. and SANTORO F. H. (1971) Identificación de acridios (Orthoptera) por las esculturas del corión. *Revta Invest. agríc. B. Aires* **8** (5) 63–82.

CANARD M. (1970) L'oophagie des larves du premier stade de chrysope (Neuroptera, Chrysopidae). *Entomologia exp. appl.* **13**, 21–36.

CAÑIZO J. DEL (1940) Las plagas de langosta en España. *Proc. 6th Int. Congr. Ent. Madrid* **2**, 845–865 (1935).

CAÑIZO J. DEL (1957) Parásitos de la langosta en España. II. Los *Trichodes* (Col. Cleridae). *Boln Patol. veg. Ent. agric.* **22**, 297–312 (1955–6).

CAPPE DE BAILLON P. (1922) Contribution anatomique et physiologique à l'étude de la reproduction chez les Locustiens et les Grilloniens. 2. La ponte et l'éclosion chez des Grillons. *Cellule* **32**, 1–193 (see id., 1920, **31**, 1–245).

CAPPE DE BAILLON P. (1927) Recherches sur la tératologie des insectes. *Encycl. ent.* **8**, 1–291.

CAPPE DE BAILLON P. (1933) La formation de la coquille de l'oeuf chez les Phasmidae. *C. r. hebd. Séanc. Acad. Sci. Paris* **196**, 809–811.

CARAYON J. (1949a) L'oothèque d'Hémiptères Plataspides de l'Afrique tropicale. *Bull. Soc. ent. Fr.* **54**, 66–69.

CARAYON J. (1949b) Observations sur la biologie des Hémiptères microphysides. *Bull. Mus. Hist. nat. Paris* **21**, 710–716.

CARAYON J. (1950) Observations sur l'accouplement, la ponte et l'éclosion chez des Hémiptères Hénicocéphalidés de l'Afrique tropicale. *Bull. Mus. Hist. nat. Paris* **22**, 739–745.

CARAYON J. (1952a) Les fécondations hémocoeliennes chez les Hémiptères Nabidés du genre *Alloeorhynchus*. *C. r. hebd. Séanc. Acad. Sci. Paris* **234**, 751–754.

CARAYON J. (1952b) Les fécondations hémocoeliennes chez les Hémiptères Nabidae du genre *Prostemma*. *C. r. hebd. Séanc. Acad. Sci. Paris* **234**, 1220–1222.

CARAYON J. (1952c) La fécondation hémocoelienne chez *Prostemma guttula* (Hémipt. Nabidae). *C. r. hebd. Séanc. Acad. Sci. Paris* **234**, 1317–1319.

CARAYON J. (1955) Tissu conducteur de spermatozoïdes et fécondation hémocoelienne chez les Hémiptères Nabidés du genre *Pagasa*. *C. r. hebd. Séanc. Acad. Sci. Paris* **240**, 357–359.

CARAYON J. (1970) Action du sperme sur la maturation des ovaires chez les Hémiptères à insémination traumatique. *Colloques int. Cent. natn. Rech. scient.* **189**, 215–247.

CARBONELL E. and FUENTES M. C. (1973) Correlaciones genéticas de la puesta del *Tribolium castaneum* en diversos medios ambientes. *An. Inst. nac. Invest. Agrarias, General* **2**, 73–82.

CARDÉ R. T., ROELOFS W. L., and DOANE C. C. (1973) Natural inhibitor of the gypsy moth sex attractant. *Nature, Lond.* **241**, 474–475.

CARDONA C. and OATMAN E. R. (1971) Biology of *Apanteles dignus* (Hymenoptera: Braconidae), a primary parasite of the tomato pinworm. *Ann. ent. Soc. Am.* **64**, 996–1007.

CARDOSO J. G. A. (1935) Dos inimigos do gafanhoto vermelho. *Docum. trim. Moçambique* **3**, 65–96.

CARFAGNA M. and LANCIERI M. (1971) Colour vision and the choice of substrate during oviposition in *Drosophila melanogaster* Meig. *Monitore zool. ital.* **5**, 215–222 (Italian summary).

CARLE P. (1966) Essai d'analyse expérimentale de facteurs conditionnant la fécondité chez la bruche du haricot (*Acanthoscelides obtectus* Say). *Annls Épiphyt.* **16**, 215–249 (1965).

CARLE P. (1975) Mise en évidence d'une attraction sécondaire d'origine sexuelle chez *Blastophagus destruens* Woll. (Col., Scolytidae). *Annls Zool. Écol. anim.* **6**, 539–550.

CARLSON O. V. and HIBBS E. T. (1970) Oviposition by *Empoasca fabae* (Homoptera: Cicadellidae). *Ann. ent. Soc. Am.* **63**, 516–519.

CARLYLE S. L., LEPPLA N. C., and MITCHELL E. R. (1975) Cabbage looper: a labor reducing oviposition cage. *J. Ga ent. Soc.* **10**, 232–234.

CARNE P. B. (1962) The characteristics and behaviour of the saw-fly *Perga affinis affinis* (Hymenoptera). *Aust. J. Zool.* **10**, 1–34.

CARNE P. B. (1966) Ecological characteristics of the eucalypt-defoliating Chrysomelid *Paropsis atomaria* Ol. *Aust. J. Zool.* **14**, 647–672.

CARNEGIE A. J. M. and HARRIS R. H. G. (1969) The introduction of Mirid egg predators (*Tytthus* spp.) into South Africa. *Proc. S. Afr. Sugar Tech. Ass.* **43D**, 113–116.

CARREL J. E., THOMPSON W., and MACLAUGHLIN M. (1975) Parental transmission of a defensive chemical (cantharidin) in blister beetles. *Am. Soc. Zool.* **13**, 1258.

CARRILLO J. L. and CALTAGIRONE L. E. (1970) Observations on the biology of *Solierella peckhami*, *S. blaisdelli* (Hym., Sphecidae) and two species of Chrysididae (Hym.). *Ann. ent. Soc. Am.* **63**, 672–681.

CARTON Y. (1971) Biologie de *Pimpla instigator* F. 1793 (Ichneumonidae, Pimplinae). I. Mode de perception de l'hôte. *Entomophaga* **16**, 285–296.

CARTON Y. (1974) Biologie de *Pimpla instigator* (Ichneumonidae: Pimplinae). III. Analyse expérimentale du processus de reconnaissance de l'hôte-chrysalide. *Entomologia exp. appl.* **17**, 265–278.

CARVALHO R. P. L. and ROSSETTO C. J. (1968) Biologia de *Zabrotes subfasciatus* (Bohemann) (Coleoptera Bruchidae). *Revta bras. Ent.* **13**, 105–117.

CASTEK K. L., BARBOUR J. F., and RUDINSKY J. A. (1967) Isolation and purification of the attractant of the striped ambrosia beetle. *J. econ. Ent.* **60**, 658–660.

CASTILLO R. L. (1945) *Los Anofelinos de la República del Ecuador*, Guayaquil, Ecuador. **1**, 172 pp.

CASTILLO R. L. (1949) *Atlas de los Anofelinos Sudamericanos*, Soc. Filantrópica del Guayas.

CATLEY A. (1963) Observations on the biology and control of the armyworm *Tiracola plagiata* Walk. (Lepidoptera: Noctuidae). *Papua New Guinea agric. J.* **15**, 105–109 (1962–3).

CATLING H. D. (1973) Notes on the biology of the South African citrus psylla *Triozo erytrae* (Del Guercio) (Homoptera: Psyllidae) (Hem.). *J. ent. Soc. S. Afr.* **36**, 299–306.

CAUDELL A. N. (1920) Zoraptera not an apterous order. *Proc. ent. Soc. Wash.* **22**, 84–97.

CAUSEY O. R., DEANE L. M., DEANE M. P., and SAMPAIO M. M. (1943) *Anopheles* (*Nyssorhynchus*) *sawyeri*, a new anopheline mosquito from Ceara, Brazil. *Ann. ent. Soc. Am.* **36**, 11–20.

CAUSEY O. R., DEANE L. M., and DEANE M. P. (1944) An illustrated key to the eggs of thirty species of Brazilian anophelines, with several new descriptions. *Am. J. Hyg.* **39**, 1–7.

CAUSSANEL C. (1966) Étude du développement larvaire de *Labidura riparia* (Derm., Labiduridae). *Annls Soc ent. Fr.* (NS) **2**, 469–498.

CAUSSANEL C. (1971) La fécondité du forficule des plages *Labidura riparia* (Insecte Dermaptère). Son cycle reproducteur près d'Arcachon (Gironde). In *Vie et Milieu, 3rd Symp. Europ. Biol. Mar.* (suppl.) **22**, 783–802.

CAVICCHI S. (1971) Lo studio della *Ceratitis capitata* in laboratorio. I. Metodi di allevamento (Diptera). *Boll. Soc. ent. Ital.* **103**, 117–124.

CAZIER M. A. (1963) The description and bionomics of a new species of *Apiocera*, with notes on other species (Diptera: Apioceridae). *Wasmann J. Biol.* **21**, 205–234.

CENTER T. D. and JOHNSON C. D. (1973) Comparative life histories of *Sennius* (Coleoptera: Bruchidae). *Envir. Ent.* **2**, 669–672.

CENTER T. D. and JOHNSON C. D. (1974) Coevolution of some seed beetles (Coleoptera: Bruchidae) and their hosts. *Ecology* **55**, 1096–1103.

CERVONE L. (1957) Sulla struttura perimicropilare dell'uovo in *Culex autogenicus* dell'Agro Pontino. *RC. Ist. sup. Sanità* **20**, 695–701.

CHABORA P. C. (1967) Hereditary behaviour variation in oviposition patterns in the parasite *Nasonia vitripennis* (Hymenoptera: Pteromalidae). *Can. Ent.* **99**, 763–765.

CHACKO M. J. (1969) Superparasitism in *Trichogramma evanescens minutum* Riley, an egg parasite of sugarcane and maize borers in India. II. Causes of superparasitism (Hymenoptera: Trichogrammatidae). *Beitr. Ent.* **19**, 637–642.

CHALAM B. S. (1927) The resistance of *Anopheles* eggs to desiccation. *Ind. J. med. Res. Calcutta* **14**, 863–866.

CHALFANT R. B. (1975) A simplified technique of rearing the lesser cornstalk borer (Lepidoptera: Phycitidae). *J. Ga ent. Soc.* **10**, 33–37.

CHALFANT R. B. and CANERDAY T. D. (1972) Feeding and oviposition of the cowpea curculio and laboratory screening of southern pea varieties for insect resistance. *J. Ga ent. Soc.* **7**, 272–277.

CHAMBERLAIN W. F. and GINGRICH A. R. (1976) Marking of horn flies with ^{32}P. *SWest. Entomol.* **1**, 174–177.

CHAMBERLIN T. R. (1941) The wheat jointworm in Oregon, with special reference to its dispersion, injury, and parasitization. *Tech. Bull. US Dep. Agric.* **784**, 1–47.

CHAMBERS V. H. (1952) The natural history of some *Pamphilius* species (Hym., Pamphiliidae). *Trans. Soc. Br. Ent.* **11**, 125–140.

CHAMPION J. P. (1968) Le comportement de ponte de *Stigmella malella* (Lep., Stigmellidae). *Annls Soc. ent. Fr.* (NS) **4**, 637–648.

CHAMPION G. C. and CHAPMAN T. A. (1901) Observations on *Orinia*, a genus of viviparous and ovo-viviparous beetles. *Trans. ent. Soc. Lond.* **1901**, 1–17.

CHAMPION H. G. (1919) A cerambycid infesting pine cones in India. *Chlorophorus strobicola* n.sp. *Entomologist's mon. Mag.* **55**, 219–224.

CHAMPLAIN R. A. and SHOLDT L. L. (1967) Life history of *Geocoris punctipes* (Hemiptera: Lygaeidae) in the laboratory. *Ann. ent. Soc. Am.* **60**, 881–883.

CHANDLER A. E. F. (1967) Oviposition responses by aphidophagous Syrphidae (Diptera). *Nature, Lond.* **213**, 736.

CHANDLER A. E. F. (1968a) Height preferences for oviposition of aphidophagous Syrphidae (Diptera). *Entomophaga* **13**, 187–195.

CHANDLER A. E. F. (1968b) A preliminary key to the eggs of some of the commoner aphidophagous Syrphidae (Diptera) occurring in Britain. *Trans. R. ent. Soc. Lond.* **120**, 199–217.

CHANG V. C. S., TANIMOTO V., and OTA A. K. (1970) Clonal resistance of sugarcane to the New Guinea sugarcane weevil (Coleoptera: Curculionidae) in Hawaii. *Hawaiian Plrs' Rec.* **58**, 103–118.

CHAPMAN J. A. (1956) Flight-muscle changes during adult life in a Scolytid. *Nature, Lond.* **177**, 1183.
CHAPMAN J. A. (1958) Studies on the physiology of the ambrosia beetle, *Tryphodendron* in relation to its ecology. *Proc. 10th Int. Congr. Ent.* **4**, 375–380.
CHAPMAN R. F. (1961) The egg pods of some tropical African grasshoppers (Orthopt: Acridoidea). Egg pods from grasshoppers collected in southern Ghana. *J. ent. Soc. S. Afr.* **24**, 259–284.
CHAPMAN R. F. and ROBERTSON I. A. D. (1958) The egg pods of some tropical African grasshoppers. *J. ent. Soc. S. Afr.* **21**, 85–112.
CHAPMAN R. N. (1928) The quantitative analysis of environmental factors. *Ecology* **9**, 111–122.
CHAPMAN R. N. (1933) The causes of fluctuations of populations of insect. *Proc. Hawaii. ent. Soc.* **8**, 279–292.
CHAPMAN R. N. and WHANG W. Y. (1934) An experimental analysis of the cause of population fluctuations. *Science, NY* **80**, 297–298.
CHAPMAN T. A. (1878) On the economy, etc., of *Bombylius*. *Entomologist's mon. Mag.* **14**, 196–200.
CHAPMAN T. A. (1911) Viviparous butterflies. *Ent. Rec.* **23**, 233–234.
CHAPMAN T. A. (1917) Notes on early stages and life history of the earwig (*Forficula aricularia*). *Ent. Rec.* **29**, 25–30, 177–180.
CHARARAS C. (1962) Étude biologique des Scolytides des conifères. *Encycl. Ent.* (A) **38**, 1–556.
CHARARAS C. and CROSASSO C. (1972) Recherches sur l'attraction de masse et l'attraction sexuelle chez *Ips sexdentatus* Boern. (Coléoptère, Scolytidae). *C. r. hebd. Séanc. Acad. Agric. Fr.* **58**, 54–61 (1972).
CHATTERJEE S. N. and RAM R. D. (1970) A technique for staining and counting leaf-hopper eggs laid in leaf tissue. *Sci. Cult.* **36**, 597–598.
CHAUDHRY H. S. and GUPTA P. C. (1972) The morphology of the reproductive system of *Bemisia gossypiperda* M. and L. (Homoptera, Aleurodidae) (Hem.). *Zool. Beitr.* **18**, 343–351.
CHAUTHANI A. R. and HAMM J. J. (1967) Biology of the exotic parasite *Drino munda* (Diptera: Tachinidae). *Ann. ent. Soc. Am.* **60**, 373–376.
CHAUVIN G. (1969) Accouplement ponte et évolution des ovaires chez *Tinea pellionella* L. (Lépidoptère, Tineidae). *C. r. hebd. Séanc. Soc. Biol. Paris* **163**, 2673–2677.
CHAUVIN G. (1971) Nature du substrat et humidité, facteurs du choix du site de ponte chez deux lépidoptères Tineidae: *Monopis rusticella* (Clerck) et *Trichophaga tapetzella* L. *Annls Zool. Écol. anim.* **3**, 319–325.
CHAUVIN G. (1972) Facteurs modifiant la fécondité des Lépidoptères. Étude de deux Tineidae; cas particulier de *Tinea pellionella* L. *Annls Zool. Ecol. anim.* **3**, 509–513 (1970).
CHAUVIN G. and BARBIER R. (1972) Perméabilité et ultrastructures des oeufs de deux lépidoptères Tineidae: *Monopis rusticella* et *Trichophaga tapetzella*. *J. Insect Physiol.* **18**, 1447–1462.
CHAUVIN G. and BARBIER R. (1974) Ultrastructure des oeufs parthénogénétiques de *Luffia ferchaultella* Steph. et de *Fumea casta* Pallas (Lepidoptera, Psychidae). *Bull. biol. Fr. Belg.* **108**, 245–252.
CHAUVIN G., BARBIER R., and BERNARD J. (1973) Ultrastructure de l'oeuf de *Triatoma infestans* Klug (Heteroptera, Reduviidae), formation des cuticules embryonnaires, rôle des enveloppes dans le transit de l'eau. *Z. Zellforsch. mikrosk. Anat.* **138**, 113–132.
CHAUVIN G., RAHN R., and BARBIER R. (1974) Comparaison des oeufs des lépidoptères *Phalera bucephala* L. (Ceruridae), *Acrolepia assectella* Z. et *Plutella maculipennis* Curt. (Plutellidae): morphologie et ultrastructures particulières du chorion au contact du support végétal. *Int. J. Insect Morph. Embryol.* **3**, 247–256.

CHAUVIN R. (1956) Les facteurs qui gouvernent la ponte chez la reine des abeilles. *Insectes soc.* **3**, 499–504.
CHEEMA P. S. (1956) Studies on the bionomics of the case-bearing clothes moth *Tinea pellionella* (L.). *Bull. ent. Res.* **47**, 167–182.
CHEMSAK J. A. (1965) A new species of *Oeme costata*, with observations on the habits of larvae and adults. *J. Kans. ent. Soc.* **38**, 351–355.
CHEMSAK J. A. and POWELL J. A. (1966) Studies on the bionomics of *Tragidion armatum* LeConte (Coleoptera: Cerambycidae). *Pan-Pacif. Ent.* **42**, 36–47.
CHEN AN-KUO, FENG WEI-HSIUNG, CHEN CHIH-HUI, and CHUNG HSIANG-CHEN (1965) Effect of high temperatures on the development and reproduction of the armyworm *Leucania separata* Walker. I. Development and hatching of the eggs. *Acta ent. sin.* **14**, 225–238 (in Chinese; English summary).
CHEN C. C. (1969) Studies on the condition of oviposition of rice green leafhoppers. *Pl. Prot. Bull. Taiwan* **11**, 83–89 (in Chinese; English summary).
CHEN P. S. and BAKER G. T. (1976) L-alanine aminotransferase in the paragonial gland of *Drosophila*. *Insect Biochem.* **6**, 441–447.
CHEN P. S. and BÜHLER R. (1970a) Paragonial substance (sex-peptide) and other free ninhydrin-positive components in male and female adults of *Drosophila melanogaster*. *J. Insect Physiol.* **16**, 615–627.
CHEN P. S. and BÜHLER R. (1970b) Isolierung und Funktion des Sexpeptids bei *Drosophila melanogaster*. *Rev. suisse Zool.* **77**, 548–554.
CHEN P. S. and DIEM C. (1961) A sex-specific ninhydrin positive substance found in the paragonia of adult males of *Drosophila melanogaster*. *J. Insect Physiol.* **7**, 289–298.
CHEN S. H. and YOUNG B. (1941) On the protective value of the egg-pedicel of Chrysopidae. *Sinensia, Nanking* **12**, 211–215.
CHEN S. H. and YOUNG B. (1943) Further remarks on the carriage of eggs on the elytra by the males of *Sphaerodema rusticum*. *Sinensia, Nanking* **14**, 49–53.
CHENG C.-H. and PATHAK M. D. (1971) Bionomics of the rice green leafhopper *Nephotettix impicticeps* Ishihara (Hem., Hom.: Cicadellidae). *Philipp. Ent.* **2**, 64–74.
CHENG H. H. and LE ROUX E. J. (1966) Preliminary life tables and notes on mortality factors of the birch leaf miner *Fenusa pusilla* (Lepeletier) (Hymenoptera: Tenthredinidae), on blue birch, *Betula caerulea grandis* Blanchard, in Quebec. *Ann. ent. Soc. Queb.* **11**, 81–104.
CHENG L. (1967) Studies on the biology of the Gerridae (Hem., Heteroptera). II. The life history of *Metrocoris tenuicornis* Esaki. *Entomologist's mon. Mag.* **102**, 273–282.
CHENG L. (1969) The life history and development of *Lypha dubia* Fall. (Dipt., Tachinidae). *Entomologist* **102**, 25–32.
CHENG L. (1972) Skaters of the seas. *Oceans* **5**, 54–55.
CHENG L. (1973a) Halobates. *Oceanogr. Mar. Biol. A. Rev.* **11**, 223–235.
CHENG L. (1973b) The ocean strider *Halobates* (Heteroptera, Gerridae) in the Atlantic Ocean. *Okeanologia* **13**, 564–570, 683–690 (in Russian; English translation).
CHENG L. (1974) Notes on the ecology of the ocean insect *Halobates*. *Mar. Fish. Rev.* **36**, 1–7.
CHEONG W. H. (1968) A note on the interesting eggs of *Aedes aurantius*. *Med. J. Malaya* **22**, 242.
CHEREVATOVA A. S. (1967) On the biology of the white willow moth. *Sborn. Rab. Mosk. lesotekhn. Inst.* **15**, 55–61 (in Russian).
CHESNUT T. L. and DOUGLAS W. A. (1971) Competitive displacement between natural populations of the maize weevil and the Angoumois grain moth in Mississippi. *J. econ. Ent.* **64**, 864–868.
CHEU S. P. (1941) Correlation of corn borer damage with growth condition of corn and its significance on corn breeding work. *Kwangsi Agric.* **2**, 126 133 (in Chinese; English summary).

CHEU S. P. and LI S. S. (1946) Further studies on the biological control of sugarcane woolly aphis (*Oregma lanigera* Zehntner) by the giant lady beetle (*Synonycha grandis* Thunb.) in Kwangsi. *Kwangsi Agric.* **6**, 26–32 (in Chinese; English summary).

CHIANG H. C. (1968) Characteristics of corn rootworm egg sampling. *Proc. N. cent. Brch Am. Ass. econ. Ent.* **23**, 19–20 (abstract).

CHIANG H. C. and HODSON A. C. (1950a) The relation of copulation to fecundity and population growth in *Drosophila melanogaster*. *Ecology* **31**, 255–259.

CHIANG H. C. and HODSON A. C. (1950b) An analytical study of population growth in *Drosophila melanogaster*. *Ecol. Monogr.* **20**, 175–206.

CHIANG H. C., SISSON V., and RASMUSSEN D. (1969) Conversion of results of concentrated samples to density estimates of egg and larval populations of the northern corn rootworm. *J. econ. Ent.* **62**, 578–583.

CHIN CHUN-TEH (1958) Studies on the locust egg. III. On the water loss and the ability to survive desiccation of the eggs of the oriental migratory locust *Locusta migratoria manilensis* Mayen. *Acta ent. sin.* **8**, 207–225 (in Chinese; English summary).

CHIN CHUN-TEH, QUO F., CHAI CHI-HUI, CHEN CHU-YING, SHA CHA-YUN, and CHAN TAK-MING (1956) Studies on the locust egg. II. Developmental changes of the locust egg during incubation and their possible physiological significances. *Acta ent. sin.* **6**, 37–60 (in Chinese; English summary).

CHIN CHUN-TEH, QUO F., CHAI CHI-HUI, and SHA CHA-YUN (1959) Studies on the locust egg. IV. The survival and embryonic development of the locust egg under water. *Acta ent. sin.* **9**, 287–305 (in Chinese; English summary).

CHIPPENDALE G. M. and BECK S. D. (1965) A method for rearing the cabbage looper *Trichoplusia ni* on a meridic diet. *J. econ. Ent.* **58**, 377–378.

CHITTENDEN F. H. (1912a) Papers on insects affecting stored products. The broadbean weevil (*Laria rufimana* Bih.). *Bull. US Bur. Ent.* **96**, 59–82.

CHITTENDEN F. H. (1912b) Papers on insects affecting stored products. The cowpea weevil (*Pachymerus chinensis* L.). *Bull. US Bur. Ent.* **96**, 83–94.

CHITTENDEN F. H. and FINK D. E. (1922) The green June beetle. *Bull. US Dep. Agric.* **891**, 1–52.

CHOCK Q. C., DAVIS C. J., and CHONG M. (1961) *Sepedon macropus* (Diptera: Sciomyzidae) introduced into Hawaii as a control for the liver fluke snail, *Lymnaea ollula*. *J. econ. Ent.* **54**, 1–4.

CHODJAI M. (1963) Étude écologique de *Ruguloscolytus mediterraneus* Eggers (Col., Scolytidae) en Iran. *Rev. Path. vég. Ent. agric. Fr.* **42**, 139–160.

CHOI S. Y., LEE H. R., LEE J.-O., and PARK J. S. (1976) Varietal differences in ovipositional preferences of the striped rice borer moths *Chilo suppressalis* W. *Korean J. Pl. Prot.* **15**, 23–27 (in Korean).

CHOI Y. H. and NAM S.-H. (1976) Notes on the early stages of *Thecla betulae* Linne. *Korean J. Ent.* **6**, 63–66 (in Korean).

CHOLODKOVSKY N. A. (1918) Miscellanea entomotomica. *Bull. Acad. Sci. Russ.* Series VI, **12** (2) 1351–1356.

CHOPARD L. (1938) *La Biologie des Orthoptères*, Paris.

CHOPARD L. (1949a) Dermaptera. In *Traité de Zoologie* **9**, Paris.

CHOPARD L. (1949b) Dictyoptera. In *Traité de Zoologie* **9**, 335–407, Paris.

CHOPARD L. (1950) Sur l'anatomie et le développement d'une blatte vivipare. *Proc. 8th int. Congr. Ent.* pp. 218–222.

CHOUDHURI J. C. B. (1956a) Observations on the oviposition behaviour of the Moroccan locust (*Dociostaurus maroccanus* Thnbg.) in Cyprus. *Saugar Univ. J.* **1** (5) 123–139.

CHOUDHURI J. C. B. (1956b) Experimental studies on the selection of oviposition sites by *Locusta migratoria migratorioides* (R. and F.). *Locusta* **4**, 23–34.

CHOUDHURI J. C. B. (1958) Experimental studies on the choice of oviposition sites by two species of *Chorthippus* (Orthoptera: Acrididae). *J. Anim. Ecol.* **27**, 201–216.

CHOUDHURI J. C. B. (1962) Experimental studies on the oviposition behaviour of *Omocestus viridulus* (Linn.) (Ord.: Orthoptera). *Agra Univ. J. Res. (Sci.)* **11**, 241–255.

CHOUDHURY A. K. S., BANDYOPADHYAY M. K., and MUKERJEE A. B. (1972) On the occurrence of some predatory mites associated with stored grain pests in India. *Entomologist's mon. Mag.* **107**, 203–204.

CHOW (SZE-CHUN) (1966) A preliminary study on *Eurydema festiva* var. *chlorotica* Horváth in Sinkiang. *Acta ent. sin.* **15**, 47–55 (in Chinese; English summary).

CHRESTIAN P. (1955) Le capnode noir des rosacées *Capnodis tenebrionis* L. (Coleoptera: Buprestidae). *Serv. Déf. Vég. Rabat* **6**, 1–141.

CHRISTENSEN C. M. and DOBSON R. C. (1977) Biological studies of *Aphodius fimetarius* (L.) (Coleoptera: Scarabaeidae). *J. Kans. ent. Soc.* **50**, 129–134.

CHRISTIANSEN E. (1971) The ecology of *Hylobius abietis* and its significance to forestry. A study of the literature. *Tidsskr. Skogbr.* **79**, 245–262 (from *Forestry Abst.* **33**).

CHRISTIANSEN E. and BAKKE A. (1968) Temperature preference in adults of *Hylobius abietis* L. (Coleoptera: Curculionidae) during feeding and oviposition. *Z. angew Ent.* **62**, 83–89.

CHRISTOPHERS S. R. (1916) An Indian tree-hole breeding *Anopheles A. baraianensis* James = *A. (Coclodiazesis) plumbeus* Haliday. *Indian J. med. Res.* **3**, 486–496.

CHRISTOPHERS S. R. (1933) *The Fauna of British India. Diptera*, Vol. IV, *Family Culicidae. Tribe Anophelini*, Taylor and Francis, London.

CHRISTOPHERS S. R. (1945) Structure of the *Culex* egg-raft in relation to function (Diptera). *Trans. R. ent. Soc. Lond.* **95**, 25–34.

CHRISTOPHERS S. R. (1960) *Aedes aegypti* (L.), *The Yellow Fever Mosquito, Its Life History, Bionomics and Structure*, Cambridge University Press.

CHRISTOPHERS S. R. and BARRAUD P. J. (1931) The eggs of Indian *Anopheles* with descriptions of the hitherto undescribed eggs of a number of species. *Rec. Malar. Surv. India* **2**, 61–192.

CHUN M. W. and SCHOONHOVEN L. M. (1973) Tarsal contact chemosensory hairs of the large white butterfly *Pieris brassicae* and their possible role in oviposition behaviour. *Entomologia exp. appl.* **16**, 343–357.

CHURCH N. S. (1967) The egg-laying behaviour of 11 species of Lyttinae (Coleoptera: Meloidae). *Can. Ent.* **99**, 752–760.

CHURCH N. S., SALKELD E. H., and REMPEL J. G. (1970) The structure of the micropyles of *Lytta nuttalli* Say and *L. viridana* Le Conte (Coleoptera: Meloidae). *Can. J. Zool.* **48**, 894–895.

CHUTTER F. M. (1972) Notes on the biology of South African Simuliidae, particularly *Simulium (Eusimulium) nigritarse* Coquillet. *Newsletter, Limnol. Soc. S. Afr.* **18**, 10–18.

CIERNIEWSKA B. (1976) Studio nad ekologia *Ephedrus persicae* Frog. (Hymenoptera, Aphidiidae) parazytoida mszycy jabloniowobabkowej, *Dysaphis plantaginea* (Pass.) (Homoptera, Aphididae). *Roczn. Nauk Rolniczych,* E **6**, 59–75.

CIESIELSKA Z. (1966) Research on the ecology of *Oryzaephilus surinamensis* (L.) (Coleoptera, Cucujidae). *Ekol. pol.* (A) **14**, 439–489.

CIESLA W. M. (1969) Forecasting population trends of an oak leaf tier, *Croesia semipurpurana*. *J. econ. Ent.* **62**, 1054–1056.

CIGLIOLI M. E. C. (1964) Tides, salinity and the breeding of *Anopheles melas* (Theobald, 1903) during the dry season in the Gambia. *Riv. Malar.* **43**, 245–263.

CIRIO U. (1970) Reperti sul meccanismo stimulo—riposta nell'ovideposizione—del *Dacus oleae* Gmelin (Diptera, Trypetidae). *Redia* **52** (3) 577–599.

CLAASSEN P. W. (1919) Life history and biological notes on *Chlaenius impunctifrons* Say (Coleoptera; Carabidae). *Ann. ent. Soc. Am.* **12**, 95–100.

CLAASSEN P. W. (1921) *Typha* insects: their ecological relationships. *Mem. Cornell Univ. agric. Exp. Stn* **57**, 459–531.

CLANCY D. W. (1946a) Natural enemies of some Arizona cotton insects. *J. econ. Ent.* **39**, 326–328.
CLANCY D. W. (1946b) The insect parasites of the Chrysopidae (Neuroptera). *Univ. Calif. Publ. Ent.* **7**, 403–496.
CLARIDGE M. F. and REYNOLDS W. J. (1972) Host plant specificity, oviposition behaviour and egg parasitism in some woodland leaf-hoppers of the genus *Oncopsis* (Hem., Hom., Cicadellidae). *Trans. R. Ent. Soc. Lond.* **124**, 149–166.
CLARIDGE M. F., REYNOLDS W. J., and WILSON M. R. (1977) Oviposition behaviour and food plant discrimination in leaf hoppers of the genus *Oncopsis*. *Ecol. Ent.* **2**, 19–25.
CLARK A. F. (1930) *Paropsis dilatata* Er. in New Zealand. Preliminary account. *NZ Jl Sci. Tech.* **12**, 114–123.
CLARK A. M. and SMITH R. E. (1967) Egg production and adult life span in two species of *Bracon* (Hymenoptera: Braconidae). *Ann. ent. Soc. Am.* **60**, 903–905.
CLARK D. P. (1965) On the sexual maturation, breeding and oviposition behaviour of the Australian plague locust, *Chorotoicetes terminifera* (Walk.). *Aust. J. Zool.* **13**, 17–45.
CLARK J. T. (1976a) The capitulum of phasmid eggs (Insecta: Phasmida). *Zool. J. Linn. Soc.* **59**, 365–375.
CLARK J. T. (1976b) The eggs of stick insects (Phasmida): a review with descriptions of the eggs of eleven species. *System. Ent.* **1**, 95–105.
CLARK J. T. (1977) A note on the European species of *Bacillus* Latreille (Phasmida) with descriptions of the eggs. *Entomologist's mon. Mag.* **112**, 63–64.
CLARK L. R. (1962) The general biology of *Cardiaspina albitextura* (Psyllidae) and its abundance in relation to weather and parasitism. *Aust. J. Zool.* **10**, 537–586.
CLARK L. R. (1963) Factors affecting the attractiveness of foliage for oviposition by *Cardiaspina albitextura* (Psyllidae). *Aust. J. Zool.* **11**, 20–34.
CLARK L. R. (1964) The intensity of parasite attack in relation to the abundance of *Cardiaspina albitextura* (Psyllidae). *Aust. J. Zool.* **12**, 150–173.
CLARK N. (1935) The effect of temperature and humidity on the eggs of the bug *Rhodnius prolixus* (Heteroptera, Reduviidae). *J. Anim. Ecol.* **4**, 82–87.
CLARK R. C. and BROWN N. R. (1958) Studies of predators of the balsam woolly aphid *Adelges piceae* (Ratz.) (Homoptera: Adelgidae). V. *Laricobius erichsonii* Rosen. (Coleoptera: Derodontidae), an introduced predator in eastern Canada. *Can. Ent.* **90**, 657–672.
CLARK R. C. and BROWN N. R. (1961) Studies of predators of the balsam woolly aphid *Adelges piceae* (Ratz.) (Homoptera: Adelgidae). IX. *Pullus impexus* (Muls.) (Coleoptera: Coccinellidae), an introduced predator in eastern Canada. *Can. Ent.* **93**, 1162–1168.
CLARK R. C. and BROWN N. R. (1962) Studies of predators of the balsam woolly aphid *Adelges piceae* (Ratz.) (Homoptera: Adelgidae). XI. *Cremifania nigrocellulata* Cz. (Diptera: Chamaemyiidae), an introduced predator in eastern Canada. *Can. Ent.* **94**, 1171–1175.
CLARKE K. U. and SARDESAI J. B. (1959) An analysis of the effects of temperature upon the growth and reproduction of *Dysdercus fasciatus* Sign. (Hemiptera, Pyrrhocoridae). I. The intrinsic rate of increase. *Bull. ent. Res.* **50**, 387–405.
CLAUSEN C. P. (1939) The effect of host size upon the sex ratio of hymenopterous parasites and its relation to methods of rearing and colonization. *Jl NY ent. Soc.* **47**, 1–9.
CLAUSEN C. P. (1940) *Entomophagous Insects*, McGraw-Hill, London.
CLAUSEN C. P. (1956) The egg–larval host relationship among the parasitic Hymenoptera. *Boll. Lab. Zool. gen. agr. Portici* **33**, 119–133.
CLAUSEN C. P., JAYNES H. A., and GARDNER T. R. (1933) Further investigations of the parasites of *Popillia japonica* in the far east. *Tech. Bull. NY Dep. Agric.* **366**, 1–58.
CLAUSEN C. P., KING J. L., and TERANISHI C. (1927) The parasites of *Popillia japonica* in Japan and Chosen (Korea) and their introduction into the United States. *Bull. US Dep. Agric.* **1429**, 1–55.

CLEMENS W. A. (1922) A parthenogenetic mayfly (*Ameletus ludens* Needham). *Can. Ent.* **54**, 77–78.

CLEMENTS A. N. (1951) On the urticating properties of adult Lymantriidae. *Proc. R. ent. Soc. Lond.* (A) **26**, 104–108.

CLEMENTS A. N. (1963) *The Physiology of Mosquitoes*, Int. Ser. Monogr. pure appl. Biol. (Zool.) **17**, 1–393, Oxford.

CLEMENTS A. N. and BENNETT F. D. (1969) The structure and biology of a new species of *Mallophora* Macq. (Diptera, Asilidae) from Trinidad, WI. *Bull. ent. Res.* **58**, 455–463.

CLEVELAND L. R., HALL S. R., SANDERS E. P., and COLLIER J. (1934) The wood-feeding roach *Cryptocercus*, its protozoa, and the symbiosis between protozoa and roach. *Mem. Am. Acad. Arts Sci.* **17**, 185–342.

CLIFFORD J. R. S. (1885) The urticating properties of the hairs of *Porthesia chrysorrhoea*. *Entomologist* **18**, 22–23.

COAD B. R. (1913) Oviposition habits of *Culex abominator* Dyar and Knab. *Can. Ent.* **45**, 265–266.

COAKER T. H. (1965) Further experiments on the effect of beetle predators on the numbers of the cabbage root fly *Erioischia brassicae* (Bouché) attacking brassica crops. *Ann. appl. Biol.* **56**, 7–20.

COAKER T. H. and DODD G. D. (1970) An apparatus for dispensing mechanically the eggs of the cabbage root fly (*Erioischia brassicae* (Bch.) (Dipt., Muscidae)). *Bull. ent. Res.* **59**, 703–705.

COAKER T. H. and WILLIAMS D. A. (1963) The importance of some Carabidae and Staphylinidae as predators of the cabbage root fly *Erioischia brassicae* (Bouché). *Entomologia exp. appl.* **6**, 156–164.

COBBEN R. H. (1965a) Das aero-mikropylare System der Homoptereneier und Evolutionstrends bei Zikadeneiern (Hom., Auchenorhyncha). *Zool. Beitr. Berl.* (NF) **11**, 13–69.

COBBEN R. H. (1965b) Egg-life and symbiont transmission in a predatory bug, *Mesovelia furcata* Ms and Rey (Heteroptera, Mesoveliidae). *Proc. 12th Int. Congr. Ent.* (sect. 2), 166–168 (1964).

COBBEN R. H. (1968) *Evolutionary Trends in Heteroptera*, Part 1, *Eggs, Architecture of the Shell, Gross Embryology and Eclosion*, Centre for Agricultural Publishing and Documentation, Wageningen.

COBBEN R. H. (1970) Morphology and taxonomy of intertidal dwarfbugs (Heteroptera: Omaniidae fam. nov.). *Tidschr. Ent.* **113**, 61–90.

COBBEN R. H. and ARNOUD BR. (1969) Anthocoridae van *Viscum, Buxus* en *Pinus* in Nederland (Heteroptera). *Meded. Lab. Ent. Wageningen* **19**, 5–16 (Dutch; English summary).

COBBEN R. H. and HENSTRA S. (1968) The egg of an assassin bug (*Rhinocoris* sp.) from the Ivory Coast. *Jeol News* **6B** (3) 2 pp.

COBBEN R. H. and WYGODZINSKY P. (1975) The Heteroptera of the Netherlands Antilles. IX. Reduviidae (assassin bugs). *Studies on the Fauna of Curaçao and Other Caribbean Islands* **48**, 1–62.

COBELLI R. (1906) A proposito del micropilo dell'uvo dei Lepidotteri. *Verh. zool.-bot. Ges. Wien* **56**, 602–604.

COCKERELL T. D. A. (1885) The urticating hairs of Lepidoptera. *Entomologist* **18**, 74–75.

COLBO M. H. and MOORHOUSE D. E. (1974) The survival of the eggs of *Austrosimulium pestilens* Mack. and Mack. (Diptera, Simuliidae). *Bull. ent. Res.* **64**, 629–632.

COLE F. R. (1919) The dipterous family Cyrtidae in North America. *Trans. Am. ent. Soc.* **45**, 1–69.

COLEMAN E. (1946) The case-moth mystery. *Vict. Nat.* **62**, 202–203.

COLEMAN L. C. (1911) The jola or Deccan grasshopper (*Colemania sphenarioides* Bol.). *Bull. Dep. Agric. Mysore (Ent.)* **2**, 1–43.

COLHOUN E. H. (1953) Notes on the stages and the biology of *Baryodma ontarionis* Casey (Coleoptera; Staphylinidae), a parasite of the cabbage maggot *Hylemya brassicae* Bouché (Diptera: Anthomyiidae). *Can. Ent.* **85**, 1–8.

COLLYER E. (1951) The separation of *Conwentzia pineticola* End. from *Conwentzia psociformis* (Curt.), and notes on their biology. *Bull. ent. Res.* **42**, 555–564.

COLLYER E. (1952) Biology of some predatory insects and mites associated with the fruit tree red spider mite (*Metatetranychus ulmi* (Koch)) in south-eastern England. I. The biology of *Blepharidopterus angulatus* (Fall.) (Hemiptera–Heteroptera, Miridae). *J. hort. Sci.* **27**, 117–129.

COLLYER E. (1953a) Biology of some predatory insects and mites associated with the fruit tree red spider mite (*Metatetranychus ulmi* (Koch)) in south-eastern England. II. Some important predators of the mite. *J. hort. Sci.* **28**, 85–97.

COLLYER E. (1953b) Biology of some predatory insects and mites associated with the fruit tree red spider mite (*Metatetranychus ulmi* (Koch)) in south-eastern England. III. Further predators of the mite. *J. hort. Sci.* **28**, 98–113.

COLTHRUP C. W. (1903) *Porthesia chrysorrhoea* in England. *Entomologist* **36**, 70.

COLTHURST I. (1930) A Himalayan hillside. *J. Darjeeling nat. Hist. Soc.* **6**, 68–72 (1931).

COLUZZI M. (1964) Morphological divergences in the *Anopheles gambiae* complex. *Riv. Malar.* **43**, 197–232.

COLUZZI M., DECO M. DI., and GIRONI A. (1975) Influenza del fotoperiodo sulla scelta del luogo di ovideposizione in *Aedes mariae* (Diptera, Culicidae). *Parassitologia* **17**, 121–130.

COLYER C. N. and HAMMOND C. O. (1951) *Flies of the British Isles*, F. Warne & Co. Ltd., London.

COMMON I. F. B. (1948) The yellow-winged locust *Gastrimargus muscius* Fabr. in central Queensland. *Qd J. agric. Sci.* **5**, 153–219.

COMMON I. F. B. (1970) Lepidoptera (moths and butterflies). In *The Insects of Australia. A Textbook for Students and Research Workers*, CSIRO, Melbourne University Press.

COMPERE H. (1939) The insect enemies of the black scale, *Saissetia oleae* (Bern.), in South America. *Univ. Calif. Publ. Ent.* **7**, 75–90.

COMSTOCK J. H. (1880) Notes on predaceous Lepidoptera. *Rep. US Comm. Agric.* **1880**, 241–242.

COMSTOCK J. H. (1887) Note on the respiration of aquatic bugs. *Am. Nat.* **21**, 577–578.

COMSTOCK J. H. (1925) *An Introduction to Entomology*, Comstock, New York.

CONCI C. (1952) L'allevamento in condizioni sperimentali dei mallofagi. *Boll. Mus. Ist. Biol. Univ. Genova* **26**, 47–70.

CONDRASHOFF S. F. (1967) An extraction method for rapid counts of insect eggs and small organisms. *Can. Ent.* **99**, 300–303.

CONIL P. A. (1881) Études sur l'*Acridium paranense* Burm., ses variétés et plusieurs insectes qui le détruisent. *Bolm Acad. Ciênc. Córdoba* **3**, 386–472.

CONNELL W. A. (1959) Estimating the abundance of corn earworm eggs. *J. econ. Ent.* **52**, 747–749.

CONNIN R. V. and JANTZ O. K. (1969) Some effects of photoperiod and cold storage on oviposition of the cereal leaf beetle *Oulema melanopus* (Coleoptera: Chrysomelidae). *Michigan Ent.* **1**, 363–366.

CONRAD M. S. (1959) The spotted lady beetle *Coleomegilla maculata* (De Geer) as a predator of European corn borer eggs. *J. econ. Ent.* **52**, 843–847.

CONSOLI R. A. G. and ESPINOLA H. N. (1973) Possíveis fatores químicos na água que influenciam as fêmeas de *Culex pipiens fatigans* para oviposição. *Revta Pat. Trop.* **1**, 49–54.

COOK F. and BUZICKY A. W. (1971) A procedure for determining the onset of diapause in *Aedes vexans* (Meigen). *Mosq. News* **31**, 116–117.

COOLING L. E. (1924) On the protracted viability of eggs of *Aedes aegypti* and *A. motoscriptus* in a desiccated condition in a state of nature. *Health* **2**, 51–52.

COOMBS C. W. (1956) Stability of grain as a factor influencing the oviposition rate of the grain weevil *Calandra granaria* (L.) (Col. Curculionidae). *Bull. ent. Res.* **47**, 737–740.

COOMBS C. W. and WOODROFFE G. E. (1962) Some factors affecting mortality of eggs and newly emerged larvae of *Ptinus tectus* Boieldieu (Col., Ptinidae). *J. Anim. Ecol.* **31**, 471–480.

COOMBS C. W. and WOODROFFE G. E. (1965) Some factors affecting the longevity and oviposition of *Ptinus tectus* Boieldieu (Coleoptera, Ptinidae) which have relevance to succession among grain beetles. *J. stored Prod. Res.* **1**, 111–127.

COOPER K. W. (1940) The genital anatomy and mating behaviour of *Boreus brumalis* Fitch (Mecoptera). *The Am. Midl. Nat.* **23**, 354–367.

COOPER K. W. (1966) Ruptor ovi, the number of moults in development, and method of exit from masoned nests. *Biology of Eumenine Wasps*, VII, *Psyche, Camb.* **73**, 238–250.

COOPER K. W. (1972) A southern California *Boreus, B. notoperates* n.sp. I. Comparative morphology and systematics (Mecoptera: Boreidae). *Psyche, Camb.* **79**, 269–283.

COOPER K. W. (1974) Sexual biology, chromosomes, development, life histories and parasites of *Boreus*, especially of *B. notoperates*. A southern California *Boreus*. II. (Mecoptera: Boreidae). *Psyche, Camb.* **81**, 84–120.

COPELAND E. L., YOUNG J. R. and LEWIS W. J. (1976) Labelling of *Trichogramma pretiosum* by rearing on eggs from ^{32}P-fed adults of *Heliothis zea*. *Ann. ent. Soc. Am.* **69**, 804–806.

COPLAND M. J. W. and KING P. E. (1971) The structures and possible function of the reproductive system in some Eulophidae and Tetracompidae. *The Entomologist* Jan. 1971, pp. 4–28.

COPONY J. A. and MORRIS C. L. (1972) Southern pine beetle suppression with frontalure and cacodylic acid treatments. *J. econ. Ent.* **65**, 754–757.

COPPEL H. C., HOUSE H. L., and MAW M. G. (1959) Studies on dipterous parasites of the spruce budworm *Choristoneura fumiferana* (Clem.) (Lepidoptera: Tortricidae). VII. *Agria affinis* (Fall.) (Diptera: Sarcophagidae). *Can. J. Zool.* **37**, 817–830.

COPPEL H. C. and SMITH B. C. (1957) Studies on dipterous parasites of the spruce budworm *Choristoneura fumiferana* (Clem.) (Lepidoptera: Tortricidae). V. *Omotoma fumiferanae* (Tot.) (Diptera: Tachinidae). *Can. J. Zool.* **35**, 581–592.

CORBET P. S. (1961) A new species of *Afronurus* (Ephemeroptera) from Tanganyika and records of *Simulium* associated with *Afronurus* larvae. *Ann. Mag. nat. Hist.* **4** (13) 573–576.

CORBET P. S. (1962a) *A Biology of Dragonflies*, H. F. and G. Witherby Ltd., London.

CORBET P. S. (1962b) Observations on the attachment of *Simulium* pupae to larvae of Odonata. *Ann. trop. Med. Parasit.* **56**, 136–140.

CORBET P. S. (1964a) Autogeny and oviposition in arctic mosquitoes. *Nature, Lond.* **203**, 669.

CORBET P. S. (1964b) Observations on mosquitoes ovipositing in small containers in Zika Forest, Uganda. *J. Anim. Ecol.* **33**, 141–164.

CORBET P. S. (1966) Diel periodicities of emergence and oviposition in riverine Trichoptera. *Can. Ent.* **98**, 1025–1034.

CORBET P. S. (1967) The diel oviposition periodicity of the black fly *Simulium vittatum*. *Can. J. Zool.* **45**, 583–584.

CORBET P. S. (1974) Habitat manipulation in the control of insects in Canada. *Proc. Tall Timbers Conf.* **5**, 147–171 (Tallahassee, Fla, 1973).

CORBET P. S., BELLAMY R. E., and KEMPSTER R. H. (1971) Automatic devices for recording oviposition periodicity of mosquitoes in the laboratory. *Mosq. News* **31**, 516–524.

CORBET P. S. and DANKS H. V. (1975) Egg-laying habits of mosquitoes in the high Arctic. *Mosq. News* **35**, 8–14.

CORBET P. S., LONGFIELD C., and MOORE N. W. (1960) *Dragonflies*, Collins, London.

CORBET S. A. (1971) Mandibular gland secretion of larvae of the flour moth *Anagasta kuehniella* contains an epideictic pheromone and elicits oviposition movements in a hymenopteran parasite. *Nature, Lond.* **232**, 481–484.

CORBET S. A. (1973) Concentration effects and the response of *Nemeritis canescens* to a secretion of its host. *J. Insect Physiol.* **19**, 2119–2128.

CORBY H. D. L. (1947) *Aphanus* (Hemiptera: Lygaeidae) in stored groundnuts. *Bull. ent. Res.* **37**, 609–617.

CORNET M., VALADE M., and DIENG P. Y. (1974) Note technique sur l'utilisation des pondoirs-piège dans une zone rurale boisée non habitée. *Cahiers ORSTOM (Ent. med. Parasit.)* **12**, 217–219.

CORNIC J. F. (1974) Élevage de *Platysma vulgare* L. (coléoptère carabique) et observations biologiques sur le développement en captivité. *Rev. Zool. agric. Path. vég.* **73**, 90–104.

CORPORAAL J. B. (1921) De Koffiebesboorder op Sumatra's oostkust en atieh. *Meded. algemeen Proefstn AVROS (Batavia)* Ser. **12**, 1–20.

CORRÊA R. R. and RAMOS A. S. (1943) Nota sistemática sobre *Anopheles (N.) rondoni* (Neiva and Pinto, 1922). Descrição do ovo (Diptera, Culicidae). *Archos. Hig. Saúde públ.* **8** (19) 133–138.

COSCARÓN S. and GIANOTTI J. F. (1960) Una nueva plaga de la fruticultura del Alto Valle del Rio Negro y Neuquen: *Eulia loxonephes* Meyrick (Lepidoptera, Tortricidae). *Rev. Invest. agric.* **14**, 229–293.

COSTA M. (1964) Descriptions of the hitherto unknown stages of *Parasitus copridis* Costa (Acari: Mesostigmata) with notes on its biology. *J. Linn. Soc. (Zool.)* **45**, 209–222.

COSTELLO R. A. and BRUST R. A. (1969) A quantitative study of uptake and loss of water by eggs of *Aedes vexans* (Diptera: Culicidae). *Can. Ent.* **101**, 1266–1269.

COSTER J. E. (1970) Production of aggregating pheromones in re-emerged parent females of the southern pine beetle. *Ann. ent. Soc. Am.* **63**, 1186–1187.

COSTER J. E. and VITÉ J. P. (1972) Effects of feeding and mating on pheromone release in the southern pine beetle. *Ann. ent. Soc. Am.* **65**, 263–266.

COTT H. B. (1940) *Adaptive Coloration in Animals*, Methuen.

COTTON R. T. (1923) Notes on the biology of the cadelle *Tenebroides mauritanicus* Linné. *J. agric. Res.* **26**, 61–68.

COTTON R. T. (1929) The meal worms. *Tech. Bull. US Dep. Agric.* **95**, 1–38.

COULSON R. N., FOLTZ J. L., MAYYASI A. M., and HAIN F. P. (1975) Quantitative evaluation of frontalure and cacodylic acid treatment effects on within-tree populations of the southern pine beetle. *J. econ. Ent.* **68**, 671–678.

COULSON R. N., OLIVERIA F. L., PAYNE T. L., and HOUSEWEART M. W. (1973a) Variables associated with use of frontalure and cacodylic acid in suppression of the southern pine beetle. 1. Factors influencing manipulation to prescribed trap trees. *J. econ. Ent.* **66**, 893–896.

COULSON R. N., OLIVERIA F. L., PAYNE T. L., and HOUSEWEART M. W. (1973b) Variables associated with use of frontalure and cacodylic acid in suppression of the southern pine beetle. 2. Brood reduction in trees treated with cacodylic acid. *J. econ. Ent.* **66**, 897–899.

COUTIN R. (1964) Le comportement de ponte chez plusieurs cécidomyies en relation avec l'état de développement chez la plante-hôte des organes recherchés pour l'oviposition. *Rev. Zool. agric. appl.* **63**, 45–55.

COUTIN R. and DUSAUSSOY G. (1956) Étude expérimentale de la ponte de *Balaninus elephas* Gyll. sur les châtaignes (Col., Curculionidae). *Bull. Soc. ent. Fr.* **61**, 62–66.

COUTTS M. P. (1965) *Sirex noctilio* and the physiology of *Pinus radiata*. Some studies of interactions between the insect, the fungus, and the tree in Tasmania. *Bull. Commonw. For. Timb. Bur. Aust.* **41**, 2–79.

COUTURIER A. and ROBERT P. (1956) Observations sur *Melolontha hippocastani* F. *Annls Épiphyt.* **7**, 431–450.

Cova-Garcia P. (1946) *Notas sobre los Anofelinos de Venezuela y su Identificación*, 12th Conferencia Sanitaria Panamericana, No. 1.

Cova-Garcia P. (1958) Resistance manifestations in the behaviour of the vectors in Venezuela. *Indian J. Malar.* **12**, 331–339.

Craig D. A. (1967) The eggs and embryology of some New Zealand Blepharoceridae (Diptera, Nematocera) with references to the embryology of other Nematocera. *Trans. R. Soc. NZ Zool.* **8**, 191–206.

Craig G. B. (1955) Preparation of the chorion of eggs of aedine mosquitoes for microscopy. *Mosq. News* **15**, 228–231.

Craig G. B. (1967) Mosquitoes: female monogamy induced by male accessory gland substance. *Science, Wash.* **156**, 1499–1501.

Craig G. B. and Horsfall W. R. (1958) Taxonomic and ecological significance of eggs of aedine mosquitoes. *Proc. 10th Int. Congr. Ent.* **3**, 853–857 (1956).

Craig G. B. and Horsfall W. R. (1960) Eggs of floodwater mosquitoes. VII. Species of *Aedes* common in the southeastern United States (Diptera: Culicidae). *Ann. ent. Soc. Am.* **53**, 11–18.

Craighead F. C. (1923) North American cerambycid larvae. *Tech. Bull. Dept. Agric. Can. (NS)* **27**, 1–239.

Cramer E. (1968) Die Tipuliden des Naturschutzparkes Hoher Vogelsberg (ein Beitrag zur Biologie, Ökologie und Entwicklung der Tipuliden sowie zur Kenntnis der Limoniinen-larven und -puppen). *Dt. ent. Z.* **15**, 133–232.

Crampton G. C. (1930) The wings of the remarkable archaic Mecopteron *Notiothauma reedi* McLachlan with remarks on their protoblattoid affinities. *Psyche, Camb.* **37**, 83–103.

Crauford-Benson H. J. (1941a) The cattle lice of Great Britain. Part I. Biology, with special reference to *Haematopinus eurystenus*. *Parasitology* **33**, 331–358.

Crauford-Benson H. J. (1941b) The cattle lice of Great Britain. Part II. Lice populations. *Parasitology* **33**, 343–358.

Crawford C. S. (1966) Photoperiod-dependent oviposition rhythm in *Crambus teterrellus* (Lepidoptera: Pyralidae: Crambinae). *Ann. ent. Soc. Am.* **59**, 1285–1288.

Crawford C. S. (1967) Oviposition rhythm studies in *Crambus topiarius* (Lepidoptera: Pyralidae: Crambinae). *Ann. ent. Soc. Am.* **60**, 1014–1018.

Crawford C. S. (1968) Oviposition rhythm in *Crambus teterrellus*: temperature–depression effects and apparent circadian periodicity. *Ann. ent. Soc. Am.* **61**, 1481–1486.

Crawford C. S. (1970) Temperature changes influencing oviposition in *Crambus teterrellus* (Lepidoptera: Pyralidae). *Jl NY ent. Soc.* **78**, 74–79.

Crawford C. S. (1971) Comparative reproduction of *Crambus harpipterus* and *Agriphila plumbifimbriella* in northern New Mexico. *Ann. ent. Soc. Am.* **64**, 52–59.

Crawford C. S. and Morrison W. P. (1969) Environmental factors influencing oviposition and egg eclosion in two crambine moths (Lepidoptera: Pyralidae). In *Physiological Systems in Semiarid Environments* (C. C. Hoff and M. L. Riesdesel, eds.), p. 165 abstr., University of New Mexico Press.

Crawshay L. R. (1903) On the life history of *Drilus flavescens* Rossi. *Trans. ent. Soc. Lond.* **51**, 39–51.

Crewe W. and Williams P. (1961) The bionomics of the Tabanid fauna of streams in the rain-forest of the southern Cameroons. I. Oviposition. *Ann. trop. Med. Parasit.* **55**, 363–378.

Criddle N. (1918) The egg-laying habits of some of the Acrididae (Orthoptera). *Can. Ent.* **50**, 145–151.

Criddle N. (1921) Some phases of the present locust outbreak in Manitoba. *Rep. ent. Soc. Ont.* **51**, 19–23 (1920).

CRISP D. J. (1964) Plastron respiration. *Rec. Prog. Surf. Sci.* **2**, 377–425.
CRISP D. J. and THORPE W. H. (1948) The water-protecting properties of insect hairs. *Discuss. Faraday Soc.* **3**, 210–220.
CRISWELL J. T., BOETHEL D. J., MORRISON R. D., and EIKENBARY R. D. (1975) Longevity, puncturing of nuts, and ovipositional activities by the pecan weevil on three cultivars of pecans. *J. econ. Ent.* **68**, 173–177.
CROCKER R. L., WHITCOMB W. H., and RAY R. M. (1975) Effects of sex, developmental stage, and temperature on predation by *Geocoris punctipes*. *Envir. Ent.* **4**, 531–534.
CROMBIE A. C. (1941) On oviposition, olfactory conditioning and host selection on *Rhizopertha dominica* (Fab.) *J. exp. Biol.* **18**, 62–79.
CROS A. (1911) *Lydus (Alosimus) viridissimus* Lucas. Ses moeurs—sa larve primaire. *Feuille jeun. Nat.* **41** (5) 191–199.
CROS A. (1912) Entomologie algérienne. *Nemognatha chrysomelina* F. Ses variétés—son évolution. *Z. wiss. Insektenbiol.* **8**, 137–141.
CROS A. (1923) Le *Leptopalpus rostratus* Fabr., ses moeurs, sa larve primaire. *Bull. Soc. Hist. nat. Afr. N.* **14**, 327–337.
CROS A. (1924) Contribution a l'étude des espèces du genre *Sitarobrachys* Reitter et plus spécialement du *Sitobrachys buigasi* Escol. *Bull. Soc. Sci. nat. Maroc.* **4**, 22–39.
CROS A. (1927a) *Zonabris circumflexa* Chevrolat—moeurs, évolution. *Bull. Soc. Sci. nat. Maroc.* **7**, 177–197.
CROS A. (1927b) *Zonabris wagneri* Chevrolat. Moeurs—évolution. *Bull. Soc. Hist. nat. Afr. N.* **18**, 103–117.
CROS A. (1927c) Le *Meloe cavensis* Petagna. Étude biologique. *Annls Sci. nat. Zool.* **10**, 347–391.
CROS A. (1929) Note sommaire sur les parasites des oothèques des sauterelles marocaines. *Bull. Soc. Hist. nat. Afr. N.* **20**, 141–142.
CROS A. (1930) *Zonabris silbermanni* Chevrolat. Étude biologique. *Bull. Soc. Hist. nat. Afr. N.* **21**, 36–42.
CROS A. (1931) *Zonabris silbermanni* Chevrolat. La larve primaire. *Bull. Soc. Hist. nat. Afr. N.* **22**, 80–88.
CROS A. (1936) Le *Meloe affinis* Lucas var. *setosus* Escherich. Étude biologique. *Bull. Soc. Hist. nat. Afr. N.* **27**, 185–196.
CROSBY C. R. and MATHESON R. (1915) An insect enemy of the four-lined leaf-bug (*Poecilocapsus lineatus* Fabr.). *Can. Ent.* **47**, 181–183.
CROSSKEY R. W. (1965) The identification of African Simuliidae (Diptera) living in phoresis with nymphal Ephemeroptera, with special reference to *Simulium berneri* Freeman. *Proc. R. ent. Soc. Lond.* (A) **40**, 118–124.
CROSSMAN S. S. (1922) *Apanteles melanoscelus*, an imported parasite of the gypsy moth. *Bull. US Dep. Agric.* **1028**, 1–25.
CROWE T. J. (1962) The biology and control of *Dirphya nigricornis* (Oliver), a pest of coffee in Kenya (Coleoptera: Cerambycidae). *J. ent. Soc. S. Afr.* **25**, 304–512.
CRUICKSHANK W. J. (1972) The formation of "accessory nuclei" and annulate lamellae in the oöcytes of the flour moth *Anagasta kühniella*. *Z. Zellforsch. mikrosk. Anat.* **130**, 181–192.
CRUZ C. (1975) Observations on pod borer oviposition and infestation of pigeonpea varieties. *J. Agric. Univ. P. Rico* **59**, 63–68.
CRYSTAL M. M. and MEYNERS H. H. (1965) Influence of mating on oviposition by screw-worm flies (Diptera: Calliphoridae). *J. med. Ent.* **2**, 214–216.
CULVER J. J. (1919) A study of *Compsilura concinnata*, an imported tachinid parasite of the gypsy moth and the brown-tail moth. *Bull. US Dep. Agric.* **766**, 1–27.
CUMMINGS M. R. (1972) Formation of the vitelline membrane and chorion in developing oöcytes of *Ephestia kühniella*. *Z. Zellforsch. mikrosk. Anat.* **127**, 175–188.

CUMMINGS M. R. and O'HALLORAN T. J. (1974) Polar aeropyles in the egg of the housefly *Musca domestica* (Diptera: Muscidae). *Trans. Am. microsc. Soc.* **93**, 277–280.

CURRIE G. A. (1932) Oviposition stimuli of the burr-seed fly *Euaresta aequalis* Loew (Dipt. Tryp.). *Bull. ent. Res.* **23**, 191–203.

CURRIE J. E. (1967) Some effects of temperature and humidity on the rates of development, mortality and oviposition of *Cryptolestes pusillus* (Schönherr) (Coleoptera, Cucujidae). *J. stored Prod. Res.* **3**, 97–108.

CUSSAC E. (1852) Moeurs et métamorphoses du *Sperchus emarginatus* et de l'*Helochares lividus*. *Ann. ent. Soc. Fr.* **10** (2) 617–627.

CUTHBERTSON A. (1929) The mating habits and oviposition of crane-flies. *Entomologist's mon. Mag.* **65**, 141–145.

CUTHBERTSON A. (1936) Biological notes on some Diptera of Southern Rhodesia. *Occ. Pap. Rhod. Mus.* **5**, 46–63.

CUTRIGHT C. R. (1923) Life history of *Micromus posticus* Wlk. *J. econ. Ent.* **16**, 448–456.

CUTTEN F. E. A. and KEVAN D. K. McE. (1970) The Nymphomyiidae (Diptera) with special reference to *Palaeodipteron walkeri* Ide and its larva in Quebec, and a description of a new genus and species from India. *Can. J. Zool.* **48**, 1–24.

CYMOREK S. (1965) Experimente mit *Lyctus*. *Holz und Organismen Int. Symp.* **1**, 391–413.

DAANJE A. (1964) Über die Ethologie und Blattrolltechnik von *Deporaus betulae* L. und ein Vergleich mit den anderen Blattrollenden Rhynchitinen und Attelabinen (Coleoptera, Attelabinae). *Verh. K. ned. Akad. Wet.* **61**, 1–215.

DAANJE A. (1975) Some special features of the leaf-rolling technique of *Byctiscus populi* L. (Coleoptera, Rhynchitini). *Behaviour* **53**, 285–316.

D'ABRERA V. ST. E. (1944) The eggs of the Ceylon anopheline mosquitoes. *J. Malar. Inst. India* **5**, 337–359.

DAFAUCE C. (1965) Combate de la *Saperda carcharias* L. (Col. Cerambycidae), insecto perforador del chopo. *Boln Serv. Plagas for.* **8** (16) 97–109.

DAFAUCE C., ASTIASO F. and BACHILLER P. (1963) Aspectos biológicos del gorgojo perforador del chopo (*Cryptorrhynchus lapathi* L. Curculionidae). *Bol. Serv. Plagas for.* **6**, 85–97.

DAHL R. (1967) Sciomyzids as a new weapon in the biological control of gastropods. *Zool. Revy* **29** 107–112 (in Swedish; English summary).

DALLAS E. D. (1933) Otro caso de dermatitis extendida producida por un lepidóptero y note sobre *Hylesia nigricans* Berg. (Lep. Bombycidae). *8th Reun. Soc. Argent. Pat. Reg.* pp. 469–474.

DAMPF A. (1925) Contribuciones a la morphologia y biologia de la *Schistocerca paranensis* Burm., la langosta devastadora de America. *Monogr. Inst. Hygiene, Mex.* **3**, 48–84.

DANILEVSKII A. S. (1950) A new genus and species of a predatory moth feeding on mealybugs, *Coccidiphila gerasimovi* Danilevsky, gen. et sp. n. (Lepidoptera, Momphidae). *Ent. Obozr.* **31**, 47–53 (in Russian).

DANKS H. V. (1974) The macrotype eggs of Tachinidae (Diptera) on *Heliothis* sp. (Lepidoptera: Noctuidae) in North Carolina. *Can. Ent.* **106**, 1277–1282.

DANTHANARAYANA W. (1966) Extraction of arthropod eggs from soil (by the Salt and Hollick technique with centrifugal flotation). *Entomologia exp. appl.* **9**, 124–125.

DARROW E. M. (1949) Factors in the elimination of the immature stages of *Anopheles quadrimaculatus* Say in a water level fluctuation cycle. *Am. J. Hyg.* **50**, 207–235.

DAS N. M. ABRAHAM C. C. and MATHEW K. P. (1974) New record of *Pheidole* sp. (Hymenoptera: Formicidae) as a predator of the rice leaf folder *Cnaphalocrocis medinalis* Guen. *Curr. Sci.* **43**, 767–768.

DAS N. M., REMAMONY K. S., and NAIR M. R. G. K. (1969) Biology of a new jassid pest of mango, *Amrasca splendens* Ghouri (Hem., Hom.). *Indian J. Ent.* **31**, 288–290.

DAUMAL J., VOEGELE J., and BRUN P. (1975) Les trichogrammes. II. Unité de production massive et quotidienne d'un hôte de substitution *Ephestia kuehniella* Zell. (Lepidoptera, Pyralidae). *Annls Zool. Écol. anim.* **7**, 45–59.

DAVATCHI A. (1956) Sur quelques insectes nuisibles au pistachier en Iran. I. Hymenoptera–Chalcidoidea. *Rev. Path. vég. ent. agric. Fr.* **35**, 17–26.

DAVEY K. G. (1958) The migration of spermatozoa in the female of *Rhodnius prolixus* Stål. *J. exp. Biol.* **35**, 694–701.

DAVEY K. G. (1960) A pharmacologically active agent in the reproductive system of insects. *Can. J. Zool.* **38**, 39–45.

DAVEY K. G. (1965) Copulation and egg production in *Rhodnius prolixus*: the role of the spermathecae. *J. exp. Biol.* **43**, 373–378.

DAVEY K. G. (1967) Some consequences of copulation in *Rhodnius prolixus*. *J. Insect Physiol.* **13**, 1629–1636.

DAVEY P. M. (1958) The groundnut bruchid, *Caryedon gonagra* (F.). *Bull. ent. Res.* **49**, 385–404.

DAVEY P. M. (1965) The susceptibility of sorghum to attack by the weevil *Sitophilus oryzae* (L.). *Bull. ent. Res.* **56**, 287–297.

DAVID J. (1970) Oviposition chez *Drosophila melanogaster*: importance des charactéristiques physiques de la surface de ponte. *Rev. Comp. anim.* **4**, 70–72.

DAVID J. (1971) Influences génétiques sur les mécanismes contrôlant la fécundité des souches sauvages de *Drosophila melanogaster* Meig. *Annls Zool. Écol. anim.* **3**, 493–500.

DAVID J. and CLAVEL M.-F. (1969) Influence de la température sur le nombre, le pourcentage d'éclosion et la taille des oeufs fondus par *Drosophila melanogaster*. *Annls Soc. ent. Fr.* (NS) **5**, 161–177.

DAVID J. and FOUILLET P. (1973) Enregistrement continu de la ponte chez *Drosophila melanogaster* et importance des conditions expérimentales pour l'étude du rhythme circadien d'oviposition. *Rev. Comp. anim.* **7**, 197–202.

DAVID J. and VAN HEREWEGE J. (1969) Action répulsive de la levure vivante sur l'oviposition de *Drosophila melanogaster* Meig. *C. r. hebd. Séanc. Acad. Sci. Paris* (D) **268**, 1778–1780.

DAVID J. and VAN HEREWEGE J. (1970) Choix d'un site de ponte chez *Drosophila melanogaster*: technique d'étude et variabilité. *Rev. Comp. anim.* **4**, 82–84.

DAVID J. and VAN HEREWEGE J. (1971) Fécondité et comportement de ponte chez *Drosophila melanogaster*: Influence de diverses qualités de levure du commerce, du volume des cages et de la surface de la nourriture. *Bull. Biol. Fr. Belg.* **55**, 345–356.

DAVID J., VAN HEREWEGE J., and FOUILLET P. (1971) Quantitative under-feeding of *Drosophila*: effects on adult longevity and fecundity. *Expl Geront.* **6**, 249–257.

DAVID J., VAN HEREWEGE J., and FOUILLET P. (1973) Influence repulsive du saccharose sur la ponte de *Drosophila melanogaster* Meig. *Rev. Comp. Anim.* **7**, 231–238.

DAVID K. (1936) Beiträge zur Anatomie und Lebensgeschichte von *Osmylus chrysops* L. *Z. Morph. Ökol. Tiere* **31**, 151–206.

DAVID M. H. and MILLS R. B. (1975) Development, oviposition, and longevity of *Ahasverus advena*. *J. econ. Ent.* **68**, 341–345.

DAVID M. H., MILLS R. B., and SAUER D. B. (1974) Development and oviposition of *Ahasverus advena* (Waltl) (Coleoptera, Silvanidae) on seven species of fungi. *J. stored Prod. Res.* **10**, 17–22.

DAVID W. A. L. and GARDINER B. O. C. (1962) Oviposition and the hatching of the eggs of *Pieris brassicae* (L.) in a laboratory culture. *Bull. ent. Res.* **53**, 91–109.

DAVIDSON W. M. (1922) Notes on certain species of *Melanostoma* (Diptera, Syrphidae). *Trans. Am. ent. Soc.* **48**, 35–47.

Davies B. R. (1976) Wind distribution of the egg masses of *Chironomus anthracinus* (Zetterstedt) (Diptera: Chironomidae) in a shallow, wind-exposed lake (Loch Leven, Kinross). *Freshwater Biol.* **6**, 421–424.

Davies D. M. and Peterson B. V. (1956) Observations on the mating, feeding, ovarian development, and oviposition of adult black flies (Simuliidae, Diptera). *Can. J. Zool.* **34**, 615–655.

Davies D. M. and Syme P. D. (1958) Three new Ontario black flies of the genus *Prosimulium* (Diptera: Simuliidae). Part II. Ecological observations and experiments. *Can. Ent.* **40**, 744–759.

Davies J. B. (1962) Egg-laying habits of *Simulium damnosum* Theobald and *Simulium medusaeforme* form *hargrevesi* Gibbins in northern Nigeria. *Nature, Lond.* **196**, 149–150.

Davis C. (1961) A study of the hatching process in aquatic invertebrates. II. Hatching in *Ranatra fusca* P. Beauvois (Hemiptera, Nepidae). *Trans. Am. Microsc. Soc.* **lxxx**, 227–234.

Davis C. C. (1964) A study of the hatching process in aquatic invertebrates. VII. Observations on hatching in *Notonecta melaena* Kirkaldy (Hemiptera, Notonectidae) and on *Ranatra absona* D. and Dec. (Hemiptera, Nepidae). VIII. The hatching process of *Amnicola* (?) *hydrobioides* (Ancey) (Prosobranchia, Hydrobiidae). *Hydrobiologia* **23**, 253–266.

Davis C. C. (1965a) A study of the hatching process in aquatic invertebrates. XVIII. *Helicopsyche borealis* (Hagen) (Trichoptera, Helicopsychidae). *Am. Midl. Nat.* **74**, 443–450.

Davis C. C. (1965b) A study of the hatching process in aquatic invertebrates. XII. The eclosion process in *Trichocorixa naias* (Kirkaldy) (Heteroptera, Corixidae). *Trans. Am. microsc. Soc.* **84**, 60–65.

Davis C. J. (1969) Notes on the grass webworm *Herpetogramma licarsisalis* (Walker) (Lepidoptera: Pyraustidae), a new pest of turfgrass in Hawaii and its enemies. *Proc. Hawaii. ent. Soc.* **20**, 311–316 (1968).

Davis E. E. (1976) A receptor sensitive to oviposition site attractants on the antennae of the mosquito *Aedes aegypti*. *J. Insect Physiol.* **22**, 1371–1376.

Davis E. G. and Wadley F. M. (1949) Grasshopper egg-pod distribution in the Northern Great Plains and its relation to egg-survey methods. *Circ. US Dep. Agric.* (NS) **816**, 1–16.

Davis J. J. (1919) Contributions to a knowledge of the natural enemies of *Phyllophaga*. *Bull. Ill., nat. Hist. Surv.* **13**, 53–138.

Davis K. C. (1903) Sialididae of North and South America. *Bull. NY State Mus.* **68**, 442–487.

Davis R. B., O'Grady J. J. Jr, and Hightower B. G. (1972) A device providing continuous stimulus to oviposition for individual screwworm flies. *J. econ. Ent.* **65**, 1214–1215.

Davis R. B., Pratt R. W., Lopez E., and Turner J. P. (1967) Oviposition by screw-worm flies in infested Mexican burros. *J. econ. Ent.* **60**, 690–691.

Dawson P. S. (1966) Developmental rate and competitive ability in *Tribolium*. *Evolution* **20**, 104–116.

Dawson P. S. (1967a) Developmental rate and competitive ability in *Tribolium*. II. Changes in competitive ability following further selection for developmental rate. *Evolution* **21**, 292–298.

Dawson P. S. (1967b) Interspecific competition, egg cannibalism and the length of larval instars in *Tribolium*. *Evolution* **21**, 857–858.

Dawson P. S. (1968) Xenocide, suicide, and cannibalism in flour beetles. *Am. Nat.* **102**, 97–105.

Dean R. L. and Hartley J. C. (1977a) Egg diapause in *Ephippiger cruciger* (Orthoptera: Tettigonidae). I. The influence, variable duration and elimination of the initial diapause. *J. exp. Biol.* **66**, 173–183.

Dean R. L. and Hartley J. C. (1977b) Egg diapause in *Ephippiger cruciger* (Orthoptera: Tettigoniidae). II. The intensity and elimination of the final egg diapause. *J. exp. Biol.* **66**, 185–195.

DEAN R. L. and HARTLEY J. C. (1977c) Egg diapause in *Ephippiger cruciger* (Orthoptera: Tettigoniidae). III. Abnormal development through the final egg diapause. *J. exp. Biol.* **66**, 197–201.

DEAN R. W. (1935) Anatomy and postpupal development of the female reproductive system in apple maggot fly *Rhagoletis pomonella* Walsh. *Tech. Bull. NY State agric. Exp. Stn* **229**, 3–31.

DEANE L. M., DEANE M. P., and CAUSEY O. R. (1943) Descricão do ovo, larva e pupa de *Anopheles (Arthuromyia) gilesi* (Neiva, 1908). *Papéis Dep. Zool. S. Paulo* **3**, 167–192.

DEANE M. P. and CAUSEY O. R. (1943) Viability of *Anopheles gambiae* eggs and morphology of unusual types found in Brazil. *Am. J. trop. Med.* **23**, 95–103.

DEBACH P. (1964) (ed.) *Biological Control of Insect Pests and Weeds*, Reinhold, London.

DEBACH P., FLESCHNER C. A., and DIETRICK E. J. (1951) A biological check method for evaluating the effectiveness of entomophagous insects. *J. econ. Ent.* **44**, 763–766.

DEBEY M. (1846) *Beiträge zur Lebens- und Entwicklungsgeschichte der Rüsselkafer aus der Familie der Attelabiden (mit Beiträge von E. Heis)*, Bonn.

DE BUCK A. (1938) Das Exochorion der *Stegomyia* Eier. *Proc. Acad. Sci. Amsterdam* **41**, 677–683.

DE BUCK A., SCHOUTE E., and SWELLENGREBEL N. H. (1932) Further investigations on the racial differentiation of *Anopheles maculipennis* in the Netherlands and its bearing on malaria. *Riv. Malar.* **11**, 1–22.

DE BUCK A. and SWELLENGREBEL N. H. (1932) On anophelism without malaria around Amsterdam. IV. The pattern of the dorsal surface of the ova in the two races of *A. maculipennis*. *Proc. K. ned. Akad. Weten. Amsterdam* **35**, 1335–1338.

DE BUCK A. and SWELLENGREBEL N. H. (1934) Further observations on the pattern of the upper surface of the ova in the Dutch varieties of *A. maculipennis*. *Proc. K. ned. Akad. Weten. Amsterdam* **37**, 578–579.

DÉCAMPS H. (1967) Écologie des Trichoptères de la vallée d'Aure (Haute-Pyrénées). *Annls Limnol.* **3**, 399–577.

DECOURSEY J. D. and WEBSTER A. P. (1952) A method of clearing the chorion of *Aedes sollicitans* (Walker) eggs and preliminary observations on their embryonic development. *Ann. ent. Soc. Am.* **45**, 625–632.

DÉDUIT Y. (1961) Études sur la ponte par autogenèse des Culicides. IV. Données numériques sur l'acte de ponte chez un groupe de femelles fécondées de *Culex pipiens autogenicus* Roubaud: recherche de l'effet de groupe. *C. r. hebd. Séanc. Soc. Biol. Paris* **154**, 1617–1619.

DEEGENER P. (1909) *Die Metamorphose der Insekten*, Leipzig, 1908.

DEEMING J. C. and KNUTSON L. V. (1966) Ecological notes on some Sphaeroceridae reared from snails, and a description of the puparium of *Copromyza (Apterina) pedestris* Meigen. *Proc. ent. Soc. Wash.* **68**, 108–112.

DE GEER C. (1771) *Mémoires pour servir à l'histoire des insectes; précédés de discours sur les insectes en général*, Vol. 2, P. Hesselberg, Stockholm.

DE GEER C. (1773) *Mémoires pour servir à l'histoire des insectes,* Vol. 3, pp. 261–266.

DEGRANGE C. (1956a) Sur les micropyles des oeufs des Ephéméroptères. *Bull. Soc. ent. Fr.* **61**, 146–148.

DEGRANGE C. (1956b) Sur l'éclosion des larves des Ephéméroptères. *C. r. hebd. Séanc. Acad. Sci. Paris* **242**, 2054–2056.

DEGRANGE C. (1957) L'oeuf et le mode d'éclosion de quelques Plécoptères. *Trav. Lab. Hydrobiol. Grenoble* **48–49**, 37–49.

DEGRANGE C. (1960) *Recherches sur la reproduction des Ephéméroptères*, Imp. Allier, Grenoble.

DEGRANGE C. (1971) L'oeuf de *Hemianax ephippiger* (Burmeister) 1839 (Odonata, Anisoptera, Aeschnidae). *Trav. Lab. Piscic. Univ. Grenoble* **62**, 131–145 (1970).

DELATTRE P. (1971) Contribution à l'étude biologique de *Zetzellia mali* Ewing (Acarina:

Stigmaeidae). *Annls Zool. Écol. anim.* **3**, 297–303.
DELCOURT A. (1907) Quelques observations sur la variabilité de *Notonecta glauca*. *C. r. hebd. Séanc. Soc. Biol. Paris* **62**, 11–13.
DELMAS H. G. (1954) Le vespère de la vigne. Techniques de lutte. *Rev. Zool. agric.* **53**, 110–120.
DELSMAN H. C. (1926) On the propagation of *Halobates*. *Treubia* **8**, 384–388.
DELUCCHI V. (1954) *Pullus impexus* (Muls.) (Coleoptera, Coccinellidae) a predator of *Adelges piceae* (Ratz.) (Hemiptera, Adelgidae), with notes on its parasites. *Bull. ent. Res.* **45**, 243–278.
DELUCCHI V., AESCHLIMANN J. P., and GRAF E. (1975) The regulating action of egg predators on the populations of *Zeiraphera diniana* Guénée (Lep. Tortricidae). *Mitt. Schweiz. ent. Ges.* **48**, 37–45.
DELUCCHI V. and PSCHORN-WALCHER H. (1954) *Cremifania nigrocellulata* Czerny (Diptera, ?Chamaemyiidae), ein Räuber an *Dreyfusia (Adelges) piceae* Ratz. (Hemiptera, Adelgidae). *Z. angew. Ent.* **36**, 84–107.
DE MEILLON B. (1934) The eggs of some South African Anophelines. *Publ. S. Afr. Inst. med. Res.* **6**, 272–287.
DE MEILLON B. (1936) The eggs of some South African Anophelines. Part II. *Publ. S. Afr. Inst. med. Res.* **7**, 131–132.
DE MEILLON B. (1937) Culicidae. 1. The eggs of some South African Anophelines. Part III. *Publ. S. Afr. Inst. med. Res.* **7**, 305.
DE MEILLON B. (1947) The Anophelini of the Ethiopian geographical region. *Publ. S. Afr. Inst. med. Res.* **49**, 1–272.
DE MEILLON B., GOLBERG L., and LAVOIPIERRE M. (1945) The nutrition of the larva of *Aedes aegypti* L. *J. exp. Biol.* **21**, 84–89.
DEMOLIN G. and DELMAS J. C. (1967) Les Éphippigères (Orthoptères, Tettigoniidae), prédateurs occasionnels mais importants de *Thaumetopoea pityocampa* Schiff. *Entomophaga* **12**, 399–401.
DEMPSTER J. P. (1957) The population dynamics of the Moroccan locust (*Dociostaurus maroccanus* Thunberg) in Cyprus. *Anti-Locust Bull.* **27**, 1–60.
DEMPSTER J. P. (1960) A quantitative study of the predators on the eggs and larvae of the broom beetle *Phytodecta olivacea* Forster using the precipitin test. *J. Anim. Ecol.* **29**, 149–167.
DENISOVA T. V. (1965) The phenolic complex of vine roots infested by *Phylloxera* as a factor in resistance. *Vest. sel'.-khoz. Nauki Mosk.* **10**, 114–118 (in Russian; English summary).
DESEÖ K. V. (1967) The role of olfactorial stimuli in the egg-laying behaviour of plum moth (*Grapholitha funebrana* Tr.). *Acta phytopath. Acad. Sci. hung.* **2**, 243–250.
DE STEFANI P. T. (1913) Cavallette, loro invasioni e lotta contro di esse in Sicilia. Osservazione fatte durante l'invasione della Provincia di Palermo negli anni 1901–1911. *Giorn. Sci. nat. Econ. Palermo* **30**, 117–199.
DETHIER V. G. (1947) The response of hymenopterous parasites to chemical stimulation of the ovipositor. *J. exp. Zool.* **105**, 199–207.
DETHIER V. G. (1959) Egg-laying habits of Lepidoptera in relation to available food. *Can. Ent.* **91**, 554–561.
DHARMARAJU E. (1968) A note on the occurrence of fruit fly maggots of *Dacus cucurbitae* Coq. in cecidomyiid galls of *Coccinia indica* in Andhra Pradesh. *Indian J. Ent.* **30**, 87.
DIAKONOFF A. (1952) Viviparity in Lepidoptera. *Trans. 9th Int. Contr. Ent.* **1**, 91–96.
DIAS B. F. DE S. (1975) Comportamento pré-social de Sinfitas do Brasil Central. I. *Themos olfersii* (Klug) (Hymenoptera, Argidae). *Studia Ent.* **18**, 401–432.
DICK J. (1937) Oviposition in certain Coleoptera. *Ann. appl. Biol.* **24**, 762–796.
DICK R. D. (1945) Ecological observations on *Oxycanus cervinata*. *NZ Jl Sci. Tech.* (A) **27**, 32–36.

DICKASON E. A. (1960) Mortality factors for the vetch bruchid *Bruchus brachialis. J. econ. Ent.* **53**, 555–558.
DICKENS J. C. and PAYNE T. L. (1977) Bark beetle olfaction: pheromone receptor system in *Dendroctonus frontalis. J. Insect Physiol.* **23**, 481–489.
DICKER G. H. L. (1951) *Agonum dorsale* Pont. (Col., Carabidae): an unusual egg-laying habit and some biological notes. *Entomologist's mon. Mag.* **87**, 33–34.
DIDLAKE M. (1926) Observations on the life-histories of two species of praying mantis (Orthop.: Mantidae). *Ent. News* **37**, 169–174.
DIMETRY N. Z. (1974) The consequences of egg cannibalism in *Adalia bipunctata* (Coleoptera: Coccinellidae). *Entomophaga* **19**, 445–451.
DIMETRY N. Z. and MANSOUR M. H. (1976) The choice of oviposition sites by the ladybird beetle *Adalia bipunctata* (L.). *Experientia* **32**, 181–182.
DINTHER J. VAN (1972) Carabids as natural enemies of the cabbage fly. *Ent. Ber.* **32** (10) 193–194 (in Dutch; English summary).
DINTHER J. B. M. VAN (1966) Laboratory experiments on the consumption capacities of some Carabidae. *Meded. LandbHoogesch. Gent.* **31**, 730–739.
DINTHER J. B. M. VAN and MENSINK F. T. (1965) Egg consumption by *Bembidion ustulatum* and *Bembidion lampros* (fam. Carabidae) in laboratory prey density experiments with house fly eggs. *Meded. LandbHoogesch. Gent.* **30,** 1542–1554.
DINTHER J. B. M. VAN and MENSINK F. T. (1971) Use of radioactive phosphorus in studying egg predation by carabids in cauliflower fields. *Meded. LandbWetensch. Gent.* **36**, 283–293.
DINULESCU G. (1932) Recherches sur la biologie des Gastérophiles. Anatomie, physiologie, cycle évolutif. *Annls Sci. nat.* **15** (10) 1–183.
DIRIMANOV M. and SENGALEVICH G. (1962) The bark borer (*Laspeyresia woeberiana* Schiff.)—morphological and biological peculiarities and possibilities for control. *Rast. Zasht.* **10** (5) 33–46.
DIRSCH V. M. (1968) The post-embryonic ontogeny of Acridomorpha. *Eos* **43**, 413–514.
DISNEY R. H. L. (1969) Phoretic association between simuliid blackflies and a prawn. *Trans. R. Soc. trop. Med. Hyg.* **63**, 292.
DISNEY R. H. L. (1971) Association between blackflies (Simuliidae) and prawns (Atyidae), with a discussion of the phoretic habit in Simuliids. *J. Anim. Ecol.* **40**, 83–92.
DISNEY R. H. L. (1973) Further observations on some blackflies (Diptera: Simuliidae) associated with mayflies (Ephemeroptera: Baetidae and Heptageniidae) in Cameroon. *J. Ent.* (A) **47**, 169–180.
DISNEY R. H. L. (1975) *Drosophila gibbinsi* larvae also eat *Simulium. Trans. R. Soc. trop. Med. Hyg.* **69**, 365–366.
DISPONS P. (1969) L'oeuf et la larve de *Stenolemus novaki* Horváth (Hemiptera, Heteroptera, Reduviidae, Emesinae). *Vie et Milieu* **20** (1C) 243–248.
DISTANT W. L. (1900) On the egg carrying habits of *Zaitha fluminea. Zoologist* **4** (4) 93–94.
DIXON T. J. (1958) An ecological study of Syrphidae (Diptera) with an account of their larvae. PhD thesis, London University.
DIXON T. J. (1959) Studies on oviposition behaviour of Syrphidae (Diptera). *Trans. R. ent. Soc. Lond.* **111**, 57–80.
DJAMIN A. and PATHAK M. D. (1967) Role of silica in resistance to Asiatic rice borer *Chilo suppressalis* (Walker) in rice varieties. *J. econ. Ent.* **60**, 347–351.
DOANE J. F. (1963) Studies on oviposition and fecundity of *Ctenicera destructor* (Brown) (Coleoptera: Elateridae). *Can. Ent.* **95**, 1145–1153.
DOANE J. F. (1966) Absorption and content of water in eggs of *Ctenicera destructor* (Brown), *Ctenicera aeripennis* (Kirby) and *Hypolithus bicolor* Eschscholtz (Coleoptera: Elateridae). *Can. Ent.* **98**, 482–486.

DOANE J. F. (1967) The influence of soil moisture and some soil physical factors on the ovipositional behaviour of the prairie grain wireworm *Ctenicera destructor*. *Entomologia exp. appl.* **10**, 275–286.

DOANE J. F. (1968) The influence of soil moisture on survival of eggs of the prairie grain wireworm *Ctenicera destructor*. *Can. Ent.* **100**, 362–373.

DOANE J. F. (1969a) A method for separating the eggs of the prairie wireworm *Ctenicera destructor* from soil. *Can. Ent.* **101**, 1002–1004.

DOANE J. F. (1969b) Effect of temperature on water absorption, development, and hatching in eggs of the prairie grain wireworm *Ctenicera destructor*. *Ann. ent. Soc. Am.* **62**, 567–572.

DOBOSH I. G. (1969) Peculiarities of ecology of oviposition of *Dolichocera* Orthoptera under conditions of the middle Dnieper area. *Dopov. Akad. Nauk. ukr. URSR* (B) **1969**, 847–849 (in Ukrainian; English summary).

DOBROVOL'SKIĬ B. V. (1950) The grape or pear leafroller (*Byctiscus betulae* L.) on the Don and in North Caucasus. *Ent. Obozr.* **31**, 41–46 (in Russian).

DOBY J.-M., RAULT B., and BEAUCOURNU-SAGUEZ F. (1967) Utilisation de rubans de plastique pour la récolte des oeufs et des stades larvaires et nymphaux de simulies (Diptères Paranématocères) et pour l'étude biologique de ceux-ci. *Annls Parasit. hum. comp.* **42**, 651–657.

DOCTERS VAN LEEUWEN W. (1910) Ueber die Lebensweise und die Entwicklung einiger holzbohrenden Cicindeliden Larven. *Tijd. Ent.* **53**, 18–40.

DODD F. P. (1904) Notes on maternal instinct in *Rhynchota*. *Trans. ent. Soc. Lond.* **1904**, 483–485.

DODGE H. R. (1938) The bark beetles of Minnesota (Coleoptera: Scolytidae). *Tech. Bull. Minn. Agric. Exp. Stn* **132**, 1–60.

DODSON M. (1937) Development of the female genital ducts in *Zygaena* (Lepidoptera). *Proc. R. ent. Soc. Lond.* (A) **12**, 61–68.

DOGIEL V. A. (1964) *General Parasitology* (revised by Yu. I. Polyanski and E. M. Kheisin), Oliver and Boyd, Edinburgh and London.

DOĬNIKOV A. (1965) The almond seed chalcid. *Zashch. Rast. Vredit. Bolez.* **1965** (8), p. 40 (in Russian).

DOLIDZE G. V. (1957) The study of problems of the ecology of the cabbage moth (*Barathra brassicae* L.) in Georgia. *Trudy. Inst. Zashch. Rast.* **12**, 79–100 (in Georgian).

DONALDSON J. M. I. (1970) Differences between top and bottom eggs and between the resulting hopper and adult populations of two strains of *Locusta migratoria migratorioides* (R. and F.). *Phytophylactica* **2**, 199–202.

DONISTHORPE H. ST. J. K. (1902) The life-history of *Clythra quadripunctata* L. *Trans. ent. Soc. Lond.* **1902**, 11–24.

DONLEY D. E. (1969) Oviposition behaviour of *Goes tigrinus* DeG. in white oak, *Quercus alba* L. *Proc. N. Cent. Brch. Am. Ass. econ. Ent.* **24**, 40.

DOOM D. (1967) Notes on *Gnathotrichus materiarius* (Col., Scolytidae), a timber beetle new to the Netherlands. *Ent. Ber.* **27**, 143–148.

D'ORCHYMONT A. (1913) Contribution à l'étude des larves Hydrophilides. *Annls biol. lacustre* **6**, 173–214.

D'ORCHYMONT A. (1933) Contribution à l'étude des Palpicornia. VIII. *Bull. A. Soc. ent. Belg.* **73**, 271–313.

DORESTE E. (1966) Ciclo biológico del cortador grande (*Agrotis repleta* Wlk.). *Mem. Sextas Jorn. agron. Soc. venez. Ingen. agrón.* **3**, 1–10.

DORIA R. C. and RAROS R. S. (1975) Varietal resistance of mungo to the bean weevil *Callosobruchus chinensis* (Linn.) and some characteristics of field infestation. *Philipp. Ent.* **2**, 399–408 (1973).

DÖRING E. (1955) *Zur Morphologie der Schmetterlingseier*, Berlin, Akademie-Verlag.

DOUGLAS J. W. (1888) Larvae of Lepidoptera feeding on Coccidae. *Entomologist's mon. Mag.* **24**, 225–228.

DOUTT R. L. (1959) The biology of parasitic Hymenoptera. *A. Rev. Ent.* **4**, 161–182.

DOUTT R. L. (1973) Maternal care of immature progeny by parasitoids. *Ann. ent. Soc. Am.* **66**, 486–487.

DOUWES P. (1968) Host selection and host finding in the egg-laying female *Cidaria albulata* L. (Lep. Geometridae). *Opusc. ent.* **33**, 233–279.

DOWDEN P. B. (1933) *Lydella nigripes* and *L. piniariae*, fly parasites of certain tree-defoliating caterpillars. *J. agric. Res.* **46**, 963–995.

DOWNE A. E. R. and WEST A. S. (1954) Progress in the use of the precipitin test in entomological studies. *Can. Ent.* **86**, 181–184.

DOWNES J. A. (1965) Adaptations of insects in the arctic. *A. Rev. Ent.* **10**, 257–274.

DRAKE C. J. and SLATER J. A. (1957) The phylogeny and systematics of the family Thaumastocoridae (Hemiptera: Heteroptera). *Ann. ent. Soc. Am.* **50**, 353–370.

DRAUDT M. (1906) Zur Kenntnis der Eupithecien-Eier. *Dt. ent. Z.* **18**, 280–320.

DREA J. J. (1969) Fecundity, hatch of eggs, and duration of oviposition of mated, isolated female alfalfa weevils. *J. econ. Ent.* **62**, 1523–1524.

DRESNER E. A. (1970) A dermestid (Coleoptera) infesting mantid egg pods. *Ann. ent. Soc. Am.* **63**, 1477–1478.

DROOZ A. T. (1965) Some relationships between host, egg potential, and pupal weight of the elm spanworm *Ennomos subsignarius* (Lepidoptera: Geometridae). *Ann. ent. Soc. Am.* **58**, 243–245.

DROOZ A. T. (1970) Rearing the elm spanworm on oak or hickory. *J. econ. Ent.* **63**, 1581–1585.

DROOZ A. T. (1975) *Mesoleius tenthredinis* and other parasites of the larch sawfly in the eastern United States. *Envir. Ent.* **4**, 645–650.

DROOZ A. T. and SOLOMON J. D. (1964) Effect of solarization on elm spanworm eggs (Lepidoptera: Geometridae). *Ann. ent. Soc. Am.* **57**, 95–98.

DRUGER M. (1962) Selection and body size in *Drosophila pseudoobscura* at different temperatures. *Genetics* **47**, 209–222.

DUBININ V. B. (1938) Changes in the parasite fauna of *Plegalis falcinellus* caused by the age and migrations of the host. *Trudỹ astrakh. Zap.* **2**, 114–212.

DU BOIS A. M. and GEIGY R. (1935) Beiträge zur Ökologie, Fortpflanzungsbiologie und Metamorphose von *Sialis lutaria* L. *Rev. suisse Zool.* **42**, 169–248.

DUBOIS J. (1966) Influence de certaines modifications récentes dans la culture de la canne à sucre sur l'évolution des populations de la cigale *Yanga guttulata* Sign. *Docum. Inst. Rech. agron. Madagascar* **63**, 1–52.

DUDLEY C. O. (1971) A sampling design for the egg and first instar larval populations of the western pine beetle *Dendroctonus brevicomis* (Coleoptera: Scolytidae). *Can. Ent.* **103**, 1291–1313.

DUERDEN J. C. and EVANS A. C. (1954) A note on the biology of *Calidea dregei* Germ. *E. Afr. agric. J.* **19**, 188–192.

DUFFEY S. S. and SCUDDER G. G. E. (1974) Cardiac glycosides in *Oncopeltus fasciatus* (Dallas) (Hemiptera: Lygaeidae). I. The uptake and distribution of natural cardenolides in the body. *Can. J. Zool.* **52**, 283–290.

DUFFY E. A. J. (1946) A contribution towards the biology of *Prionus coriarius* L. (Coleoptera, Cerambycidae). *Trans. R. ent. Soc. Lond.* **97**, 419–442.

DUFFY E. A. J. (1953) *A Monograph of the Immature Stages of British and Imported Timber Beetles (Cerambycidae)*, British Museum (Nat. Hist.), London.

DUFFY E. A. J. (1957) *A Monograph of the Immature Stages of African Timber Beetles (Cerambycidae)*, British Museum (Nat. Hist.), London.
DUFOUR L. (1858) Histoire des métamorphoses du *Bombylius major*. *Annls Soc. ent. Fr.* **6**, 503–511.
DULIZIBARIĆ T. (1966) The possibilities of chemical control of the maize stem borer *O. nubilalis* with regard to its bionomics, ecology and the new method of cultivating maize in Yugoslavia. *Zašt. Bilja* **17** (89–90), 1–180 (English summary).
DUMBLETON L. J. (1940) *Oncodes brunneus* Hutton: a dipterous spider parasite. *NZ Jl Sci. Tech.* **22**, 97A–102A.
DU MERLE P. (1966) Modèle de cage permettant d'obtenir la ponte d'un Diptère Bombyliidae *Villa quinquefasciata* Wied. AP Meig. *Entomophaga* **11**, 325–330.
DUNBAR D. M. and BACON O. G. (1972) Feeding, development, and reproduction of *Geocoris punctipes* (Heteroptera: Lygaeidae) on eight diets. *Ann. ent. Soc. Am.* **65**, 892–895.
DU PLESSIS C. (1937) The occurrence of the brown and red locust in the Union during the seasons 1934–35 and 1935–36. *Sci. Bull. Dep. Agric. S. Afr.* **164**, 1–17.
DUPREE J. W. and MORGAN H. A. (1902) Mosquito development and hibernation. *Science, NY* **16**, 1036–1038.
DUPUIS C. (1960) Expériences sur l'oviparité, le comportement de ponte et l'incubation chez quelques Diptères Phasiinae. *C. r. hebd. Séanc. Acad. Sci. Paris* **250**, 1744–1746.
DUPUIS C. (1963) Essai monographique sur les Phasiinae (Diptères Tachinaires, parasites d'Hétéroptères). *Mem. Mus. Hist. nat. (Zool.), Paris* (NS) **26**, 1–461.
DURCHON M. (1946) Les adultes de Doryphore peuvent s'attaquer à leurs propres oeufs. *C. r. hebd. Séanc. Acad. Sci. Paris* **222**, 340–342.
DÜRR H. J. R. (1956) The morphology and bionomics of the European houseborer *Hylotrupes bajulus* (Coleoptera: Cerambycidae). *Ent. Mem. Dep. Agric. S. Afr.* **4**, 1–136.
DUSTAN G. G. (1964) Mating behaviour of the oriental fruit moth *Grapholitha molesta* (Busck) (Lepidoptera: Olethreutidae). *Can. Ent.* **96**, 1087–1093.
DYAR H. G. (1929) A new beneficial moth from Panama and a scavenger (Lepidoptera, Pyralidae, Phycitinae). *Proc. ent. Soc. Wash.* **31**, 16–17.
DYER E. D. A. (1973) Spruce beetle aggregated by synthetic pheromone frontalin. *Can. J. For. Res.* **3**, 486–494.
DYER E. D. A. (1975) Frontalin attractant in stands infested by the spruce beetle *Dendroctonus rufipennis* (Coleoptera: Scolytidae). *Can. Ent.* **107**, 979–988.
DYER E. D. A. and CHAPMAN J. A. (1971) Attack by the spruce beetle, induced by frontalin or billets with burrowing females. *Bi-mon. Res. Notes* **27**, 10–11.
DYER E. D. A., HALL P. M., and SAFRANYIK L. (1975) Numbers of *Dendroctonus rufipennis* (Kirby) and *Thanasimus undatulus* Say at pheromone-baited poisoned and unpoisoned trees. *J. ent. Soc. Br. Columb.* **72**, 20–22.
DZHIBLADZE K. N. (1969) A new species of predacious mite (Hemisarcoptidae) attacking the sour-orange scale *Cornuaspis beckii* Newman (Homoptera, Coccoidea) in western Georgia. *Ent. Obozr.* **48**, 689–691 (English translation: *Ent. Rev.* **48**, 435–436).

EASTHAM J. W. (1909) Some enemies of Ontario Coccidae. *Ann. Rep. ent. Soc. Ont.* **39**, 54–56 (1908).
EASTHAM L. E. S. and MCCULLY S. B. (1943) The oviposition responses of *Calandra granaria* Linn. *J. exp. Biol.* **20**, 35–42.
EATON A. E. (1865) Occurrence of the female imago under submerged stones. *Entomologist's mon. Mag.* **2**, 14.

EBERHARD W. G. (1975) The ecology and behaviour of a subsocial pentatomid bug and two scelionid wasps: strategy and counterstrategy in a host and its parasites. *Smithson. Contr. Zool.* **205**, 1–39.

EBERHARDT G. (1930) *Chortophila cicicrura* Rond., a new parasite of the migratory locust in Dagestan. *Zasch. Rast. Vredit.* **6**, 813–814 (in Russian).

ECKBLAD J. W. (1973) Experimental predation studies of malacophagous larvae of *Sepedon fuscipennis* (Diptera: Sciomyzidae) and aquatic snails. *Expl Parasit.* **33**, 331–342.

ECKBLAD J. W. and BERG C. O. (1972) Population dynamics of *Sepedon fuscipennis* (Diptera: Sciomyzidae). *Can. Ent.* **104**, 1735–1742.

ECKENRODE C. J., HARMAN G. E., and WEBB D. R. (1975) Seed-borne microorganisms stimulate seedcorn maggot egg laying. *Nature, Lond.* **256**, 487–488.

EDDY G. W., DEVANEY J. A., and HANDKE B. D. (1975) Response of the adult screwworm (Diptera: Calliphoridae) to bacteria-inoculated and incubated bovine blood in olfactometer and oviposition tests. *J. med. Ent.* **12**, 379–381.

EDEL'MAN N. M. (1956) The biology of *Lymantria dispar* (L.) under conditions of the Kuba district of the Azerbaijan SSR. *Zool. Zh.* **35**, 572–582 (in Russian; English summary).

EDNEY E. B. (1945) Laboratory bionomics on the rat fleas *Xenopsylla brasiliensis* Baker and *X. cheopis* Roths. I. Certain effects of light, temperature and humidity on the rate of development and on adult longevity. *Bull. ent. Res.* **35**, 399–416.

EDNEY E. B. (1957) *The Water Relations of Terrestrial Arthropods*, Cambridge University Press.

EDWARDS C. A. (1964) The bionomics of swift moths. I. The ghost swift moth *Hepialus humuli* (L.). *Bull. ent. Res.* **55**, 147–160.

EDWARDS D. K. (1961) Influence of electrical field on pupation and oviposition in *Nepytia phatasmaria* Stkr. (Lepidoptera: Geometridae). *Nature, Lond.* **191**, 976–993.

EDWARDS D. K. (1964) Activity rhythms of lepidopterous defoliators. I. Techniques for recording activity, eclosion, and emergence. *Can. J. Zool.* **42**, 923–937.

EDWARDS F. W. (1921) A revision of the mosquitoes of the Palaearctic Region. *Bull. ent. Res.* **12**, 263–351.

EDWARDS J. S. (1962) Observations on the development and predatory habit of two Reduviid Heteroptera *Rhinocoris carmelita* Stål and *Platymeris rhadamanthus* Gerst. *Proc. R. ent. Soc. Lond.* (A) **37**, 89–98.

EDWARDS J. S. (1966) Observations on the life history and predatory behaviour of *Zelus exsanguis* (Stål) (Heteroptera: Reduviidae). *Proc. R. ent. Soc. Lond.* (A) **41**, 21–24.

EDWARDS R. L. (1954) The hostfinding and oviposition behaviour of *Mormoniella vitripennis* (Walker) (Hym., Pteromalidae), a parasite of Muscoid flies. *Behaviour* **7**, 88–112.

EDWARDS R. L. and EPP H. T. (1965) The influence of soil moisture and soil type on the oviposition behaviour of the migratory grasshopper *Melanoplus sanguinipes* (Fabricius). *Can. ent.* **97**, 401–409.

EDWARDS W. D., GRAY K., and MOTE D. C. (1934) Observations on the life habits of *Cnephasia longana* Haw. *Mon. Bull. Dep. Agric. Calif.* **23**, 328–333.

EGE R. (1915) On the respiratory function of the air stores carried by some aquatic insects. *Z. allg. Physiol.* **17**, 81–124.

EGGER A. (1974) Zur Biologie und wirtschaftlichen Bedeutung von *Chrysopa carnea* Steph. (Neuropt., Planip., Chrysopidae). *Anz. Schädlingsk.* **47** (12) 183–189.

EGGLISHAW H. J. (1960a) Studies on the family Coelopidae (Diptera). *Trans. R. ent. Soc. Lond.* **112**, 109–140.

EGGLISHAW H. J. (1960b) The life-history of *Fucellia maritima* (Haliday) (Diptera, Muscidae). *Entomologist* **95**, 225–231.

EGGLISHAW H. J. (1961) The life-history of *Thoracochaeta zosterae* (Hal.) (Dipt., Sphaeroceridae). *Entomologist's mon. Mag.* **96**, 124–128.

EGOROV N. N. and SOLOZHÉNIKINA T. N. (1963) *Cacoecia crataegana* Hb.—a mass pest of oak stands in the Voronezh region. *Zool. Zh.* **42**, 1501–1512 (in Russian; English summary).
EGUAGIE W. E. (1972) Effects of temperature and humidity on the development and hatching of eggs of the thistle lacebug *Tingis ampliata* (Heteroptera, Tingidae). *Entomologia exp. appl.* **15**, 183–189.
EGUAGIE W. E. (1973) Mating and oviposition of *Tingis ampliata* H.-S. (Het., Tingidae) (Hem.). *Entomologia scand.* **4**, 285–298.
EGUAGIE W. E. (1977) Observations on the ecology of the cocoa shield bug *Bathycoelia thalassina* (H.-S.) (Heteroptera) in Nigeria (abstract). Pub. in Kumar (ed.), *Proc. 4th Conf. W. African Cocoa Entomologists, Ghana,* 1974.
EICHHORN O., KRIEGL M., and SECHSER B. (1968) Investigations on the distribution and the predator complex of *Adelges* (=*Dreyfusia*) *prelli* Gros. on *Abies nordmanniana* Spach. in central Europe. *Tech. Bull. Commonw. Inst. biol. Control* **10**, 27–32.
EICHLER W. (1937) Einige Bemerkungen zur Ernährung und Eiablage der Mallophagen. *Sber. Ges. naturf. Freunde Berlin* **1937**, 80–111.
EICHLER W. (1938) Lebensraum und Lebensgeschichte der Dahlemer Plamenhausheuschrecke *Phlugiola dahlemica* nov. spec. (Orthoptera, Tettigodiid). Inaug. diss. Fak. Friedrich-Wilhelms-Univ., Berlin, pp. 4–79.
EICHLER W. (1940) Topographische Spezialisation bei Ektoparasiten. *Z. ParasitKde.* **11**, 205–214.
EICHLER W. (1946) Parthenogenese und Ovoviviparie als Entwicklungseigentümlichkeiten bei Läusen und Federlingen. *Tierärztl. Umsch.* **1**, 10.
EICHLER W. (1952) *Behandlungstechnik parasitärer Insekten,* Leipzig.
EICHLER W. (1963) Arthropoda. Insecta. Phthiraptera 1. Mallophaga. In *Bronn's Klassen and Ordnungen des Tierreichs, Leipzig,* **5**, III, Abt. 7, Buch (b), 1–290.
EIDMANN H. (1929) Morphologische und physiologische Untersuchungen am weiblichen Genitalapparat der Lepidopteren. I. Morphologischer Teil. *Z. angew. Ent.* **15**, 1–66.
EIDMANN H. (1930) Ueber den taxonomischen Wert des weiblichen Genitalapparates der Lepidopteren (nach Untersuchungen an der Gattung *Papilio* L.). *Zool. Anz.* **92**, 113–122.
EIDT D. C. and CAMERON M. D. (1970) Pretreatment of spruce budworm eggs for counting. *Bi-mon. Res. Notes* **26** (5) 46–47.
EKBLOM T. (1926) Morphological and biological studies of the Swedish families of Hemiptera–Heteroptera. Part I. The families Saldidae, Nabidae, Lygaeidae, Hydrometridae, Veliidae, and Gerridae. *Zool. Bidr.* **10**, 29–179.
EKKENS D. (1972) Peruvian treehopper behaviour (Homoptera: Membracidae). *Ent. News* **83**, 257–271.
ELBADRY E. (1965) Some effects of gamma radiation of the potato tuberworm *Gnorimoschema operculella* (Lepidoptera: Gelechiidae). *Ann. ent. Soc. Am.* **58**, 206–209.
ELBADRY E. A. and TAWFIK M. F. S. (1966) Life-cycle of the mite *Adactylidium* sp. (Acarina: Pyemotidae), a predator of thrips eggs in the United Arab Republic. *Ann. ent. Soc. Am.* **59**, 458–461.
EL-BOROLLOSY F. M., WAFA A. K., and EL-HEFNY A. M. (1972) Studies on the biology of *Philanthus triangulum* F. Abdel Kader (Hymenoptera: Sphecidae). *Bull. Soc. ent. Égypte* **56**, 287–295.
ELDEFRAWI M. E., HANBAL I., and HAMMAD S. M. (1967) Biology and control of leopard moth on pear trees in the United Arab Republic. *Pl. Prot. Bull. FAO* **15** (4) 70–76.
EL-HALFAWY M. (1977) A method for staining the egg-plugs of the granary and rice weevils (*Sitophilus* spp.) on certain grains. *Agric. Res. Rev.* **53**, 113.
EL-HELALY M. S., EL-SHAZLI A. Y., and EL-GAYAR (1971) Morphological studies on immature stages of *Bemista tabaci* Gennadius (Hem. Hom. Aleyrodidae). *Z. angew. Ent.* **68**, 403–408.

EL-HELALY M. S., EL-SHAZLI A. Y., and EL-GAYAR F. H. (1972) Morphological studies on a new pest in Egypt, *Aleyrodes proletella* L. (Hom., Aleyrodidae) Hem. *Z. angew. Ent.* **71**, 12–26.

EL-KADY E., ZAZOU H., EL-DEEB A., and HAMMAD S. M. (1963) The biology of the dried-fruit beetle *Carpophilus hemipterus* L. (Coleoptera: Nitidulidae). *Bull. Soc. ent. Égypte* **46**, 97–118 (1962).

EL KHIDIR E. (1969) A contribution to the biology of *Epilachna chrysomelina* F., the melon ladybird beetle in the Sudan (Col., Coccinellidae). *Sudan agric. J.* **4** (2) 32–37.

ELKIND A. (1915) *Les Tubes ovariques et l'ovogénèse chez* Carausius hilari *Br.* Lausanne.

ELLENBERGER J. S. and CAMERON E. A. (1977) The spacial distribution of oak leafroller egg masses on primary host trees. *Envir. Ent.* **6**, 101–106.

ELLIOTT E. W., LANIER G. N. and SIMEONE J. B. (1975) Termination of aggregation by the European elm bark beetle *Scolytus multistriatus*. *J. chem. Ecol.* **1**, 283–289.

ELLIOTT J. M. (1969) Life history and biology of *Sericostoma personatum* Spence (Trichoptera). *Oikos* **20**, 110–118.

ELLIOTT K. R. and WONG H. R. (1966) An important predator of the aspen leaf beetle *Chrysomela crotchi* Brown in Manitoba and Saskatchewan. *Bi-mon. Res. Notes* **22** (5) 3–4.

EL-MINSHAWY A. M., EL-SAWAF S. K., HAMMAD S. M., and DONIA A. (1971) The biology of *Asterolecanium pustulans* Cockerell in Alexandria district (Egypt) (Hem., Hom.: Asterolecaniidae). *Bull. Soc. ent. Égypte* **55**, 441–446.

EL-SAADANY G. and ABDEL-FATTAH M. I. (1976) Contributions to the ecology of cotton-pests in Egypt. I. Studies on the egg-laying preferences of the cotton leaf worm moth *Spodoptera littoralis* Boisd. (Noctuidae). *Z. angew. Ent.* **80**, 31–35.

EL-SAWAF S. K. (1956) Some factors affecting the longevity, oviposition, and rate of development in the southern cowpea weevil *Callosobruchus maculatus* F. (Coleoptera: Bruchidae). *Bull. Soc. ent. Égypte* **40**, 29–95.

ELSEY K. D. (1971) Stilt bug predation on artificial infestation of tobacco hornworm eggs. *J. econ. Ent.* **64**, 772–773.

ELSEY K. D. (1972a) Defenses of eggs of *Manduca sexta* against predation by *Jalysus spinosus*. *Ann. ent. Soc. Am.* **65**, 896–897.

ELSEY K. D. (1972b) Predation of eggs of *Heliothis* spp. on tobacco. *Envir. Ent.* **1**, 433–438.

ELSEY K. D. (1973) *Jalysus spinosus*: spring biology and factors that influence occurrence of the predator on tobacco in North Carolina. *Envir. Ent.* **2**, 421–425.

ELSEY K. D. (1974) *Jalysus spinosus*: effect of age, starvation, host plant, and photoperiod on flight activity. *Envir. Ent.* **3**, 653–655.

ELSEY K. D. and RABB R. L. (1970) Biology of *Voria ruralis* (Diptera: Tachinidae). *Ann. ent. Soc. Am.* **63**, 216–222.

ELSEY K. D. and STINNER R. E. (1971) Biology of *Jalysus spinosus*, an insect predator found on tobacco. *Ann. ent. Soc. Am.* **64**, 779–783.

EL-SHAZLY N. Z. (1972a) Der Einfluss von Ernährung und Alter des Muttertieres auf die hämocytäre Abwehrreaktion von *Neomyzus circumflexus* (Buck.). *Entomophaga* **17**, 203–209.

EL-SHAZLY N. Z. (1972b) Der Einfluss äusserer Faktoren auf die hämocytäre Abwehrreaktion von *Neomyzus circumflexus* (Buckton) (Homoptera: Aphididae). *Z. angew. Ent.* **70**, 414–436.

EL SHERIF A. R. A. (1961) Preliminary biological studies on the potato tuber worm in UAR *Gnorimoschema operculella* (Zeller) (Lepidoptera: Tineidae). *Agric. Res. Rev.* **39**, 288–298.

EL-TITI A. (1973) Einflüsse von Beutedichte und Morphologie der Wirtspflanze auf die Eiablage von *Aphidoletes aphidimyza* (Rond.) (Diptera: Itonididae). *Z. angew. Ent.* **72**, 400–415.

El-Titi A. (1974) Zur Auslösung der Eiablage bei der aphidophagen Gallmücke *Aphidoletes aphidimyza* (Diptera : Cecidomyiidae). *Entomologia exp. appl.* **17**, 9–21.

Eltringham H. (1909) Curious place chosen by *Triphena pronuba* for ovipositing. *Entomologist* **42**, 236–237.

Eltringham H. (1912) (no title) *Proc. ent. Soc. Lond.* **1912**, lxxviii–lxxxi.

Eltringham H. (1914) On the urticating properties of *Porthesia similis* Fuess. *Trans. ent. Soc. Lond.* **1913**, 423–427.

Eltringham H. (1923) *Butterfly Lore*, Clarendon Press, Oxford.

Eluwa M. C. (1972) The egg-laying habits of *Euthypoda acutipennis* Karsch (Orthoptera : Tettigoniidae). *Zool. J. Linn. Soc.* **51**, 97–103.

Elzghari M. (1969) Elevage de la teigne de la farine *Anagasta kuhniella* Zeller et utilisation de la ponte comme nourriture de Coccinelles prédatrices. *Awamia* **33**, 25–43.

Embleton A. L. (1904) On the anatomy and development of *Comys infelix* Embleton, a hymenopterous parasite of *Lecanium hemisphaericum*. *Trans. Linn. Soc. Lond.* **9** (2) 231–254.

Embree D. G. and Sisojevic P. (1965) The bionomics and population density of *Cyzenis albicans* (Fall.) (Tachinidae : Diptera) in Nova Scotia. *Can. Ent.* **97**, 631–639.

Emden F. van (1931) Zur Kenntnis der Morphologie und Ökologie des Brotkafer-Parasiten *Cephalonomia quadridentata* Duchaussoy. *Z. Morph. Ökol. Tiere* **23**, 425–574.

Emden F. I. van (1946) Egg-bursters in some more families of polyphagous beetles and some general remarks on egg-bursters. *Proc. R. ent. Soc. Lond.* (A) **21**, 89–97.

Emden F. I. van (1948) A *Trox* larva feeding on locust eggs in Somalia. *Proc. R. ent. Soc. Lond.* (B) **17**, 145–148.

Emschermann F. (1969) Zur Morphologie, Biologie und Ökologie von *Chrysocharis seiuncta* Delucchi (Hymenoptera, Chalcidoidea, Eulophidae, Entedontinae), einem Larvenparasiten der Sattelmücke *Haplodiplosis equestris* Wagner (Diptera, Cecidomyiidae). I. Teil. *Z. angew. Ent.* **63**, 132–155 : II. Teil. 237–262.

Ene J. C. (1962) Parasitization of mantid oöthecae in West Africa. *Proc. 11th Int. Congr. Ent.* **2**, 725–727.

Ene J. C. (1964) The distribution and post-embryonic development of *Tarachodes afzelii* (Stål) (Mantodea : Eremiaphilidae). *Ann. Mag. nat. Hist.* **7** (13) 493–511.

Engel H. (1935) Biologie und Ökologie von *Cassida viridis* L. *Z. Morph. Ökol. Tiere* **30**, 41–96.

Engelmann F. (1957) Bau und Funktion des weiblichen Geschlechtsapparates bei der ovoviviparen Schabe *Leucophaea maderae*, und einige Beobachtungen über die Entwicklung. *Biol. Zbl.* **76**, 722–740.

Engelmann F. (1970) *The Physiology of Insect Reproduction*, Pergamon Press, Oxford.

Englert D. C. and Thomas W. H. (1970) The influence of the homeotic mutation, "antennapedia", on egg cannibalism in *Tribolium castaneum*. *Trans. Ill. State Acad. Sci.* **63**, 51–56.

English K. M. I. (1946) Notes on morphology and biology of *Apiocera maritima* Hardy (Diptera; Apioceridae). *Proc. Linn. Soc. NSW* **71**, 296–302.

Enock F. (1899) Oviposition of *Nepa cinerea* and *Notonecta*. *Proc. ent. Soc. Lond.* 1899, xiv.

Entwistle P. F. (1963) A note on *Eulophonotus myrmeleon* Fldr. (Lepidoptera, Cossidae), a stem borer of cocoa in West Africa. *Bull. ent. Res.* **54**, 1–3.

Ercelik T. M., Adams T. S., Holt G. G., and Nelson D. R. (1973) Differentiation of eggs with dominant lethal mutations from parthenogenic, unfertilized and viable eggs in *Trichoplusia ni* (Lepidoptera : Noctuidae). *Ann. ent. Soc. Am.* **66**, 292–297.

Esaki T. and Miyamoto S. (1959) A new or little known *Hypselosoma* from Amami-Oshima and Japan, with the proposal of a new tribe for the genus (Hemiptera). *Sieboldia* **2**, 109–120.

Esau K. (1961) *Plants, Viruses and Insects*. Cambridge, Mass.

ESKAFI F. M. (1975) Increased efficiency in the mass production of *Hippelates collusor* (Diptera: Chloropidae). *J. med. Ent.* **12**, 575–576.

ESPÍNOLA H. N. (1965) Ritmo de oviposicão em *Triatoma infestans* (Klug, 1834) e *Panstrongylus megistus* (Burmeister, 1835) (Hemiptera, Reduviidae). *Cienc. Cult. S. Paulo* **17**, 227–228.

ESSELBAUGH C. O. (1946) A study of the eggs of the Pentatomidae (Hemiptera). *Ann. ent. Soc. Am.* **39**, 667–691.

ESSIG E. O. (1910) The natural enemies of the citrus mealybug. 1. *Sympherobius angustatus* Banks. *Pomona Coll. J. Ent. Zool.* **2**, 143–146.

ESSIG E. O. (1916) A coccid-feeding moth, *Holcocera iceryaella* Riley. *J. agric. Res.* **9**, 369–370.

ESSIG E. O. (1926) *Insects of Western North America*, Macmillan, New York.

ETCHEVERRY M. and RETAMAL T. (1965) Biologia de *Helicoverpa zea* (Boddie) 1850 (Lepidoptera, Noctuidae). *Publnes Cent. Estud. ent. Univ. Chile* **7**, 49–56.

ÉTIENNE J. (1973) Élevage permanent de *Pardalaspis cyanescens* (Dipt., Trypetidae) sur hôte végétal de remplacement. *Annls Soc. ent. Fr.* **9**, 853–860.

EUW J. VON, FISHELSON L., PARSONS J. A., REICHSTEIN T., and ROTHSCHILD M. (1967) Cardenolides (heart poisons) in a grasshopper feeding on milkweeds. *Nature, Lond.* **214**, 35–39.

EUW J. VON, REICHSTEIN T., and ROTHSCHILD M. (1971) Heart poisons (cardiac glycosides) in the lygaeid bugs *Caenocoris nerii* and *Spilostethus pandurus*. *Insect Biochem.* **1**, 373–384.

EVANS A. C. (1934) Studies on the influence of the environment on the sheep blow-fly *Lucilia sericata* Meig. I. The influence of humidity and temperature on the egg. *Parasitology* **26**, 366–377.

EVANS A. C. (1936) Studies on the influence of the environment of sheep blow-fly *Lucilia sericata* Meig. IV. The indirect effect of temperature and humidity acting through certain competing species of blow-flies. *Parasitology* **28**, 431–439.

EVANS A. M. (1936) The *funestus* series of *Anopheles* at Kisumu and a coastal locality in Kenya. *Ann. trop. Med. Parasit.* **30**, 511–520.

EVANS A. M. and LEESON H. S. (1935) The *funestus* series of *Anopheles* in Southern Rhodesia, with description of a new variety. *Ann. trop. Med. Parasit.* **29**, 33–47.

EVANS B. R. and BEVIER G. A. (1969) Measurement of field populations of *Aedes aegypti* with the ovitrap in 1968. *Mosq. News* **29**, 347–353.

EVANS H. E. (1958) The evolution of social life in wasps. *Proc. 10th Int. Congr. Ent.* **2**, 449–457.

EVANS H. F. (1976) The effect of prey density and host plant characteristics on oviposition and the fertility in *Anthocoris confusus* (Reuter). *Ecol. Ent.* **1**, 157–161.

EVENDEN J. C. (1943) The mountain pine beetle, an important enemy of western pines. *Circ. US Dep. Ent.* **664**, 1–25.

EVERETT T. R. (1964a) Feeding and oviposition reaction of boll weevils to cotton, althea, and okra flower buds. *J. econ. Ent.* **57**, 165–166.

EVERETT T. R. (1964b) An inherited behavioral variant in the boll weevil. *J. econ. Ent.* **57**, 760–761.

EVERETT T. R. and EARLE N. W. (1964) Boll weevil oviposition responses in cotton squares and various other substrates. *J. econ. Ent.* **57**, 651–656.

EVERETT T. R. and RAY J. O. (1964) Observations of puncturing and oviposition behavior of boll weevils. *J. econ. Ent.* **57**, 121–123.

EVERETT T. R. and TRAHAN G. (1967) Oviposition by rice weevils in Louisiana. *J. econ. Ent.* **60**, 305–307.

EWER D. W. and EWER R. F. (1942) The biology and behaviour of *Ptinus tectus* Boie. (Coleoptera, Ptinidae), a pest of stored products. *J. exp. Biol.* **18**, 290–305.

EWER R. F. (1945) The effect of grain size on the oviposition of *Calandra granaria* Linn. (Coleoptera, Curculionidae). *Proc. R. ent. Soc. Lond.* (A) **20**, 57–63.

EYLES A. C. (1963) Fecundity and oviposition rhythms in *Nysius huttoni* White (Heteroptera: Lygaeidae). *NZ Jl Sci.* **6**, 186–207.

EZZAT Y. M. (1957) Egg-laying and hatching habits of *Magicicada septendecim* (Linné) (Hemiptera–Homoptera: Cicadidae). *Bull. Soc. ent. Égypte* **41**, 365–370.

FABRE J. H. (1897) *Souvenirs entomologiques*, 5th ser., Delagrave (ed.), Paris.

FABRE J. H. (1923) *Souvenirs entomologiques*, 8.

FALKENSTRÖM G. (1933) Die Metamorphose von *Deronectes depressus sensu* Fabr. und *latescens* Falk. nebst Zugen aus ihrer Ökologie. *Int. Rev. Hydrobiol.* **29**, 161–220.

FALLERONI D. (1926) Fauna anofelica italiana e suo habitat. Metodi di lotta contro la malaria. *Riv. Malar.* **5**, 553–593.

FANG Y. C. (1976) A preliminary study on the biology and control of the geometric *Euctenurapteryx nigrociliaria* Leech. *Acta ent. sin.* **19**, 318–324.

FARES F. M. (1973) The effect of cold storage on the hatchability of the Mediterranean fruit fly eggs. *Agric. Res. Rev.* **51**, 57–58.

FARNER D. S. (1960) Digestion and the digestive system. In *Biology and Comparative Physiology of Birds* (A. J. Marshall, ed.), London.

FARQUHARSON C. O. (1922) Five years' observations (1914–1918) on the bionomics of Southern Nigerian insects, chiefly directed to the investigation of Lycaenid life-histories and to the relation of Lycaenidae, Diptera, and other insects to ants. *Trans. ent. Soc. Lond.* **1921**, 319–448.

FARROW R. A. (1974) The larva of *Laius villosus* (Coleoptera: Melyridae) feeding on the egg pods of the Australian plague locust, *Chortoicetes terminifera* (Orthoptera: Acrididae). *J. Aust. ent. Soc.* **13**, 185–188.

FATZINGER C. W. (1970) Rearing successive generations of *Dioryctria abietella* (Lepidoptera: Pyralidae (Phycitinae)) on artificial media. *Ann. ent. Soc. Am.* **63**, 809–814.

FAURE J. C. (1923) The life-history of the brown locust. *J. Dept. Agric. Pretoria TUC Bull.* **4**, 1–30.

FAURE J. C. (1935) The life history of the red locust (*Nomadacris septemfasciata* (Serville)). *Bull. S. Afr. Dep. Agric.* **144**, 1–32.

FAURE J. C. (1940) Maternal care displayed by mantids (Orthoptera). *J. ent. Soc. S. Afr.* **3**, 139–150.

FAVRELLE M. (1936) Notes sur l'habitat et la biologie de l'*Eugaster nigripes* (Orth., Tettigoniidae). *Annls Soc. ent. Fr.* **105**, 369–374.

FAVRELLE M. (1938) Étude du *Phalces longiscaphus* (Orthopt., Phasmidae). *Annls Soc. ent. Fr.* **107**, 197–211.

FEDEROV S. M. (1927) Studies in the copulation and oviposition of *Anacridium aegyptium* L. (Orthoptera, Acrididae). *Trans. ent. Soc. Lond.* **75**, 53–60.

FELLOWS A. G. (1967) Egg-case of mantis. *Victorian Nat.* **84**, 237.

FELT E. P. (1894) The scorpion flies. *NY ent. Rep.* **10**, 463–480.

FENARD G. (1896) Sur les annexes internes de l'appareil génital femelle des Orthoptères, *C. r. hebd. Séanc. Acad. Sci. Paris* **122**, 1137–1139.

FENEMORE P. G. and ALLEN V. A. L. (1969) Oviposition preference and larval survival in *Wiseana cervinata* (Walker) (Hepialidae). *NZ Jl agric. Res.* **12**, 146–161.

FERNALD C. H. and KIRKLAND A. H. (1903) *The Brown-tail Moth (Euproctis chrysorrhoea L.)*, Mass. State Bd. Agric., Boston.

FERNANDO C. H. and LEONG C. Y. (1963) Miscellaneous notes on the biology of Malayan Corixidae (Hem.: Heteroptera) and a study of the life histories of two species, *Micronecta quadristrigata* Bredd. and *Agraptocorixa hyalinipennis* (F.). *Ann. Mag. nat. Hist.* **13**, 545–558.

FERNANDO W. (1935) The early embryology of a viviparous Psocid. *Q. Jl microsc. Sci.* **77**, 99–119.
FÉRON M. (1957) Le comportement de ponte de *Ceratitis capitata* Wied. : influence de la lumière. *Rev. Path. vég. Ent. agric. Fr.* **36**, 127–143.
FÉRON M. (1958) Mise en évidence d'un stimulus significatif dans le comportement de ponte de *Ceratitis capitata* Wied. (Dipt., Trypetidae). *C. r. hebd. Séanc. Acad. Sci. Paris* **246**, 1590–1592.
FÉRON M. (1962) L'instinct de reproduction chez la mouche méditerranéenne des fruits *Ceratitis capitata*. Comportement sexuel. Comportement de ponte. *Rev. Path. vég. Ent. agric. Fr.* **41**, 1–129.
FÉRON M. and AUDEMARD H. (1956) Notes sur *Hydrellia griseola* (Dipt., Ephydridae), mouche mineuse du riz en France. *Annls Épiphyt.* **7**, 421–430.
FÉRON M., DELANOUE P., and SORIA F. (1958) L'élevage massif artificiel de *Ceratitis capitata* Wied. *Entomophaga* **3**, 45–53.
FÉRON M. and VIDAUD J. (1960) La mouche du carthame *Acanthiophilus helianthi* Rossi (Dipt., Trypetidae) en France. *Rev. Path. vég. Ent. agric. Fr.* **39**, 1–12.
FERRIS C. D. (1973) Life history of *Callophrys s. sheridanii* (Lycaenidae) (Lep.) and notes on other species. *J. Lep. Soc.* **27**, 279–283.
FERRIS G. F. (1931) The louse of elephants, *Haematomyzus elephantis* Piaget (Mallophaga: Haematomyzidae). *Parasitology, Cambridge* **23**, 112–127.
FERRIS G. F. (1951) The suckling lice. *Mem. Pacif. Coast ent. Soc.* **1**, 1–320.
FERTON C. (1923) *La vie des abeilles et des guêpes*, Paris.
FEWKES D. W. (1963) The effect of exposure to dry conditions on the eggs of *Aeneolamia varia saccharina* (Homoptera: Cercopidae). *Ann. ent. Soc. Am.* **56**, 719–720.
FIEBRIG K. (1910) Cassiden und Cryptocephaliden Paraguays. Ihre Entwicklungsstadien und Chutzvorrichtungen. *Zool. Jb.* (Suppl.) **12**, 161–264.
FIEDLER O. G. H. (1950) Entomologisches aus Afrika (Beobachtungen über Kaffeeschädlinge). *Z. angew. Ent.* **32**, 289–306.
FIELDING J. W. (1919) Notes on the bionomics of *Stegomyia fasciata* Fabr. Part I. *Ann. trop. Med. Parasit.* **13**, 259–296.
FILIPEK P. (1962) Studies on the bionomics and ecology of *Acanthoscelides obtectus* in laboratory conditions. *Pr. nauk. Inst. Ochr. Rośl.* **4**, 177–200 (in Russian; English summary).
FILIPPONI A. (1955) Sulla natura dell'associazione tra *Macrocheles muscaedomesticae* e *Musca domestica*. *Riv. Parassit.* **16**, 83–102.
FILIPPONI A. (1964) The feasibility of mass producing macrochelid mites for field trials against houseflies. *Bull. Wld Hlth Org.* **31**, 499–501.
FILIPPONI A., MOSNA B., and PETRELLI G. (1971) L'ottimo di temperatura di *Macrocheles muscaedomesticae* (Scopoli), come attributo di popolazione. *Riv. Parassit.* **32**, 193–218.
FINCH S. and COAKER T. H. (1969) A method for the continuous rearing of the cabbage root fly *Erioischia brassicae* (Bch.) and some observations on its biology. *Bull. ent. Res.* **58**, 619–627.
FINCH S., SKINNER B., and FREEMAN G. H. (1975) The distribution and analysis of cabbage root fly egg populations. *Ann. appl. Biol.* **79**, 1–18.
FINK D. E. (1915) The eggplant lace-bug. *Bull. US Bur. Ent.* **239**, 1–7.
FINLAYSON L. H. (1949) The life-history and anatomy of *Lepinotus patruelis* Pearman. *Proc. zool. Soc. Lond.* **119**, 301–323.
FINLAYSON L. R. (1933) Some notes on the biology and life-history of Psocids. *Ann. Rep. ent. Soc. Ont.* **63**, 56–58 (1932).
FINNEGAN R. J. (1967) Notes on the biology of the pitted ambrosia beetle *Corthylus punctatissimus* (Coleoptera: Scolytidae) in Ontario and Quebec. *Can. Ent.* **99**, 49–54.
FINNEY G. L. (1950) Mass-culturing *Chrysopa californica* to obtain eggs for field distribution. *J. econ. Ent.* **43**, 97–100.

FIORI G. (1951) Contributi alla conoscenza morphologica ed etologica die Coleotteri. V. *Coptocephala küsteri* Kraatz e *Cryptocephalus frenatus* Laich. (Chrysomelidae). *Boll. Ist. Ent. Univ. Bologna* **18**, 182–196.
FISCHER M. (1954) Untersuchungen über den kleinen Holzbohrer (*Xyleborinus saxeseni* Ratz.). *Pfl. Schutzberichte* **12**, 137–180.
FISCHER O. VON (1928) Die Entwicklung von *Periplaneta americana*. *Mitt. naturf. Ges. Bern* 1927.
FISHER R. C. (1961) A study in insect multiparasitism. I. Host selection and oviposition. *J. exp. Biol.* **38**, 267–275.
FISHER R. C. (1971) Aspects of the physiology of endoparasitic Hymenoptera. *Biol. Rev.* **46**, 243–278.
FISHER T. W. (1963) Mass culture of *Cryptolaemus* and *Leptomastix*—natural enemies of *Citrus* mealybug. *Bull. Calif. agric. Exp. Stn* **797**, 1–39.
FISHER T. W. and ORTH R. E. (1964) Biology and immature stages of *Antichaeta testacea* Melander (Diptera: Sciomyzidae). *Hilgardia* **36**, 1–29.
FITSCH E. A. (1880) Insects bred from *Cynips kollari* galls. *Entomologist* **28**, 252–263.
FITZGERALD T. D. and NAGEL W. P. (1972) Oviposition and larval bark-surface orientation of *Medetera aldrichii* (Diptera: Dolichopodidae): response to a prey-liberated plant terpene. *Ann. ent. Soc. Am.* **65**, 328–330.
FLANDERS S. E. (1930) Mass production of egg parasites of the genus *Trichogramma*. *Hilgardia* **4**, 465–501.
FLANDERS S. E. (1937) Ovipositional instincts and developmental sex differences in the genus *Coccophagus*. *Univ. Calif. Publs Ent.* **6**, 401–422.
FLANDERS S. E. (1939) Environmental control of sex in hymenopterous insects. *Ann. ent. Soc. Am.* **32**, 11–26.
FLANDERS S. E. (1942) Oosorption and ovulation in relation to oviposition in the parasitic Hymenoptera. *Ann. ent. Soc. Am.* **35**, 251–266.
FLANDERS S. E. (1943) The Argentine ant versus the parasites of the black scale. *Calif. Citrogr.* **28**, 117, 128, 137.
FLANDERS S. E. (1945a) The bisexuality of uniparental Hymenoptera, a function of the environment. *Am. Nat.* **lxxix**, 122–141.
FLANDERS S. E. (1945b) Coincident infestations of *Aonidiella citrina* and *Coccus hesperidum*, a result of ant activity. *J. econ. Ent.* **38**, 711–712.
FLANDERS S. E. (1951) Mass culture of California red scale and its golden chalcid parasites. *Hilgardia* **21**, 1–42.
FLANDERS S. E. (1959) Differential host relations of the sexes in parasitic Hymenoptera. *Entomologia exp. appl.* **2**, 125–142.
FLANDERS S. E. (1968) Mechanisms of population homeostasis in *Anagasta* ecosystems. *Hilgardia* **39**, 367–404.
FLANDERS S. E. (1969) Alternative differential mortality of the sexes concomitant to sex differentiation in host relations. *Entomophaga* **14**, 335–346.
FLANDERS S. E. and BADGLEY M. E. (1963) Prey–predator interactions in self-balanced laboratory populations. *Hilgardia* **35**, 145–183.
FLANDERS S. E., BARTLETT B. R., and FISHER T. W. (1961) *Coccophagus basalis* (Hymenoptera: Aphelinidae): its introduction into California with studies of its biology. *Ann. ent. Soc. Am.* **54**, 227–236.
FLETCHER B. S. and WATSON C. A. (1974) The ovipositional response of the Tephritid fruit fly *Dacus tryoni* to 2-chloro-ethanol in laboratory bioassays. *Ann. ent. Soc. Am.* **67**, 21–23.
FLETCHER L. W. and LONG J. S. (1971) Influence of food odors on oviposition by the cigarette beetle on nonfood materials. *J. econ. Ent.* **64**, 770–771.

FLETCHER L. W., TURNER J. P., and HUSMAN C. N. (1973) Surface temperature as a factor in selection of ovipositional sites by three strains of the screwworm. *J. econ. Ent.* **66**, 422–423.
FLETCHER T. B. (1916) One hundred notes on Indian insects. *Bull. agric. Res. Inst. Pusa* **59**, 1–39.
FLETCHER T. B. (1920) Life-histories of Indian insects, Microlepidoptera. *Mem. Dep. Agric. India (Ent.)* **6**, 1–217.
FLETCHER T. B. (1933) Life histories of Indian Microlepidoptera. Cosmopterygidae to Neopseustidae. *Sci. Monogr. Imp. Council Agric. Res. Delhi* **4**, 1–58.
FLITTERS N. E. (1964) The effect of photoperiod, light intensity, and temperature on copulation, oviposition, and fertility of the Mexican fruit fly. *J. econ. Ent.* **57**, 811–813.
FLORENCE L. (1921) The hog louse *Haematopinus suis* Linne. Its biology, anatomy, and histology. *Mem. Cornell Univ. agric. Exp. Stn* **51**, 635–743.
FOGLE T. A. and ENGLERT D. C. (1976) Differential egg cannibalism among larvae of *Tribolium castaneum* as influenced by the antennapedia mutation. *Can. J. Genet. Cytol.* **18**, 179–187.
FONSECA J. P. C. DA (1964) Studies on the larval competition of the Bruchid beetle *Caryedon gonagra* (Fab.). *Garcia de Orta* **12**, 633–643.
FONSECA J. P. C. DA (1965) Oviposition and length of adult life in *Caryedon gonagra* (F.) (Col., Bruchidae). *Bull. ent. Res.* **55**, 697–707.
FONSECA J. P. C. DA and AUTUORI M. (1938) O pulgao branco das laranjeiras. *Inst. Biol. (São Paulo)* **88**, 1–11.
FOOTE B. A. (1959) Biology and life history of the snail-killing flies belonging to the genus *Sciomyza* Fallén (Diptera, Sciomyzidae). *Ann. ent. Soc. Am.* **52**, 31–43.
FOOTE B. A. (1967) Biology and immature stages of fruit flies: the genus *Icterica* (Diptera: Tephritidae). *Ann. ent. Soc. Am.* **60**, 1295–1305.
FOOTE B. A. (1971) Biology of *Hedria mixta* (Diptera: Sciomyzidae) (preying on aquatic snails in the United States). *Ann. ent. Soc. Am.* **64**, 931–941.
FOOTE B. A. and KNUTSON L. V. (1970) Clam-killing fly larvae. *Nature, Lond.* **226**, 466.
FOOTE B. A., NEFF S. E., and BERG C. O. (1960) Biology and immature stages of *Atrichomelina pubera* (Diptera: Sciomyzidae). *Ann. ent. Soc. Am.* **53**, 192–199.
FOOTT W. H. (1968) Laboratory rearing of the pepper maggot *Zonosemata electa* (Say) (Diptera: Tephritidae). *Proc. ent. Soc. Ont.* **98**, 18–21 (1967).
FORATTINI O. P. and BARATA J. M. S. (1974) Nota sobre a diferenciação de ovos de *Rhodnius neglectus* e *R. prolixus. Rev. Saúde Públ.* **8**, 447–450.
FORBES W. T. M. (1923) The Lepidoptera of New York and neighbouring states. *Mem. Cornell Univ. agric. Exp. Stn* **68**, 1–729.
FORCE D. C. (1966) Reactions of the three-lined potato beetle *Lema trilineata* (Coleoptera: Chrysomelidae) to its host and certain nonhost plants. *Ann. ent. Soc. Am.* **59**, 1112–1119.
FORISTER G. W. and JOHNSON C. D. (1970) Bionomics of *Merobruchus julianus* (Coleoptera: Bruchidae). *Coleopt. Bull.* **24**, 84–87.
FOSTER C. H., RENAUD G. D., and HAYS K. L. (1973) Some effects of the environment on oviposition by *Chrysops* (Diptera: Tabanidae). *Envir. Ent.* **2**, 1048–1050.
FOSTER W. A. and LEA A. C. (1975) Renewable fecundity of male *Aedes·aegypti* following replenishment of seminal vesicles and accessory glands. *J. Insect Physiol.* **21**, 1085–1090.
FOURIE G. J. J. (1967) The influence of certain factors on the fecundity of *Sitophilus oryzae* (L.). *S. Afr. J. agric. Sci.* **10**, 331–344.
FOX A. S., MEAD C. G., and MUNYON I. L. (1959) Sex-peptide of *Drosophila melanogaster*. *Science, Wash.* **129**, 1489–1490.
FOX C. J. S. (1973) Influence of vegetation on the distribution of wireworms in grassland: observations on *Agriotes obscurus* (L.) (Col.: Elateridae). *Phytoprotection* **54**, 69–71.
FOX C. J. S. (1974) Note on the ovipositing preferences of the imported cabbageworm *Pieris rapae* (L.) on cole crops. *Phytoprotection* **55**, 36–37.

Fox H. (1939) The egg content and nymphal production and emergence in oothecae of two introduced species of Asiatic mantids (Orthoptera: Mantidae). *Ann. ent. Soc. Am.* **32**, 549–560.

Fox H. (1943) Further studies on oothecae of introduced Asiatic mantids (Orthoptera: Mantidae). *Ann. ent. Soc. Am.* **36**, 25–33.

Fox L. and Landis B. J. (1973) Notes on the predaceous habits of the gray field slug *Deroceras laeve. Envir. Ent.* **2**, 306–307.

Fox R. C. and Griffith K. H. (1976) Predation of pine cinaron aphids by spiders. *J. Ga Ent. Soc.* **11**, 241–243.

Fraenkel G. (1969) Evaluation of our thoughts on secondary plant substances. *Entomologia exp. appl.* **12**, 473–486.

Fraenkel G. and Blewett M. (1943) The utilization of metabolic water in insects. *Bull. ent. Res.* **35**, 127–139.

Francis E. (1907) Observation on the life cycle of *Stegomyia calopus. Publ. Hlth Rep. Wash.* **22**, 381–383.

Francke-Grosmann H. (1953a) Über die Brutgewohnheiten von *Rhynchites (Merhynchites) germanicus* Hbst. *Dt. EntTag Hamburg* **1953**, 114–128.

Francke-Grosmann H. (1953b) Über die Brutfürsorge einiger an Kulturweiden lebender triebstechender Rüssler (Curculionidae) und ihre phytopathologische Bedeutung. *Beitr. Ent.* **3**, 468–478.

Francke-Grosmann H. (1967) Ectosymbiosis in wood-inhabiting insects. In *Symbiosis* (S. M. Henry, ed.), pp. 141–205, Academic Press, New York.

François J. (1975) L'encapsulation hémocytaire expérimentale chez le Lépisme *Thermobia domestica. J. Insect Physiol.* **21**, 1535–1546.

Frank J. H. (1967) A serological method used in the investigation of the predators of the pupal stage of the winter moth *Operophtera brumata* (L.) (Hydriomenidae). *Quaest. ent.* **3**, 95–105.

Frank J. H. (1971) Carabidae (Coleoptera) as predators of the red-backed cutworm (Lepidoptera: Noctuidae) in central Alberta. *Can. Ent.* **103**, 1039–1044.

Frankenberg G. (1940) Ein brutpflegender und musizierender Wasserkäfer (*Spercheus emarginatus*). *Natur Volk, Frankfurt* **70**, 79–89.

Frankie G. W. and Koehler C. S. (1967) Cypress bark moth on Monterey cypress. *Calif. Agric.* **21**, 6–7.

Franssen C. J. H. (1956) Biology and control of the common bean weevil. *Meded. Dir. Tuinb.* **19**, 797–809 (in Dutch; English summary).

Franz J. (1958) Studies on *Laricobius erichsonii* Rosenh. (Coleoptera: Derondontidae): a predator on chermesids. *Entomophaga* **3**, 109–196.

Franz J. and Szmidt A. (1960) Beobachtungen beim Züchten von *Perillus bioculatus* (Fabr.) (Heteropt., Pentatomidae), einem aus Nordamerika importierten Räuber des Kartoffelkäfers. *Entomophaga* **5**, 87–110.

Fraser F. C. (1933) *The Fauna of British India, Including Ceylon and Burma. Odonata* **1**, 1–423, Taylor and Francis, London.

Fraser F. C. (1934) *The Fauna of British India, Including Ceylon and Burma. Odonata,* **2**, 1–398, Taylor and Francis, London.

Fraser F. C. (1936) *The Fauna of British India, Including Ceylon and Burma. Odonata,* **3**, 1–461, Taylor and Francis, London.

Fraser F. C. (1952) Methods of exophytic oviposition in Odonata. *Entomologist's mon. Mag.* **88**, 261–262.

Fraser F. C. (1953) Methods of exophytic oviposition in Odonata. *Entomologist's mon. Mag.* **89**, 252–254.

FREDEEN F. J. H. (1959) Collection, extraction, sterilization and low-temperature storage of black-fly eggs (Diptera: Simuliidae). *Can. Ent.* **91**, 450–453.

FREDEEN F. J. H., REMPEL J. G., and ARNASON A. P. (1951) Egg-laying habits, overwintering stages, and life-cycle of *Simulium arcticum* Mall. (Diptera: Simuliidae). *Can. Ent.* **83**, 73–76.

FREDEEN F. J. H. and TAYLOR M. E. (1964) Borborids (Diptera: Sphaeroceridae) infesting sewage disposal tanks, with notes on the life cycle, behaviour and control of *Leptocera (Leptocera) caenosa* (Rondani). *Can. Ent.* **96**, 801–808.

FREEMAN B. E. (1967) The biology of the white clover seed weevil *Apion dichroum* Bedel (Col. Curculionidae). *J. appl. Ecol.* **4**, 535–552.

FREEMAN P. (1954) A new African species of *Simulium* (Diptera, Simuliidae) in phoretic association with mayfly nymphs. *Ann. Mag. nat. Hist.* **7** (12) 113–115.

FRICK K. E. (1970) Behavior of adult *Hylemya seneciella*, an Anthomyiid (Diptera) used for the biological control of tansy ragwort. *Ann. ent. Soc. Am.* **63**, 184–187.

FRICK K. E. and ANDRES L. A. (1967) Host specificity of the ragwort seed fly. *J. econ. Ent.* **60**, 457–463.

FRIEDEL T. and GILLOTT C. (1976) Male accessory gland substance of *Melanoplus sanguinipes*: an oviposition stimulant under the control of the corpus allatum. *J. Insect Physiol.* **22**, 489–495.

FRILLI F. (1968) Allevamento sperimentale di *Carpocapsa pomonella* L. e del suo parassita *Ascogaster quadridentatus* Wesm. Studi del gruppo di lavoro del CNR per la lotta integrata contro i nemici animali delle piante: XXV. *Boll. Zool. agr. Bachic.* **8** (2) 181–190 (1966–7).

FRISON T. H. (1929) Fall and winter stoneflies or Plecoptera of Illinois. *Bull. Ill. nat. Hist. Surv.* **18**, 340–409.

FRITZE A. (1915) Widerstandsfähigkeit der Eierkokons der Fangheuschrecken (Mantodeen). *Z. wiss. Insektenbiol.* **11**, 275–276.

FROESCHNER R. C. (1960) Cydnidae of the Western Hemisphere. *Proc. US nat. Mus.* **111**, 337–680.

FROGGATT W. W. (1901) The pear and cherry slug (*Eriocampa limacina* Retz), generally known as *Selandria cerasi*, with notes on Australian sawflies. *Agric. Gaz. NSW* **5**, 1063–1073.

FROGGATT W. W. (1910) Scale-eating moths. *Agric. Gaz. NSW* **21**, 801.

FRÖHLICH G. (1956) Zur Frage der biologischen Abhängigkeit der Kohlschoten-Gallmücke (*Dasyneura brassicae* Winn.) vom Kohlschotenrüssler (*Ceuthorrhynchus assimilis* Payk.). *Beitr. Ent.* **6**, 100–110.

FRONK W. D. (1947) The southern pine beetle—its life history. *Tech. Bull. Va agric. Exp. Stn* **108**, 1–12.

FROST S. W. (1973) Hosts and eggs of *Blepharida dorothea* (Coleoptera: Chrysomelidae). *Fla Ent.* **56**, 120–122.

FROST S. W. and HABER V. R. (1944) A case of parental care in the Heteroptera. *Ann. ent. Soc. Am.* **37**, 161–166.

FUCHS M. S. and CRAIG G. B. (1967) Extraction and purification of male accessory gland substance in *Aedes aegypti*. *Bull. ent. Soc. Am.* **13**, 201 (abstract).

FUCHS M. S., CRAIG G. B., and DESPOMMIER D. D. (1969) The protein nature of the substance inducing female monogamy in *Aedes aegypti*. *J. Insect Physiol.* **15**, 701–709.

FUCHS M. S., CRAIG G. B., and HISS E. A. (1968) The biochemical basis of female monogamy in mosquitoes. I. Extraction of the active principle from *Aedes aegypti*. *Life Sci.* **7**, 835–839.

FUCHS M. S. and HISS E. A. (1970) The partial purification and separation of the protein components of matrone from *Aedes aegypti*. *J. Insect Physiol.* **16**, 931–939.

FUCHS M. S. and SCHLAEGER D. A. (1973) The stimulation of dopa decarboxylase activity by ecdysone and its enhancement by cyclic AMP in adult mosquitoes. *Biochem. Biophys. Res. Comm.* **54**, 784–789.

FUENTES M. DEL C. and OROZCO F. (1970) Influencia de diversos factores en la puesta del *Tribolium castaneum*. II. Influencia de una amplia gama de temperaturas en la puesta. *An. Inst. nac. Invest. agron.* **19**, 373–401.

FÜHRER E. (1975) Physiological specificity of the polyphagous pupal parasite *Pimpla turionellae* and its ecological significance. *Zentbl. ges. Forstw.* **92**, 218–227.

FUJISAKI K. (1973) The spatial distribution pattern of the winter cherry bug *Acanthocoris sordidus* Thunberg in relation to its aggregation behaviour. *Jap. J. appl. Ent. Zool.* **17**, 31–38.

FULDNER D. (1960) Beiträge zur Morphologie und Biologie von *Aleochara bilineata* Gyll. und *A. bipustulata* L. (Coleoptera: Staphylinidae). *Z. Morph. Ökol. Tiere* **49**, 312–386.

FULLER M. E. (1938) Notes on *Trichopsidea oestracea* (Nemestrinidae) and *Cyrtomorpha flaviscutellaris* (Bombyliidae)—two dipterous enemies of grasshoppers. *Proc. Linn. Soc. NSW* **63**, 95–104.

FULTON B. B. (1915) The tree crickets of New York: life history and bionomics. *Tech. Bull. NY St. agric. Exp. Stn* **42**, 1–47.

FULTON B. B. (1924) Some habits of earwigs. *Ann. ent. Soc. Am.* **17**, 357–367.

FULTON B. B. (1933) Notes on *Habrocytus cerealellae*, parasite of the angoumois grain moth. *Ann. ent. Soc. Am.* **26**, 536–553.

FUNKHOUSER W. D. (1917) Biology of the Membracidae of the Cayuga Lake basin. *Mem. Cornell Univ. agric. Exp. Stn* **11**, 177–445.

FURNEAUX P. J. S. (1970) *o*-Phosphoserine as a hydrolysis product and amino acid analysis of shells of new laid eggs of the house cricket *Acheta domesticus* L. *Biochim. biophys. Acta* **215**, 52–56.

FURNEAUX P. J. S., JAMES C. R., and POTTER S. A. (1969) The egg shell of the house cricket (*Acheta domesticus*): an electron microscopic study. *J. Cell Sci.* **5**, 227–249.

FURNEAUX P. J. S. and MACKAY A. L. (1970) Periodic protein structure in insect egg shells. *Proc. 7th Congr. Int. Micr. Electr.*, Grenoble.

FURNEAUX P. J. S. and MACKAY A. L. (1972) Crystalline protein in the chorion of insect egg shells. *J. ultrastr. Res.* **38**, 343–359.

FURNEAUX P. J. S. and McFARLANE J. E. (1965a) Identification, estimation, and localization of catecholamines in eggs of the house cricket *Acheta domesticus* (L.). *J. Insect Physiol.* **11**, 591–600.

FURNEAUX P. J. S. and McFARLANE J. E. (1965b) A possible relationship between the occurrence of catecholamines and water absorption in insect eggs. *J. Insect Physiol.* **11**, 631–635.

FURNISS M. M., DATERMAN G. E., KLINE L. N., McGREGOR M. D., TROSTLE G. C., PETTINGER L. F., and RUDINSKY J. A. (1974) Effectiveness of the Douglas-fir beetle antiaggregative pheromone methylcyclohexenone at three concentrations and spacings around felled host trees. *Can. Ent.* **106**, 381–392.

FURNISS M. M., KLINE L. N., SCHMITZ R. F., and RUDINSKY J. A. (1972) Tests of three pheromones to induce or disrupt aggregation of Douglas-fir beetles (Coleoptera: Scolytidae) on live trees. *Ann. ent. Soc. Am.* **65**, 1227–1232.

FURNISS M. M. and SCHMITZ R. F. (1971) Comparative attraction of Douglas-fir beetle to frontalin and tree volatiles. *US For. Serv. Res. Pap.* INT-96, 16 pp.

FUSEINI B. A. and KUMAR R. (1973) The accessory glands of some female mantids. *Entomologist's mon. Mag.* **108**, 98–101.

GABALDON A. (1939) A method for mounting Anopheline eggs. *J. Parasit.* **25**, 281.

GADD C. H. (1946) *Macrocentrus homonae*—a polyemoryonic parasite of tea Tortrix (*Homona coffearia*). *Ceylon J. Sci.* (B) **23**, 67–79.

GADEAU DE KERVILLE H. (1907) *Accouplement, oeufs, et amour maternel des Forficulides*, Rouen.
GADEAU DE KERVILLE H. (1930) Observations sur la *Pseudochelidura sinuata* Germ. *Bull. Soc. ent. Fr.* **1930**, 61–63.
GADEAU DE KERVILLE H. (1931) Sur les oeufs et l'instinct maternel du *Pseudochelidura sinuata* Germ. (Dermaptera). *Bull. Soc. ent. Fr.* **1931**, 119–120.
GAGNEPAIN C. (1973a) Comportement de ponte des Oscinies et mise au point d'un système de pondoir (Dipt., Chloropidae). *Bull. Soc. ent. Fr.* **78**, 239–243.
GAGNEPAIN C. (1973b) Étude de la fécondité chez des couples isolés d'*Oscinella pusilla* (Dipt., Chloropidae). *Entomologia exp. appl.* **16**, 223–231.
GALFORD J. R. (1971) Improved techniques for rearing the smaller European elm bark beetle on artificial media. *J. econ. Ent.* **64**, 1327–1328.
GALINDO P. (1957) A note on the oviposition behaviour of *Sabethes (Sabethoides) chloropterus* Humboldt. *Proc. ent. Soc. Wash.* **59**, 287–288.
GALINDO P. (1958) Bionomics of *Sabethes chloropterus* Humboldt, a vector of sylvan yellow fever in middle America. *Am. J. trop. Med. Hyg.* **7**, 429–440.
GALLARD L. (1914) Notes on *Psychopsis newmani*. *Aust. Nat.* **3**, 29–32.
GALLARD L. (1935) Notes on the life history of the yellow lacewing, *Nymphes myremeleonoides*. *Aust. Nat.* **9**, 118–119.
GALLIKER P. (1958) Morphologie und Systematik der präimaginalen Stadien des schweizerischen *Solenobia arten* (Lep., Psychidae). Prom. Nr. 2716 Eidgenössischen Technischen Hochschule in Zürich.
GALVÃO A. L. A. (1941a) Notas sôbre alguns Anofelinos do sub-genero *Nyssorhynchus* do Notre do Brasil. *Rev. Biol. Higiene* **11**, 92–96.
GALVÃO A. L. A. (1941b) Cotribuição ao conhecimento das especies de *Myzorhynchella* (Diptera, Culicidae). *Rev. Mus. paul.* **25**, 505–576.
GALVÃO A. L. A. and DAMASCENO R. G. (1944) Observações sobre *Anofelinos* do complexo *albitarsis* (Diptera Culicidae). *An. Fac. Med. Univ. S. Paulo* **20**, 73–87.
GALVÃO A. L. A. and LANE J. (1937a) Nota sobre os *Nyssorhynchus* de S. Paulo. VII. Estudo sobre as variedades deste grupo com a descrição de *Anopheles (Nyssorhynchus) albitarsis* Arrib., 1878 var. *limai* n. var. *An. Fac. Med. Univ. S. Paulo* **13**, 211–238.
GALVÃO A. L. A. and LANE J. (1937b) Notas sobre os *Nyssorhynchus* de São Paulo (Diptera Culicidae). *Rev. Mus. paul.* **23**, 25–27.
GALVÃO A. L. A. and LANE J. (1938) Notas sobre os *Nyssorhynchus* de S. Paulo. VI. Revalidação de *Anopheles (Nyssorhynchus) oswaldoi* Peryassú, 1922 e discussão sobre *Anopheles (Nyssorhynchus) tarsimaculatus* Goeldi, 1905. *Livr. Jub. Prof. Travassos* **3**, 169–178.
GALVÃO A. L. A., LANE J., and CORREIA R. (1937) Notas sobre os *Nyssorhynchus* de S. Paulo. V. Sobre os *Nyssorhynchus* de Novo Oriente. *Revta Biol. Hyg.* **8**, 37–45.
GAMBLES R. M. (1956) Eggs of *Lestinogomphus africanus* Fraser. *Nature, Lond.* **177**, 663.
GAMBLES R. M. and GARDNER A. E. (1960) The egg and early stages of *Lestinogomphus africanus* (Fraser) (Odonata: Gomphidae). *Proc. R. ent. Soc. Lond.* (A) **35**, 12–16.
GAMEEL O. I. (1971) The whitefly eggs and first larval stages as prey for certain Phytoseiid mites. *Rev. Zool. Bot. afr.* **84**, 79–82.
GAMEEL O. I. (1974) Some aspects of the mating and oviposition behaviour of the cotton whitefly *Bemisia tabaci* (Genn.). *Rev. zool. afr.* **88**, 784–788.
GANDER R. (1951) Experimentelle und ökologische Untersuchungen über das Schlüpfvermögen der Larven von *Aedes aegypti* (L.). *Rev. suisse Zool.* **58**, 215–278.
GANESALINGAM V. K. (1972) Anatomy and histology of the sense organs of the ovipositor of the ichneumonid wasp *Devorgilla canescens*. *J. Insect Physiol.* **18**, 1857–1867.
GANGRADE G. A. (1964a) Female reproductive organs of *Necroscia sparaxes* Westwood (Phasmidae: Phasmida). *Entomologist* **97**, 201–208.

GANGRADE G. A. (1964b) Accessory glands of female phasmid *Necroscia sparaxes* Westwood. *Ann. Mag. nat. Hist.* **7** (13) 739–742.
GARCIA M. F. (1970) Bioecologia del taladrillo de los frutales de carozo *Scolytus rugulosus* Ratzeburg. *Revta Invest Agropec.* **7**, 11–19.
GARCIA R. (1973) Harvesting eggs from wild *Notonecta* populations. *Proc. and Papers 41st Ann. Conf. Califn. Mosquito Control Assoc.* (Peck and Mulhern, eds.).
GARCIA R. and LAING J. (1970) Breath, CO_2 and human handling of the host as factors affecting the feeding behavior of adult *Aedes aegypti* (L.) on Lepidoptera larvae. *J. med. Ent.* **5**, 601–604.
GARCIA-BELLIDO A. (1964) Das Sekret der Paragonien als Stimulus der Fekundität bei Weibehen von *Drosophila melanogaster. Z. Naturforsch.* **196**, 491–495.
GARDINER B. O. C. (1963) Notes on the breeding and biology of *Papilio machaon* L. (Lepidoptera: Papilionidae). *Proc. R. ent. Soc. Lond.* (A) **38**, 206–211.
GARDINER L. M. (1966a) A photographic record of oviposition by *Rhyssa lineolata* (Kirby) (Hymenoptera: Ichneumonidae). *Can. Ent.* **98**, 95–97.
GARDINER L. M. (1966b) Egg bursters and hatching in the Cerambycidae (Coleoptera). *Can. J. Zool.* **44**, 199–212.
GARDNER A. E. (1951) The early stages of Odonata. *Proc. Trans. S. Lond. ent. nat. Hist. Soc.* **1950–1**, 83–88.
GARDNER A. E. (1953) Further notes on exophytic oviposition in Odonata. *Entomologist's mon. Mag.* **89**, 98–99.
GARDNER A. E. (1954) The life-history of *Coenagrion hastulatum* (Charp.) (Odonata: Coenagriidae). *Ent. Gaz.* **5**, 17–40.
GARDNER A. E. (1955) The egg and mature larva of *Aeschna isosceles* (Mueller) (Odonata: Aeshnidae). *Ent. Gaz.* **6**, 13–20.
GARDNER A. E. (1956) The biology of dragonflies. *Proc. Trans. S. Lond. ent. nat. Hist. Soc.* **1954–5**, 109–134.
GARDNER A. E. (1963) Report on the insects collected by the E. W. Classey and A. E. Gardner expedition to Madeira in December 1957. *Proc. S. Lond. ent. nat. Hist. Soc.* **1962**, 62–84.
GARDNER J. C. M. (1947) A note on the larva of *Trox procerus* Har. (Scarabaeidae, Col.). *Indian J. Ent.* **8**, 31–32.
GARDOŠ J. (1976) Effect of amount of blood-meal size and temperature of environment and fertility and oogenesis of mosquitoes *Aedes aegypti* L. *Biológia Czechoslovakia* **31**, 839–844.
GARHAM P. (1927) Guide to the insects of Connecticut. Part V. The Odonata or dragonflies of Connecticut. *Bull. Conn. geol. nat. Hist. Surv.* **39**, 1–331.
GARMAN H. (1881) The egg-case and larva of *Hydrophilus triangularis. Am. Nat.* **15**, 660–663.
GARMAN H. and JEWETT H. H. (1914) The lace-wing fly (*Chrysopa oculata*). In The life-history and habits of the corn-ear worm (*Chloidea obsoleta*). *Bull. Ky agric. Exp. Stn* **187**, 589–591.
GARMS R. (1972) Vorkommen phoretischer Simulien in Liberia. *Z. Tropenmed. Parasit.* **23**, 302–307.
GARNETT W. B. and FOOTE B. A. (1967) Biology and immature stages of *Pseudolaria crassata* (Diptera: Heleomyzidae). *Ann. ent. Soc. Am.* **60**, 126–134.
GARRETT-JONES C. (1951) The Congo floor maggot *Auchmeromyia luteola* (F.) in a laboratory culture. *Bull. ent. Res.* **41**, 679–708.
GASSNER G. (1963) Notes on the biology and immature stages of *Panorpa nuptialis* Gerstaecker (Mecoptera: Panorpidae). *Texas J. Sci.* **15**, 33–34.
GAST R. T. (1961) Some shortcuts in laboratory rearing of boll weevils. *J. econ. Ent.* **54**, 395–396.
GAST R. T. (1966a) Oviposition and fecundity of boll weevils in mass-rearing laboratory cultures. *J. econ. Ent.* **59**, 173–176.

GAST R. T. (1966b) A spray technique for implanting boll weevil eggs on artificial diets. *J. econ. Ent.* **59**, 239–240.

GAST R. T. and LANDIN M. (1966) Adult boll weevils and eggs marked with dye fed in larval diet. *J. econ. Ent.* **59**, 474–475.

GAUMONT R. (1957) Études sur la biologie de quelques Chermesidae. *Annls Inst. nat. agron. Paris* **43**, 1–230.

GEBERT S. (1937) Notes on the viability of *Anopheles costalis* ova subjected to natural desiccation. *Trans. R. Soc. trop. Med. Hyg.* **31**, 115–117.

GEER B. W. (1967) Dietary choline requirements for sperm motility and normal mating activity in *Drosophila melanogaster*. *Biol. Bull. Woods Hole* **133**, 548–566.

GEHRING R. D. and MADSEN H. F. (1963) Some aspects of the mating and oviposition behavior of the codling moth *Carpocapsa pomonella*. *J. econ. Ent.* **56**, 140–143.

GEIGY R. and DU BOIS A. M. (1935) Sinnesphysiologische Beobachtungen über die Begattung von *Sialis lutaria* L. *Rev. suisse Zool.* **42**, 447–457.

GEIGY R. and GANDER R. (1949) Äussere Einwirkungen beim Schlüpfen von *Aëdes aegypti* aus dem Ei. *Acta tropica Basle* **6**, 97–104.

GEISSLER K. (1966) Untersuchungen zur Morphologie und Ökologie der Erbsengallmücke *Contarinia pisi* Winn. *Arch. PflSchutz* **2**, 39–75.

GEISTHARDT G. (1937) Über die ökologische Valenz zweier Wansenarten mit verschiedenem Verbreitungsgebiet. *Z. ParasitKde* **9**, 151–202.

GENCHEV N. (1968) An ecological study of *Agrotis segetum* Schiff. (Lepidoptera, Noctuidae). *RastVŭd. Nauki* **5**, 127–137 (in Bulgarian; English summary).

GENIEYS P. (1924) Contributions à l'étude des Evaniidae: *Zeuxevania splendidula* Costa. *Bull. biol. Fr. Belg.* **58**, 482–494.

GENSITSKIĬ I. P. and BOGACH A. V. (1972) A photometric method for determining the number of eggs laid by *Hyphantria cunea* Drury at each oviposition. *Vest. Zool.* 1972 (3) 78–79 (in Russian; English summary).

GEORGE C. J. (1928) The morphology and development of the genitalia and genital ducts of Homoptera and Zygoptera as shown in the life histories of *Philaenus* and *Agrion*. *Q. Jl microsc. Sci.* **72**, 447–485.

GERBER G. H. and CHURCH N. S. (1976) The reproductive cycles of male and female *Lytta nuttalli* (Coleoptera: Meloidae). *Can. Ent.* **108**, 1125–1136.

GERBERG E. J. (1970) Manual for mosquito rearing and experimental techniques. *Bull. Am. Mosq. Control Ass.* **5**, viii + 1–109.

GERLING D. (1966) Biological studies on *Encarsia formosa* (Hymenoptera: Aphelinidae). *Ann. ent. Soc. Am.* **59**, 142–143.

GERLING D. (1969) Host acceptance and oviposition by *Podagrion meridionale* Masi. *Entomophaga* **14**, 329–334.

GERMAIN M., GRENIER P., and MOUCHET J. (1966) Une simulie nouvelle du Cameroun Occidental: *Simulium rickenbachi* n.sp. (Diptera, Simuliidae) associée à des larves d'éphémères (*Afronurus*). *Bull. Soc. Path. exot.* **59**, 133–144.

GERRITY R. G., REMPEL J. G., SWEENEY P. R., and CHURCH N. S. (1967) The embryology of *Lytta viridana* Le Conte (Coleoptera: Meloidae). 2. The structure of the vitelline membrane. *Can. J. Zool.* **45**, 497–503.

GETZIN L. W. (1962) Mass rearing of virus-free cabbage loopers on an artificial diet. *J. Insect Pathol.* **4**, 486–488.

GEYER J. W. C. (1947a) A study of the biology and ecology of *Exochomus flavipes* Thunb. (Coccinellidae, Coleoptera). Pt. I. *J. ent. Soc. S. Afr.* **9**, 219–234.

GEYER J. W. C. (1947b) A study of the biology and ecology of *Exochomus flavipes* Thunb. (Coccinellidae, Coleoptera). Pt. II. *J. ent. Soc. S. Afr.* **10**, 64–109.

GHANI M. A. and SWEETMAN H. L. (1951) Ecological studies of the book louse *Liposcelis divinatorius* (Mull.). *Ecology* **32**, 230–244.

GHENT A. W. (1959) Row-type oviposition in *Neodiprion* sawflies as exemplified by the European pine sawfly *N. sertifer* (Geoff.). *Can. J. Zool.* **37**, 267–281.

GHENT A. W. and WALLACE D. R. (1958) Oviposition behaviour of the Swaine jack-pine sawfly. *For. Sci. Wash.* **4**, 264–272.

GHOSE S. K. and PAUL P. K. (1972) Observations on the biology of the mealybug *Ferrisia virgata* (Cockrell) (Pseudococcidae: Hemiptera) (Hom.). *Proc. zool. Soc. Calcutta* **25**, 39–48.

GHOSH C. C. (1910) Entomological notes—*Croce filipennis* Westw. *J. Bombay nat. Hist. Soc.* **20**, 530–532.

GHOSH C. C. (1912) The big brown cricket (*Brachypterus achatinus* Stoll.). *Mem. Dep. Agric. India* **4**, 161–182.

GHOSH C. C. (1913) Life-history of *Helicomitus dicax* Walker. *J. Bombay nat. Hist. Soc.* **22**, 643–648.

GIARDINA A. (1898) Sul nido dell *Mantis religiosa*. *Nat. sicil.* **2** (2) 141–149.

GIARDINA A. (1899) Sulla biologia della mantidi. *G. Sci. nat. econ. Palermo* **22**, 287–328.

GIARDINA A. (1901) Funzionamento dell'armatura gebitale femminile e considerazioni intorno alle ooteche degli Acridii. *G. Sci. nat. econ. Palermo* **23**, 54–61.

GIBBINS E. G. (1933) Eggs of some Ethiopian *Anopheles* mosquitoes. *Bull. ent. Res.* **24**, 257–262.

GIBSON A. and TREHERNE R. C. (1916) The cabbage root maggot and its control in Canada, with notes on the imported onion maggot and the seed-corn maggot. *Bull. Can. Dep. Agric.* **12**, 1–58.

GIBSON R. W. (1971) Glandular hairs providing resistance to aphids in certain wild potato species. *Ann. appl. Biol.* **68**, 113–119.

GIBSON W. P. and BERBERET R. C. (1974) Histological studies on encapsulation of *Bathyplectes curculionis* eggs by larvae of the alfalfa weevil. *Ann. ent. Soc. Am.* **67**, 588–590.

GIFFORD J. R. and TRAHAN G. B. (1969) Staining techniques for eggs of rice water weevils oviposited intracellularly in the tissue of the leaf sheaths of rice. *J. econ. Ent.* **62**, 740–741.

GILBERT B. L., BAKER J. E., and NORRIS D. M. (1967) Juglone (5-hydroxy-1,4-naphthoquinone) from *Carya ovata*, a deterrent to feeding by *Scolytus multistriatus*. *J. Insect Physiol.* **13**, 1453–1459.

GILBERT B. L. and NORRIS D. M. (1968) A chemical basis for bark beetle (*Scolytus*) distinction between host and non-host trees. *J. Insect Physiol.* **14**, 1063–1068.

GILBERT L. E. (1975) Ecological consequences of a coevolved mutualism between butterflies and moths. In *Coevolution of Animals and Plants* (L. E. Gilbert and P. H. Raven, eds.), pp. 210–240, University of Texas Press, Austin.

GILES E. T. (1953) The biology of *Anisolabis littorea* (White) (Dermaptera: Labiduridae). *Trans. R. Soc. NZ* **80**, 383–398.

GILES E. T. (1961) The female reproductive organs and genital segments of *Anisolabis littorea* (White) (Dermaptera: Labiduridae). *Trans. Roy. Soc. NZ Zool.* **1**, 293–302.

GILLETT J. D. (1955a) Variation in the hatching response of *Aedes* eggs (Diptera: Culicidae). *Bull. ent. Res.* **46**, 241–254.

GILLETT J. D. (1955b) Behaviour differences in two strains of *Aedes aegypti*. *Nature, Lond.* **176**, 124–125.

GILLETT J. D., HADDOW A. J., and CORBET P. S. (1959) Observations on the oviposition cycle of *Aedes (Stegomyia) aegypti* (Linnaeus). II. *Ann. trop. Med. Parasit.* **53**, 35–41.

GILLETT J. D., ROMAN E. A., and PHILLIPS V. (1977) Erratic hatching in *Aedes* eggs: a new interpretation. *Proc. R. Soc. Lond.* (B) **196**, 223–232.

GILLETT S. D. (1975a) Changes in the social behaviour of the desert locust *Schistocerca gregaria* in response to the gregarizing pheromone. *Animal Behav.* **23**, 494–503.

GILLETT S. D. (1975b) The action of the gregarisation pheromone on five non-behavioural characters of phase polymorphism of the desert locust. *Schistocerca gregaria* (Forskål). *Acrida* **4**, 137–149.

GILLETTE C. P. (1924) A peculiar egg-laying experience with the spider parasite *Oncodes costatus* Loew. *Circ. Colo. State Ent.* **43**, 49–51.

GILLIES M. T. (1955) Notes on the eggs of some East African *Anopheles*. *Ann. trop. Med. Parasit.* **49**, 158–160.

GILLIES M. T. and DE MEILLON B. (1968) The Anophelinae of Africa south of the Sahara (Ethiopian zoogeographical region). *Publs S. Afr. Inst. med. Res.* **54**, 1–343.

GILLIES M. T. and WILKES T. J. (1972) The range of attraction of animal baits and carbon dioxide for mosquitoes. Studies in a freshwater area of West Africa. *Bull. ent. Res.* **61**, 389–404.

GILLON Y. and ROY R. (1968) Les mantes de Lamto et des savanes de Côte d'Ivoire. *Bull. Inst. fr. Afr. noire* **30**, 1038–1151.

GILMER P. M. (1925) A comparative study of the poison apparatus of certain lepidopterous larvae. *Ann. ent. Soc. Am.* **18**, 203–239.

GIORGI D. (1936) Studio sulle ghiandole mucipare del *Bombyx mori*. *Boll. ent. Oss. Fitopath. Milano* **6**, 85–91.

GIRAULT A. A. (1907) Standards of the number of eggs laid by insects. 5. *Ent. News* **18**, 89.

GIRAULT A. A. (1913) Notes on plague of locusts in North Queensland and its relation to sugar cane. *J. Soc. Ent.* **28**, 45–48.

GJULLIN C. M. (1938) A machine for separating mosquito eggs from soil. *Circ. US Dept. Agric. (Bur. Ent.) ET* 135.

GJULLIN C. M., HEGARTY C. P., and BOLLEN W. B. (1941) The necessity of a low oxygen concentration for the hatching of *Aedes* mosquito eggs. *J. cell. comp. Physiol.* **17**, 193–202.

GJULLIN C. M., YATES W. W., and STAGE H. H. (1939) The effect of certain chemicals on the hatching of mosquito eggs. *Science, NY* **89**, 539–540.

GJULLIN C. M., YATES W. W., and STAGE H. H. (1950) Studies on *Aedes vexans* (Meig.) and *Aedes sticticus* (Meig.) floodwater mosquitoes in the lower Columbia River Valley. *Ann. ent. Soc. Am.* **43**, 262–275.

GLASGOW J. P. (1936) The bionomics of *Hydropsyche colonica* McL. and *H. philpotti* Till. *Proc. R. ent. Soc. Lond.* (A) **11**, 122–128.

GLEN D. M. (1975) The effects of predators on the eggs of codling moth *Cydia pomonella* in a cider-apple orchard in south-west England. *Ann. appl. Biol.* **80**, 115–119.

GLEN D. M. (1977) Ecology of the parasites of a predatory bug *Blepharidopterus angulatus* (Fall.). *Ecol. Ent.* **2**, 47–55.

GLENDENNING R. (1924) The satin moth in British Columbia. *Pamphlet Dep. Agric.* **50**, 3–14.

GLOVER P. M. (1937) *Lac Cultivation in India*, Bihar.

GLOVER P. M. and NEGI P. S. (1935) Specificity of parasitism by *Eublemma amabilis*. *Curr. Sci.* **3**, 426–427.

GNINENKO YU. I. (1974) The biology of *Lophopteryx camelina* in birch forests in Transuralia. *Zool. Zh.* **53**, 1495–1501 (in Russian; English summary).

GOANTSA I. K. (1966) The parasites of the bituberculate soft scale (*Palaeolecanium bituberculatum* Targ.) and the spherical apple soft scale (*Eulecanium mali* Schr.) in the Moldavian SSR. *Trudy moldav. nauchno-issled. Inst. Sadov. Vinogr. Vinod.* **13**, 59–69 (in Russian).

GODDEN D. H. (1973) A re-examination of circadian rhythmicity in *Carausius morosus*. *J. Insect Physiol.* **19**, 1377–1386.

GODWIN P. A. and ODELL T. M. (1965) The life-history of the white-pine cone beetle *Conophthorus coniperda*. *Ann. ent. Soc. Am.* **58**, 213–219.

GOEDEN R. D. and RICKER D. W. (1967) *Geocoris pallens* found to be predaceous on *Microlarinus* spp. introduced to California for the biological control of puncture vine, *Tribulus terrestris*. *J. econ. Ent.* **60**, 725–729.

GOEDEN R. D. and RICKER D. W. (1971) Biology of *Zonosemata vittigera* relative to silverleaf nightshade. *J. econ. Ent.* **64**, 417–421.

GOELDI E. A. (1886) Die Eier zweier brasilianischen Gespenstheuschrecken. *Zool. Jb. (Syst.)* **1**, 724–729.

GOELDI E. A. (1905) Os mosquitos do Pare. *Mem. Mus. Goeldi (paraense)* **4**, 1–154.

GOIDANICH A. (1956) Gregarismi od individualismi larvali e cure materne nei Crisomelidi (Col. Chrysomelidae). *Mem. Soc. ent. ital.* **35**, 151–182.

GOINY H., SOMEREN E. C. C. VAN, and HEISCH R. B. (1957) The eggs of *Aedes (Skusea) pembaensis* Theobald discovered on crabs. *E. Afr. med. J.* **34**, 1–2.

GOLEBIOWSKA Z. (1968) Development and fertility of the Mediterranean flour moth *Anagasta (Ephestia) kühniella* Zeller (Lepidoptera, Pyralidae) in differently coloured vessels. *Ekol. pol. (A)* **16**, 485–504.

GOLINI V. I. (1975) Relative response to coloured substrates by ovipositing blackflies (Diptera: Simuliidae). III. Oviposition by *Simulium (Psilozia) vittatum* Zetterstedt. *Proc. ent. Soc. Ontario* **105**, 48–55 (1974).

GOLINI V. I. and DAVIES D. M. (1975a) Relative response to coloured substrates by ovipositing black-flies (Diptera, Simuliidae). II. Oviposition by *Simulium (Odagmia) ornatum* Meigen. *Norw. J. Ent.* **22**, 89–94.

GOLINI V. I. and DAVIES D. M. (1975b) Relative response to coloured substrates by ovipositing blackflies (Diptera: Simuliidae). I. Oviposition by *Simulium (Simulium) verecundum* Stone and Jamnback. *Can. J. Zool.* **53**, 521–535.

GOLUBENKO N. N. (1969) A test of integrated control of the plum fruit moth *Laspeyresia funebrana* Tr. (Lepidoptera, Tortricidae). *Ent. Obozr.* **48**, 502–506 (English translation: *Ent. Rev.* **48**, 318–321).

GOMA L. K. H. (1963) Oviposition on dry surfaces by *Anopheles (Cellia) gambiae* Giles. *Nature, Lond.* **200**, 1232–1233.

GOMAA A. A. (1973) Biological studies on the Eri silkworm *Attacus ricini* Boisd. (Lepidoptera: Saturniidae). *Z. angew. Ent.* **74**, 120–126.

GONZÁLEZ B. J. E. (1956) Algunas observaciones ecológicas sobre los tortricidos enrolladores de la hoja del algodonero en el Perú. *Bolm trim. Exp. agropec.* **5** (3) 2–6.

GOODWIN J. A. and MADSEN H. F. (1964) Mating and oviposition behavior of the navel orangeworm *Paramyelois transitella* (Walker). *Hilgardia* **35**, 507–525.

GOONEWARDENE H. F. and TOWNSHEND B. G. (1974) Hatch of eggs of the Japanese beetle after storage at reduced temperatures. *Ann. ent. Soc. Am.* **67**, 867–870.

GOONEWARDENE H. F. and TOWNSHEND B. G. (1975) Pattern of hatch of eggs in the laboratory from field-trapped female Japanese beetles. *Ann. ent. Soc. Am.* **68**, 185–186.

GOOT P. VAN DER (1929) Einige dierlijke vijanden van *Vigna hosei* en *Calopogonium mucunoides*. *Landbk. Tijdschr. Ned. Ind.* **4**, 753.

GOPINATH K. (1966) Observations on the symbiotic ambrosia fungus of *Xylosandrus campactus* (Eichhoff) (*Xyleborus morstatti* Hagedorn). *Proc. Indian Sci. Congr.* **53**, 359.

GOPINATH K. (1967) Mating and oviposition in the ambrosia beetle *Xylosandrus compactus* Eichhoff (*Xyleborus morstatti* Hagedorn) (Coleoptera: Scolytidae). *Proc. Indian Sci. Congr.* **54**, 466.

GORDH G. and DEBACH P. (1976) Male inseminative potential in *Aphytis lingnanensis* (Hymenoptera: Aphelinidae). *Can. Ent.* **108**, 583–589.

GORDH G. and EVANS H. E. (1976) A new species of *Goniozus* imported into California from Ethiopia for the biological control of pink bollworm and some notes on the taxonomic status of *Parasierola* and *Goniozus* (Hymenoptera: Bethylidae). *Proc. ent. Soc. Washington* **78**, 479–489.

GORDON H. T. and BANDAL S. K. (1967) Effect of mating on egg production by the large milkweed bug *Oncopeltus fasciatus* (Hemiptera: Lygaeidae). *Ann. ent. Soc. Am.* **60**, 1099–1102.

GORDON H. T. and LOHER W. (1968) Egg production and male activation in new laboratory strains of the large milkweed bug *Oncopeltus fasciatus*. *Ann. ent. Soc. Am.* **61**, 1573–1578.

GORE W. E., PEARCE G. T., and SILVERSTEIN R. M. (1976) Mass spectrometric studies of the dioxabicyclo (3.2.1) octanes multistriatin, frontalin, and brevicomin. *J. org. Chem.* **41**, 607–610.

GÖRG J. (1959) Untersuchungen am Keim von *Hierodula (Rhombodera) crassa* Giglio Tos, ein Beitrag zur Embryologie der Mantiden. *Dt. ent. Z.* (NF) **6**, 389–450.

GORYUNOVA Z. S. (1964) Intraspecific differentiation in *Prospaltella (Prospaltella perniciosi* Tow.), the parasite of the San José scale. *Trudy vses. Inst. Zashch. Rast.* **21**, 40–55 (in Russian; English summary).

GOSECO F. P. (1933) *Elassogaster sepsoides* Walker, a parasite of locust eggs. *Sug. News* **14**, 185.

GOTHILF S. (1969) Natural enemies of the carob moth *Ectomyelois ceratoniae* (Zeller). *Entomophaga* **14**, 195–202.

GOTHILF S., LEVY E. C., COOPER R., and LAVIE D. (1975) Oviposition stimulants of the moth *Ectomyelois ceratoniae*: the effect of short-chain alcohols. *J. chem. Ecol.* **1**, 457–464.

GOTZ R. (1949) Der Einfluss von Tageszeit und Witterung auf Ausschlüpfen, Begattung und Eiablage des Springwurmwicklers *Sparganothis pilleriana* Schiff. *Z. angew. Ent.* **31**, 261–274.

GOWDEY C. C. (1912) On the utilisation of an indigenous African silk worm (*Anaphe infracta* Whm.) in Uganda. *Bull. ent. Res.* **3**, 269–274.

GOWER A. M. (1966) The life cycle of *Drusus annulatus* Steph. (Trich., Limnephilidae) in watercress beds. *Entomologist's mon. Mag.* **101**, 133–141.

GOWER A. M. (1967) A study of *Limnephilus lunatus* Curtis (Trichoptera: Limnephilidae) with reference to its life cycle in watercress beds. *Trans. R. ent. Soc. Lond.* **119**, 283–302.

GRABER V. (1872) Anatomisch-physiologische Studien über *Phthirius inguinalis* Leach. *Z. wiss. Zool.* **22**, 137–167.

GRAHAM H. M. and MANGUM C. L. (1971) Larval diets containing dyes for tagging pink bollworm moths internally. *J. econ. Ent.* **64**, 376–379.

GRAHAM M. W. R. DE V. (1960) Oviposition of *Tetrastichus pallipes* (Dalm.) and host of *T. lycidas* (Walk.) (Hym., Eulophidae). *Entomologist's mon. Mag.* **96**, 183.

GRANDI M. (1941) Contributi allo studio degli Efemerotteri italiani. III. *Cloën dipterum*. *Boll. Ist. Ent. Bologna* **13**, 29–71.

GRANDI M. (1947) Contributi allo studio degli Efemeroidei italiani. IX. *Oligoneuriella rhenana* Imh. *Boll. Ist. Ent. Bologna* **16**, 176–218.

GRASSÉ P. P. (1922) Étude biologique sur le criquet égyptien *Orthacanthacris aegyptia* (L.). *Bull. biol. Fr. Belg.* **56**, 545–578.

GRASSÉ P. P. (1924a) Étude biologique sur *Phanoptera 4-punctata* Br. et *P. falcata* Scop. *Bull. biol. Fr. Belg.* **58**, 454–472.

GRASSÉ P.-P. (1924b) Les ennemis des acridiens ravageurs français. *Rev. Zool. agric.* **23**, 1–15.

GRASSÉ P. P. and VICHET G. DE (1924) Sur la ponte des *Phanoptera quadripunctata* Br. et *P. falcata* Scop. *Bull. Soc. ent. Fr.* **1924**, 186–187.

GRASSIA A. and HARDY R. J. (1970) Application of the inflated poisson distribution to eggs per floret of clover seed moth *Coleophora alcyonipennella* Kollar. *J. Aust. ent. Soc.* **9**, 162–164.

GRAY B., BILLINGS R. F., GARA R. I., and JOHNSEY R. L. (1972) On the emergence and initial flight behaviour of the mountain pine beetle *Dendroctonus ponderosae* in eastern Washington. *Z. angew. Ent.* **71**, 250–259.

Gray T. G., Shepherd R. F., and Wood C. S. (1973) A hot-water technique to remove insect eggs from foliage. *Bi-mon. Res. Notes* **29**, 29.
Grbić V. (1964) Proučvanje života i suzbijanja *Otiorhynchus lavandus* Germ. *Zašt. Bilja* **15**, 463–564 (English summary).
Greany P. D. and Oatman E. R. (1972a) Demonstration of host discrimination in the parasite *Orgilus lepidus* (Hymenoptera: Braconidae). *Ann. ent. Soc. Am.* **65**, 375–376.
Greany P. D. and Oatman E. R. (1972b) Analysis of host discrimination in the parasite *Orgilus lepidus* (Hymenoptera: Braconidae). *Ann. ent. Soc. Am.* **65**, 377–383.
Greany P. D. and Oatman E. R. (1972c) Oviposition by an isolated parasite abdomen. *Ann. ent. Soc. Am.* **65**, 991–992.
Greathead D. J. (1958a) A new species of *Systoechus* (Dipt., Bombyliidae), a predator on egg-pods of the desert locust *Schistocerca gregaria* (Forskål). *Entomologist's mon. Mag.* **94**, 22–23.
Greathead D. J. (1958b) Observations on two species of *Systoechus* (Diptera: Bombyliidae) preying on the desert locust *Schistocerca gregaria* (Forskål) in eastern Africa. *Entomophaga* **3**, 3–22.
Greathead D. J. (1958c) Notes on the life history of *Symmictus flavopilosus* Bigot (Diptera: Nemestrinidae) as a parasite of *Schistocerca gregaria* (Forskål) (Orthoptera: Acrididae). *Proc. R. ent. Soc. Lond.* (A) **33**, 107–119.
Greathead D. J. (1958d) Notes on the larva and life history of *Cyrtonotum cuthbertsoni* (Dipt., Drosophilidae), a fly associated with the desert locust *Schistocerca gregaria* (Forskål). *Entomologist's mon. Mag.* **94**, 36–37.
Greathead D. J. (1962) The biology of *Stomorhina lunata* (Fabricius) (Diptera: Calliphoridae), predator of the eggs of Acrididae. *Proc. zool. Soc. Lond.* **139**, 139–180.
Greathead D. J. (1963) A review of the insect enemies of Acridoidea (Orthoptera). *Trans. R. ent. Soc. Lond.* **114**, 437–517.
Greathead D. J. (1970) White sugar-cane scales (*Aulacaspis* spp.) (Diaspididae: Hemiptera) in East Africa with notes on their natural enemies. *E. Afr. agric. For. J.* **36**, 70–76.
Green E. E. (1903) Short-hole borer. *Circ. agric. J. R. bot. Gdns, Pevadeniya Ceylon* **2**, 141–156.
Greenberg B. (1954) A method for the sterile culture of housefly larvae *Musca domestica* L. *Can. Ent.* **86**, 527–528.
Greene G. L. (1968) Distribution of *Trichoplusia ni* eggs and larvae on cabbage plants as a basis for sampling efficiency. *J. econ. Ent.* **61**, 1648–1650.
Greene G. L., Reid J. C., Blount V. N., and Riddle T. C. (1973) Mating and oviposition behaviour of the velvetbean caterpillar in soybeans. *Envir. Ent.* **2**, 1113–1115.
Greene G. L. and Thurston R. (1971) Ovipositional preference of *Heliothis virescens* for *Nicotiana* species. *J. econ. Ent.* **64**, 641–643.
Greenfield M. D. and Karandinos M. G. (1976a) Oviposition rhythm of *Synanthedon pictipes* under a 16:8 L:D photoperiod and various thermoperiods. *Envir. Ent.* **5**, 712–713.
Greenfield M. D. and Karandinos M. G. (1976b) Fecundity and longevity of *Synathedon pictipes* under constant and fluctuating temperatures. *Envir. Ent.* **5**, 883–887.
Grellet P. (1969a) Modalités de l'absorption d'eau par l'oeuf de *Scapsipedus marginatus* Afz. et Br. (orthoptère, gryllide). *C. r. hebd. Séanc. Acad. Sci. Paris* **269**, 2011–2013.
Grellet P. (1969b) Perméabilité des enveloppes de l'oeuf de *Scapsipedus marginatus* (orthoptère, gryllide). *C. r. hebd. Séanc. Acad. Sci. Paris* **269**, 2236–2239.
Grellet P. (1969c) Étude expérimentale de l'absorption d'eau par l'oeuf de *Scapsipedus marginatus* Afz. et Br. (orthoptère, gryllide). *C. r. hebd. Séanc. Acad. Sci. Paris* **269**, 2417–2420.
Grellet P. (1971a) Structure et perméabilité des enveloppes de l'oeuf de *Scapsipedus marginatus* (orthoptère, gryllide). *J. Insect Physiol.* **17**, 1275–1293.

GRELLET P. (1971b) Analyse expérimentale de l'absorption d'eau dans l'oeuf de *Scapsipedus marginatus* (orthoptère, gryllide). *J. Insect Physiol.* **17**, 1533–1553.
GRENACHER H. (1868) Beiträge zur Kenntnis des Eies der Ephemeriden. *Z. wiss. Zool.* **18**, 95–98.
GRENIER P. and MOUCHET J. (1959) Note complémentaire sur la morphologie et la biologie de *S. ovazzae* Grenier et Mouchet 1959 (Diptera, Simuliidae), espèce associée au crabe *Potamonautes chaperi* M.-Edw., dans l'Ouest africain. *Bull. Soc. Path. exot.* **52**, 373–385.
GRESSITT J. L. (1953) The coconut rhinoceros beetle (*Oryctes rhinoceros*) with particular reference to the Palau Islands. *Bull. Bishop Mus.* **212**, viii + 1–157.
GREWAL S. S. and ATWAL A. S. (1969) The influence of temperature and humidity on the development of *Sitotroga cerealella* Oliver (Gelechiidae, Lepidoptera). *J. Res. Punjab agric. Univ.* **6**, 353–358.
GRIFFIN J. G. and LINDIG O. H. (1975) Granular materials used in implanting boll weevil eggs on artificial diet. *J. econ. Ent.* **68**, 433–434.
GRIFFITH M. E. (1945) The environment, life history and structure of the waterboatman *Rhampocorixa acuminata* (Uhler) (Hemiptera, Corixidae). *Kans. Univ. Sci. Bull.* **30**, 241–366.
GRIFFITHS D. C. (1960) The behaviour and specificity of *Monoctonus paludum* Marshall (Hym., Braconidae), a parasite of *Nasonovia ribis-nigri* (Mosley) on lettuce. *Bull. ent. Res.* **51**, 303–319.
GRIFFITHS D. C. (1961) The development of *Monoctonus paludum* Marshall (Hym., Braconidae) in *Nasonovia ribis-nigri* on lettuce, and immunity reactions in other lettuce aphids. *Bull. ent. Res.* **52**, 147–163.
GRIFFITHS K. J. (1972) Discrimination between parasitized and unparasitized hosts by *Pleolophus basizonus* (Hymenoptera: Ichneumonidae). *Proc. ent. Soc. Ontario* **102**, 83–91.
GRIGARICK A. A. (1959) Bionomics of the rice leaf miner *Hydrellia griseola* (Fallen) in California (Diptera: Ephydridae). *Hilgardia* **29**, 1–80.
GRIGARICK A. A. and BEARDS G. W. (1965) Ovipositional habits of the rice water weevil in California as related to a greenhouse evaluation of seed treatments. *J. econ. Ent.* **58**, 1053–1056.
GRIGOLO A., SACCHI L., GASPERI G., and CAPROTTI M. (1974) Alcuni aspetti della monogamia delle femmine di *Piophila casei* L. *Riv. Parassit.* **35**, 213–225.
GRIGOROV S. and GOSPODINOV G. (1964) The noxious Pentatomid (*Eurygaster integriceps* Put.). *Rastit. Zasht.* **12**, 3–10 (in Bulgarian).
GRIMBLE D. G., NORD J. C., and KNIGHT F. B. (1969) Oviposition characteristics and early larval mortality of *Saperda inornata* and *Oberea schaumii* in Michigan aspen. *Ann. ent. Soc. Am.* **62**, 308–315.
GRIMPE G. (1921) Beitrage zur Biologie von *Phyllium bioculatum* G. R. Gray (Phasmidae). *Zool. Jb. (Syst.)* **44**, 227–266.
GRIMSTONE A. V., ROTHERAM S., and SALT G. (1967) An electron-microscope study of capsule formation by insect blood cells. *J. Cell Sci.* **2**, 281–292.
GRIOT M. (1955) Observaciones sobre algunos parásitos de *Ceroplastes grandis* Hempel. *Rev. Fac. Agron.* **13**, 491–504.
GRISON P. (1957) Les facteurs alimentaires de la fécondité chez le doryphore (*Leptinotarsa decemlineata* Say) (Col., Chrysomelidae). *Annls Épiphyt.* **8**, 305–381.
GRISON P. and RITTER R. (1961) Effets du groupement sur l'activité et la ponte du doryphore *Leptinotarsa decemlineata* Say (Col. Chrysomelidae). *Insectes soc.* **8**, 109–123.
GRISON P., SILVESTRE R. DE S., and GALICHET P. F. (1951) La processionnaire du pin (*Thaumaetopoea pityocampa* Schiff.). Moeurs dégats—moyens de lutte. *Rev. Zool. agric. appl.* **1–3, 4–6**, 1–16.

GRISWOLD G. H. (1941) Studies on the biology of four common carpet beetles. I. The black carpet beetle (*Attagenus piceus* Oliv.), the varied carpet beetle (*Anthrenus verbasci* L.), and the furniture carpet beetle (*Anthrenus vorax* Waterh.). *Mem. Cornell Univ. agric. Exp. Stn* **240**, 3–57; 70–75.

GRISWOLD G. H. (1944) Studies on the biology of the webbing clothes moth (*Tineola bisselliella* Hum.). *Mem. Cornell Univ. agric. Exp. Stn* **262**, 1–59.

GRIVANOV K. P. and ANTONENKO O. P. (1970) A study of predators of *Eurygaster integriceps* during the active period of its life by means of labelling with ^{14}C. *Zool. Zh.* **49**, 1563–1569 (in Russian).

GROMOVA A. A. and ROGACHEVA T. V. (1976) The food specialization of the gold tail. *Zashch. Rast.* 12, 76, p. 51.

GROMOVAYA E. F. and SMIRNOVA I. M. (1964) Control of the apple Aegeriid. *Zashch. Rast. Vredit. Bolez.* **1964**, (4) 30–31 (in Russian).

GROSS F. (1930) Odonata. *Biologie der Tiere Deutschlands*, pp. 1–78, Berlin.

GROSS H. R. JR and BARTLETT F. J. (1972) Improved technique for collecting eggs of whitefringed beetles. *J. econ. Ent.* **65**, 611–612.

GROSS H. R. JR, HARRELL E. A., and PERKINS W. D. (1975) *Heliothis zea*: oviposition chamber and egg collection technique for mass production. *J. econ. Ent.* **68**, 630–632.

GROSS H. R. JR, MITCHELL J. A., SHAW Z. A., and PADGETT G. R. (1972) Extended storage of eggs of whitefringed beetles. *J. econ. Ent.* **65**, 731–733.

GROSS J. (1906) Untersuchungen über die Ovarien von Mallophagen und Pediculiden. *Zool. Jb. (Anat.)* **22**, 347–386.

GROVER P. and PRASAD S. N. (1968) *Golanudiplosis japonicus*—a new midge attacking Comstock mealybug in Japan (Diptera: Itonididae). *Beitr. Ent.* **18**, 213–220.

GRUBER F. and DYSART R. J. (1974) *Peridesmia discus*, an egg predator of *Hypera postica* in Europe. *Envir. Ent.* **3**, 789–792.

GRUNER L. (1968) Contribution à l'étude de l'ovogénèse d'un dynastide *Phyllognathus silenus* F. Essais préliminaires de chimiostérilisation. *Annls Épiphyt.* **19**, 267–304.

GRUNIN K. YA. (1953) Viviparity in coprobionts in the order Diptera. *Trudȳ zool. Inst. Akad. Nauk SSSR* **13**, 387–389.

GRUWEZ G., HOSTE C., LINTS C. V., and LINTS F. A. (1971) Oviposition rhythm in *Drosophila melanogaster* and its alteration by a change in the photoperiodicity. *Experientia* **27**, 1414–1416.

GRUYS P. (1970) Growth in *Bupalus piniarius* (Lepidoptera: Geometridae) in relation to larval population density. I. The influence of some abiotic factors on growth. II. The effect of larval density. *Verh. Rijksinst. Natuurbeheer* **1**, 1–127.

GUBB D. C. (1976) Helicoidal architecture and the control of morphogenesis in extracellular matrices. Thesis, University of Bristol.

GUBLER D. J. (1971) Studies on the comparative oviposition behaviour of *Aedes (Stegomyia) albopictus* and *Aedes (Stegomyia) polynesiensis* Marks. *J. med. Ent.* **8**, 675–682.

GUENNELON G. and AUDEMARD H. (1960) La pyrale du mais *Ostrinia (Pyrausta) nubilalis* Hbn. (lépidoptères Pyralidae) dans la basse vallée du Rhône: observations écologiques; incidences économiques. *Annls Épiphyt.* **11**, 337–396.

GUENNELON G. and TORT M. J. (1958a) Influence de la lumière sur la ponte d'*Archips rosana* Linné (Lep., Tortricidae). *Bull. Soc. ent. Fr.* **63**, 59–62.

GUENNELON G. and TORT M. J. (1958b) Sur les facteurs de réduction naturelle des populations hivernantes d'*Archips rosana* Linné dans la Basse Vallée du Rhône (Lep. Tortricidae). *Bull. Soc. ent. Fr.* **63**, 117–122.

GUERRITORE A., RAMPONI G., and BACCETTI B. (1958) Sulla defosforilazione enzimatica dell'ATP nelle ghiandole di Insetti. I. Ghiandole colleteriche di *Acrida bicolor* Thbg.

Monitore zool. ital. **46**, 49–52.

GUERRITORE A., RAMPONI G., and BACCETTI B. (1960) Acylphosphatase in organs of a grasshopper *Acrida bicolor* Thunberg. *J. Insect Physiol.* **5**, 213–215.

GUIDICELLI J. (1964) L'oviposition chez les Blépharocérides (Diptera). *Rev. fr. Ent.* **31**, 116–119.

GUILDING L. (1827) Communication on *Ascalaphus macleayanus*. *Trans. Linn. Soc. Lond.* **15**, 509–512.

GUILLOT F. S. and VINSON S. B. (1972) Sources of substances which elicit a behavioural response from the insect parasitoid *Campoletis perdistinctus*. *Nature, Lond.* **235**, 169–170.

GÜNTHER K. (1933) Funktionell-anatomische Untersuchungen über die Bursa copulatrix, den Ovipositor und den mannlichen Kopulationsapparat bei Phasmiden. *Z. Naturwiss.* **68**, 403–462.

GÜNTHER K. and HERTER K. (1974) Dermaptera (Ohrwurmer). *Handbuch der Zoologie* 4 (2.2).

GUPPY P. L. (1913) Life-history of the syrphid fly predaceous on frog-hopper nymphs. *Bull. Dep. Agric. Trin. Tobago* **12**, 159–161.

GUPPY P. L. (1914) Breeding and colonizing the syrphid. *Circ. Dep. Agric. Trin. Tobago* **10**, 217–226.

GUPPY R. (1950) Biology of *Anisolabis maritima* (Géné), the seaside earwig, on Vancouver Island (Dermaptera: Labiduridae). *Proc. ent. Soc. Br. Columb.* **46**, 14–18.

GUPTA A. P. and RILEY R. C. (1967) Female reproductive system and histology of the ovariole of the asparagus beetle *Crioceris asparagi* (Coleoptera: Chrysomelidae). *Ann. ent. Soc. Am.* **60**, 980–988.

GUPTA B. P. (1966) Bionomics of the cotton grey weevil *Myllocerus maculosus* Desb. (Curculionidae-Coleoptera). *Labdev J. Sci. Technol.* **4**, 124–127.

GUPTA P. D. (1950) On the structure, development and homology of the female reproductive organs in orthopteroid insects. *Indian J. Ent.* **10**, 75–123 (1948).

GUPTA P. D. and THORSTEINSON A. J. (1960) Food plant relationships of the diamond-back moth (*Plutella maculipennis* (Curt.)). II. Sensory regulation of oviposition of the adult female. *Entomologia exp. appl.* **3**, 305–314.

GURNEY A. B. (1938) A synopsis of the order Zoraptera with notes on the biology of *Zorotypus hubbardi* Caudell. *Proc. ent. Soc. Wash.* **40**, 57–87.

GURNEY A. B. (1950) Praying mantids of the United States: native and introduced. *Smithson. Rep.* **1950**, 339–362.

GURNEY A. B. (1955) Further notes on *Iris oratoria* in California. *Pan-Pacif. Ent.* **31**, 67–72.

GURR L. (1957) Observations on the distribution, life-history, and economic importance of *Nysius huttoni* (Lygaeidae: Hemiptera). *NZ Jl Sci. Tech.* **38** (A) 710–714.

GUR'YANOVA T. M. (1972) The distribution of eggs in *Exenterus abruptorius* (Hymenoptera, Ichneumonidae) at different population levels of *Neodiprion sertifer* (Diprionidae). *Zool. Zh.* **51**, 845–854 (in Russian).

GWADZ R. W. (1972) Neuro-hormonal regulation of sexual receptivity in female *Aedes aegypti*. *J. Insect Physiol.* **18**, 259–266.

GWADZ R. W., CRAIG G. B., and HICKEY W. A. (1971) Female sexual behaviour as the mechanism rendering *Aedes aegypti* refractory to insemination. *Biol. Bull. Woods Hole* **140**, 201–214.

GYRISCO G. G., WHITCOMB W. H., BURRAGE R. H., LOGOTHETIS C., and SCHWARDT H. H. (1953) Biology of the European chafer *Amphimallon majalis*, Razoumowsky (Scarabaeidae). *Mem. Cornell Univ. agric. Exp. Stn* **328**, 1–35.

HABIB A., BADAWI A., and HERAKLY F. (1972) Biological studies on certain species of leafhoppers (Hem., Cicadellidae) (Hom.) in Egypt. *Z. angew. Ent.* **71**, 172–178.

HACKETT L. W., MARTINI E., and MISSIROLI A. (1932) The races of *A. maculipennis*. *Am. J. Hyg.* **16**, 137–162.
HACKETT L. W. and MISSIROLI A. (1935) The varieties of *Anopheles maculipennis* and their relation to the distribution of malaria in Europe. *Riv. Malar.* **14**, 45–109.
HACKMAN R. H. and GOLDBERG M. (1960) Composition of the oötheca of three Orthoptera. *J. Insect Physiol.* **5**, 73–78.
HACKMAN R. H. and GOLDBERG M. (1963) Phenolic compounds in the cockroach oötheca. *Biochim. biophys. Acta* **71**, 738–740.
HACKMAN W. (1963) On the dipterous fauna of rodent burrows in northern Lapland. *Notul. ent. Helsingf.* **43**, 121–131.
HADAWAY A. B. (1956) The biology of the dermestid beetles *Trogoderma granarium* Everts and *Trogoderma versicolor* (Creutz.). *Bull. ent. Res.* **46**, 781–796.
HADDOW A. J. (1961) Entomological studies from a high tower in Mpanga forest, Uganda. VII. The biting behaviour of mosquitoes and tabanids. *Trans. R. ent. Soc. Lond.* **113**, 315–335.
HADDOW A. J., GILLETT J. D., and CORBET P. S. (1961) Observations on the oviposition-cycle of *Aedes (Stegomyia) aegypti* (Linnaeus). V. *Ann. trop. Med. Parasit.* **55**, 343–356.
HADLINGTON P. and SHIPP E. (1962) Diapause and parthenogenesis in the eggs of three species of Phasmatodea. *Proc. Linn. Soc. NSW* **86**, 268–279.
HADWEN S. (1923) Insects affecting livestock. *Bull. ent. Br. Can.* **24**, 1–32.
HAEGERMARK U. (1956) Investigations on the frequency of perforation of ovaries and pods in two fields of autumn rape. *Växtskyddsnotiser* **3**, 45–47 (in Swedish).
HAFEEZ M. A. (1971) Physiological effects of food renewal on the rate of oviposition of *Latheticus oryzea* (Waterhouse) (Coleoptera, Tenebrionidae). *Agric. Res. Rev.* **49**, 40–45.
HAFEZ M. (1939) The life history of *Philonthus quisquiliarius* Gyllh. (Coleoptera : Staphylinidae). *Bull. Soc. Fouad I Ent. Cairo* **23**, 302–311.
HAFEZ M. (1949) Observations on the biology of some coprophagous Borboridae (Diptera). *Proc. R. ent. Soc. Lond.* (A) **24**, 1–5.
HAFEZ M. and EL-MOURSY A. A. (1956) Studies on desert insects in Egypt. II. On the general biology of *Vermileo vermileo* L. (Diptera : Rhagionidae). *Bull. Soc. ent. Égypte* **40**, 333–348.
HAFEZ M., EZZAT M. A., FARES F., and AWADALLAH A. M. (1967) Mass rearing of the Mediterranean fruit fly *Ceratitis capitata* (Wied.) on artificial medium in UAR. *Agric. Res. Rev. Cairo* **45** (2) 77–90.
HAFEZ M. and GAMAL-EDDIN F. M. (1959) Ecological studies on *Stomoxys calcitrans* L. and *sitiens* Rond. in Egypt, with suggestions on their control. *Bull. Soc. ent. Égypte* **43**, 245–283.
HAFEZ M. and MAKKY A. M. M. (1959) Studies on desert insects in Egypt. III. On the bionomics of *Adesmia bicarinata* Klug (Coleoptera : Tenebrionidae). *Bull. Soc. ent. Égypte* **43**, 89–113.
HAFEZ M. and OSMAN F. H. (1956) Biological studies on *Bruchidius trifolii* (Motsch.) and *Bruchidius alfierii* Pic., in Egypt. *Bull. Soc. ent. Égypte* **40**, 231–277.
HAGEMANN-MEURER U. and THALENHORST W. (1964) Vergleichende Untersuchungen über die Puppengewicht-Eizahl-Relation der Forleule *Panolis flammea* (Schiff.). *Z. angew. Ent.* **54**, 290–299.
HAGEN H. R. (1941) The general morphology of the female reproductive system of a viviparous roach, *Diploptera dytiscoides*. *Psyche, Camb.* **48**, 1–9.
HAGEN K. S. (1962) Biology and ecology of predaceous Coccinellidae. *A. Rev. Ent.* **7**, 289–326.
HAGLEY E. A. C. (1965) On the life history and habits of the palm weevil *Rhynchophorus palmarum*. *Ann. ent. Soc. Am.* **58**, 22–28.
HAGLEY E. A. C. (1976) Effect of rainfall and temperature on codling moth oviposition. *Envir. Ent.* **5**, 967–969.
HAGSTRUM D. W. (1973) Infestation of flour by *Tribolium castaneum*: relationship between

distribution of eggs and adults in a dispersing population. *Ann. ent. Soc. Am.* **66**, 587–591.

HAGSTRUM D. W. and GUNSTREAM S. E. (1971) Salinity, pH, and organic nitrogen of water in relation to presence of mosquito larvae. *Ann. ent. Soc. Am.* **64**, 465–467.

HAGSTRUM D. W. and TOMBLIN C. F. (1973) Oviposition by the almond moth *Cadra cautella* in response to falling temperature and onset of darkness. *Ann. ent. Soc. Am.* **66**, 809–812.

HAHN J. (1923) Morfologie vajička indické pakobylky. *Dixippus (Carausius) morosus* Br. et Redt. *Mém. Soc. r. Sci. Bohème Cl. Sci.* **1921–2** (Pt. 8), 1–12.

HALE H. M. (1924) Notes on the eggs, habits and migration of some Australian aquatic bugs (Corixidae and Notonectidae). *S. Aust. Nat.* **5**, 133–135.

HALFFTER G., HALFFTER V., and LOPEZ I. G. (1974) *Phanaeus* behaviour: food transportation and bisexual cooperation. *Envir. Ent.* **3**, 341–345.

HALFFTER G. and LOPEZ, Y. G. (1977) Development of the ovary and mating behaviour in *Phanaeus. Ann. ent. Soc. Am.* **70**, 203–212.

HALFFTER G. and MATTHEWS E. G. (1966) The natural history of dung beetles of the subfamily Scarabaeinae (Coleoptera, Scarabaeidae). *Folia ent. mex.* **12–14**, 1–312.

HALL D. W., ANDREADIS T. G., FLANAGAN T. R., and KACZOR W. J. (1975) Melanotic encapsulation of the nematode *Neoaplectana carpocapsae* by *Aedes aegypti* larvae concurrently parasitized by the nematode *Reesimermis nielseni. J. Invert. Path.* **26**, 269–270.

HALL R. R., DOWNE A. E. R., MACLELLAN C. R. and WEST A. S. (1953) Evaluation of insect predator–prey relationships by preciptin test studies. *Mosq. News* **13**, 199–204.

HALLEZ P. (1885) Orientation de l'embryon et formation du cocoon chez *Periplaneta orientalis. C. r. hebd. Séanc. Acad. Sci. Paris* **101**, 444–446.

HALLEZ P. (1909) Sur les cristaux de la Blatte. *C. r. hebd. Séanc. Acad. Sci. Paris* **148**, 317–318.

HALSTEAD D. G. H. (1967) Biological studies on species of *Palorus* and *Coelopalorus* with comparative notes on *Tribolium* and *Latheticus* (Coleoptera: Tenebrionidae). *J. stored Prod. Res.* **2**, 273–313.

HALSTEAD D. G. H. (1968) Observations on the biology of *Marmidius ovalis* (Beck) (Coleoptera: Cerylonidae). *J. stored Prod. Res.* **4**, 13–21.

HAMID A. (1966) Sex ratio and population studies on *Clerodendron* caseworm, *Clania cameri* Hamps and its parasite, *Brachycorhyphus nursei* Cam. *Agriculture Pakist.* **17**, 91–95.

HAMILTON A. (1940) The New Zealand Dobson-fly (*Archichauliodes diversus* Walk.): life-history and bionomics. *NZ Jl Sci. Tech.* **22**, 44–55.

HAMILTON A. G. (1936) The relation of humidity and temperature to the development of three species of African locusts—*Locusta migratoria migratoriodes* (R. and F.), *Schistocerca gregaria* (Forsk.), *Nomadacris septemfasciata* (Serv.). *Trans. R. ent. Soc. Lond.* **85**, 1–60.

HAMILTON M. A. (1931) The morphology of the water-scorpion *Nepa cinerea* Linn. (Rhynchota, Hemiptera). *Proc. zool. Soc. Lond.* **1931**, 1067–1136.

HAMM A. H. (1919) A ribbon-making fly: the oviposition of *Ceratopogon nitidus* Macq. *Entomologist's mon. Mag.* **5** (3) 66–67.

HAMMAD S. M. (1953) The immature stages of *Metophthalmus serripennis* Broun (Coleoptera: Lathridiidae). *Proc. R. ent. Soc. Lond.* (A) **28**, 133–138.

HAMMAD S. M., ARMANIUS N. E., and EL-DEEB A. A. (1972) Some biological aspects of *Oxycarene hyalinipennis* Costa (Hemiptera: Lygaeidae) (Het.). *Bull. Soc. ent. Égypte* **56**, 33–38.

HAMMER O. (1941) Biological and ecological investigations on flies associated with pasturing cattle and their excrement. *Vidensk. Meddr dansk naturh. Foren.* **105**, 141–393.

HAMON C. (1972) Formation du chorion et pénétration des symbiontes dans l'oeuf d'un insecte Homoptère Auchenorrhynchae, *Ulopa reticulata* Fab. *Z. Zellforsch. mikrosk. Anat.* **123**, 112–120.

HAMPSON G. F. (1910) *Catalogue of the Lepidoptera Phalaenae in the British Museum* **10**, British Museum (NH), London.

HANCOCK J. L. (1902) *The Tetrigidae of North America*, Chicago.

HANCOCK J. L. (1904) The oviposition and carnivorous habits of the green meadow grasshopper (*Orchelimum glaberrimum* Burmeister). *Psyche, Camb.* **11**, 69–71.

HANDLIRSCH A. (1882) Die Metamorphose und Lebensweise von *Hirmoneura obscura* Meig., einem Vertreter der Dipterenfamilie Nemestrinidae. *Wien. ent. Z.* **1**, 224–228 (also 1883, **2**, 11–15).

HANITSCH R. (1933) Notes by Dr R. Hanitsch, PhD, on female cockroaches (Blattidae) which carry their young. *Proc. R. ent. Soc. Lond.* **8**, 18–19.

HANNA H. M. (1961) Observations on the egg-laying of some British caddis flies and on case-building by newly hatched larvae. *Proc. R. ent. Soc. Lond.* (A) **36**, 57–62.

HANSELL M. H. (1969) A field study on the attraction of female *Anopheles pharoensis* Theobald to carbon dioxide. *Rev. Comp. anim.* **3**, 65–68.

HANSEN G. (1964) Utah's economic future—a fly in the ointment. *Proc. 22nd A. Meet. Utah Mosq. Abatement Ass.* **1964** (M. K. Benge and B. Rosay, eds.), pp. 18–20.

HARADA F., MORIYA K., and YABE T. (1967) Observations on the habits of feeding and oviposition of *Culex tritaeniorhynchus* Giles. *Jap. J. appl. Ent. Zool.* **11**, 83–89.

HARADA F., MORIYA K., and YABE T. (1970) Observations on the habits of feeding and oviposition of *Culex tritaeniorhynchus summorosus* Hyar (III). *Jap. J. sanit. Zool.* **21**, 46–54.

HARAKLY F. A. (1966) External morphology and life-history of the egg-plant stem borer *Euzophera osseatella* Treit. (Lepidoptera: Pyralidae). *Bull. Soc. ent. Égypte* **49**, 259–276.

HARCHARAN SINGH (1965) *Pachymerus chinensis* Linn., its biology and extent of damage to gram. *Pl. Prot. Bull.* **16**, 23–28 (1964).

HARDEN F. W., POOLSON B. J., BENNETT L. W., and GASKIN R. C. (1970) Analysis of CO_2 supplemented mosquito adult landing rate counts. *Mosq. News* **30**, 369–374.

HARDING J. A. (1976) *Heliothis* spp.: parasitism and parasites plus host plants and parasites of the beet armyworm, diamondback moth and two tortricids in the Lower Rio Grande Valley of Texas. *Envir. Ent.* **5**, 669–671.

HARDWICK D. F. (1971) The life history of *Schinia separata* (Noctuidae). *J. Lepid. Soc.* **25**, 177–180.

HARDY D. E. (1949) Studies in Hawaiian fruit flies (Diptera, Tephritidae). *Proc. ent. Soc. Wash.* **51**, 181–205.

HARDY G. A. and PREECE W. H. A. (1926) Notes on some species of Cerambycidae (Col.) from the southern portion of Vancouver Island, BC. *Pan-Pacif. Ent.* **3**, 34–40.

HARDY J. (1848) On the manner in which the female of *Psocus quadripunctatus* constructs a web to protect her egg. *Newmann's Zoologist* **6**, 2221–2222.

HARDY Y. J. and ALLEN D. C. (1975) Selection of ovipositional sites by the European pine sawfly *Neodiprion sertifer* (Hymenoptera: Diprionidae) on Scots pine. *Can. Ent.* **107**, 761–768.

HARGREAVES H. (1940) Report of the work on the Entomological Section. (i) Report of the senior entomologist. *Rep. Dep. Agric. Uganda* II. **1939**, 5–9.

HARLAN D. P. and ENNIS W. R. (1966) Surface-printing of grasshopper eggs for identification. *Ann. ent. Soc. Am.* **59**, 1018–1020.

HARMAN D. M. and WALLACE J. B. (1971) Description of the immature stages of *Lonchaea corticis*, with notes on its role as a predator of the white pine weevil *Pissodes strobi*. *Ann. ent. Soc. Am.* **64**, 1221–1226.

HARMAN M. T. (1925) The reproductive system of *Apotettix eurycephalus* Hancock. *J. Morph.* **41**, 217–238.

HARRELL E. A., BURTON R. L., and SPARKS A. N. (1970) A machine to manipulate corn earworm (*Heliothis zea* (Boddie)) eggs in a mass-rearing program. *J. econ. Ent.* **63**, 1362–1363.

HARRELL E. A., SPARKS A. N., PERKINS W. D., and HARE W. W. (1974) Equipment to place insect eggs in cells on a form-fill-seal machine. *US Dep. Agric.* 1–4.

HARRIS K. M. (1968) A systematic revision and biological review of the cecidomyiid predators (Diptera: Cecidomyiidae) on world Coccoidea (Hemiptera: Homoptera). *Trans. R. ent. Soc. Lond.* **119**, 401–494.

HARRIS M. K. (1976) Pecan weevil adult emergence, onset of oviposition and larval emergence from the nut as affected by the phenology of the pecan. *J. econ. Ent.* **69**, 167–170.

HARRIS W. V. (1971) *Termites. Their Recognition and Control*, 2nd edn., Longmans, London.

HARRISON F. P. (1960) Corn earworm oviposition and the effect of DDT on the egg predator complex in corn silk. *J. econ. Ent.* **53**, 1088–1094.

HARTLEY J. C. (1961a) The shell of Acridid eggs. *Q. Jl microsc. Sci.* **102**, 249–255.

HARTLEY J. C. (1961b) A taxonomic account of the larvae of some British Syrphidae. *Proc. zool. Soc. Lond.* **136**, 505–573.

HARTLEY J. C. (1962) The egg of *Tetrix* (Tetrigidae, Orthoptera), with a discussion on the probable significance of the anterior horn. *Q. Jl microsc. Sci.* **103**, 253–259.

HARTLEY J. C. (1964) The structure of the eggs of the British Tettigoniidae (Orthoptera). *Proc. R. ent. Soc. Lond.* (A) **39**, 111–117.

HARTLEY J. C. (1965) The structure and function of the egg-shell of *Deraeocoris ruber* L. (Heteroptera, Miridae). *J. Insect Physiol.* **11**, 103–109.

HARTLEY J. C. (1967) Some notes on hatching the eggs of *Pholidoptera griseoaptera* (Degeer) (Orth., Tettigoniidae). *Entomologist's mon. Mag.* **103**, 123–124.

HARTLEY J. C. (1971) The respiratory system of the egg-shell of *Homorocoryphus nitidulus vicinus* (Orthoptera, Tettigoniidae). *J. exp. Biol.* **55**, 165–176.

HARTZELL F. Z. (1917) The cherry leaf-beetle. *Bull. NY agric. Exp. Stn* **444**, 747–820.

HARVEY G. W. (1907) A ferocious water bug. *Can. Ent.* **39**, 17–21 (see also *Proc. ent. Soc. Wash.* **8**, 72–75).

HARWOOD W. G. and HORSFALL W. R. (1959) Development, structure, and function of coverings of eggs of floodwater mosquitoes. III. Functions of coverings. *Ann. ent. Soc. Am.* **52**, 113–116.

HARWOOD W. G. and RUDINSKY J. A. (1966) The flight and olfactory behavior of checkered beetles (Coleoptera: Cleridae) predatory on the Douglas-fir beetle. *Tech. Bull. Ore. agric. Exp. Stn* **95**, 1–36.

HARZ K. (1957) *Die Geradflügler Mitteleuropas*, Jena.

HARZ K. (1958) Orthopterologische Beiträge. *NachrBl. bayer. Ent.* **7**, 38–40, 47–48.

HARZ K. (1960) Geradflüger oder Orthoptera (Blattodea, Mantodea, Saltatoria, Dermaptera). *Tierwelt. Dtl.* **46**, 1–232.

HASE A. (1954) Über Menschenläuse (Anoplura). Bibliographien, Geschichte der Forschungen, Eier und Mikropylapparat. *Z. ParasitKde* **16**, 145–190.

HASEGAWA T. and CHIBA T. (1975) Ovipositional behaviour of *Tabanus sapporoenus* and *Tabanus katoi* (Diptera: Tabanidae). *Bull. Tohoku natn. agric. exp. Stn* **51**, 61–72 (in Japanese; English summary).

HASKELL P. T. (1958) Stridulation and associated behaviour in certain Orthoptera. 2. Stridulation of females and their behaviour with males. *Anim. Behav.* **6**, 27–42.

HASSAN A. I. (1939) The biology of some British Delphacidae (Homopt.) and their parasites with special reference to the Strepsiptera. *Trans. R. ent. Soc. Lond.* **89**, 345–384.

HASSAN S. A. (1969) Observations on the effect of insecticides on Coleopterous predators of *Erioischia brassicae* (Diptera: Anthomyiidae). *Entomologia exp. appl.* **12**, 157–168.

HASSANEIN M. H. (1958) Biological studies on the diamond-back moth *Plutella maculipennis* Curtis (Lepidoptera: Plutellidae). *Bull. Soc. ent. Égypte* **42**, 325–337.

Hassell M. P. (1969) A study of the mortality factors acting upon *Cyzenis albicans* (Fall.), a tachinid parasite of the winter moth (*Operophtera brumata* (L.)). *J. Anim. Ecol.* **38**, 329–339.

Hathaway D. O. (1966) Laboratory and field cage studies of the effects of gamma radiation on codling moths. *J. econ. Ent.* **59**, 35–37.

Hathaway D. O., Schoenleber L. G., and Lydin L. V. (1972) Codling moths: plastic pellets or waxed paper as oviposition substrates. *J. econ. Ent.* **65**, 1756–1757.

Hausermann W. and Nijhout H. F. (1975) Permanent loss of male fecundity following sperm depletion in *Aedes aegypti* (L.). *J. med. Ent.* **11**, 707–715.

Haviland M. D. (1925) The Membracidae of Kartabo, Bartica District, British Guiana. *Zoologica, NY* **6**, 229–290.

Hawkes R. B. (1972) A fluorescent dye technique for marking insect eggs in predation studies. *J. econ. Ent.* **65**, 1477–1478.

Haynes D. L. and Butcher J. W. (1962) Studies on host preference and its influence on European pine shoot moth success and development. *Can. Ent.* **94**, 690–706.

Hays D. B. and Vinson S. B. (1971) Acceptance of *Heliothis virescens* (F.) (Lepidoptera, Noctuidae) as a host by the parasite *Cardiochiles nigriceps* Viereck (Hymenoptera, Braconidae). *Anim. Behav.* **19**, 344–352.

Hayslip N. C. (1943) Notes on biological studies of male crickets at Plant City, Florida. *Fla Ent.* **36**, 33–46.

Hayward K. J. (1936) Contribución al conocimiento de la langosta *Schistocerca paranensis* Burm. y sus enemigos naturales. *Mem. Com. cent. Invest. Langosta, 1934*, pp. 217–229.

Hazard E. I., Mayer M. S., and Savage K. E. (1967) Attraction and oviposition stimulation of gravid female mosquitoes by bacteria isolated from hay infusions. *Mosq. News* **27**, 133–136.

Hearle E. (1932) Warble flies and their control in Canada. *Pamphl. Dep. Agric. Can. ent. Br.* (NS) **147**, 1–9.

Heberday R. F. (1931) Zur Entwicklungsgeschichte, vergleichenden Anatomie und Physiologie der weiblichen Geschlechtsausfürwege der Insekten. *Z. Morph. Ökol. Tiere* **22**, 416–586.

Hedden R., Vité J. P., and Mori K. (1976) Synergistic effect of a pheromone and a kairomone on host selection and colonisation by *Ips avulsus*. *Nature, Lond.* **261**, 696–697.

Hegdekar B. M. and Arthur A. P. (1973) Host heamolymph chemicals that induce oviposition in the parasite *Itoplectis conquisitor* (Hymenoptera: Ichneumonidae). *Can. Ent.* **105**, 787–793.

Heidemann O. (1901) Eggs of *Halobates*. In *Papers from the Hopkins Stanford Galapagos Expedition 1898*, p. 362 (see also *Proc. Wash. Acad. Sci.* **3**, 364).

Heie O. E. (1972) Aphids on birch (*Betula*) in Denmark (Hom., Aphidoidea) (Hem.). *Ent. Medd.* **40**, 81–105.

Helms T. J. and Raun E. S. (1971) Perennial laboratory culture of disease-free insects. In *Microbial Control of Insects and Mites* (H. D. Burges and N. W. Hussey, eds.), pp. 639–653.

Helson G. A. H. (1935) The hatching and early instars of *Stenoperla prasina* Newman. *Trans. R. Soc. NZ, Dunedin* **65**, 11–16.

Hemmingsen A. M. (1947) Plant bug guarding eggs and offspring and shooting anal jets (*Physomerus grossipes* F., Coreidae). *Ent. Medd.* **25**, 200.

Hemmingsen A. M. (1952) The oviposition of some crane-fly species (*Tipulidae*) from different types of localities. *Vidensk. Meddr dansk naturh. Foren.* **114**, 366–430.

Hemmingsen A. M. (1956a) Deep-boring ovipository instincts of some crane-fly species (*Tipulidae*) of the subgenera *Vestiplex* Bezzi and *Oreomyza* Pok. and some associated phenomena. *Vidensk. Meddr dansk naturh. Foren.* **118**, 244–315.

HEMMINGSEN A. M. (1956b) Convergent methods of oviposition in short-horned grasshoppers (Acridiidae) and some craneflies (Tipulidae) compared with other types of convergent evolution. *Proc. XIVth Int. Congr. Zool. Copenhagen, 1953.*

HEMMINGSEN A. M. (1960) The function of some remarkable crane-fly ovipositors. *Ent. Medd.* **29**, 221–247.

HEMMINGSEN A. M. (1968) The role of *Trigona trisulcata* (Schummel) (Diptera, Tipulidae, Cylindrotominae) in the adaptive radiation of the Cylindrotominae. *Folia limnol. scand.* **15**, 1–30.

HEMMINGSEN A. M. and JENSEN B. (1960) Relative wing length and abdominal prolongation in some crane-fly species (Tipulidae) with deep-boring instincts. *Vidensk. Meddr dansk naturh. Foren.* **123**, 81–110.

HEMMINGSEN A. M. and JENSEN B. (1972) Egg characteristics and body size in crane-flies (Diptera: Tipulidae), with comparative notes on birds and other organisms. *Vidensk. Meddr dansk naturh. Foren.* **135**, 85–127.

HEMMINGSEN A. M. and NIELSEN B. R. (1971) Species differences in ovipository instincts within the Vermileoninae (Diptera Brachycera, Rhagionidae = Leptidae). *Vidensk. Meddr dansk naturh. Foren.* **134**, 149–203.

HEMMINGSEN A. M. and NØRREVANG A. (1966) Cinematographic and other studies of ovipository mechanisms in crane-flies. *Vidensk. Meddr dansk naturh. Foren.* **129**, 261–274.

HEMMINGSEN A. M. and THEISEN B. F. (1956) The inheritance of terminal egg-filaments in fertile hybrids of *Tipula paludosa* Meigen and *Tipula czizeki* de Jong. *Vidensk. Meddr dansk naturh. Foren.* **118**, 15–32.

HENDRICKS D. E. (1971) Oil-soluble blue dye in larval diet marks adults, eggs, and first-stage F_1 larvae of the pink bollworm. *J. econ. Ent.* **64**, 1404–1406.

HENDRICKS D. E. and GRAHAM H. M. (1970) Oil-soluble dye in larval diet for tagging moths, eggs, and spermatophores of tobacco budworms. *J. econ. Ent.* **63**, 1019–1020.

HENDRICKS D. E., LEAL M. P., ROBINSON S. H., and HERNANDEZ N. S. (1971) Oil-soluble black dye in larval diet marks adults and eggs of tobacco budworm and pink bollworm. *J. econ. Ent.* **64**, 1399–1401.

HENDRICKS D. E. and SHAVER T. N. (1975) Tobacco budworm: male pheromone suppressed emission of sex pheromone by the female. *Envir. Ent.* **4**, 555–558.

HENKING H. (1890) Untersuchungen über die ersten Entwickelungsvorgänge in den Eiern der Insekten: I. Das Ei von *Pieris brassicae* L. nebst Bemerkungen über Samen und Samenbildung. *Z. wiss. Zool.* **49**, 503–564.

HENNEBERG B. (1927) Viviparität bei *Phytodecta rufipes* Fbr. (Coleopt. Chrysomelid.). *Ber. oberhess. Ges. Nat. Heilk. Giessen* **11**, 17–20.

HENNEBERRY T. J. (1963) Effects of gamma radiation on the fertility and longevity of *Drosophila melanogaster*. *J. econ. Ent.* **56**, 279–281.

HENNEBERRY T. J. and KISHABA A. N. (1967) Mating and oviposition of the cabbage looper in the laboratory. *J. econ. Ent.* **60**, 692–696.

HENNEGUY L. F. (1890) Note sur la structure de l'enveloppe de l'oeuf des Phyllies. *Bull. Soc. philom. Paris* **2**, 18–25.

HENRY C. S. (1972) Eggs and rapagula of *Ululodes* and *Ascaloptynx* (Neuroptera: Ascalaphidae): a comparative study. *Psyche, Camb.* **79**, 1–22.

HENSLEY S. D. and HAMMOND A. M. (1968) Laboratory techniques for rearing the sugarcane borer on an artificial diet. *J. econ. Ent.* **61**, 1742–1743.

HEREWEGE J. VAN (1970) Intervention de stimulus olfactifs et gustatifs dans l'oviposition de *Drosophila melanogaster*. *C. r. hebd. Séanc. Acad. Sci. Paris* **271**, 108–110.

HERFS A. (1936) Ökologisch–physiologische Studien an *Anthrenus fasciatus* Herbst. *Zoologica, Stuttgart* **34**, 1–96.

HERMS W. B. (1932) Deterrent effect of artificial light on the codling moth. *Hilgardia* **7**, 263–280.
HERMS W. B. and FREEBORN S. B. (1920) The egg-laying habits of Californian anophelines. *J. Parasit.* **7**, 69–79.
HERMS W. B. and FROST F. M. (1932) A comparative study of the eggs of Californian anophelines. *J. Parasit.* **18**, 240–244.
HERNANDEZ R. F. (1973) Estudios sobre la mosquita blanca *Trialeurodes vaporariorum* (Westwood) en el Estado de Morelos. *Agricultura téc. Méx.* **3**, 165–172 (1972).
HERON R. J. (1968) Vital dyes as markers for behavioral and population studies of the larch sawfly *Pristiphora erichsonii* (Hymenoptera: Tenthredinidae). *Can. Ent.* **100**, 470–475.
HERREBOUT W. M. and VEER J. VAN DER (1969) Habitat selection in *Eucarcelia rutilla* Vill. (Diptera: Tachinidae). III. Preliminary results of olfactometer experiments with females of known age. *Z. angew. Ent.* **64**, 55–61.
HERRICK G. W. (1926) *Insects Injurious to the Household and Annoying to Man*, New York.
HERRICK G. W. and HADLEY C. H. (1916) The lesser migratory locust. *Bull. Cornell agric. Exp. Stn* **378**, 1–45.
HERRING J. L. (1961) The genus *Halobates* (Hemiptera: Gerridae). *Pac. Insects* **3**, 223–305.
HERTEL G. D., HAIN F. P., and ANDERSON R. F. (1969) Response of *Ips grandicollis* (Coleoptera: Scolytidae) to the attractant produced by attacking male beetles. *Can. Ent.* **101**, 1084–1091.
HERTEL R. (1955) Zur Kenntnis der Systematik, Biologie und Morphologie von *Neides tipularius* L. (Heteroptera: Neididae). *Abh. Mus. Tierk.* **22**, 11–183.
HERTER K. (1943) Zur Fortpflanzungsbiologie eines lebendgebärenden Ohrwurmes. *Z. Morph. Ökol. Tiere* **40**, 158–180.
HERTER K. (1960) Zur Fortpflanzungsbiologie des Meerstrand-Ohrwurmes *Anisolabis maritima* (Géné). *Zool. Beitr.* (NF) **5**, 199–239.
HERTER K. (1963) Weitere Beobachtungen über die Fortpflanzungsbiologie des Meerstrand-Ohrwurmes *Anisolabis maritima* (Géné). *Sber. Ges. naturf. Freunde Berlin* (NF) **3**, 103–116.
HERTER K. (1964) Zur Fortpflanzungsbiologie des Ohrwurmes *Forficula pubescens* (Géné). *Zool. Beitr.* (NF) **10**, 1–28.
HESSE A. J. (1938) A revision of the Bombyliidae (Diptera) of southern Africa. I. *Ann. S. Afr. Mus.* **34**, 1–1053.
HETRICK L. A. (1949) The oviposition of the two-striped walkingstick *Anisomorpha buprestoides* (Stoll) (Orthoptera, Phasmidae). *Proc. ent. Soc. Wash.* **51**, 103–104.
HEWER H. R. (1932) Studies in *Zygaena* (Lepidoptera). Part I (A). The female genitalia; (B) The male genitalia. *Proc. zool. Soc. Lond.* **1932**, 33–75.
HEYMER A. (1967) Contribution à l'étude du comportement de ponte du genre *Platycnemis* Burmeister, 1839 (Odonata: Zygoptera). *Z. Tierpsychol.* **24**, 645–650.
HEYMONS R. (1895) *Die Embryonalentwicklung von Dermapteren und Orthopteren*, Jena.
HEYMONS R. (1896) Grundzüge der Entwickelung und des Köperbaues von Odonaten und Ephemeriden. *Abh. preuss. Akad. Wiss. Berlin* **1896**, 1–66.
HEYMONS R. (1929) Über die Biologie der Passaluskafer. *Z. Morph. Ökol. Tiere* **16**, 74–100.
HEYMONS R., LENGERKEN H. VON, and BAYER M. (1932) Studien über die Lebenserscheinungen der Silphini. VIII. *Ablattaria laevigata* F. *Z. Morph. Ökol. Tiere* **24**, 259–287.
HIBBS E. T., CARLSON O. V., and DE LEON R. R. (1969) Selective oviposition of the potato leafhopper on solanaceous plants. *Proc. N. cent. Brch Am. Ass. econ. Ent.* **24**, 25.
HILL A. R. (1961) The biology of *Anthocoris sarothamni* Douglas and Scott in Scotland (Hemiptera: Anthococidae). *Trans. R. ent. Soc. Lond.* **113**, 41–54.
HILL D. L., BELL V. A., and CHADWICK L. E. (1947) Rearing of the blowfly *Phormia regina* Meigen on a sterile synthetic diet. *Ann. ent. Soc. Am.* **40**, 213–216.

HILL M. A. (1947) The life-cycle and habits of *Culicoides impunctatus* Goetghebuer and *Culicoides obsoletus* Meigen, together with some observations on the life cycle of *Culicoides odibilis* Austen, *Culicoides pallidicornis* Kieffer, *Culicoides cubitalis* Edwards, and *Culicoides chiopterus* Meigen. *Ann. trop. Med. Parasit.* **41**, 55–115.

HILL W. R., RUBENSTEIN A. D., and KOVACS J. (1948) Dermatitis resulting from contact with moths (genus *Hylesia*). *J. Am. med. Ass.* **138**, 737–740.

HILLHOUSE T. L. and PITRE H. N. (1976) Oviposition by *Heliothis* on soybeans and cotton. *J. econ. Ent.* **69**, 144–146.

HILLIARD J. R. (1959) The specificity of acridian egg pods and eggs, with biological notes. Unpubl. thesis, University of Texas.

HILLYER R. J. and THORSTEINSON A. J. (1969) The influence of the host plant or males on ovarian development or oviposition in the diamondback moth *Plutella maculipennis* (Curt.). *Can. J. Zool.* **47**, 805–816.

HILLYER R. J. and THORSTEINSON A. J. (1971) Influence of the host plant or males on programming of oviposition in the diamondback moth *Plutella maculipennis* (Curt.) Lepidoptera. *Can. J. Zool.* **49**, 983–990.

HILSENHOFF W. L. (1966) The biology of *Chironomus plumosus* (Diptera, Chironomidae) in Lake Winnebayo, Wisconsin. *Ann. ent. Soc. Am.* **59**, 465–473.

HINCKLEY A. D. (1963) The rice leafroller *Susumia exigua* (Butler) in Fiji. *J. econ. Ent.* **56**, 112–113.

HINCKS W. D. (1948) Preliminary notes on Mauritian earwigs (Dermaptera). *Ann. Mag. nat. Hist.* **14** (11) 517–540.

HINES B. M., HARRIS F. A., and MITLIN N. (1973) *Heliothis virescens*: assimilation and retention of radioactivity in eggs and larvae from ^{32}P-treated adults. *J. econ. Ent.* **66**, 1071–1073.

HINKO E. (1952) Practical method of transporting mosquito eggs. *Higijena (Yugoslavia)* **2** (4) 319–321 (English translation: Defence Sci. Inf. Serv. DRB Canada).

HINKS C. F. (1971) Observations on larval behaviour and avoidance of encapsulation of *Perilampus hyalinus* (Hymenoptera: Perilampidae) parasitic in *Neodiprion lecontei* (Hymenoptera: Diprionidae). *Can. Ent.* **103**, 182–187.

HINMAN E. H. (1932) The presence of bacteria within the eggs of mosquitoes. *Science, NY* **76**, 106–107.

HINTON H. E. (1940) A monographic revision of the Mexican water beetles of the family Elmidae. *Novit. Zool.* **42**, 217–396.

HINTON H. E. (1944) Some general remarks on sub-social beetles, with notes on the biology of the staphylinid *Platystethus arenarius* (Fourcroy). *Proc. R. ent. Soc. Lond.* (A) **19**, 115–128.

HINTON H. E. (1945a) *A Monograph of the Beetles Associated with Stored Products*, **1**, British Museum (NH), London.

HINTON H. E. (1945b) The Histeridae associated with stored products. *Bull. ent. Res.* **35**, 309–340.

HINTON H. E. (1946a) On the homology and nomenclature of the setae of lepidopterous larvae, with some notes on the phylogeny of the Lepidoptera. *Trans. R. ent. Soc. Lond.* **97**, 1–37.

HINTON H. E. (1946b) A new classification of insect pupae. *Proc. zool. Soc. Lond.* **116**, 282–328.

HINTON H. E. (1947) The gills of some aquatic beetle pupae (Coleoptera, Psephenidae). *Proc. R. ent. Soc. Lond.* (A) **22**, 52–60.

HINTON H. E. (1948a) On the origin and function of the pupal stage. *Trans. R. ent. Soc. Lond.* **99**, 395–409.

HINTON H. E. (1948b) A synopsis of the genus *Tribolium* Macleay, with some remarks on the evolution of its species groups (Coleoptera, Tenebrionidae). *Bull. ent. Res.* **39**, 13–56.

HINTON H. E. (1951) Myrmecophilous Lycaenidae and other Lepidoptera—a summary. *Proc. S. Lond. ent. nat. Hist. Soc.* **1949–50**, 111–175.

HINTON H. E. (1953) Some adaptations of insects to environments that are alternately dry and flooded, with some notes on the habits of the Stratiomyidae. *Trans. Soc. Br. Ent.* **11**, 209–227.

HINTON H. E. (1954a) On the structure and function of the respiratory horns of the pupae of the genus *Pseudolimnophila* (Diptera: Tipulidae). *Proc. R. ent. Soc. Lond.* (A) **29**, 135–140.

HINTON H. E. (1954b) Resistance of the dry eggs of *Artemia salina* L. to high temperatures. *Ann. Mag. nat. Hist.* **7** (12) 158–160.

HINTON H. E. (1955a) On the respiratory adaptations, biology, and taxonomy of the Psephenidae, with notes on some related families (Coleoptera). *Proc. zool. Soc. Lond.* **125**, 543–568.

HINTON H. E. (1955b) On the structure, function, and distribution of the prolegs of the Panorpoidea, with a criticism of the Berlese–Imms theory. *Trans. R. ent. Soc. Lond.* **106**, 455–556.

HINTON H. E. (1955c) Protective devices of endopterygote pupae. *Trans. Soc. Br. Ent.* **12**, 49–92.

HINTON H. E. (1957a) The structure and function of the spiracular gill of the fly *Taphrophila vitripennis*. *Proc. R. Soc.* (B) **147**, 90–120.

HINTON H. E. (1957b) Some little known respiratory adaptations. *Sci. Progr. Lond.* **45**, 692–700.

HINTON H. E. (1957c) Biological control of pests. Some considerations. *Sci. Progr. Lond.* **45**, 11–26.

HINTON H. E. (1957d) Some aspects of diapause. *Sci. Progr. Lond.* **45**, 307–320.

HINTON H. E. (1958) The phylogeny of the Panorpoid orders. *A. Rev. Ent.* **3**, 181–206.

HINTON H. E. (1959a) Plastron respiration in the eggs of *Drosophila* and other flies. *Nature, Lond.* **184**, 280–281.

HINTON H. E. (1959b) Origin of indirect flight muscles in primitive flies. *Nature, Lond.* **183**, 557–558.

HINTON H. E. (1960a) The structure and function of the respiratory horns of the eggs of some flies. *Phil. Trans. R. Soc.* (B) **243**, 45–73.

HINTON H. E. (1960b) The chorionic plastron and its role in the eggs of the Muscinae (Diptera). *Q. Jl microsc. Sci.* **101**, 313–332.

HINTON H. E. (1960c) Plastron respiration in the eggs of blowflies. *J. Insect Physiol.* **4**, 176–183.

HINTON H. E. (1961a) How some insects, especially the egg stages, avoid drowning when it rains. *Proc. S. Lond. ent. nat. Hist. Soc.* **1960**, 138–154.

HINTON H. E. (1961b) The structure and function of the egg-shell in the Nepidae (Hemiptera). *J. Insect Physiol.* **7**, 224–257.

HINTON H. E. (1962a) A key to the eggs of the Nepidae (Hemiptera). *Proc. R. ent. Soc.* (A) **37**, 65–68.

HINTON H. E. (1962b) The structure of the shell and respiratory system of the eggs of *Helopeltis* and related genera (Hemiptera, Miridae). *Proc. zool. Soc. Lond.* **139**, 482–488.

HINTON H. E. (1962c) The fine structure and biology of the egg-shell of the wheat bulb fly *Leptohylemyia coarctata*. *Q. Jl microsc. Sci.* **103**, 243–251.

HINTON H. E. (1962d) Respiratory systems of insect egg-shells. *Sci. Prog.* **50**, 96–113.

HINTON H. E. (1963a) The origin and function of the pupal stage. *Proc. R. ent. Soc. Lond.* (A) **38**, 77–85.

HINTON H. E. (1963b) The respiratory system of the egg-shell of the blowfly *Calliphora erythrocephala* Meig. as seen with the electron microscope. *J. Insect Physiol.* **9**, 121–129.

HINTON H. E. (1964a) The respiratory efficiency of the spiracular gill of *Simulium*. *J. Insect Physiol.* **10**, 73–80.

HINTON H. E. (1964b) Sperm transfer in insects and the evolution of haemocoelic insemination. *Symp. R. ent. Soc. Lond.* **2**, 95–107.

HINTON H. E. (1966a) Respiratory adaptations of the pupae of beetles of the family Psephenidae. *Phil. Trans. R. Soc.* (B) **251**, 211–245.
HINTON H. E. (1966b) The spiracular gill of the fly *Eutanyderus* (Tanyderidae). *Aust. J. Zool.* **14**, 365–369.
HINTON H. E. (1967) The respiratory system of the egg-shell of the common housefly. *J. Insect Physiol.* **13**, 647–651.
HINTON H. E. (1968a) Structure and protective devices of the egg of the mosquito *Culex pipiens. J. Insect Physiol.* **14**, 145–161.
HINTON H. E. (1968b) Observations on the biology and taxonomy of the eggs of *Anopheles* mosquitoes. *Bull. ent. Res.* **57**, 495–508.
HINTON H. E. (1968c) Reversible suspension of metabolism and the origin of life. *Proc. R. Soc.* (B) **171**, 43–47.
HINTON H. E. (1968d) Spiracular gills. *Adv. Insect Physiol.* **5**, 65–162.
HINTON H. E. (1969) Respiratory systems of insect egg shells. *A. Rev. Ent.* **14**, 343–368.
HINTON H. E. (1970a) Insect eggshells. *Scient. Am.* **223** (2) 84–91.
HINTON H. E. (1970b) Algunas pequeñas estructuras de insectos observadas con microscopio electrónico explorador. *Acta politec. méx.* **10**, 181–201 (1969).
HINTON H. E. (1970c) Some structures of insects as seen with the scanning electron microscope. *Micron* **1**, 84–108 (1969).
HINTON H. E. (1971a) Plastron respiration in the mite *Platyseius italicus. J. Insect Physiol.* **17**, 1185–1199.
HINTON H. E. (1971b) Polyphyletic evolution of respiratory systems of eggshells, with a discussion of structure and density-independent and density-dependent selective pressures. In *Scanning Electron Microscopy* (V. H. Heywood, ed.), pp. 17–36, London.
HINTON H. E. (1971c) Some neglected phases in metamorphosis. *Proc. R. ent. Soc. Lond.* (C) **35**, 55–64.
HINTON H. E. (1973) Neglected phases in metamorphosis: a reply to V. B. Wigglesworth. *J. Ent.* (A) **48**, 57–68.
HINTON H. E. (1974) Accessory functions of seminal fluid. *J. med. Ent.* **11**, 19–25.
HINTON H. E. (1976a) Maternal care in the Membracidae. *Proc. R. ent. Soc. Lond.* (C) **40**, 33.
HINTON H. E. (1976b) Respiratory adaptations in marine insects. In *Marine Insects* (L. Cheng, ed.), pp. 43–78, North-Holland.
HINTON H. E. (1976c) Enabling mechanisms. *Proc. 15th Int. Congr. Ent. Washington*, pp. 71–83.
HINTON H. E. (1976d) Notes on neglected phases in metamorphosis, and a reply to J. M. Whitten. *Ann. ent. Soc. Am.* **69**, 560–566.
HINTON H. E. (1977a) Subsocial behaviour and biology of some Mexican membracid bugs. *Ecol. Ent.* **2**, 61–79.
HINTON H. E. (1977b) Functions of shell structures of pig louse and how egg maintains a low equilibrium temperature in direct sunlight. *J. Insect Physiol.* **23**, 785–800.
HINTON H. E. and COLE S. (1965) The structure of the egg-shell of the cabbage root fly *Erioischia brassicae. Ann. appl. Biol.* **56**, 1–6.
HINTON H. E. and SERVICE M. W. (1969) The surface structure of aedine eggs as seen with the scanning electron microscope. *Ann. trop. Med. Parasit.* **63**, 409–411.
HINTON T. (1955) The genetic basis of a nutritional requirement in *Drosophila. Genetics* **40**, 224–234.
HIRAKOSO S. (1964) Studies on oviposition site of *Culex pipiens* (L.). *Jap. J. sanit. Zool.* **15**, 150–165 (in Japanese; English summary).
HIRATA S. (1956) On the bimodality of oviposition curve observed in the first brood of cabbage armyworm *Barathra brassicae* L. *Ōyō-Kontyū, Tokyo* **12**, 195–201 (in Japanese; English summary).

HIRATA S. (1962) On the phase variation of cabbage armyworm *Mamestra (Barathra) brassicae* (L.). VI. Phase variation in larval characters under different conditions of density and temperature. *Jap. J. Ecol.* **12**, 223–228.

HIROSE Y., KIMOTO H., and HIEHATA K. (1976) The effect of host aggregation on parasitism by *Trichogramma papilionis* Nagarkatti (Hymenoptera: Trichogrammatidae) as egg parasitoid of *Papilio xuthus* Linné (Lepidoptera: Papilionidae). *Appl. Ent. Zoo.* **11**, 116–125.

HISS E. A. (1971) The biochemical basis of female monogamy in *Aedes aegypti* (L.), PhD thesis, University of Notre Dame, Indiana, USA.

HISS E. A. and FUCHS M. S. (1972) The effect of matrone on oviposition in the mosquito *Aedes aegypti*. *J. Insect Physiol.* **18**, 2217–2227.

HITCHCOCK S. W. (1961) Egg parasitism and larval habits of the orange-striped oakworm. *J. econ. Ent.* **54**, 502–503.

HO F. K. and DAWSON P. S. (1966) Egg cannibalism by *Tribolium* larvae. *Ecology* **47**, 318–322.

HOBBY B. M. and SMITH K. G. V. (1962) The larva of the viviparous fly *Ocydromia glabricula* (Fln.) (Dipt., Empididae). *Entomologist's mon. Mag.* **98**, 49–50.

HOCHMUT R. (1959) A contribution to the study of the morphology, bionomics and population dynamics of *Archips crataeganus*. *Práce výzk. Úst. lesn. ČSR* **16**, 23–58 (in Czech; summary in German).

HOCKING H. (1968) Studies on the biology of *Rhyssa persuasoria* (L.) (Hymenoptera: Ichneumonidae) incorporating an X-ray technique. *J. Aust. ent. Soc.* **7**, 1–5.

HODEK I. (1970) Coccinellids and the modern pest management. *BioScience* **20**, 543–552.

HODEK I. (1973) *Biology of Coccinellidae with Keys for Identification of Larvae by Co-authors*, The Hague, Netherlands.

HODGES J. D. and PICKARD L. S. (1971) Lightning in the ecology of the southern pine beetle *Dendroctonus frontalis* (Coleoptera: Scolytidae). *Can. Ent.* **103**, 44–51.

HODJAT S. H. (1970) Effects of crowding on colour, size and larval activity of *Spodoptera littoralis* (Lepidoptera: Noctuidae). *Entomologia exp. appl.* **13**, 97–106.

HODSON A. C. and WEINMAN C. J. (1945) Factors affecting recovery from diapause and hatching of eggs of the forest tent caterpillar *Malacosoma disstria* Hbn. *Tech. Bull. Minn. agric. Exp. Stn* **170**, 1–31.

HOFFMAN B. L. and KILLINGWORTH B. F. (1967) The egg-laying habits of *Aedes aegypti* (Linnaeus) in central Texas. *Mosq. News* **27**, 466–469.

HOFFMAN C. H. (1936) Notes on *Climaciella brunnea* var. *occidentalis* Banks. *Bull. Brooklyn ent. Soc.* **31**, 202–203.

HOFFMAN J. D., ERTLE L. R., BROWN J. B., and LAWSON F. R. (1970) Techniques for collecting, holding, and determining parasitism of lepidopterous eggs. *J. econ. Ent.* **63**, 1367–1369.

HOFFMAN J. D., IGNOFFO C. M., and DICKERSON W. A. (1975) *In vitro* rearing of the endoparasitic wasp *Trichogramma pretiosum*. *Ann. ent. Soc. Am.* **68**, 335–336.

HOFFMAN J. D., LONG S. H., and DICKERSON W. A. (1968) Teflon and cornstarch-coated paper as substrates for oviposition by cabbage looper moths. *J. econ. Ent.* **61**, 1760.

HOFFMANN W. E. (1936a) The life history of *Limnogonus fossarum* (Fabr.) in Canton (Hemiptera: Gerridae). *Lingnam Sci. J.* **15**, 289–299.

HOFFMANN W. E. (1936b) Life history notes on *Rhagadotarsus kraepelini* Breddin (Hemiptera: Gerridae) in Canton. *Lingnam Sci. J.* **15**, 477–482.

HOFFMEYER S. (1970) Dispersal of *Pachythelia* species (Lep.: Psychidae). *Entomologist's Rec. J. Var.* **82**, 33.

HOGAN T. W. (1967) The influence of diapause on the resistance to desiccation of eggs of *Teleogryllus commodus* (Walk.) (Orthoptera: Gryllidae). *Proc. R. Soc. Vict.* **80**, 37–41.

HOHORST W. (1938) Die Mallophagen des Haushuhnes und ihre Eigelege. *Vet.-med. Nachr.* **4**, 1–88.

HOKAMA Y. and JUDSON C. L. (1963) A new bleaching technique with possible general use in entomology. *Ann. ent. Soc. Am.* **56**, 407–408.

HOKYO N., KIRITANI K., NAKASUJI F., and SHIGA M. (1966) Comparative biology of the two scelionid egg parasites of *Nezara viridula* L. (Hemiptera: Pentatomidae). *Appl. Ent. Zool.* **1**, 94–102.

HOLDAWAY F. G. (1932) An experimental study of the growth of populations of the "flour beetle" *Tribolium confusum* Duval as affected by atmospheric moisture. *Ecol. Monogr.* **2**, 263–304.

HÖLLER G. (1962) Untersuchungen zur Ökologie der Attagenuskäfer. *Z. angew. Ent.* **50**, 46–52.

HOLLING C. S. (1961) Principles of insect predation. *Rev. Ent.* **6**, 163–182.

HOLLINGER A. H. and PARKS H. B. (1919) *Euclemensia bassettella* (Clemens), the *Kermes* parasite. *Ent. News* **30**, 91–100.

HOLST CHRISTENSEN P. J. (1937) Zur Histologie und Embryologie der über winterten Eier von *Orgyia antiqua* L. *Zool. Jb.* **62**, 567–582.

HOLT G. G. and NORTH D. T. (1970) Effects of gamma irradiation on the mechanisms of sperm transfer in *Trichoplusia ni*. *J. Insect Physiol.* **16**, 2211–2222.

HONDA K. (1971) Some observations on the ovipositing habit and the larval food plants of *Polygonia c-aureum* Linné. *Butterfl. Moths* **22**, 49–52.

HOOD J. D. (1950) Thrips that "talk". *Proc. ent. Soc. Wash.* **52**, 42–43.

HOOPER R. L., PITTS C. W., and WESTFALL J. A. (1972) Sense organs on the ovipositor of the face fly *Musca autumnalis*. *Ann. ent. Soc. Am.* **65**, 577–586.

HOPKINS A. D. (1909) Contributions toward a monograph of the scolytid beetles. I. The genus *Dendroctonus*. *Tech. Ser. US Bur. Ent.* **17**, 1–164.

HOPKINS A. D. (1915) Contributions toward a monograph of the scolytid beetles. II. Preliminary classification of the superfamily Scolytoidea. *Tech. Ser. US Bur. Ent.* **17**, 165–232.

HOPKINS C. R. (1964) The histochemistry and fine structure of the accessory nuclei in the oocyte of *Bombus terrestris*. *Q. Jl microsc. Sci.* **105**, 475–480.

HOPKINS C. R. and KING P. E. (1964a) Occurrence of microvilli and micropinocytosis in trophocyte of *Bombus*. *Nature, Lond.* **204**, 298–299.

HOPKINS C. R. and KING P. E. (1964b) Egg resorption in *Nansonia vitripennis* (Walker) (Hymenoptera: Pteromalidae). *Proc. R. ent. Soc. Lond.* (A) **39**, 101–107.

HOPKINS C. R. and KING P. E. (1966) An electron-microscopical and histological study of the oocyte periphery in *Bombus terrestris* during vitellogenesis. *J. Cell Sci.* **1**, 201–216.

HOPKINS D. E. (1970) *In vitro* colonization of the sheep biting louse, *Bovicola ovis*. *Ann. ent. Soc. Am.* **63**, 1196–1197.

HOPKINS D. E. and CHAMBERLAIN W. F. (1972) *In vitro* colonization of the cattle biting louse *Bovicola bovis*. *Ann. ent. Soc. Am.* **65**, 771–772.

HORMCHAN P., SCHUSTER M. F., and HEPNER L. W. (1976) Biology of *Tropiconabis capsiformis*. *Ann. ent. Soc. Am.* **69**, 1016–1018.

HORN D. J. (1970) Oviposition behaviour of *Tetrastichus incertus*, a parasite of the alfalfa weevil. *J. econ. Ent.* **63**, 303–304.

HORSBURGH R. L. and ASQUITH D. (1968) The eggs and oviposition sites of *Hyaliodes vitripennis* on apple trees (Miridae: Hemiptera). *Can. Ent.* **100**, 199–201.

HORSBURGH R. L. and ASQUITH D. (1970) The eggs and oviposition sites of *Diaphnidia capitata* (Hemiptera: Miridae) on apple trees. *Can. Ent.* **102**, 1316–1319.

HORSFALL W. R. (1943) Biology and control of common blister beetles in Arkansas. *Bull. Ark. agric. Exp. Stn* **436**, 1–55.

HORSFALL W. R. (1949) Hatching eggs of floodwater mosquitoes in media that promote plant growth. *Science, Lancaster Pa,* **110**, 504–506.

Horsfall W. R. (1956) Eggs of floodwater mosquitoes. III (Diptera, Culicidae). Conditioning and hatching of *Aedes vexans. Ann. ent. Soc. Am.* **49**, 66–71.

Horsfall W. R. and Anderson J. F. (1964) Thermal stress and anomalous development of mosquitoes (Diptera: Culicidae). II. Effect of alternating temperatures on dimorphism of adults of *Aedes stimulans. J. exp. Zool.* **156**, 61–89.

Horsfall W. R., Anderson J. F., and Brust R. A. (1964) Thermal stress and anomalous development of mosquitoes (Diptera: Culicidae) III. *Aedes sierrensis. Can. Ent.* **96**, 1369–1372.

Horsfall W. R. and Craig G. B. (1956) Eggs of floodwater mosquitoes. IV. Species of *Aedes* common in Illinois (Diptera: Culicidae). *Ann. ent. Soc. Am.* **49**, 368–374.

Horsfall W. R. and Craig G. B. (1960) Eggs of floodwater mosquitoes. VII. Species of *Aedes* common in the southeastern United States (Diptera: Culicidae). *Ann. ent. Soc. Am.* **53**, 11–18.

Horsfall W. R. and Fowler H. W. (1961) Eggs of floodwater mosquitoes. VIII. Effect of serial temperatures on conditioning of eggs of *Aedes stimulans* Walker (Diptera: Culicidae). *Ann. ent. Soc. Am.* **54**, 664–666.

Horsfall W. R., Lum P. T. M., and Henderson L. M. (1958) Eggs of floodwater mosquitoes (Diptera: Culicidae). V. Effect of oxygen on hatching of intact eggs. *Ann. ent. Soc. Am.* **51**, 209–213.

Horsfall W. R., Miles R. C., and Sokatch J. T. (1952) Eggs of floodwater mosquitoes. I. Species of *Psorophora* (Diptera: Culicidae). *Ann. ent. Soc. Am.* **45**, 618–624.

Horsfall W. R. and Trpiš M. (1967) Eggs of floodwater mosquitoes. X. Conditioning and hatching of winterized eggs of *Aedes sticticus* (Diptera: Culicidae). *Ann. ent. Soc. Am.* **60**, 1021–1025.

Horsfall W. R. and Voorhees F. R. (1972) Eggs of floodwater mosquitoes. XIV. Northern *Aedes* (Diptera: Culicidae). *Ann. ent. Soc. Am.* **65**, 123–126.

Horton J. R. (1922) A swallow-tail butterfly injurious to California orange trees. *Mon. Bull. Dep. Agric. Calif.* **11**, 377–387.

Hosny M. M. and Kotby F. A. (1960) Host-plants favoured by the cotton leaf-worm moth *Prodenia litura* F. for egg-laying, and their value as egg-mass trap-crops (Lepidoptera: Agrotidae–Zenobiinae). *Bull. Soc. ent. Égypte* **44**, 223–234.

Hosoi M. (1939) The life-history of *Hydrous acuminatus* Motschulsky. *Bot. Zool. Tokyo* **7**, 1867–1874 (in Japanese).

Hosoi T. (1954) Egg production in *Culex pipiens pallens* Coquillett. IV. Influence of breeding conditions on wing length, body weight and follicle production. *Jap. J. med. Sci. Biol.* **7**, 129–134.

Houllier M. (1965) Influence de l'éclairement du cacaoyer sur le comportement de ponte du miridé *Sahlbergella singularis* Hagl. *Rev. Path. vég. Ent. agric. Fr.* **43**, 209–211 (1964).

Houser J. S. and Balduf W. V. (1926) The striped cucumber beetle. *Bull. Ohio agric. Exp. Stn* **388**, 241–264.

Houseweart M. W. and Kulman H. M. (1976) Fecundity and parthenogenesis of the yellowheaded spruce sawfly *Pikonema alaskensis. Ann. ent. Soc. Am.* **69**, 748–750.

Houyez P. (1970) Technique: comment conserver les oeufs des Lépidoptères. *Lambillionea* **70**, 22–23.

Hovanitz W. and Chang V. C. S. (1964) Adult oviposition responses in *Pieris rapae. J. Res. Lepidoptera* **3**, 159–172.

Howard L. O. (1896) A coleopterous enemy of *Corydalis cornutus. Proc. ent. Soc. Wash.* **3**, 310–313.

Howe R. W. (1950) The development of *Rhizopertha dominica* (F.) (Col., Bostrichidae) under constant conditions. *Entomologist's mon. Mag.* **86**, 1–5.

Howe R. W. (1952) The biology of the rice weevil *Calandra oryzae* (L.). *Ann. appl. Biol.* **39**, 168–180.

Howe R. W. (1955) Studies on beetles of the family Ptinidae. 12. The biology of *Tipnus unicolor* Pill. and Mitt. *Entomologist's mon. Mag.* **91**, 253–257.

Howe R. W. (1957) A laboratory study of the cigarette beetle *Lasioderma serricorne* (F.) (Col., Anobiidae) with a critical review of the literature on its biology. *Bull. ent. Res.* **48**, 9–56.

Howe R. W. (1959) Studies on beetles of the family Ptinidae. XVII. Conclusions and additional remarks. *Bull. ent. Res.* **50**, 287–326.

Howe R. W. (1960) The effects of temperature and humidity on the rate of development and the mortality of *Tribolium confusum* Duval (Coleoptera, Tenebrionidae). *Ann. appl. Biol.* **48**, 363–376.

Howe R. W. (1962) The effects of temperature and humidity on the oviposition rate of *Tribolium castaneum* (Hbst.) (Coleoptera, Tenebrionidae). *Bull. ent. Res.* **53**, 301–310.

Howe R. W. and Bull J. O. (1956) Studies on beetle of the family Ptinidae. 13. The oviposition rate of *Pseudeurostus hilleri* (Reittl). *Entomologist's mon. Mag.* **92**, 113–115.

Howe R. W. and Burges H. D. (1951) Studies on beetles of the family Ptinidae. VI. The biology of *Ptinus fur* (L.), *P. sexpunctatus* Panzer. *Bull. ent. Res.* **42**, 499–511.

Howe R. W. and Burges H. D. (1952) Studies on beetles of the family Ptinidae. VII. The biology of five ptinid species found in stored products. *Bull. ent. Res.* **43**, 153–186.

Howe R. W. and Burges H. D. (1953a) Studies on beetles of the family Ptinidae. IX. A laboratory study of the biology of *Ptinus tectus* Boield. *Bull. ent. Res.* **44**, 461–516.

Howe R. W. and Burges H. D. (1953b) Studies on beetles of the family Ptinidae. 10. The biology of *Mezium affine* Boieldieu. *Entomologist's mon. Mag.* **89**, 217–220.

Howe R. W. and Currie J. E. (1964) Some laboratory observations on the rates of development, mortality and oviposition of several species of Bruchidae breeding in stored pulses. *Bull. ent. Res.* **55**, 437–477.

Howe W. L. and Zdarkova E. (1971) A simple method for continuous rearing of the striped cucumber beetle. *J. econ. Ent.* **64**, 1337.

Howell J. F. and Clift A. E. (1972) Rearing codling moths on an artificial diet in trays. *J. econ. Ent.* **65**, 888–890.

Hower A. A. Jr and Ferrer F. R. (1970) An artificial oviposition technique for the alfalfa weevil. *J. econ. Ent.* **63**, 761–764.

Hoyt C. P. (1963) Investigations of rhinoceros beetles in West Africa. *Pacif. Sci.* **17**, 444–451.

Hrdý I., Jasič J., Novák K., Růžička Z., Vallo V., Weismann L., and Zelený J. (1969) The sugar-cane borer *Diatraea saccharalis* (Lepidoptera, Pyralidae) in Cuba. II. Natural limitation, biological control and contribution to the methods of population density estimation. *Acta ent. bohemoslov.* **66**, 255–265.

Hsiao T. H. and Fraenkel G. (1968a) The role of secondary plant substances in the food specificity of the Colorado potato beetle. *Ann. ent. Soc. Am.* **61**, 485–493.

Hsiao T. H. and Fraenkel G. (1968b) Selection and specificity of the Colorado potato beetle for solanaceous and nonsolanaceous plants. *Ann. ent. Soc. Am.* **61**, 493–503.

Hsiao T. H., Holdaway F. G., and Chiang H. C. (1966) Ecological and physiological adaptations in insect parasitism. *Entomologia exp. appl.* **9**, 113–123.

Hubbard H. C. (1897) The ambrosia beetles of the United States. *Bull. US Bur. Ent.* **7**, 9–35.

Hudon M. and Perron J. P. (1957) Méthode pour l'obtention et la manipulation de grandes quantités de masses d'oeufs de la pyrale du maïs *Pyrausta nubilalis* (Hbn.) (Lépidoptères: Pyralidae). *Ann. ent. Soc. Quebec* **2**, 76–80 (1956).

Hudson A. (1970) Factors affecting egg maturation and oviposition by autogenous *Aedes atropalpus* (Diptera: Culicidae). *Can. Ent.* **102**, 939–949.

Hudson A. and McLintock J. (1967) A chemical factor that stimulates oviposition by *Culex*

tarsalis Coquillet (Diptera: Culicidae). *Anim. Behav.* **15**, 336–341.
HUDSON G. V. (1928) *The Butterflies and Moths of New Zealand*, Wellington.
HUGHES P. R. (1973a) Effect of α-pinene exposure on *trans*-verbenol synthesis in *Dendroctonus ponderosae* Hopk. *Naturwissenschaften* **60** (5) 261–262.
HUGHES P. R. (1973b) *Dendroctonus*: production of pheromones and related compounds in response to host monoterpenes. *Z. angew. Ent.* **73**, 294–312.
HUGHES P. R. (1974) Myrcene: a precursor of pheromones in *Ips* beetles. *J. Insect Physiol.* **20**, 1271–1275.
HUGHES P. R. (1975) Pheromones of *Dendroctonus*: origin of α-pinene oxidation products present in emergent adults. *J. Insect Physiol.* **21**, 687–691.
HUGHES P. R. (1976) Response of female southern pine beetles to the aggregation pheromone frontalin. *Z. angew. Ent.* **80**, 280–284.
HUGHES R. D. (1959) The natural mortality of *Erioischia brassicae* (Bouché) (Diptera, Anthomyiidae) during the egg stage of the first generation. *J. Anim. Ecol.* **28**, 343–357.
HUIE L. H. (1916a) The bionomics of the tiger-beetle (*Cicindela campestris*). *Proc. R. phys. Soc. Edinb.* **20**, 1–11.
HUIE L. H. (1916b) Observations on the hatching of *Stenopsocus cruciatus*. *Scot. Nat.* **1916**, 61–65.
HUIGNARD J. (1971) Variations de l'activité reproductrice des males d'*Acanthoscelides obtectus*. *J. Insect Physiol.* **17**, 1245–1255.
HUNGERFORD H. B. (1917) The egg laying habits of a backswimmer, *Buenoa margaritacea*. *Ent. News* **38**, 177–183.
HUNGERFORD H. B. (1918) Concerning the oviposition of *Notonecta*. *Ent. News* **29**, 241–245.
HUNGERFORD H. B. (1920) The biology and ecology of aquatic and semiaquatic Hemiptera. *Kans. Univ. Sci. Bull.* **11**, 1–341 (1919).
HUNGERFORD H. B. (1922) The life history of the toad bug *Gelastocoris oculatus* Fabr. *Kans. Univ. Sci. Bull.* **14**, 145–171.
HUNGERFORD H. B. (1923) Notes on the eggs of Corixidae. *Bull. Brooklyn ent. Soc.* **18**, 13–16.
HUNGERFORD H. B. (1925) *Lethocerus americanus*, habits. *Psyche, Camb.* **32**, 88–91.
HUNGERFORD H. B. (1931) Concerning the egg of *Polystoechotes punctatus* Fabr. *Bull. Brooklyn ent. Soc.* **26**, 22–23.
HUNGERFORD H. B. (1933) The genus *Notonecta* of the World. *Kans. Univ. Sci. Bull.* **34**, 5–195.
HUNGERFORD H. B. (1936) The Mantispidae of the Douglas Lake, Michigan Region, with some biological observations (Neurop.). *Ent. News* **47**, 69–72, 85–88.
HUNGERFORD H. B. (1948) The eggs of Corixidae (Hemiptera). *J. Kans. ent. Soc.* **21**, 141–146.
HUNT T. N. and FARRIER M. H. (1974) Oviposition and feeding preference of pales weevil (Coleoptera: Curculionidae) for five types of loblolly pine bark. *Ann. ent. Soc. Am.* **67**, 407–408.
HUNTER P. E. and MOSER J. C. (1968) *Pseudoparasitus thatcheri* n.sp. (Acarina: Dermanyssidae, Laelapinae) associated with southern pine beetles (*Dendroctonus frontalis* Zimm. in *Pinus taeda* in Louisiana). *Fla Ent.* **51**, 119–123.
HUNTER-JONES P. (1964) Egg development in the desert locust (*Schistocerca gregaria* Forsk.) in relation to the availability of water. *Proc. R. ent. Soc. Lond.* (A) **39**, 25–33.
HUNTER-JONES P. and LAMBERT J. (1961) Egg development of *Humbe tenuicornis* Schaum (Orthoptera: Acrididae) in relation to availability of water. *Proc. R. ent. Soc. Lond.* (A) **36**, 75–80.
HUNTER-JONES P. and WARD V. K. (1960) The life-history of *Gastrimargus africanus* Saussure (Orth., Acrididae) in the laboratory. *Entomologist's mon. Mag.* **95**, 169–172.
HURLBUT H. S. (1938) Further notes on the overwintering of the eggs of *Anopheles walkeri* Theobald, with a description of the eggs. *J. Parisit.* **24**, 521–526.

HURPIN B. (1969) Sur la biologie de *Oryctes boas* F. *Oléagineux* **24**, 673–678.
HURPIN B. and FRESNEAU M. (1969) Contribution à l'étude de *Oryctes elegans* (Col. Dynastidae). *Annls Soc. ent. Fr.* (NS) **5**, 595–612.
HURPIN B. and MARIAU D. (1966) Contribution à la lutte contre les *Oryctes* nuisibles aux palmiers. Mise au point d'un élevage permanent en laboratoire. *C. r. hebd. Séanc. Acad. Agric. Fr.* **52**, 178–186.
HURTER J., KATSOYANNOS B., BOLLER E. F., and WIRZ P. (1976) Beitrag zur Anreicherung und teilweisen Reinigung des eiablageverhindernden Pheromons der Kirschenfliege, *Rhagoletis cerasi* L. (Dipt., Trypetidae). *Z. angew. Ent.* **80**, 50–56.
HUSAIN M. A. and ROONWAL M. L. (1933) Studies on *Schistocerca gregaria* Forsk. I. The micropyle in *Schistocerca gregaria* and some other Acrididae. *Indian J. agric. Sci.* **3**, 639–645.
HUSBANDS R. C. (1952) Some techniques used in the study of *Aedes* eggs in irrigated pastures in California. *Mosq. News* **12**, 145–150.
HUSSEINY M. M. and MADSEN H. F. (1964) Sterilization of the navel orangeworm *Paramyelois transitella* (Walker) by gamma radiation (Lepidoptera: Phycitidae). *Hilgardia* **36**, 113–117.
HUSSEY N. W. (1957) Effects of the physical environment on the development of the pine looper *Bupalus piniarius*. *Rep. For. Res.* **1956–7**, 111–128.
HUSSEY N. W. (1960) Biology of mushroom phorids. *Mushroom Sci.* **4**, 260–270.
HUSSEY R. F. (1934) Observations on *Pachycoris torridus* (Scop.), with remarks on parental care in other Hemiptera. *Bull. Brooklyn ent. Soc.* **29**, 133–145.
HYDE G. E. (1961) Egg-laying habits of British moths. *New Scient.* **9**, 158–160.
HYNES H. B. N. (1941) The taxonomy and ecology of the nymphs of British Plecoptera with notes on the adults and eggs. *Trans. R. ent. Soc. Lond.* **91**, 459–557.
HYNES H. B. N. (1947) Observations on *Systoechus somali* (Diptera, Bombyliidae) attacking the eggs of the desert locust (*Schistocerca gregaria* (Forskål)) in Somalia. *Proc. R. ent. Soc. Lond.* (A) **22**, 79–85.
HYNES H. B. N. (1952) The Neoperlinae of the Ethiopian Region (Plecoptera, Perlidae). *Trans. R. ent. Soc. Lond.* **103**, 85–108.
HYNES H. B. N. (1970) *The Ecology of Running Waters*, Liverpool University Press.
HYNES H. B. N. (1974) Observations on the adults and eggs of Australian Plecoptera. *Aust. J. Zool.* (Suppl. Ser.) **29**, 37–52.
HYNES H. B. N., WILLIAMS T. R., and KERSHAW W. E. (1961) Freshwater crabs and *Simulium neavei* in East Africa. I. Preliminary observations made on the slopes of Mount Elgon in December 1960 and January 1961. *Ann. trop. Med. Parasit.* **55**, 197–201.

IBBOTSON A. (1960) Observations on the oviposition behaviour of frit fly (*Oscinella frit* L. Dipt., Chloropidae). *Entomologia exp. appl.* **3**, 84–92.
IBBOTSON A. and EDWARDS C. A. T. (1954) The biology and control of *Otiorrhynchus clavipes* Bonsd. (Rhyn., Coleop.), a pest of strawberries. *Ann. appl. Biol.* **41**, 520–535.
IBRAHIM M. M. (1970a) Studies on the ecology and biology of *Pyrgomorpha conica* Olivier (Orthoptera: Pyrgomorphidae). *Bull. Soc. ent. Égypte* **53**, 137–146 (1969).
IBRAHIM M. M. (1970b) Biological studies on the immature forms of *Pyrgomorpha conica* Olivier (Orthoptera: Pyrgomorphidae). *Bull. Soc. ent. Égypte* **53**, 147–160 (1969).
IDRIS B. E. M. (1960a) Die Entwicklung im geschnürten Ei von *Culex pipiens* L. (Diptera). *Arch. EntwMech. Organ.* **152**, 230–262.
IDRIS B. E. M. (1960b) Die Entwicklung im normalen Ei von *Culex pipiens* L. (Diptera). *Z. Morph. Ökol. Tiere* **49**, 387–429.
IGNOFFO C. M. (1963) A successful technique for mass-rearing cabbage loopers on a semisynthetic diet. *Ann. ent. Soc. Am.* **56**, 178–182.

IGNOFFO C. M. (1964) Production and virulence of a nuclear-polyhedrosis virus from larvae of *Trichoplusia ni* (Hübner) reared on a semisynthetic diet. *J. Insect Pathol.* **6**, 318–326.

IHERING R. VON (1909) As especies brasileiras do gen *Phloea*. *Ent. brasil* **2**, 129–133.

IJIMA H. (1930) Habits of "Kawatobi-Kera". *Kontyû Tokyo* **4**, 19–22 (in Japanese).

IKESHOJI T. (1966a) Studies on mosquito attractants and stimulants. Part 1. Chemical factors determining the choice of oviposition site by *Culex pipiens fatigans* and *pallens*. *Jap. J. exp. Med.* **36**, 49–59.

IKESHOJI T. (1966b) Studies on mosquito attractants and stimulants. Part 3. The presence in mosquito breeding waters of a factor which stimulates oviposition. *Jap. J. exp. Med.* **36**, 67–72.

IKESHOJI T. (1975) Chemical analysis of woodcreosote for species-specific attraction of mosquito oviposition. *Appl. Ent. Zool.* **10**, 302–308.

IKESHOJI T. and MULLA M. S. (1970) Oviposition attractants for four species of mosquitoes in natural breeding waters. *Ann. ent. Soc. Am.* **63**, 1322–1327.

IKESHOJI T., SAITO K., and YANO A. (1975) Bacterial production of the ovipositional attractants for mosquitoes on fatty acid substrates. *Appl. Ent. Zool.* **10**, 239–242.

ILTIS W. G. and ZWEIG G. (1962) Surfactant in apical drop of eggs of some culicine mosquitoes. *Ann. ent. Soc. Am.* **55**, 409–415.

IMMS A. D. (1911) On the life history of *Croce filipennis* Westw. *Trans. Linn. Soc. Lond.* **11**, 151–160.

IMMS A. D. (1913) Contributions to a knowledge of the structure and biology of some Indian insects. II. On *Embia major* sp.nov. from the Himalayas. *Trans. Linn. Soc. Zool.* **11**, 167–195.

IMMS A. D. (1931) *Recent Advances in Entomology*, London.

IMMS A. D. (1937) *Recent Advances in Entomology*, 2nd edn., London.

IMMS A. D. (1942) On *Braula coeca* Nitsch and its affinities. *Parasitology* **34**, 88–100.

IMMS A. D. (1957) *A General Textbook of Entomology Including the Anatomy, Physiology, Development and Classification of Insects*, 9th edn. (revised by O. W. Richards and R. G. Davis), Methuen, London.

IMMS A. D. and CHATTERJEE N. C. (1915) On the structure and biology of *Tachardia lacca* Kerr, with observations on certain insects predaceous or parasitic upon it. *Ind. For. Mem. (Zool.)* **3**, 1–41.

INGRAM J. W., BYNUM E. K., MATHOS R., HALEY W. E., and CHARPENTIER L. J. (1951) Pests of sugarcane and their control. *Circ. US Dep. Agric.* **878**, 1–38.

INGRAM M. J., STAY B., and CAIN G. D. (1977) Composition of milk from the viviparous cockroach *Diploptera punctata*. *Insect Biochem.* **7**, 257–267.

IPERTI G. (1966) The choice of oviposition sites in aphidophagous Coccinellidae. In *Ecology of Phytophagous Insects* (review) (I. Hodek, ed.), pp. 121–122, The Hague.

IRWIN M. E. (1973) A new genus of the *Xestomyza*-group from the western coast of South Africa, based on two new species with flightless females (Diptera: Therevidae). *Ann. Natal Mus.* **21**, 533–556.

IRWIN M. E. (1976) Morphology of the terminalia and known ovipositing behaviour of female Therevidae (Diptera: Asiloidea) with an account of correlated adaptations and comments on phylogenetic relationships. *Ann. Natal Mus.* **22**, 913–935.

IRWIN M. E., GILL R. W., and GONZALEZ D. (1974) Field-cage studies of native egg predators of the pink bollworm in southern California cotton. *J. econ. Ent.* **67**, 193–196.

IRWIN M. E. and STUCKENBERG B. R. (1972) A description of the female, egg and first-instar larva of *Tongamya miranda*, with notes on oviposition and the habitat of the species (Diptera: Apioceridae). *Ann. Natal Mus.* **21**, 439–453.

ISAAC P. V. (1925a) Papers on Indian Tabanidae. VIII. The bionomics and life-histories of some of the common Tabanidae of Pusa. *Mem. Dep. Agric. India, Ent.* **9**, 21–28.

ISAAC P. V. (1925b) Papers on Indian Tabanidae. VII. Notes on the life-history of *Tabanus striatus* Fabr. (= *hilaris* Wlk.). *Mem. Dep. Agric. India, Ent.* **8**, 108–109.

ISAACSON D. L. (1973) A life table for the cinnabar moth *Tyria jacobaeae* in Oregon. *Entomophaga* **18**, 291–303.

ISAEV V. A. (1976) The effect of temperature on the embryonic diapause development and egg reactivation in *Culicoides pulicaris punctatus* (Diptera, Ceratopogonidae). *Zool. Zh.* **55**, 1172–1177.

ISART J. (1966) *Lixus junci* Boh. y su importancia como plaga de la remolacha. *Boln R. Soc. esp. Hist. nat. (Biol.)* **64**, 63–86.

ISHII G. (1937a) Morphology and ecology of *Megatoma varius* Matsumura and Yokoyama (Dermestidae). *Bull. seric. Exp. Stn Japan* **9** (3) 125–150 (in Japanese; English summary).

ISHII G. (1937b) The life history and habits of *Orphiloides ovivorus* Matsumura and Yokoyama (Dermestidae). *Bull seric. Exp. Stn Japan* **9** (3) 151–165 (in Japanese; English summary).

ISHII S. (1952) Studies on the host preference of the cowpea weevil (*Callosobruchus chinensis* L.). *Bull. natn. Inst. agric. Sci. Tokyo* (C) **1**, 185–256.

IŚHII S., HIRANO C., IWATA Y., NAKASAWA M., and MIYAGAWA H. (1962) Isolation of benzoic and salicylic acids from the rice plant as growth-inhibiting factors for the rice stem borer (*Chilo suppressalis* Walker) and some rice plant fungus pathogens. *Jap. J. appl. Ent. Zool.* **6**, 281–288 (in Japanese; English summary).

ISLE D. (1937) New observations on responses to colours in egg laying butterflies. *Nature, Lond.* **140**, 544.

ITO H. (1924) Contribution histologique et physiologique à l'étude des annexes des organes génitaux des Orthoptères. *Archs Anat. microsc.* **20**, 342–460.

ITÔ Y. (1967) Population dynamics of the chestnut gallwasp *Dryocosmus kuriphilus* Yasumatsu (Hymenoptera: Cynipidae). 4. Further analysis of the distribution of eggs and young larvae in buds using the truncated negative binomial series. *Researches Popul. Ecol. Kyoto Univ.* **9**, 177–191.

ITÔ Y. (1973) The effects of nematode feeding on the predatory efficiency for house fly eggs and reproduction rate of *Macrocheles muscaedomesticae* (Acarina: Mesostigmata). *Jap. J. sanit. Zool.* **23**, 209–213.

ITÔ Y., YAMANAKA H., NAKASUJI F., and KIRITANI K. (1972) Determination of predator–prey relationship with an activable tracer, europium-151. *Kontyû Tokyo* **40**, 278–283.

IVES W. G. H. (1955) Estimation of egg populations of the larch sawfly *Pristiphora erichsonii* (Htg.). *Can. J. Zool.* **33**, 370–388.

IVES W. G. H. (1967) Relations between invertebrate predators and prey associated with larch sawfly eggs and larvae on tamarack. *Can. Ent.* **99**, 607–622.

IYENGAR M. O. T. (1938) Egg of *Anopheles leucosphyrus* Dön. *J. Malar. Inst. India* **1**, 353–354.

JACKLIN S. W., RICHARDSON E. G., and YONCE C. E. (1970) Substerilizing doses of gamma irradiation to produce population suppression in plum curculio. *J. econ. Ent.* **63**, 1053–1057.

JACKSON C. G., BRYAN D. E., BUTLER G. D., and PATANA R. (1970) Development, fecundity, and longevity of *Leschenaultia adusta*, a Tachinid parasite of the salt-marsh caterpillar. *J. econ. Ent.* **63**, 1396–1397.

JACKSON D. J. (1928) The biology of *Dinocampus (Perilitus) rutilis* Nees, a braconid parasite of *Sitonia lineata* L. Part I. *Proc. zool. Soc. Lond.* **1928**, 597–630.

JACKSON D. J. (1956) Notes on hymenopterous parasitoids bred from eggs of Dytiscidae in Fife. *J. Soc. Br. Ent.* **5**, 144–149.

Jackson D. J. (1957) A note on the embryonic cuticle shed on hatching by the larvae of *A abus bipustulatus* L. and *Dytiscus marginalis* L. (Coleoptera: Dytiscidae). *Proc. R. ent. Soc. Lond.* (A) **32**, 115–118.

Jackson D. J. (1958a) Egg-laying and egg-hatching in *Agabus bipustulatus* L., with notes on oviposition in other species of *Agabus* (Coleoptera: Dytiscidae). *Trans. R. ent. Soc. Lond.* **110**, 53–80.

Jackson D. J. (1958b) Observations on the biology of *Caraphractus cinctus* Walker (Hymenoptera: Mymaridae), a parasitoid of the eggs of Dytiscidae. *Trans. R. ent. Soc. Lond.* **110**, 533–554.

Jackson D. J. (1959) The association of a slime bacterium with the inner envelope of the egg of *Dytiscus marginalis* (Coleoptera), and the less common occurrence of a similar bacterium on the egg of *D. semisulcatus*. *Q. Jl microsc. Sci.* **100**, 433–443.

Jackson D. J. (1960) Observations on egg-laying in *Ilybius fuliginosus* Fabricus and *I. ater* Degeer (Coleoptera: Dytiscidae), with an account of the female genitalia. *Trans. R. ent. Soc. Lond.* **112**, 37–52.

Jackson D. J. (1961) Observations on the biology of *Caraphractus cinctus* Walker (Hymenoptera: Mymaridae), a parasitoid of the eggs of Dytiscidae (Coleoptera). *Parasitology* **51**, 269–294.

Jackson D. J. (1964) Observations on the life-history of *Mestocharis bimacularis* (Dalman) (Hym., Eulophidae), a parasitoid of the eggs of Dytiscidae. *Opusc. ent.* **29**, 81–97.

Jackson D. J. (1966) Observations on the biology of *Caraphractus cinctus* Walker (Hymenoptera: Myrmaridae), a parasitoid of the eggs of Dytiscidae (Coleoptera). III. The adult life and sex ratio. *Trans. R. ent. Soc. Lond.* **118**, 23–49.

Jackson D. J. (1969) Observations on the female reproductive organs and the poison apparatus of *Caraphractus cinctus* Walker (Hymenoptera: Mymaridae). *Zool. J. Linn. Soc.* **48**, 49–81.

Jackson H. B. and Eikenbary R. D. (1971) Bionomics of *Aphelinus asychis* (Hymenoptera: Eulophidae), an introduced parasite of the sorghum greenbug. *Ann. ent. Soc. Am.* **64**, 81–85.

Jacob S. A. and Mohan M. S. (1975) Predation on certain stored product insects by red flour beetle. *Indian J. Ent.* **35**, 95–98 (1973).

Jacobs W. (1949) Die Eiablage von *Euthystira brachyptera* (Osck.) (Orth., Acrid.). *Entomon, Munich* **1**, 198–200.

Jacobson E. (1913) Biological notes on the Heterocera: *Eublemma rubra* (Hampson), *Cactoblemma sumbarensis* (Hampson), and *Eublemma versicolor* (Walker). *Tijd. Ent.* **56**, 165–173.

Jacobson L. A. (1965) Mating and oviposition of the pale western cutworm *Agrotis orthogona* Morrison (Lepidoptera: Noctuidae) in the laboratory. *Can. Ent.* **97**, 994–1000.

Jacobson L. A. and Blakeley P. E. (1959) Development and behavior of the army cutworm in the laboratory. *Ann. ent. Soc. Am.* **52**, 100–105.

Jacobson L. A. and Blakeley P. E. (1960) Survival, development, and fecundity of the pale western cutworm *Agrotis orthogonia* Morr. (Lepidoptera: Noctuidae) after starvation. *Can. Ent.* **92**, 184–188.

Jakob W. L. and Bevier G. A. (1969) Evaluation of ovitraps in the US *Aedes aegypti* eradication program. *Mosq. News* **29**, 650–653.

Jakob W. L., Fay R. W., Windeguth D. L. von, and Schoof H. F. (1970) Evaluation of materials for ovitrap paddles in *Aedes aegypti* surveillance. *J. econ. Ent.* **63**, 1013–1014.

Jalil M. and Rodriguez J. G. (1970a) Studies of behaviour of *Macrocheles muscaedomesticae* (Acarina: Macrochelidae) with emphasis on its attraction to the house fly. *Ann. ent. Soc. Am.* **63**, 738–744.

Jalil M. and Rodriguez J. G. (1970b) Biology of and odor perception by *Fuscuropoda vegetans*

(Acarina: Uropodidae), a predator of the house fly. *Ann. ent. Soc. Am.* **63**, 935–938.

JAMES H. G. (1958) Egg development, hatching, and prey taken by the European mantis *Mantis religiosa* L. in several habitats. *Rep. ent. Soc. Ont.* **89**, 50–55.

JAMES H. G. (1966) Location of univoltine *Aedes* eggs in woodland pool areas and experimental exposure to predators. *Mosq. News* **26**, 59–63.

JAMES S. P. (1922) *Trans. R. Soc. trop. Med. Hyg.* **16**, 267–269.

JAMES S. P. (1923) Eggs of *Finlaya geniculata* and of other English mosquitoes illuminated with the aid of Lieberkühn reflectors. *Trans. R. Soc. trop. Med. Hyg.* **17**, 8–9.

JANCKE O. (1938) Flöhe oder Aphaniptera (Scutoria). Laüse oder Anoplura (Siphunculata). *Tierwelt Dtl.* **35**, 1–78.

JANISCH E. (1930) Experimentelle Untersuchungen über die Wirkung der Umweltfaktoren auf Insekten. I. Die Massenvermehrung der Baumwolleule *Prodenia littoralis* im Ägypten. *Z. Morph. Ökol. Tiere* **17**, 339–416.

JANNONE G. (1934) Osservazioni ecologiche e biologiche sul *Dociostaurus maroccanus* Thunb., *Calliptamus italicus* L. e loro parassiti in Provincia di Napoli (primo contributo). *Boll. Lab. Zool. gen. agr. Portici* **28**, 75–151.

JANNONE G. (1939) Studio morfologico, anatomico e istologico del *Dociostaurus maroccanus* (Thunb.) nelle sue fasi *transiens congregans, gregaria* e *solitaria*. (Terzo contributo.) *Boll. Lab. Ent. Agr. Portici* **4**, 1–443.

JANSEN D. H. (1971) Escape of *Cassia grandis* beans from predators in time and space. *Ecology* **52**, 964–979.

JANSEN D. H. (1975) Interactions of seeds and their insects predators/parasitoids in a tropical deciduous forest. In *Evolutionary Strategies of Parasitic Insects and Mites* (P. W. Price, ed.), pp. 154–186, New York.

JANSON O. E. (1913) Mantid oothecae. *Trans. ent. Soc. Lond.* **1912**, cxxi–cxxii.

JANTZ O. K. and RUDINSKY J. A. (1966) Studies of the olfactory behavior of the Douglas-fir beetle *Dendroctonus pseudotsugae* Hopkins. *Tech. Bull. Ore. agric. Exp. Stn* **94**, 1–38.

JANVIER H. (1933) Étude biologique de quelques Hyménoptères du Chili. *Ann. Sci. nat. (Zool.)* **16**, 210–256.

JANVIER H. (1956) L'anesthésie des proies, l'oophagie et la ponte chez *Tiphia femorata* F. C. r. hebd. Séanc. Acad. Sci. Paris **243**, 872–874.

JARA P. B. (1970) Nuevos dispositivos de oviposición para Crianza Masal de *Anastrepha fraterculus* (Wied.). *Revta peru. Ent. agríc.* **13**, 90–92.

JASIČ J. and BÍROVÁ H. (1958) Plodnost spriadača amerického (*Hyphantria cunea* Drury) a jej stanovenie. I. *Biológia* **13**, 793–809.

JEANNEL R. (1942) Les Hénicocéphalidés. Monographie d'un group d'Hémiptères hématophages. *Annls Soc. ent. Fr.* **110**, 273–366.

JENNINGS D. T. (1975) Distribution of *Rhyacionia neomexicana* eggs within ponderosa pine trees. *Ann. ent. Soc. Am.* **68**, 1008–1010.

JENNINGS D. T. and ADDY N. D. (1968) A staining technique for jack-pine budworm egg masses. *J. econ. Ent.* **61**, 1766.

JENSEN-HAARUP A. C. (1916) On the parental care of *Elasmosthethus griseus*. *Flora og Fauna København* **1916**, 124–126.

JENSEN-HAARUP A. C. (1917) Brutpflege bei einer Wanze (*Elasmostethus griseus* L.). *Ent. Mitt.* **6**, 187.

JENTSCH S. (1936) Ovovivipiarie bei einer einheimischen Copeognathenart (*Hyperetes guestfalicus*). *Zool. Anz.* **116**, 287.

JEPSON W. F. (1939) Entomological division. *Rep. Dep. Agric. Mauritius*, **1938**, 40–51.

JERMY T. (ed.) (1976) *The Host Plant in Relation to Insect Behaviour and Reproduction*, New York, USA.

JOBLING B. (1935) The effect of light and darkness on oviposition in mosquitoes. *Trans. R. Soc. trop. Med. Hyg.* **29**, 157–166.

JOBLING B. (1953) On the blood-sucking midge *Culicoides vexans* Stager, including a description of its eggs and the first-stage larva. *Parasitology* **43**, 148–159.

JOHANNSEN O. A. (1905) Aquatic nematocerus Diptera. Part II. *Bull. NY State Mus.* **86**, 76–331.

JOHANNSEN O. A. (1937a) Aquatic Diptera. Part III. Chironomidae: subfamilies Tanypodinae, Diamesinae, and Orthocladiinae. *Mem. Cornell Univ. agric. Exp. Stn* **205**, 1–84.

JOHANNSEN O. A. (1937b) Aquatic Diptera. Part IV. Chironomidae: subfamily Chironominae. *Mem. Cornell Univ. agric. Exp. Stn* **210**, 1–56.

JOHN O. (1923) Fakultative Viviparität bei Thysanopteren. *Ent. Mitt.* **12**, 227–232.

JOHNSON A. W. and HAYS S. B. (1969) Laboratory mating behaviour of the plum curculio. *J. econ. Ent.* **62**, 438–440.

JOHNSON A. W. and HAYS K. L. (1973) Some predators of immature Tabanidae (Diptera) in Alabama. *Envir. Ent.* **2**, 1116–1117.

JOHNSON B. (1953) The injurious effects of the hooked epidermal hairs of french beans (*Phaseolus vulgaris* L.) on *Aphis craccivora* Koch. *Bull. ent. Res.* **44**, 779–788.

JOHNSON B. (1958) Embryonic cuticle in aphids. *Aust. J. Sci.* **21**, 146.

JOHNSON C. (1961) Breeding behaviour and oviposition in *Hetareina americana* Fabricius and *H. tita* (Drury) (Odonata: Agriidae). *Can. Ent.* **93**, 260–266.

JOHNSON C. (1965) Mating and oviposition of damselflies in the laboratory. *Can. Ent.* **97**, 321–326.

JOHNSON C. D. and KINGSOLVER J. M. (1975) Ecology and redescription of the Arizona grape bruchid *Amblycerus vitis* (Coleoptera). *Coleopt. Bull.* **29**, 321–331.

JOHNSON C. G. (1934) On the eggs of *Notostira erratica* L. (Hemiptera, Capsidae). *Trans. Soc. Br. Ent.* **1**, 1–32.

JOHNSON C. G. (1936) The biology of *Leptobyrsa rhododendri* Horváth (Hemiptera: Tingitidae), the rhododendron lace-bug. *Ann. appl. Biol.* **23**, 342–368.

JOHNSON C. G. (1937) The absorption of water and the associated volume changes occurring in the eggs of *Notostira erratica* L. (Hemiptera, Capsidae) during embryonic development under experimental conditions. *J. exp. Biol.* **14**, 413–421.

JOHNSON C. G. (1940) Development, hatching, and mortality of the eggs of *Cimex lectularis* (Hemipt.) in relation to climate, with observations on the effects of preconditioning to temperature. *Parasitology* **32**, 127–173.

JOHNSON M. W., STINNER R. E., and RABB R. L. (1975) Ovipositional response of *Heliothis zea* (Boddie) to its major hosts in North Carolina. *Envir. Ent.* **4**, 291–297.

JÖHNSSEN A. (1930) Beiträge zur Entwicklungs- und Ernährungsbiologie einheimischer Coccinelliden unter besonderer Berücksichtigung von *Coccinella septempunctata* L. *Z. angew. Ent.* **16**, 87–158.

JOHNSTON A. F. (1915) Asparagus-beetle egg parasite. *J. agric. Res.* **4**, 303–312.

JOLIVET P. (1951) Contribution à l'étude de genre *Gastrophysa* Chevolat (Coléoptèra, Chrysomelidae), (1ʳᵉ Note). *Bull. Inst. Sci. nat. Belg.* **27**, 1–11.

JOLLY M. S., SUBBA RAO S., and KRISHNASWAMI S. (1964) Effect of plane of inclination on egg laying by *Bombyx mori* L. *Indian J. exp. Biol.* **2**, 165–166.

JOLY N. (1876) Études sur l'embryogénie des Éphémères, notamment chez le *Palingerria virgo*. *Mém. Acad. Toulouse* **7**, 243–254.

JONES B. J. (1906) Catalogue of Ephydridae with bibliography and description of new species. *Univ. Calif. Publ. Ent.* **1**, 153–198.

JONES B. M. (1956) Endocrine activity during insect embryogenesis, function of the ventral head glands in locust embryos (*Locustana pardalina* and *Locusta migratoria*, Orthoptera). *J. exp. Biol.* **33**, 174–185.

Jones B. M. (1958) Enzymatic oxidation of protein as a rate-determining step in the formation of highly stable surface membranes. *Proc. R. ent. Soc.* (B) **148**, 263–277.

Jones C. R. (1922) A contribution to our knowledge of the Syrphidae of Colorado. *Colo. agric. Bull.* **269**.

Jones D. A., Parsons J., and Rothschild M. (1962) Release of hydrocyanic acid from crushed tissues of all stages in the life-cycle of species of Zygaeninae (Lepidoptera). *Nature, Lond.* **193**, 52–53.

Jones E. L. (1968) Bloodworms. Pests of rice. *Agric. Gaz. NSW* **79**, 477–478.

Jones F. M. (1933) Another oriental mantis well established in the United States (*Tenodera agustipennis* Saussure. Orthoptera: Mantidae). *Ent. News* **44**, 1–3.

Jones J. C. (1972) Sexual activities during single and multiple co-habitations in *Aedes aegypti* mosquitoes. *J. Ent.* (A), **48** (2), 185–191 (1974).

Jones J. C. (1973) A study of the fecundity of male *Aedes aegypti*. *J. Insect Physiol.* **19**, 435–439.

Jones J. C. and Wheeler R. E. (1965) Studies on spermathecal filling in *Aedes aegypti* (Linnaeus). II. Experimental. *Biol. Bull. Woods Hole* **129**, 532–545.

Jones M. G. (1968) The effect of moving carabids on oviposition by frit fly (*Oscinella frit* L.). *Entomologist's mon. Mag.* **104**, 85–87.

Jones M. G. (1969a) The effect of weather on frit fly (*Oscinella frit* L.) and its predators. *J. appl. Ecol.* **6**, 425–441.

Jones M. G. (1969b) Oviposition of frit fly (*Oscinella frit* L.) on oat seedlings and subsequent larval development. *J. appl. Ecol.* **6**, 411–424.

Jones M. G. (1972a) The egg-shell of the wheat bulb fly *Leptohylemyia coarctata* (Fall.). *Pl. Path.* **21**, 83.

Jones M. G. (1972b) The egg-shell of the cabbage root fly *Erioischia brassicae* (Bouché). *Pl. Path.* **21**, 84–85.

Jones R. L., Burton R. L., McGovern T. P., and Beroza M. (1973a) Potential oviposition inducers for corn earworm. *Ann. ent. Soc. Am.* **66**, 921–925.

Jones R. L., Lewis W. J., Beroza M., Bierl B. A., and Sparks A. N. (1973b) Host-seeking stimulants (kairomones) for the egg parasite *Trichogramma evanescens*. *Envir. Ent.* **2**, 593–596.

Jones S. L., Lingren P. D., and Bee M. J. (1977) Diel periodicity of feeding, mating, and oviposition of adult *Chrysopa carnea*. *Ann. ent. Soc. Am.* **70**, 43–47.

Jordan K. H. C. (1932) Zur Kenntnis des Eies und der Larven von *Microvelia schneideri*. *Z. wiss. InsektBiol.* **127**, 18–22.

Jordan K. H. C. (1935) Beitrag zur Lebensweise der Wanzen auf feuchten Böden (Heteroptera). *Stett. ent. Ztg.* **96**, 1–26.

Jordan K. H. C. and Wendt A. (1938) Zur Biologie von *Salda littoralis* L. (Hem.: Het.). *Stett. ent. Ztg.* **99**, 273–292.

Jörg M. E. (1933) Nota previa sobre el principio activo urticante de *Hylesia nigricans* (Lepidopt. Hemileucidae) y las dermatitis provocadas por el mismo. *8th Reun. Soc. Argent. Patol. Reg.* pp. 482–495.

Jörg M. E. (1939) Dermatosis lepidopterianas (segunda nota). *9th Reun. Soc. Argent. Patol. Reg.* **3**, 1617–1639.

Jörg M. E. (1969) Subcutaneous inflammatory nodule by spicules of *Hylesia fulviventris* (Lepidoptera). *Bol. chil. Parasit.* **24** (3–4) 146–150 (in Spanish; English summary).

Joshi B. G. (1963) Laboratory studies on the biology of the tobacco ground beetle *Mesomorphus villiger* Blanch. *Indian J. Ent.* **24**, 205–210 (1962).

Joubert P. C. (1964) The reproductive system of *Sitotroga cerealella* Olivier (Lepidoptera: Gelechiidae). 1. Development of the female reproductive system. *S. Afr. J. agric. Sci.* **7**, 65–77.

JOURNET P. (1952) Contribution à l'étude de la sésie du groseillier. *Phytoma* **5**, 24–25.
JOYCE R. J. V. (1952) The ecology of grasshoppers in East Central Sudan. *Anti-Locust Bull.* **11**, 1–97.
JUAREZ E., SILVA E. P. DE C. E., and ZUCAS S. M. (1971) Marcação de ovos de *Triatoma infestans* com diversos corantes. *Rev. Saúde Públ.* **5**, 291–293.
JUDENKO E. (1958) The appearance of adult shot-hole borers (*Xyleborus fornicatus* Eich.) outside their galleries under natural conditions. *Tea Q.* **29** (2) 104–111.
JUDGE P. J. (1953) A new method for determining the transition temperature of insect cuticular waxes. *Nature, Lond.* **172**, 405–406.
JUDSON C. L. (1960) The physiology of hatching aedine mosquito eggs: hatching stimulus. *Ann. ent. Soc. Am.* **53**, 688–691.
JUDSON C. L. (1963) The physiology of hatching of aedine mosquito eggs: carbon monoxide stimulation and ethyl chloride inhibition of hatching of *Aedes aegypti* (L.) egg. *J. Insect Physiol.* **9**, 787–792.
JUDSON C. L. (1967) Feeding and oviposition behaviour in the mosquito *Aedes aegypti* (L.). 1. Preliminary studies of physiological control mechanisms. *Biol. Bull. Woods Hole* **133**, 369–377.
JUDSON C. L., HOKAMA Y., and BRAY A. D. (1962) The effects of various chemicals on eggs of the yellow fever mosquito *Aedes aegypti*. *J. econ. Ent.* **55**, 805–807.
JUDSON C. L., HOKAMA Y., and HAYDOCK I. (1965) The physiology of hatching of aedine mosquito eggs: some larval responses to the hatching stimulus. *J. Insect Physiol.* **11**, 1169–1177.
JUDSON C. L., HOKAMA Y., and KLIEWER J. W. (1966) Embryology and hatching of *Aedes sierrensis* eggs (Diptera, Culicidae). *Ann. ent. Soc. Am.* **59**, 1181–1184.
JUDULIEN F. (1899) Quelques notes sur plusieurs Coprophages de Buenos Aires. *Revta Mus. La Plata* **9**, 371–380.
JUILLET J. A. (1961) Observations on arthropod predators of the European pine shoot moth *Rhyacionia buoliana* (Schiff.) (Lepidoptera: Olethreutidae) in Ontario. *Can. Ent.* **93**, 195–198.
JUNNIKKALA E. (1960) Life history and insect enemies of *Hyponomeuta malinellus* Zell. (Lep., Hyponomeutidae) in Finland. *Ann. Zool. Soc. "Vanamo"* **21**, 1–44.
JURA C. (1959) The early developmental stages of the ooviviparous scale insect *Quadraspidiotus ostreaeformis* (Curt.) (Homoptera: Coccidae, Aspidiotini). *Zool. polon.* **9**, 17–34 (1958).
JURA C. (1972) Development of apterygote insects. In *Developmental Systems: Insects* (S. J. Counce and C. H. Waddington, eds.) **1**, 49–94.

KABANOV V. A. (1969) The biology and ecology of the wireworms *Agriotes ponticus* Stepanov and *Melanotus fusciceps* Gyll. (Coleoptera, Elateridae) in the conditions of the Krasnodar region. *Ent. Obozr.* **48**, 486–492 (in Russian; English translation: *Ent. Rev.* **48**, 307–311).
KABIR A. K. M. F. and GIESE R. L. (1966) The Columbian timber beetle *Corthylus columbianus* (Coleoptera: Scolytidae). I. Biology of the beetle. *Ann. ent. Soc. Am.* **59**, 883–894.
KABOS W. J. (1943) Eischaal structuren bij Syrphiden. *Tijd. Ent.* **86**, 43–44.
KADYI H. (1879) Beitrag zur Kenntnis der Vorgänge beim Eierlegen der *Blatta orientalis*. *Zool. Anz.* **2**, 632–636.
KAHN N. H. (= KHAN) (1952) Oviposition and hatching in some species of Tabanidae (Diptera). *Ann. ent. Soc. Am.* **45**, 550–552.
KAHN N. H. (1953) Oviposition in Tabanidae (Diptera). *Indian J. Ent.* **15**, 39–44.
KAIN W. M. and BURTON G. (1975) Effect of ground cover and pasture height on oviposition and establishment of soldier fly larvae. *Proc. 28th NZ Weed and Pest Control Conference 1975*, pp. 237–241.

KAJITA H. (1966) Studies on the utilization of natural enemies as "biotic insecticides". On the number of eggs deposited by *Pseudaphycus malinus* Gahan and its host selection. *Sci. Bull. Fac. Agric. Kyushu Univ.* **22**, 319–324 (in Japanese; English summary).

KALANDADZE L. (1927) Über die Biologie des Museumskäfers *Anthrenus verbasci* L. und seine Bekämpfung. *Z. angew. Ent.* **13**, 301–311.

KALPAGE K. S. P. and BRUST R. A. (1968) Mosquitoes of Manitoba. I. Descriptions and a key to *Aedes* eggs (Diptera: Culicidae). *Can. J. Zool.* **46**, 699–718.

KALPAGE K. S. P. and BRUST R. A. (1973) Oviposition attractant produced by immature *Aedes atropalpus*. *Envir. Ent.* **2**, 729–730.

KALSHOVEN L. G. E. (1940) Observations on the red branchborer *Zeuzera coffeae* Nietn. (Lep., Coss.). *Ent. Med. Ned.-Ind.* **6**, 50–54.

KALSHOVEN L. G. E. (1958a) Studies on the biology of Indonesian Scolytoidea. 1. *Xyleborus fornicatus* Eichh. as a primary and secondary shot-hole borer in Java and Sumatra. *Ent. Ber.* **18**, 147–160.

KALSHOVEN L. G. E. (1958b) Studies on the biology of Indonesian Scolytoidea. 2. A case of primary infestation of *Glochidion* by *Xyleborus xanthopus* Eichh. *Ent. Ber.* **18**, 190–193.

KALTENBACH A. (1968) Embioptera (Spinnfüsser). *Handbuch der Zoologie* **4**, 1–29.

KAMAL M. (1938a) *Brachymeria femorata* Panz. (Hymenoptera–Chalcididae) a primary parasite of the cabbage worm *Pieris rapae* L. *Bull. Soc. ent. Égypte* **21**, 5–25.

KAMAL M. (1938b) The cotton green bug *Nezara viridula* L. and its important egg parasite *Microphanurus megacephalus* (Ashmead). *Bull. Soc. ent. Égypte* **21**, 175–207.

KANEKO T. and TAKAGI K. (1966) Biology of some scolytid ambrosia beetles attacking tea plants. VI. A comparative study of two ambrosia fungi associated with *Xyleborus compactus* Eichhoff and *Xyleborus germanus* Blandford (Coleoptera: Scolytidae). *Appl. Ent. Zool.* **1**, 173–176.

KANERVO V. (1946) Tutkimuksia lepän lehtikuoriaisen *Melasoma aenea* L. (Col., Chrysomelidae), luontaisista vihollisista. *Ann. zool. Soc. zool.-bot. fenn.* **12**, 1–206.

KANGAS E., OKSANEN H., and PERTTUNEN V. (1970) Responses of *Blastophagus piniperda* L. (Col. Scolytidae) to transverbenol, cis-verbenol, and verbenome, known to be population pheromones of some American bark beetles. *Ann. ent. fenn.* **36**, 75–83.

KANGAS E., PERTUNNEN V., and OKSANEN H. (1967) Studies on the olfactory stimuli guiding the bark beetle *Blastophagus piniperda* L. (Coleoptera, Scolytidae) to its host tree. *Suom. hyönt. Aikak.* **33**, 181–211.

KAPIL R. P. and BHANOT J. P. (1972) *Acaropsis docta* (Berlese), a predacious mite on *Trogoderma granarium* Everts. *Indian J. Ent.* **33**, 457.

KAPIL R. P. and BHANOT J. P. (1973) Feeding behaviour of the predatory mite *Acaropsis docta* (Berlese). *J. stored Prod. Res.* **9**, 1–6.

KAPLAN W. D., OJIMA Y., TANAKA K. K., TANAKA T., and LEE H. (1957) Snail mantle tissue as a matrix for processing insect eggs and other small objects. *Stain Tech.* **32**, 153–155.

KARADZHOV S. (1973) Dynamics of the phytophagous and predacious mites in a biological system of pest control on apple. *Gradinar. Loz. Nauka* **10** (5) 51–62 (in Bulgarian; English summary).

KARANDINOS M. G. and AXTELL R. C. (1974) Age-related changes in the rate of oviposition of three species of *Hippelates* (Diptera: Chloropidae) eye gnats: experimental and mathematical analysis. *Ann. ent. Soc. Am.* **67**, 669–677.

KARPENKO C. P. and NORTH D. T. (1973) Ovipositional response elicited by normal, irradiated, F_1 male progeny, or castrated male *Trichoplusia ni* (Lepidoptera: Noctuidae). *Ann. ent. Soc. Am.* **66**, 1278–1280.

KARTMAN L. and REPASS R. P. (1952) The effects of desiccation on the eggs of *Anopheles quadrimaculatus* Say. *Mosq. News* **12**, 107–110.

KASCHEF A. H. (1961) *Gibbium psylloides* Czemp. (Col., Ptinidae) new host of *Lariophagus distinguendus* Först. (Hym., Pteromalidae). *Z. ParasitKde* **21**, 65–70.

KÄSTNER A. (1934) Zur Lebensweise der Kamelhalsfliegen (Raphidiina). *Zool. Anz.* **108**, 1–11.

KATANYUKUL W. and THURSTON R. (1973) Seasonal parasitism and predation of eggs of the tobacco hornworm on various host plants in Kentucky. *Envir. Ent.* **2**, 939–945.

KATAYAMA E. (1965) Studies on the development of the broods of *Bombus diversus* Smith (Hymenoptera, Apidae). I. On the egg-laying habits. *Kontyû, Tokyo* **33**, 291–298.

KATIYAR K. N. (1960) Ecology of oviposition and the structure of egg-pods and eggs in some Indian Acrididae. *Rec. Indian Mus.* **55**, 29–68 (1957).

KATO K. and KUBOMURA K. (1956) Composition of the egg-packet of the praying mantis. *Sci. Rep. Saitama Univ. Urawa* (B) **2**, 165–182.

KATÔ M. and TORIUMI M. (1956) Emergence of larvae of *Aedes vexans* in the ground pool and the relation to the micro-organism community. *Ecol. Rev. Tôhoku* **14**, 163–167 (in Japanese; English summary).

KATO S. (1955) Investigation on *Hydrellia griseola* Fallén, the smaller rice leaf-miner. *Soc. Pl. Prot. N. Japan*, Special Rep. no. 3 (in Japanese).

KATSOYANNOS B. I. (1975) Oviposition-deterring, male-arresting, fruit-marking pheromone in *Rhagoletis cerasi*. *Envir. Ent.* **4**, 801–807.

KATSOYANNOS B. I. and BOLLER E. F. (1976) First field application of oviposition-deterring marking pheromone of European cherry fruit fly. *Envir. Ent.* **5**, 151–152.

KAUFMANN T. (1968) Observations on the biology and behavior of the evergreen bagworm moth *Thyridopteryx ephemeraeformis* (Lepidoptera: Psychidae). *Ann. ent. Soc. Am.* **61**, 38–44.

KAUP J. J. (1871) Ueber die Eier der Phasmiden. *Berl. ent. Z.* **15**, 17–24.

KAWAHARA S. (1959) The influences of temperature upon the growth and reproduction of the case-bearing clothes moth *Tinea pellionella* (L.). *Botyu-kagaku* **24**, 191–199 (in Japanese; English summary).

KAWAHARA S. (1960) The influences of relative humidity upon the growth and reproduction of the case-bearing clothes moth *Tinea pellionella* (L.). *Botyu-kagaku* **25**, 78–84 (in Japanese; English summary).

KAWAHARA S. (1962) The influences of population density of the adults upon the growth and reproduction of the case-bearing clothes moth *Tinea pellionella*. *Botyu-kagaku* **27**, 113–119 (in Japanese; English summary).

KAWASAKI H., SATO H., SUZUKI M., and OJIMA N. (1969) Conversion of serine to glycine during the formation of egg-shells in the silkworm *Bombyx mori*. *J. Insect Physiol.* **15**, 25–32.

KAWASAKI H., SATO H., and SUZUKI M. (1971a) Structural proteins in the silkworm egg-shells. *Insect Biochem.* **1**, 130–148.

KAWASAKI H., SATO H., and SUZUKI M. (1971b) Structural proteins in the egg-shell of the oriental garden cricket *Gryllus mitratus*. *Biochem. J.* **125**, 495–505.

KAWASAKI H., SATO H., and SUZUKI M. (1972) Structural proteins in the egg-shell of silkworms *Bombyx mandarina* and *Antheraea mylitta*. *Insect Biochem.* **2**, 53–57.

KAWASAKI H., SATO H., and SUZUKI M. (1974) Structural proteins in the egg envelopes of the dragonflies *Sympetrum infuscatum* and *S. frequens*. *Insect Biochem.* **4**, 99–111.

KAWASAKI H., SATO H., and SUZUKI M. (1975) Structural proteins in the egg envelopes of the mealworm beetle *Tenebrio molitor*. *Insect Biochem.* **5**, 25–34.

KAWASAKI H., SATO H., and YAGO M. (1976) Structural proteins in the egg-shell of the horned beetle *Xylotrupes dichotomus*. *Insect Biochem.* **6**, 43–52.

KAY, D., ROTHSCHILD M., and ALPIN R. (1969) Particles present in the haemolymph and defensive secretions of insects. *J. Cell Sci.* **4**, 369–380.

KEARNS C. W. (1934) A hymenopterous parasite (*Cephalonomia gallicola* Ashm.) new to the cigarette beetle (*Lasioderma serricorne* Fab.). *J. econ. Ent.* **27**, 801–806.

KEHAT M. and WYNDHAM M. (1971) The influence of temperature on development, longevity, and fecundity in the Rutherglen bug *Nysius vinitor* (Hemiptera: Lygaeidae). *Aust. J. Zool.* **20**, 67–78.

KEILIN D. (1917) Recherches sur les Anthomyides à larves carnivores. *Parasitology* **9**, 325–450.

KEILIN D. and NUTTALL G. H. F. (1930) Iconographic studies of *Pediculus humanus*. *Parasitology* **22**, 1–10.

KEISER I., CHAMBERS D. L., and SCHNEIDER E. L. (1972) Modified commercial containers as laboratory cages, watering devices, and egging receptacles for fruit flies. *J. econ. Ent.* **65**, 1514–1516.

KEISTER M. and BUCK J. (1964) Respiration: some exogenous and endogenous effects on rate of respiration. In the *Physiology of the Insecta* (M. Rockstein, ed.) **3**, 617–658.

KÉLER S. VON (1942) Ein Beitrag zur Kenntnis der Mallophagen. 5. Über dei Eier einiger Mallophagenarten. *Arb. morph. taxon. Ent.* **9**, 174–180.

KÉLER S. VON (1969) Mallophaga (Federlinge und Haarlinge). *Handbuch der Zoologie* **4**, 1–72.

KELLEN W. R. (1959) Notes on the biology of *Halovelia marianarum* Usinger in Samoa (Velidae: Heteroptera). *Ann. ent. Soc. Am.* **52**, 53–62.

KELLEN W. R. (1960) A new species of *Omania* from Samoa, with notes on its biology (Heteroptera: Saldidae). *Ann. ent. Soc. Am.* **53**, 494–499.

KELLOGG F. E. (1970) Water vapour and carbon dioxide receptors in *Aedes aegypti*. *J. Insect. Physiol.* **16**, 99–108.

KELNER-PILLAULT S. (1960) Biologie, écologie d'*Enoicyla pusilla* Burm. (Trichoptères Limnophilides. *Ann. Biol.* **36** (3) 51–99.

KELSEY J. M. (1957) Oviposition preference by *Costelytra zealandica* (White). *NZ Jl Sci. Tech.* (A) **38**, 633–637.

KELSEY J. M. (1968) Oviposition preference by *Costelytra zealandica* (White). II. *NZ Jl agric. Res.* **11**, 206–210.

KEMNER N. A. (1926) Die Lebensweise und die parasitische Entwicklung der echten Aleochariden. *Ent. Tidskr.* **47**, 133–170.

KEMPTON R. A., BARDNER R., FLETCHER K. E., JONES M. G., and MASKELL F. E. (1974) Fluctuations in wheat bulb fly egg population in eastern England. *Ann. appl. Biol.* **77**, 102–107.

KENCHINGTON W. (1969) The hatching thread of praying mantids: an unusual chitinous structure. *J. Morph.* **129**, 307–316.

KENCHINGTON W. (1974) Experimentally induced *in vitro* increase in size of calcium citrate crystals in praying mantids. *J. Insect Physiol.* **20**, 2043–2047.

KENCHINGTON W. and FLOWER N. E. (1969) Studies on insect fibrous proteins: the structural protein of the ootheca in the praying mantis *Sphodromantis centralis* Rehn. *J. Microsc.* **89**, 263–281.

KENNEDY C. H. (1915) Notes on the life history and ecology of the dragonflies of Washington and Oregon. *Proc. US natn. Mus.* **49**, 259–345.

KENNEDY C. H. (1917) Notes on the life history and ecology of the dragonflies of Central California and Nevada. *Proc. US natn. Mus.* **52**, 483–635.

KENNEDY G. G. (1971) Reproduction of *Scaphytopius delongi* (Homoptera: Cicadellidae) in relation to age of males and females. *Ann. ent. Soc. Am.* **64**, 1180–1182.

KENNEDY J. S. (1949) A preliminary analysis of oviposition behaviour by *Locusta* (Orthoptera: Acrididae) in relation to moisture. *Proc. R. ent. Soc. Lond.* (A) **24**, 83–89.

KENT P. W. and BRUNET P. C. J. (1959) The occurrence of protocatechuic acid and its 4-0-β-D-glucoside in *Blatta* and *Periplaneta*. *Tetrahedron* **7**, 252–256.

Kephart C. F. (1914) The poison glands of the larva of the brown-tail moth (*Euproctis chrysorrhoea* Linn.). *J. Parasit.* **1**, 95–103.
Kerenski J. (1930) Beobachtungen über die Entwicklung der Eier von *Anisoplia austriaca* Reitt. *Z. angew. Ent.* **16**, 178–188.
Kershaw J. C. (1910) The formation of the ootheca of a Chinese mantis *Hierodula saussurii*. *Psyche, Camb.* **17**, 136–141.
Kessel E. (1940) Gelege und Ei eines Federlings. *Natur Volk* **70**, 27–32.
Kessel E. (1942) Von Haarlingen und Federlingen. *Mikrokosmos* **35**, 79–84.
Kettle D. S. and Sellick G. (1947) The duration of the egg stage in the races of *Anopheles maculipennis* Meigen (Diptera, Culicidae). *J. Anim. Ecol.* 16, **1**, 38–43.
Keuchenius P. E. (1915) Onderzoekingen en Beschovwingen over eeniger Schadelyke Schildluizen van de Koffiekultur op Java. *Meded. Besoekisch Proefsta Djember* **16**, 1–63.
Kevan D. K. McE. (1944) The bionomics of the neotropical cornstalk borer *Diatraea lineolata* Wlk. (Lep., Pyral.) in Trinidad, BWI. *Bull. ent. Res.* **35**, 23–30.
Keys J. H. (1895) Some remarks on the habits of *Aepophilus bonnairii* Sign. *Entomologist's mon. Mag.* **31**, 135–137.
Keys J. H. (1914) Some further remarks on *Aepophilus bonnairii* Sign. *Entomologist's mon. Mag.* **50**, 284–285.
Khalifa A. (1950) Spermatophore production and egg-laying behaviour in *Rhodnius prolixus* Stål (Hemiptera: Reduviidae). *Parasitology* **40**, 283–289.
Khalifa A. (1956) The egg-pods of some Egyptian grasshoppers and the preference of females for soils of different moisture contents (Orthoptera: Acrididae). *Bull. Soc. ent. Égypte* **40**, 175–186.
Khalifa A. (1957) The development of eggs of some Egyptian species of grasshoppers, with a special reference to the incidence of diapause in the eggs of *Euprepocnemis plorans* Charp. *Bull. Soc. ent. Égypte* **41**, 299–330.
Khalsa H. G., Nigam B. S., and Agarwal P. N. (1966) Resistance of coniferous timbers to *Lyctus* attack. *Indian J. Ent.* **27**, 377–388 (1965).
Khan A. A. and Maibach H. I. (1972) Effect of human breath on mosquito attraction to man. *Mosq. News* **32**, 11–15.
Kharizanov A. and Dimitrov A. (1972) Some biological characteristics of *Chrysopa carnea*. *Rast. Zasht.* **20**, 36–38.
Khitsova L. N. and Golub V. B. (1972) Morphological differences in the eggs and young larvae of three species of *Gymnosoma* Mg. (Diptera, Tachinidae). *Zool. Zh.* **51**, 458–461.
Kikuchi T. and Ogura K. (1976) A three-binding site model for aggregation pheromone activities of the bark beetle *Ips confusus*. *Insect Biochem.* **6**, 115–122.
Kilincer N. (1975) Untersuchungen über die hämacytäre Abwehrreaktion der Puppe von *Galleria mellonella* L. (Lepidoptera) und über ihre Hemmung durch den Puppenparasiten *Pimpla turionellae* L. (Hym., Ichneumonidae). *Z. angew. Ent.* **78**, 340–370.
Killick-Kendrick R., Leaney A. J., and Ready P. D. (1977) The establishment, maintenance and productivity of a laboratory colony of *Lutzomyia longipalpis* (Diptera: Psychodidae). *J. med. Ent.* **13**, 429–440.
Killington F. J. (1930) A synopsis of British Neuroptera. *Trans. ent. Soc. S. Engl.* **1929**, 1–36.
Killington F. J. (1934) On the life histories of some British Hemerobiidae. *Trans. Soc. Br. Ent.* **1**, 119–134.
Killington F. J. (1936) *A Monograph of the British Neuroptera* **1**, i–xix, 1–269, Ray Society, London.
Kim C. W. and Kim B. K. (1975) Evaluation of the predators on the larvae of the pine-needle gall-midge *Thecodiplosis japonensis* Uchida et Inouye by the precipitin test. *Korean J. Ent.* **5**, 1–5.

King A. B. S. (1975) The extraction, distribution and sampling of the eggs of the sugar-cane froghopper *Aeneolamia varia saccharina* (Dist.) (Homoptera, Cercopidae). *Bull. ent. Res.* **65**, 157–164.

King J. L. (1916) Observations on the life history of *Pterodonta flavipes* Gray. *Ann. ent. Soc. Am.* **9**, 308–321.

King J. L. (1919) Notes on the biology of the carabid genera *Brachinus*, *Galerita* and *Chlaenius*. *Ann. ent. Soc. Am.* **12**, 382–388.

King P. E. (1962a) The effect of resorbing eggs upon the sex ratio of the offspring in *Nasonia vitripennis* (Hymenoptera, Pteromalidae). *J. exp. Biol.* **39**, 161–165.

King P. E. (1962b) Structure of the micropyle of eggs of *Nasonia vitripennis*. *Nature, Lond.* **195**, 829–830.

King P. E. (1962c) The muscular structure of the ovipositor and its mode of function in *Nasonia vitripennis* (Walker) (Hymenoptera: Pteromalidae). *Proc. R. ent. Soc. Lond.* (A) 121–128.

King P. E. (1963) The rate of egg resorption in *Nasonia vitripennis* (Walker) (Hymenoptera: Pteromalidae) deprived of hosts. *Proc. R. ent. Soc. Lond.* (A) **38**, 98–100.

King P. E. and Copland M. J. W. (1969) The structure of the female reproductive system in the Mymaridae (Chalcidoidea: Hymenoptera). *J. nat. Hist.* **3**, 349–365.

King P. E. and Fordy M. R. (1970) The external morphology of the "pore" structures on the tip of the ovipositor in Hymenoptera. *Entomologist's mon. Mag.* **106**, 65–66.

King P. E. and Hopkins C. R. (1963) Length of life of the sexes in *Nasonia vitripennis* (Walker) (Hymenoptera, Pteromalidae) under conditions of starvation. *J. exp. Biol.* **40**, 751–761.

King P. E. and Rafai J. (1970) Host discrimination in a gregarious parasitoid *Nasonia vitripennis* (Walker) (Hymenoptera: Pteromalidae). *J. exp. Biol.* **53**, 245–254.

King P. E. and Ratcliffe N. A. (1969) The structure and possible mode of functioning of the female reproductive system in *Nasonia vitripennis* (Hymenoptera: Pteromalidae). *J. Zool. Lond.* **157**, 319–344.

King P. E., Ratcliffe N. A., and Copland M. J. W. (1969) The structure of the egg membranes in *Apanteles glomeratus* (L.) (Hymenoptera: Braconidae). *Proc. R. ent. Soc. Lond.* (A) **44**, 137–142.

King P. E. and Richards J. G. (1968) Accessory nuclei and annulate lamellae in Hymenopteran oocytes. *Nature, Lond.* **218**, 488.

King P. E., Richards J. G., and Copland M. J. W. (1968) The structure of the chorion and its possible significance during oviposition in *Nasonia vitripennis* (Walker) (Hymenoptera: Pteromalidae) and other chalcids. *Proc. R. ent. Soc. Lond.* (A) **43**, 13–20.

King P. E. and Stabins V. (1971) Aspects of the biology of a strand-living beetle *Eurynebria complanata* (L.). *J. nat. Hist.* **5**, 17–28.

King R. C. (1963) Studies on early stages of insect oögenesis. *Roy. ent. Soc. Symp. Insect Repr.* **2**, 13–25.

King W. V., Bradley G. H., Smith C. N., and McDuffie W. C. (1960) *A Handbook of the Mosquitoes of the Southeastern United States*, US Dep. Agric. Handbook no. 173, 1–188.

Kingsolver J. G. and Norris D. M. (1977) The interaction of the female ambrosia beetle *Xyleborus ferrugineus* (Coleoptera: Scolytidae) with her eggs in relation to the morphology of the gallery system. *Entomologia exp. appl.* **21**, 9–13.

Kinn D. N. (1966) Predation by the mite *Macrocheles muscaedomesticae* (Acarina: Macrochelidae) on three species of flies. *J. med. Ent.* **3**, 155–158.

Kinn D. N. (1971) The life cycle and behavior of *Cercoleipus coelonotus* (Acarina: Mesostigmata). Including a survey of phoretic mite associates of California Scolytidae. *Univ. Calif. Publ. Ent.* **65**, 1–66.

Kinzer G. W., Fentiman A. F., Foltz R. L., and Rudinsky J. A. (1971) Bark beetle attractants: 2-methyl-2-cyclohexane-1-one isolated from *Dendroctonus pseudotsugae*. *J. econ. Ent.* **64**,

970–971.

Kinzer G. W., Fentiman A. F., Page T. F., Foltz R. L., Vité J. P., and Pitman G. B. (1969) Bark beetle attractants: identification, synthesis and field bioassay of a new compound isolated from *Dendroctonus*. *Nature, Lond.* **221**, 477–478.

Kinzer H. G., Ridgill B. J., and Watts J. G. (1970) Biology and cone attack behavior of *Conophthorus ponderosae* in southern New Mexico (Coleoptera: Scolytidae). *Ann. ent. Soc. Am.* **63**, 795–798.

Kirichenko A. (1926) A study of the ecology and biology of *Calliptamus italicus* L. in the steppe zone of the Ukraine. *Odessa Reg. agric. Exp. Stn Ent. Dep.* **1**, 1–47 (in Russian).

Kiritani K. (1959) Egg laying habit of *Thaumaglossa ovivorus* (Col., Dermestidae). *Insect Ecol.* **7** (3) 111–116.

Kiritani K. (1963) Oviposition habit and effect of parental age upon the post-embryonic development in the southern green stink bug *Nezara viridula*. *Jap. J. Ecol.* **13**, 88–96.

Kiritani K. and Hokyo N. (1965) Variation of egg mass size in relation to the oviposition in Pentatomidae. *Kontyû, Tokyo* **33**, 427–433.

Kiritani K. and Kawahara S. (1963) Effect of adult diet on the longevity, fecundity oviposition period and phototaxis in the black carpet beetle *Attagenus megatoma* (F.). *Jap. J. Ecol.* **13**, 21–28.

Kiritani K. and Kimura K. (1965) The effect of population density during nymphal and adult stages on fecundity and other reproductive performances. *Jap. J. Ecol.* **15**, 233–236.

Kirk V. M., Calkins C. O., and Post F. J. (1968) Oviposition preferences of western corn rootworms for various soil surface conditions. *J. econ. Ent.* **61**, 1322–1324.

Kirkaldy G. W. (1896) *Notonecta*, oviposition. *Proc. ent. Soc. Lond.* **1896**, xxvii.

Kirkpatrick T. W. (1937) Studies on the ecology of coffee plantations in East Africa. II. The autecology of *Antestia* spp. (Pentatomidae) with a particular account of a strepsipterous parasite. *Trans. R. ent. Soc. Lond.* **86**, 247–343.

Kirkpatrick T. W. (1947) Notes on a species of Epipyropidae (Lepidoptera) parasitic on *Metaphaena* species (Hemiptera: Fulgoridae) at Amani, Tanganyika. *Proc. R. ent. Soc. Lond.* (A) **22**, 61–64.

Kirkpatrick T. W. (1953) Insect pests of cacao and insect vectors of cacao virus disease. *Rep. Cacao Res. Trinidad 1945–51*, pp. 122–125.

Kirkpatrick T. (1957) *Insect Life in the Tropics*, Longmans, London.

Kishi Y. (1970a) *Mimemodes japonus* Reitter (Coleoptera: Rhizophagidae), an egg predator of the pine bark beetle *Cryphalus fulvus* Niijima (Coleoptera: Ipidae). *Kontyû, Tokyo* **38**, 195–197.

Kishi Y. (1970b) Difference in the sex ratio of the pine bark weevil parasite *Dolichomitus* sp. (Hymenoptera: Ichneumonidae) emerging from different host species. *Appl. Ent. Zool.* **5**, 126–132.

Kitamura B. (1966) Some observations on the ecology of a papilionid butterfly *Luehdorfia puziloi inexpecta* Sheljuzhko. III. The egg laying place and the number of eggs in each batch. *New Ent. Ueda* **15** (4) 1–4.

Kitano H. (1969a) Defensive ability of *Apanteles glomeratus* L. (Hymenoptera: Braconidae) to the hemocytic reaction of *Pieris rapae crucivora* Boisduval (Lepidoptera; Pieridae). *Appl. Ent. Zool.* **4**, 51–55.

Kitano H. (1969b) The parasitism of endophagous parasites with special reference to non-specific cellular defence reactions of their hosts. *Zool. Mag. Tokyo* **78**, 463–475 (in Japanese).

Kitano H. (1974) Effects of the parasitization of a braconid, *Apanteles*, on the blood of its host, *Pieris*. *J. Insect Physiol.* **20**, 315–328.

Kitching R. L. (1974) The immature stages of *Sextus virescens* (Fairmaire) (Homoptera:

Membracidae). *J. Aust. ent. Soc.* **13**, 55–60.
KLATT B. (1920) Beiträge zur Sexualphysiologie der Schwammspinners. *Biol. Zbl.* **40**, 539–558.
KLAUSNITZER B. (1969) Zur Unterscheidung der Eier mitteleuropäischer Coccinellidae. *Acta ent. bohemoslov.* **66**, 146–149.
KLAUSNITZER B. and FÖRSTER G. (1971) Zur Eimorphologie einiger mitteleuropäischer Chrysomelidae (Coleoptera). *Polskie Pismo ent.* **41**, 429–437 (in Polish; German summary).
KLEIN M. G. and COPPEL H. C. (1966) Oviposition habits of a white pine chrysomelid *Glyptoscelis pubescens* (Fabr.). *Proc. N. cent. Brch Am. Ass. econ. Ent.* **20**, 140–141 (1965).
KLEIN-KRAUTHEIM F. (1936) Über das Chorion der Eier einiger Syrphiden (Diptera). *Biol. Zbl.* **56**, 323–329.
KLEINE R. (1916) *Cassida nebulosa* L. und ihr Frassbild. Eine biologische Betrachtung. *Stett. ent. Ztg.* **77**, 187–216.
KLEMPERER H.-G. and BOULTON R. (1976) Brood burrow construction and brood care by *Heliocopris japetus* (Klug) and *Heliocopris hamadryas* (Fabricius) (Coleoptera, Scarabaeidae). *Ecol. Ent.* **1**, 19–29.
KLINGLER J. (1959) Biologische Beobachtungen über den gefurchten Dickmaulrüssler (*Otiorrhynchus sulcatus* Fabr.) während seines Massenauftretens der letzten Jahre auf Reben der deutschen Schweiz. *Landw. Jb. Schweiz.* **73**, 409–438.
KLINGLER J., VOGEL W., and WILLE H. (1958) Der Einfluss der Temperatur auf die Eiablage des Apfelwicklers. *Schweiz. Z. Obst- u. Weinb.* **67**, 256–262.
KLINGSTEDT H. (1926) Beobachtungen über die Biologie, insbesondere das Eierlegen von *Limnophilus decipiens* Kol. (Trich.). *Not. Ent.* **6**, 118–120.
KLOMP H. (1958) Larval density and adult fecundity in a natural population of the pine looper (*Bupalus piniarius* L.). *Arch. néerl. Zool.* **13** (suppl. 1) 319–334.
KLUG W.-S., CAMPBELL D., and CUMMINGS M. R. (1974) External morphology of the egg of *Drosophila melanogaster* Meigen (Diptera: Drosophilidae). *Int. J. Insect Morph. Embryol.* **3**, 33–40.
KLUN J. A. and BRINDLEY T. A. (1966) Role of 6-methoxybenzoxazolinone in inbred resistance of host plant (maize) to first-brood larvae of European corn borer. *J. econ. Ent.* **59**, 711–718.
KNAB F. (1904) The eggs of *Culex territans* Walker. *Jl NY ent. Soc.* **12**, 246–248.
KNABKE J. J. and GRIGARICK A. A. (1971) Biology of the African earwig *Euborellia cincticollis* (Gerstaecker) in California and comparative notes on *Euborellia annulipes* (Lucas). *Hilgardia* **41**, 157–194.
KNIGHT A. W., NEBEKER A. V., and GAUFIN A. R. (1965a) Description of the eggs of common Plecoptera of western United States. *Ent. News* **76**, 105–111.
KNIGHT A. W., NEBEKER A. V., and GAUFIN A. R. (1965b) Further descriptions of the eggs of Plecoptera of western United States. *Ent. News* **76**, 233–239.
KNIGHT F. B. (1961) Variations in the life history of the Engelmann spruce beetle. *Ann. ent. Soc. Am.* **54**, 209–214.
KNIGHT F. B. (1969) Egg production by the Engelmann spruce beetle *Dendroctonus obesus* in relation to status of infestation. *Ann. ent. Soc. Am.* **62**, 448.
KNIGHT G. H. (1970) The life-history of the aphid or red-campion (*Brachycaudus klugisti* Hem., Hom. Aphididae). *Entomologist's mon. Mag.* **106**, 63–64.
KNOLL F. (1922) Insekten und Blumen. Experimentelle Arbeiten zur Vertiefung unserer Kenntnisse über die Wechselbeziehungen zwischen Pflanzen und Tieren. III. Lichtsinn und Blumenbesuch der Falters von *Macroglossum stellatarum*. *Abh. zool-bot. ges. Wien.* **12**, 123–377.

KNOPF J. A. E. and PITMAN G. B. (1972) Aggregation pheromone for manipulation of the Douglas-fir beetle. *J. econ. Ent.* **65**, 723–726.

KNOTT C. M., LAWSON F. R., and HOBGOOD J. M. (1966) Oviposition cage for the tobacco budworm and the corn earworm. *J. econ. Ent.* **59**, 1290.

KNOX P. C. and HAYS K. L. (1972) Attraction of *Tabanus* spp. (Diptera: Tabanidae) to traps baited with carbon dioxide and other chemicals. *Envir. Ent.* **1**, 323–326.

KNUTSON L. V. (1966) Biology and immature stages of malacophagous flies: *Antichaeta analis, A. atriseta, A. brevipennis* and *A. obliviosa* (Diptera: Sciomyzidae). *Trans. Am. ent. Soc.* **92**, 67–101.

KNUTSON L. V. (1970a) Biology and immature stages of *Tetanura pallidiventris*, a parasitoid of terrestrial snails (Dipt., Sciomyzidae). *Ent. scand.* **1**, 81–89.

KNUTSON L. V. (1970b) Biology of snail-killing flies in Sweden (Dipt., Sciomyzidae). *Ent. scand.* **1**, 307–314.

KNUTSON L. V. and BERG C. O. (1963) Biology and immature stages of a snail-killing fly. *Hydromya dorsalis* (Fabricius) (Diptera: Sciomyzidae). *Proc. R. ent. Soc. Lond.* (A) **38**, 45–58.

KNUTSON L. V. and BERG C. O. (1964) Biology and immature stages of snail-killing flies: the genus *Elgiva* (Diptera: Sciomyzidae). *Ann. ent. Soc. Am.* **57**, 173–192.

KNUTSON L. V. and BERG C. O. (1966) Parasitoid development in snail-killing sciomyzid flies. *Trans. Am. microsc. Soc.* **85**, 164–165.

KNUTSON L. V. and BERG C. O. (1967) Biology and immature stages of malacophagous Diptera of the genus *Knutsonia* Verbeke (Sciomyzidae). *Bull. Inst. R. Sci. nat. Belg.* **43** (7) 1–60.

KNUTSON L. V., NEFF S. E., and BERG C. O. (1967) Biology of snail-killing flies from Africa and southern Spain (Sciomyzidae: *Sepedon*). *Parasitology* **57**, 487–505.

KNUTSON L. V., STEPHENSON J. W., and BERG C. O. (1965) Biology of a slug-killing fly *Tetanocera elata* (Diptera: Sciomyzidae). *Proc. malac. Soc. Lond.* **36**, 213–220.

KNUTSON L. V., STEPHENSON J. W., and BERG C. O. (1970) Biosystematic studies of *Salticella fasciata* (Meigen), a snail-killing fly (Diptera: Sciomyzidae). *Trans. R. ent. Soc. Lond.* **122**, 81–100.

KOBAYASHI S. (1960) Studies on the distribution pattern of the eggs of the common cabbage butterfly *Pieris rapae crucivora* in a cabbage farm and the factors affecting its concentrating trend. *Jap. J. Ecol.* **10**, 154–160 (in Japanese; English summary).

KOBAYASHI S. (1965) Influence of adult density upon the oviposition site in the cabbage butterfly *Pieris rapae crucivora*. *Jap. J. Ecol.* **15**, 35–38.

KOCHETOVA N. I. (1968) The sex ratio in *Anastatus disparis* Rusch. (Hymenoptera, Eupelmidae) when developing in different hosts. *Zool. Zh.* **47**, 1572–1574 (in Russian; English summary).

KOCHLAR R. D., DIXIT R. S., and SOMAYA C. I. (1972) A study of oviposition of *Aedes* mosquitoes. *Mosq. News* **32**, 114–115.

KOHNO M. (1943) Notes on the eggs of some species of the family Perlidae (Plecoptera). *Mushi, Fukuoka* **15**, 54–59 (in Japanese).

KOHNO M. (1946) Observations on the Plecopteran eggs. *Collecting Breed. Tokyo* **8**, 194–196 (in Japanese).

KOIDE T. (1962) Observations on the feeding habit of the larvae of *Coccinella septempunctata bruchii* Mulsant. The feeding behaviour and number of prey fed under different temperatures. *Kontyû, Tokyo* **30**, 236–241.

KOIZUMI K. (1955) Effect of constant temperature upon the development of the potato tuber moth *Gnorimoschema operculella* (Zeller). *Sci. Rep. Fac. Agric. Okayama Univ.* **7**, 36–45 (in Japanese; English summary).

KOJIMA T. (1932) Beiträge zur Kenntnis von *Lyctus linearis* Goeze. *Z. angew. Ent.* **19**, 325–356.
KOJIMA T. (1935) Effect of temperature and moisture upon the hatching of *Dendrolimus spectabilis* Butl. *Oyo Dobuts. Zasshi, Tokyo* **7**, 211–244 (in Japanese).
KOKHMANYUK F. S. (1964) On the oviposition of *Ocneria (Lymantria) dispar* L. *Zool. Zh.* **43**, 290–291 (in Russian; English summary).
KOLMAKOVA V. D. (1958) On the biology of Siberian fruit moths of the genus *Grapholitha* (Lepidoptera, Tortricidae), injurious to fruit trees in Transbaikalia. *Rev. Ent. URSS* **37**, 134–150 (in Russian; English summary).
KOLMAKOVA V. D. (1965) The use of the local form of Trichogramma (*Trichogramma embryophagum* Htg.) in the orchards of Transbaikalia. *Trudȳ vses. Inst. Zashch. Rast.* **24**, 203–210 (in Russian; English summary).
KOMÁREK J. (1955) Mutterpflege bei *Molops piceus* Panz. (Col., Carabidae). *Acta Soc. ent. Bohem. Čsl. Prague* **51**, 132–134 (1954) (in Czech; German summary).
KOMP W. H. W. (1941) The species of *Nyssorhynchus* confused under *Tarsimaculatus goeldi*, and a new name, *A. emilianus*, for one species found in Para, Brazil (Diptera, Culicidae). *Ann. ent. Soc. Am.* **34**, 791–807.
KONDO S. (1960) On the ovipositing behavior of *Periplaneta fuliginosa* S. *Sanit. injur. Insects* **5**, 71–77.
KONIG A. (1894) Ueber die Larvae von *Oncodes*. *Verh. zool.-bot. Ges. Wien* **44**, 163–166.
KÔNO H. (1929a) Über zwei neue Gattungen von Rhynchitien und ihre Lebensweise. *Trans. Sapporo nat. Hist. Soc.* **10**, 122–137.
KÔNO H. (1929b) Biologische Bemerkungen über einige Attelabinen aus Hokkaido, mit den Beschreibungen der neuen Arten. *Trans. Sapporo nat. Hist. Soc.* **11**, 48–58.
KORMONDY E. J. (1959) The systematics of *Tetragoneuria*, based on ecological, life history, and morphological evidence (Odonata : Corduliidae). *Misc. Publ. Mus. Zool. Univ. Mich.* **107**, 1–79.
KORSAKOFF M. N. (1949) Notes sur l'oothèque et le comportement des *Blepharopsis mendica* Fab. *Bull. Soc. Sci. nat. Maroc* **25–27**, 213–226 (1945–7).
KORSCHELT E. (1884a) Ueber die Bildung des Chorions und der Mikropylen bei den Insecteneiern. *Zool. Anz.* **7**, 394–398, 420–424.
KORSCHELT E. (1884b) Die Bildung des Chorions bei einigen Wasserwanzen. *Zool. Anz.* **7**, 500–504.
KORSCHELT E. (1887) Zur Bildung der Eihüllen, der Mikropylen und Chorionanhänge bei den Insekten. *Nova Acta Acad. Caesar. Leop. Carol.* **51**, 181–252.
KORSCHELT E. (1923) Bearbeitung einheimischer Tiere, Erste Monographie: Der Gelbrand, *Dytiscus marginalis* L. **1**, 1–863; **2** (1924), 1–964 (Leipzig).
KOSS R. W. (1968) Morphology and taxonomic use of Ephemeroptera eggs. *Ann. ent. Soc. Am.* **61**, 296–721.
KOSS R. W. (1970) Ephemeroptera eggs: sperm guide morphology and adhesive layer formation. *Trans. Am. microsc. Soc.* **89**, 295–299.
KOSS R. W. and EDMUNDS G. F. (1974) Ephemeroptera eggs and their contribution to phylogenetic studies of the order. *Zool. J. Linn. Soc.* **55**, 267–349.
KOTBY F. A. (1977) A brief note on the effect of temperature and washing on hatchability of mulberry silkworm eggs. *Agric. Res. Rev.* **53**, 217–220.
KOVAL' YU. V. (1968) Predators of the Colorado beetle. *Zashch. Rast.* **13**, 52 (in Russian).
KOVALEVA M. F. (1957) The effectiveness of *Trichogramma* in the control of the codling moth. *Zool. Zh.* **36**, 225–229 (in Russian; English summary).
KOVTUN I. V. (1966) The biological characteristics of the development of the Colorado beetle in the western regions of the Ukrainian SSR. *Zakhȳst Roslȳn* **3**, 3–7 (in Ukrainian; Russian summary).

Kowalska T. (1962) A method of rearing larvae of *Agrotis c-nigrum* L. and *A. exclamationis* L. (Lepidoptera, Noctuidae) in laboratory conditions. *Biul. Inst. Ochr. Rośl.* **14**, 35–42 (in Polish; English summary).
Koyama K. (1972) Experiments on the preservation of eggs of *Inazuma (Recilia) dorsalis* (Hemiptera: Deltocephalidae) at low temperatures. *Jap. J. appl. Ent. Zool.* **16**, 50–51 (in Japanese).
Koyama T. (1964) Bionomics and control of *Chilotraea polychrysa* (Meyr.) in Malaya. *Bull. Div. Agric. Minist. Agric. Co-op. Malaysia* **115** (3+) 51 pp.
Kozhanchikov I. V. (1969) Lepidoptera. Psychidae. *Fauna of the USSR* **3** (2).
Kozulina O. V. (1949) The mechanism of egg-laying in the clothes louse (*Pediculus humanus corporis* De Geer) and its anomaly. *Ent. Obozr.* **30**, 235–238 (in Russian).
Kozulina O. V. (1957) On the morphology and biology of *Pediculus humanus corporis* De Geer (Anoplura, Pediculidae). *Ent. Obozr.* **26**, 577–596.
Kozulina O. V. (1958) The effect of feeding on the oviposition and the methods of increasing the proportion of hatching eggs in the body-lice *Pediculus humanus corporis* De Geer (Anoplura, Pediculidae) under the conditions of mass breeding. *Ent. Obozr.* **37**, 580–588 (in Russian; English summary).
Kraemer G. D. (1950) Der grosse Tannenborkenkäfer, unter Berücksichtigung seiner beiden Verwandten und der Brutbaumdisposition *Pityokteines curvidens* Germ., *vorontzowi* Jakobs und *spinidens* Reitt. *Z. angew. Ent.* **31**, 349–430.
Kramer K. J., Ong J., and Law J. H. (1973) Oöthecal proteins of the oriental praying mantid *Tenodera sinensis*. *Insect Biochem.* **3**, 297–302.
Kramer P. (1869) Beitrage zur Anatomie und Physiologie der Gattung *Philopterus* (Nitzsch.). *Z. wiss. Zool.* **19**, 452–468.
Kramer S. (1956) Notes and observations on the biology and rearing of *Creophilus maxillosus* (L.) (Coleoptera, Staphylinidae). *Ann. ent. Soc. Am.* **48**, 375–380.
Kreyenberg J. (1928) Experimentell-biologische Untersuchungen über *Dermestes lardarius* L. und *Dermestes vulpinus* F. Ein Beitrag zur Frage nach der Inkonstanz der Häutungszahlen bei Coleopteren. *Z. angew. Ent.* **14**, 140–188.
Krieg A. (1961) *Grundlagen der Insektenpathologie. Viren-, Rickettsien- und Bakterien-Infektionen.* Wissensch. Forschungsb. (Naturwiss. Reihe) Band 69. Darmstadt, Dr. Dietrich Steinkopff Verlag.
Krnjaić S. (1968) The effect of some factors upon oviposition and life span of *Acanthoscelides obtectus* Say. *Zašt. Bilja* **19**, 179–185 (in Serbian; English summary).
Kugler O. E., Frankenstein P. W., and Rafferty K. A. (1956) Histochemical localization of alkaline phosphatase, glycogen and nucleic acids in the female reproductive organs of the cockroach *Periplaneta americana*. *J. Morph.* **98**, 235–249.
Kühlow (=Kühlhorn) F. (1962a) Studies on the bionomics and the morphology of the saltwater breeding *Anopheles gambiae* on the coast of Tanganyika. *Malaria* **41**, 187–197.
Kühlow (=Kühlhorn) F. (1962b) Dipterologische Studien in Niedersachsen. I. Über *Anopheles*-Vorkommen (Diptera: Culicidae) und die Milieuverhältnisse verschiedener Brutbiotope im Raum Göttingen-Northeim. *Beitr. Naturk. Niedersach.* **15**, 84–104.
Kullenberg B. (1942) Die Eier der schwedischen Capsiden (Rhynchota). I. *Ark. Zool.* **33**, 1–16.
Kullenberg B. (1943) Die Eier der schwedischen Capsiden (Rhynchota). II. *Ark. Zool.* **34**, 1–8.
Kullenberg B. (1946) Studien über die Biologie der Capsiden. *Zool. Bidr.* **23**, 1–522.
Kullenberg B. (1947) Über Morphologie und Funktion des Kopulations-apparats der Capsiden und Nabiden. *Zool. Bidr.* **24**, 217–418.
Kulman H. M. (1965) Oviposition habits of *Trichogramma minutum* on artificial concentrations of eggs of the European pine shoot moth. *Ann. ent. Soc. Am.* **58**, 241–243.
Kulshrestha S. K. (1969a) Observations on the ovulation and oviposition with reference to

corpus luteum formation in *Musca domestica neobulo* Fabr. (Muscidae: Diptera). *J. nat. ist.* **3**, 561–570.

KULSHRESTHA S. K. (1969b) Observations on the ovulation and oviposition with reference to corpus luteum formation in *Epilachna vigintioctopunctata* Fabr. (Coccinellidae: Coleoptera). *Zool. Anz.* **182**, 276–285.

KUMAR R. (1964) Anatomy and relationships of Thaumastocoridae (Hemiptera: Cimicoidea). *J. ent. Soc. Q.* **3**, 48–51.

KUMAR R. (1966) Studies on the biology, immature stages, and relative growth of some Australian bugs of the superfamily Coreoidea (Hemiptera: Heteroptera). *Aust. J. Zool.* **14**, 895–991.

KUMAR R. (1973) The biology of some Ghanaian mantids (Dictyoptera: Mantodea). *Bull. Inst. fr. Afr. noire* **35**, 551–578.

KUMAR R. and ANSARI A. K. (1974) Biology, immature stages and rearing of cocoa-capsids (Miridae: Heteroptera). *Zool. J. Linn. Soc.* **54**, 1–29.

KUMAR R. and BARNOR J. L. (1974) On some substances produced by the colleterial glands of certain Orthopteroid insects. *Ann. ent. Soc. Am.* **67**, 753–755.

KUMM H. W. (1941) The eggs of some Costa Rican Anophelines. *Am. J. trop. Med.* **21**, 91–102.

KUMMER H. (1960) Experimentelle Untersuchungen zur Wirkung von Fortpflanzungsfaktoren und die Lebensdauer von *Drosophila melanogaster*-Weibchen. *Z. vergl. Physiol.* **43**, 642–679.

KUNCKEL D'HERCULAIS J. (1893–1905) *Invasions des Acridiens vulgo Sauterelles en Algérie*, Algiers.

KUNDU G. G. and KISHORE P. (1971) Biology of the sorghum shoot fly *Atherigona varia soccata* Rond. (Anthomyiidae: Diptera). *Indian J. Ent.* **32**, 215–217 (1970).

KUNG K. S. (1955) The banana stem-borer weevil *Odoiporus longicollis* Oliv. in Taiwan. *J. Agric. For. Taiwan* **4**, 80–113 (in Chinese; English summary).

KUNIKE G. (1939) Neue Ergebnisse über die Eiablage und Generations-folge der *Anthrenus*-Arten. Vorläufige Mitteilung. *Anz. Schädlingsk.* **15**, 80–84.

KUNITSKAYA N. T. and PROKOP'EV V. N. (1976) On the number of eggtubes in fleas (Aphaniptera). *Ent. Obozr.* **55**, 800–807.

KUNZ S. E., BERRY I. L. and FOERSTER K. W. (1977) The development of the immature forms of *Stomoxys calcitrans*. *Ann. ent. Soc. Am.* **70**, 169–172.

KUNZ S. E., BLUME R. R., HOGAN B. F., and MATTER J. J. (1970) Biological and ecological investigations of hornflies in central Texas: influence of time of manure deposition on oviposition. *J. econ. Ent.* **63**, 930–933.

KUPERSHTEIN M. L. (1974) Use of the precipitin test for quantitative estimation of the influence of *Pterostichus crenuliger* (Coleoptera, Carabidae) on the population dynamics of *Eurygaster integriceps* (Hemiptera, Scutelleridae). *Zool. Zh.* **53**, 557–562.

KURBANOVA D. D. (1967) The bionomics of the apple moth in Azerbaijan and its differences from the fruit moth. *Zool. Zh.* **46**, 551–555 (in Russian; English summary).

KURIR A. (1975) Eiablage der blassen Kiefernbuschhornblattwespe *Diprion pallidum* Klug (Diprionidae, Hymenoptera) bei der 1. Generation im Freiland. *Z. angew. Ent.* **78**, 66–75.

KUROKO H. (1961) On the eggs and first-instar larvae of two species of Mantispidae. *Esakia, Hikosan* **3**, 25–32.

KUROSAWA T. (1937) Morphology and ecology of *Trinodes hirtus* Fabricius (Dermestidae). *Bull. seric. Exp. Stn* **9**, 185–204 (in Japanese; English summary).

KURSTAK E. (1964) Some data on the bionomics and ecology of the Ichneumonid *Devorgilla canescens*. *Pr. nauk. Inst. Ochr. Rośl.* **6**, 173–187 (in Polish; English summary).

KUSNEZOV N. J. (1910) On the probable viviparity in some Danaid, i.e. Pierid, butterflies. *Horae Soc. ent. ross.* **39**, 634–651 (in Russian; English summary).

KUSUI K. (1934) Über die chemische Zusammensetzung des Larvensacks von Sackträgern

(*Psyche*). *J. Biochem.* **21**, 453–455.

KUWAYAMA S. (1924) Morphological and ecological studies on the eggs of Chrysopidae. *Zool. Mag. Tokyo* **36**, 1–30 (in Japanese).

KUWAYAMA S. (1934) On the life-history of two species of Leptocerid caddis-flies injurious to the rice plant. *Trans. Sapporo nat. Hist. Soc.* **13**, 266–274.

KUWAYAMA S. (1955) Investigations on *Hydrellia griseola* Fallén, the smaller rice leaf-miner. *Soc. Pl. Prot. N. Japan*, Special Rep. no. 3 (in Japanese).

KUWAYAMA S. (1958) The smaller rice leaf-miner *Hydrellia griseola* Fallén in Japan. *Proc. 10th Int. Congr. Ent.* **3**, 399–405 (1956).

KUWAYAMA S. (1971) Observations on the biology of the mulberry sucker, with special reference to the influence of its parasitism on the growth of silkworms. *Jap. J. appl. Ent. Zool.* **15**, 115–120 (in Japanese; English summary).

KUZNETZOVA YU. I. (1970) A study of the possibility of storage of the eggs of *Chrysopa carnea* Steph. (Neuroptera, Chrysopidae) under low temperatures. *Zool. Zh.* **49**, 1505–1514 (in Russian; English summary).

LAABS A. (1939) Brutfürsorge und Brutpflege einiger Hydrophiliden mit Berücksichtigung des Spinnapparates, seines äusseren Baues und seiner Tätigkeit. *Z. Morph. Ökol. Tiere* **36**, 123–178.

LABEYRIE V. (1956) Observations sur la teigne du poireau (*Acrolepia assectella* Zeller). *Rev. Zool. agric.* **55**, 8–13.

LABEYRIE V. (1957a) Observations sur le comportement de ponte de la mouche du céleri (*Philophylla heraclei* L.). *Annls Épiphyt.* **8**, 171–183.

LABEYRIE V. (1957b) Sur l'attractivité de diverses alliacées vis-à-vis de la mouche de l'oignon: *Hylemyia antiqua* Meigen (Dipt., Muscidae). *Bull. Soc. ent. Fr.* **62**, 13–15.

LABEYRIE V. (1962) Mise en évidence d'influences multiples de la plante hôte sur la stimulation de la ponte chez *Acanthoscelides obtectus* Say. *C. r. Séanc. Soc. Biol.* **156**, 1473–1477.

LABEYRIE V. (1967) Résultat de la sélection d'*Acanthoscelides obtectus* répondant immédiatement par l'émission d'oeufs à une introduction différée de grains de *Phaseolus vulgaris*. *C. r. Séanc. Soc. Biol.* **160**, 1696–1699 (1966).

LABEYRIE V. (1969) Longévité et capacité reproductrice de lignées d'*Acanthoscelides obtectus* sélectionnées en fonction de la réponse aux stimuli de pont. *C. r. Séanc. Soc. Biol.* **162**, 2203–2206.

LABEYRIE V. (1970) Influence déterminante du lieu de ponte sur la rencontre des sexes chez *Acanthoscelides obtectus* Say (Coléoptère, Bruchidae). *C. r. hebd. Séanc. Acad. Sci. Paris* (D) **271**, 1578–1581.

LABINE P. A. (1968) The population biology of the butterfly *Euphydryas editha*. VIII. Oviposition and its relation to patterns of oviposition in other butterflies. *Evolution* **22**, 799–805.

LACHMAJER J. (1971) Host selection by *Anopheles labranchiae atroparvus* v. Tiel 1927 (Diptera, subfamilia Culicidae) in Gdańsk environment. *Biul. Inst. Med. morsk. Gdańsku* **22**, 41–48.

LACROIX J. L. (1925) Note détachée sur les oeufs des Chrysopides (Neur.). *Bull. Soc. ent. Fr.* 1925, 227–323.

LADDUWAHETTY A. M. (1967) Oviposition cycles and the stimulus for oviposition in *Dermestes maculatus* DeGeer (Coleoptera: Dermestidae). *Ceylon J. Sci.* (B) **7**, 128–138.

LAIDLAW W. B. R. (1936) The brown lacewing flies; their importance as controls of *Adelges cooleyi* Gillette. *Entomologist's mon. Mag.* **72**, 164–174.

LAING D. R. and CALTAGIRONE L. E. (1969) Biology of *Habrobracon lineatellae* (Hymenoptera: Braconidae). *Can. Ent.* **101**, 135–142.

LAKE C. R. and MILLS R. R. (1975) *In vitro* biosynthesis of oöthecal sclerotization agents from tyrosine by haemolymph of *Periplaneta americana*. *Insect Biochem.* **5**, 659–669.

LAKE C. R., MILLS R. R., and KOEPPE J. K. (1975) *In vivo* conversion of noradrenalin to 3-hydroxy-4-0-β-D glucosidobenzoic acid by the American cockroach. *Insect Biochem.* **5**, 223–229.

LAKER A. (1881) The cocoon of *Hydrophilus piceus* and *Hydrobius fuscipes*. *Entomologist* **14**, 82–84.

LAL K. B. (1934) The biology of Scottish Psyllidae. *Trans. R. ent. Soc. Lond.* **82**, 363–385.

LAL L., KATIYAR O. P., SINGH J., and MUKHARJI S. P. (1973) A new host of *Tyrophagus putrescentiae* (Schrank) (Tyroglyphidae: Acarina) at Varanasi. *Bull. Grain Tech.* **11**, 68–69.

LAL R. (1953) Notes on the effect of temperature on the developmental stages of *Anopheles subpictus* Grassi and *Anopheles stephensi* Liston. *Indian J. Ent.* **15**, 97–106.

LAL R. and GUPTA S. B. L. (1953) Morphology of the immature stages of *Sphaerophoria scutellaris* (Fabr.) (Syrphidae: Diptera) with notes on its biology. *Indian J. Ent.* **15**, 207–218.

LALL B. S. (1962) On the biology of *Pediobius foveolatus* (Crawford) (Eulophidae: Hymenoptera). *Indian J. Ent.* **23**, 268–273 (1961).

LAMB R. J. (1976) Parental behaviour in the Dermaptera with special reference to *Forficula auricularia* (Dermaptera: Forficulidae). *Can. Ent.* **108**, 609–619.

LAMBORN W. A. (1914) On the relationship between certain West African insects, especially ants, Lycaenidae and Homoptera. *Trans. ent. Soc. Lond.* **1913**, 438–498.

LANDIS B. J., WALLIS R. L., and REDMOND R. D. (1967) *Psilopa leucostoma*, a new leaf miner of sugar beets in the United States. *J. econ. Ent.* **60**, 115–118.

LANE C. (1962) Differences in the egg-laying habits of the five spot burnet (*Zygaena longicerae* von Schev.) and the six-spot burnet (*Z. filipendulae* L.) (Lep., Zygaenidae). *Entomologist's Gaz.* **13**, 11–12.

LANGE W. H. (1950) Biology and systematics of plume moths of the genus *Platyptilia* in California. *Hilgardia* **19**, 561–668.

LANIER G. N., BIRCH M. C., SCHMITZ R. F., and FURNISS M. M. (1972) Pheromones of *Ips pini* (Coleoptera: Scolytidae): variation in response among three populations. *Can. Ent.* **104**, 1917–1923.

LANIER G. N. and BURKHOLDER W. E. (1974) Pheromones in speciation of Coleoptera. In *Pheromones. Frontiers of Biology* **32**, 161–189 (M. C. Birch, ed.), Amsterdam and London.

LANIER G. N. and WOOD D. L. (1975) Specificity of response to pheromones in the genus *Ips* (Coleoptera: Scolytidae). *J. chem. Ecol.* **1**, 9–23.

LAPIE G. E. (1923) Les chenilles venimeuses et les accidents éruciques. Thesis published by Nancy Societé d'impressions typographiques, pp. 193.

LARA E. F. (1965) The banana stalk borer *Castniomera humboldti* (Boisduval) in La Estrella Valley, Costa Rica. II. Bionomics. *Turrialba* **14**, 188–195.

LARSEN E. B. (1936) Biologische Studien über die tunnelgraben Käfer auf Skallingen. *Vidensk. Meddr dansk naturh. Foren.* **100**, 1–231.

LARSEN E. B. (1943) The influence of humidity on life and development of insects. *Vidensk. Meddr dansk naturh. Foren.* **107**, 127–184.

LARSEN E. B. (1953) Studies on the soil fauna of Skallingen. Qualitative and quantitative studies on alterations in the beetle fauna during five years' natural development of some sand and salt-marsh biotopes. *Oikos* **3**, 166–192 (1951).

LARSÉN O. (1927) Über die Entwicklung und Biologie von *Aphelocheirus aestivalis* Fabr. *Ent. Tidskr.* **48**, 181–206.

LARSON A. O. (1927) The host-selection principle as applied to *Bruchus quadrimaculatus* Fab. *Ann. ent. Soc. Am.* **20**, 37–79.

LARSON A. O., BRINDLEY T. A., and HINMAN F. G. (1938) Biology of the pea weevil in the

Pacific northwest with suggestions for its control on seed peas. *Tech. Bull. US Dep. Agric.* **599**, 1–48.

LARSON A. O. and FISHER C. K. (1924) Longevity and fecundity of *Bruchus quadrimaculatus* Fab. as influenced by different foods. *J. agric. Res.* **29**, 297–305.

LATTIN J. D. (1955) The eggs of *Corimelaena virilis* (McAtee and Malloch) (Hemiptera: Cydnidae: Corimalaeninae). *Pan-Pacif. Ent.* **31**, 63–66.

LAUBMANN M. (1959) Der Einfluss von Länge, Breite und Materialbeschaffenheit von Spalten auf die Eiablage des Getreidekapuziners, *Rhizopertha dominica* Fab. *Anz. Schädlingsk.* **32**, 161–166.

LAUCK D. and MENKE A. S. (1961) The higher classification of the Belostomatidae (Hemiptera). *Ann. ent. Soc. Am.* **54**, 644–657.

LAUDANI U. and BIANCHI U. (1973) Comportamento discriminante di *Anopheles atroparvus* durante l'ovodeposizione. *Riv. Parassit.* **34**, 73–81.

LAUGHLIN R. (1953) Absorption of water by the egg of the garden chafer *Phyllopertha horticola* L. *Nature, Lond.* **171**, 577.

LAUGHLIN R. (1957) Absorption of water by the egg of the garden chafer *Phyllopertha horticola* L. *J. exp. Biol.* **34**, 226–236.

LAUGHLIN R. (1958a) Desiccation of eggs of the crane fly (*Tipula oleracea* L.). *Nature, Lond.* **182**, 613.

LAUGHLIN R. (1958b) The rearing of crane flies (Tipulidae). *Entomologia exp. appl.* **1**, 241–245.

LAURENCE B. R. (1954) The larval inhabitants of cow pats. *J. Anim. Ecol.* **23**, 234–260.

LAURENCE B. R. (1959) Oviposition by *Mansonioides* mosquitoes in the Gambia, West Africa. *Proc. R. ent. Soc. Lond.* (A) **34**, 161–170.

LAURENCE B. R. (1960) The biology of two species of mosquito, *Mansonia africana* (Theobald) and *Mansonia uniformis* (Theobald), belonging to the subgenus *Mansonioides* (Diptera, Cilicidae). *Bull. ent. Res.* **51**, 491–517.

LAURENCE B. R. and SAMARAWICKREMA W. A. (1970) Aggregation by ovipositing *Mansonioides* mosquitoes. *J. med. Ent.* **7**, 594–600.

LAUVERJAT S. (1965) Données histologiques et histochimiques sur les voies génitales femelles et sur la sécrétion de l'oothèque chez quelques Acridiens (Orth., Acridoidea). *Annls Soc. ent. Fr.* (NS) **1**, 879–935.

LAVABRE E. M. (1958–1959) Le scolyte des branchettes du caféier robuste *Xyleborus morstatti* Haged. *Café, Cacao, Thé* **2**, 119–130; **3**, 21–33.

LAVIGNE R. J. (1963) Notes on the behaviour of *Stenopogon coyote* Bromley with a description of the eggs (Diptera: Asilidae). *Pan-Pacif. Ent.* **39**, 103–107.

LAVOIPIERRE M. M. J. and REYNAUD P. (1953) A simple way of counting mosquito eggs. *Annls trop. Med. Parasit.* **47**, 307–308.

LAWKO C. M. and DYER E. D. A. (1974) Flight ability of spruce beetle emerging after attacking frontalin-baited trees. *Bi-mon. Res. Notes* **30**, 17.

LAWLOR W. K. (1940) Notes on a variation in the eggs of *Anopheles punctipennis* Say. *Publ. Hlth Rep. Wash.* **55**, 371–373.

LAWSON D. E. and WEEKMAN G. T. (1966) A method of recovering eggs of the western corn rootworm from the soil. *J. econ. Ent.* **59**, 657–659.

LAWSON F. A. (1951) Structural features of the oothecae of certain species of cockroaches (Orthoptera, Blattidae). *Ann. ent. Soc. Am.* **44**, 269–285.

LAWSON F. A. (1952) Structural features of cockroach egg capsules. II. The ootheca of *Cariblatta lutea lutea* (Orthoptera: Blattidae). *Ohio J. Sci.* **52**, 296–300.

LAWSON F. A. (1953) Structural features of cockroach egg capsules. III. The ootheca of *Euryotis floridana* (Orthoptera: Blattidae). *J. Tenn. Acad. Sci.* **28**, 28–33.

LAWSON F. A. (1954) Structural features of cockroach egg capsules. IV. The ootheca of

Parcoblatta uhleriana (Orthoptera: Blattidae). *J. Kans. ent. Soc.* **27**, 14–20.

LAWSON F. A. (1967) Structural features of cockroach egg capsules. V. The ootheca of *Lamproblatta albipalpus* Hebard (Orthoptera: Blattidae). *J. Kans. ent. Soc.* **40**, 601–607.

LAWSON F. A. (1976) Egg and larval case formation by *Pachybrachis bivittatus*. *Annls. ent. Soc. Am.* **69**, 942–944.

LAWSON F. R. (1959) The natural enemies of the hornworms on tobacco (Lepidoptera: Sphingidae). *Ann. ent. Soc. Am.* **52**, 741–755.

LAZAREVIĆ B. (1958) Jabukin svrdlaš. *Zashch. Rast.* **43**, 29–53 (1957) (English summary).

LAZAREVIĆ B. (1963) A contribution to knowledge of *Rhynchites aequatus* L. *Zashch. Rast.* **72**, 211–222 (in Yugoslavian; English summary).

LEA A. O. and EDMAN J. D. (1972) Sexual behaviour of mosquitoes. 3. Age dependence of insemination of *Culex nigripalpus* and *C. pipiens quinquefasciatus* in nature. *Ann. ent. Soc. Am.* **65**, 290–293.

LEA A. O. and EVANS D. G. (1972) Sexual behaviour of mosquitoes. 1. Age dependence of copulation and insemination in the *Culex pipiens* complex and *Aedes taeniorhynchus* in the laboratory. *Ann. ent. Soc. Am.* **65**, 285–289.

LEAHY M. G. (1962) Male accessory gland substance as a stimulant for oviposition in mosquitoes (abstract). *Bull. ent. Soc. Am.* **8**, 163.

LEAHY M. G. (1966) Egg deposition in *D. melanogaster* increased by transplant of male paragonia. *Drosoph. Inf. Serv.* **41**, 145–146.

LEAHY M. G. (1967) Non-specificity of the male factor enhancing egg-laying in Diptera. *J. Insect Physiol.* **13**, 1283–1292.

LEAHY M. G. (1970) Effect of the male accessory gland secretion on mosquitoes and fruit flies. *Colloques int. Cent. natn. Rech. scient.* **189**, 297–309.

LEAHY M. G. (1973a) Oviposition of virgin *Schistocerca gregaria* (Forskål) (Orthoptera: Acrididae) after implant of the male accessory gland complex. *J. Ent.* (A) **48**, 69–78.

LEAHY M. G. (1973b) Oviposition of *Schistocerca gregaria* (Forskål) (Orthoptera: Acrididae) mated with males unable to transfer spermatophores. *J. Ent.* (A) **48**, 79–84.

LEAHY M. G. and CRAIG G. B. (1965) Male accessory gland substance as a stimulant for oviposition in *Aedes aegypti* and *A. albopictus*. *Mosq. News* **25**, 448–452.

LEAHY M. G. and CRAIG G. B. JR (1967) Barriers to hybridization between *Aedes aegypti* and *Aedes albopictus* (Diptera: Culicidae). *Evolution* **21**, 41–58.

LEAHY M. G. and LOWE M. L. (1966) Extraction of a substance from male accessory glands that enhances oviposition in Diptera. *Bull. ent. Soc. Am.* (abstract) **12**, 302.

LEAHY M. G. and LOWE M. L. (1967) Purification of the male factor increasing egg deposition in *D. melanogaster*. *Life Sci.* **6**, 151–156.

LE BERRE J. R. and JAILLET I. (1967) Inhibition du comportement de ponte du criquet migrateur (*Locusta migratoria* L.) provoquée par la salinité du sol. *C. r. hebd. Séanc. Acad. Sci. Paris* **265D**, 430–433.

LE BERRE J. R., LAUNOIS M., and LUONG H. (1969) Étude de l'influence de congénères sur le comportement de ponte du criquet *Locusta migratoria cinerascens* (Fab.). *C. r. hebd. Séanc. Soc. Biol. Paris* **163**, 1078–1082.

LE BERRE J. R. and PORTIER G. (1963) Utilisation d'un hétéroptère Pentatomidae *Perillus bioculatus* (Fabr.) dans la lutte contre le doryphore *Leptinotarsa decemlineata* (Say): premiers résultats obtenus en France. *Entomophaga* **8**, 183–190.

LEBRUN D. (1960) Recherches sur la biologie et l'éthologie de quelques Hétéroptères aquatiques. *Annls Soc. ent. Fr.* **129**, 179–199.

LÉCAILLON A. (1898) Sur les enveloppes ovulaires de quelques Chrysomelides. *Archs Anat. microsc.* **2**, 89–117.

LECATO G. L. (1975) Predation by red flour beetle on sawtoothed grain beetle. *Envir. Ent.* **4**,

504–506.
LeCato G. L. and Collins J. M. (1976) *Xylocoris flavipes*: maximum kill of *Tribolium castaneum* and minimum kill required for survival of the predator. *Envir. Ent.* **5**, 1059–1061.
LeCato G. L. and Pienkowski R. L. (1970) Effects of temperature and presence of males on laboratory oviposition by the alfalfa weevil. *J. econ. Ent.* **63**, 897–900.
LeCato G. L. and Pienkowski R. L. (1972a) Reproductive efficiency of the alfalfa weevil *Hypera postica* at constant and alternating temperatures. *Envir. Ent.* **1**, 166–169.
LeCato G. L. and Pienkowski R. L. (1972b) High- or low-temperature treatments affecting alfalfa weevil fecundity, egg fertility, and longevity. *J. econ. Ent.* **65**, 146–148.
LeCato G. L. and Pienkowski R. L. (1972c) Alfalfa weevil oviposition: influence of sperm stored in the spermatheca. *Ann. ent. Soc. Am.* **65**, 979–980.
LeCato G. L. and Pienkowski R. L. (1972d) Fecundity, egg fertility, duration of oviposition, and longevity of alfalfa weevils from eight mating and storage conditions. *Ann. ent. Soc. Am.* **65**, 319–323.
Leclercq J. (1940) Les problèmes de la ponte chez les Hyménoptères. Aperçu critique. *Lambillionea* **40**, 19–25.
Leclercq J. (1946) Effect of atmospheric humidity on the eggs of a phasmid *Carausius (Dixippus) morosus* Br. *Proc. R. ent. Soc. Lond.* (A) **21**, 3–5.
Leclercq J. (1948) Influence des conditions hygrométriques sur les oeufs de *Melasoma populi* L. (Col., Chrysomelidae). *Annls ent. Soc. Belg.* **84**, 26–27.
Leclercq-Smekens M. (1976) Organogénèse et différentiation des voies génitales femelles d'*Euproctis chrysorrhea* L. (Lépidoptère: Lymantriidae). *Int. J. Insect Morph. Embryol.* **5**, 241–252.
Ledesma L. (1971) Notas relativas a la distribución y predación de puestas de procesionaria del pino (*Thaumetopoea pityocampa* Schiff.) sobre pies en edades de monte bravo y latizal de pino negral (*Pinas laricio* Poir.). *Boln Serv. Plagas for.* **14** (27) 71–80.
Ledoux A. (1949) La ponte des ouvrières de la fourmie fileuse (*Oecophylla longinoda* Latr.). *C. r. hebd. Séanc. Acad. Sci. Paris* **228**, 1154–1155.
Ledoux A. (1958) Biologie et comportement de l'Embioptère *Monotyla ramburi* Rims.-Kors. *Annls Sci. nat.* **12** (11) 515–532.
Lee D. J. and Woodhill A. R. (1944) The anopheline mosquitoes of the Australasian Region. *Publ. Univ. Sydney, Zool.* **2**, 1–209.
Lee R. D. (1954) Oviposition by the poultry bug. *J. econ. Ent.* **47**, 224–226.
Lees A. D. (1964) The location of the photoperiodic receptors in the aphid *Megoura viciae* Buckton. *J. exp. Biol.* **41**, 119–133.
Lees A. H. (1916) Some observations on the egg of *Psylla mali*. *Ann. appl. Biol.* **2**, 251–257.
Leeson H. S. (1941) The effect of temperature upon the hatching of the eggs of *Pediculus humanus corporis* De Geer (Anoplura). *Parasitology* **33**, 243–249.
Le Faucheux M. (1961) Contribution à l'étude du cycle biologique de *Vermelio degeeri* Macquart (Diptère Rhagionidae): Ponte et éclosion des jeunes larves. *Bull. Soc. sci. Bretagne* **36**, 133–141.
Lefkovitch L. P. (1957) The biology of *Cryptolestes ugandae* Steel and Howe (Coleoptera, Cucujidae), a pest of stored products in Africa. *Proc. zool. Soc. Lond.* **128**, 419–429.
Lefkovitch L. P. (1967) A laboratory study of *Stegobium paniceum* (L.) (Coleoptera: Anobiidae). *J. stored Prod. Res.* **3**, 235–249.
Lefkovitch L. P. and Brust R. A. (1968) Locating the eggs of *Aedes vexans* (Mg.) (Diptera: Culicidae). *Bull. ent. Res.* **58**, 119–122.
Lefkovitch L. P. and Currie J. E. (1967) Factors influencing fecundity in *Lasioderma serricorne* (F.) (Coleoptera, Anobiidae). *J. stored Prod. Res.* **3**, 199–212.
Lefroy H. M. (1909) *Indian Insect Life*, Calcutta.

LEGAY J. M. (1976) Effets de la température sur la morphogénèse des ovocytes de *Bombyx mori* L. C. r. hebd. Séanc. Acad. Sci. Paris (D) **282**, 1313–1316.

LEGER M. and MOUZELS P. (1918) Dermatose provigineuse determinée par des papillons saturnides du genre *Hylesia*. Bull. Soc. Path. exot. **11**, 104–107.

LEGNER E. F. (1968) Parasite activity related to ovipositional responses in *Hippelates collusor*. J. econ. Ent. **61**, 1160–1163.

LEHMENSICK R. and LIEBERS R. (1937) Die Oberflächenstruktur von Motteneiern als Bestimmungsmerkmal. Z. angew. Ent. **24**, 436–447.

LEKANDER B. (1959) Der doppeläugige Fichtenbastkäfer *Polygraphus poligraphus* L. Ein Beitrag zur Kenntnis seiner Morphologie, Anatomie, Biologie und Bekämpfung. Medd. Skogsforskn Inst. **48** (9) 1–127 (with a summary in Swedish).

LEKANDER B. (1963) *Xleborus cryptographus* (Col., Ipidae). Ein Beitrag zur Kenntnis seiner Verbreitung und Biologie. Ent. Tidskr. **84**, 96–109.

LEKIĆ M. (1966) Prilog proucavanju ekologije populacija *Coroebus rubi* L. na malinki kao hraniteljki. Arh. poljopr. Nauke Teh. **19** (65) 132–144 (English summary).

LEKIĆ M. (1970) *Cicadetta dimissa* Hagen-nova štetočina maline. Arh. poljopr. Nauke Teh. **20**, 3–14 (English summary).

LENGERKEN H. VON (1922) Eisprenger bei Carabidenlarven. Zool. Anz. **54**, 18–21.

LENGERKEN H. VON (1929) Die Blattschnittmethode des Ahornblattrollers (*Deporaus tristis* F.) (Coleopt.). Biol. Zbl. **49**, 469–490.

LENGERKEN H. VON (1951a) Zur Brutbiologie des Pappelblattrollers (*Byctiscus populi* L.). Z. angew. Ent. **32**, 599–603.

LENGERKEN H. VON (1951b) Der Ahornblattroller (*Deporaus tristis* F.) auch an Rotbuche? Zool. Anz. **146**, 27–30.

LENGERKEN H. VON (1951c) Zur Brutbiologie des spanischen Mondhornkäfers (*Copris hispanus* L.). Biol. Zbl. **70**, 418–432.

LENGERKEN H. VON (1953) Die Brutbiologie von *Copris hispanus* L. Trans. 11th Int. Congr. Ent. **2** (Symp.) 117–124.

LENGERKEN H. VON (1954) *Die Brutfürsorge- und Brutpflegeinstinkte der Käfer*, 2nd edn., Leipzig, Akad. Verlagsges., Geest and Portig, 383 pp.

LENGERKEN H. VON (1955) Brutpflege-Instinkte beim Mondkornkäfer. Leopoldina **1** (3) 37–40.

LENGERKEN H. VON (1961) Zum Brutfürsorgeverhalten des Ahornblattrollers (*Deporaus tristis* F.), Curculionidae, Coleoptera. Zool. Anz. **167**, 442–448.

LENZ F. (1921) Die Eiablage von *Cylindrotoma distinctissima* (Mg.). Arch. Naturgesch. **87** (A, 7) 128–135.

LEONARD D. E., BIERL B. A., and BEROZA M. (1975) Gypsy moth kairomones influencing behavior of the parasitoids *Brachymeria intermedia* and *Apanteles melanoscelus*. Envir. Ent. **4**, 929–930.

LEONARD D. E. and DOANE C. C. (1966) An artificial diet for the gypsy moth *Porthetria dispar* (Lepidoptera: Lymantriidae). Ann. ent. Soc. Am. **59**, 462–464.

LEONARD D. E., SIMMONS G. A., and VANDERWERKER G. K. (1974) Spruce budworm: techniques to improve counting of eggs. J. econ. Ent. **66**, 992 (1973).

LEONG C. Y. (1962) The life-history of *Anisops breddini* Kirkaldy (Hemiptera, Notonectidae). Ann. Mag. nat. Hist. **5** (13) 377–383.

LEONG J. K. L. and OATMAN E. R. (1968) The biology of *Campoplex haywardi* (Hymenoptera: Ichneumonidae), a primary parasite of the potato tuberworm. Ann. ent. Soc. Am. **61**, 26–36.

LÉONIDE J. C. (1964) Contribution à l'étude de la biologie du *Symmictus costratus* Loew, Diptère, Némestrinidé Acridiophage. IV. La ponte et l'infestation de l'hôte. Bull. Soc. zool. Fr. **89**, 135–142.

LEOPOLD R. A. (1970) Cytological and cytochemical studies on the ejaculatory duct and accessory secretion in *Musca domestica* L. *J. Insect Physiol.* **16**, 1859–1872.

LEOPOLD R. A. (1976) The role of male accessory glands in insect reproduction. *A. Rev. Ent.* **21**, 199–221.

LEOPOLD R. A. and TERRANOVA A. C. (1970) The distribution and persistence of ^{3}H-labelled male accessory secretion in mated female houseflies. *Proc. N. Dak. Acad. Sci.* **24**, 20.

LEOPOLD R. A., TERRANOVA A. C., and SWILLEY E. M. (1970) Studies on the biosynthetic mechanisms concerned with mating refusal in the housefly. *Proc. N. cent. Brch Am. Ass. econ. Ent.* **25**, 33 (Abstract).

LEOPOLD R. A., TERRANOVA A. C., and SWILLEY E. M. (1971a) Mating refusal in *Musca domestica*: Effects of repeated mating and decerebration upon frequency and duration of copulation. *J. exp. Zool.* **176**, 353–359.

LEOPOLD R. A., TERRANOVA A. C., THORSON B. J., and DEGRUGILLIER M. E. (1971b) The biosynthesis of the male housefly accessory secretion and its fate in the mated female. *J. Insect Physiol.* **17**, 987–1003.

LÉPINEY J. DE (1930) Contribution à l'étude du complexe biologique de *Lymantria dispar*. *Mém. Soc. Sci. nat. Maroc* **23**, 1–100.

LÉPINEY J. DE (1933) Le rôle de la Direction des Eaux et Forêts du Maroc et de l'Institut Scientifique Chérifien dans la lutte biologique entreprise contre *Lymantria dispar* à l'aide de *Schedius kuwanae*. *Int. Congr. Ent.* **5** (2) 807–812.

LÉPINEY J. DE and MIMEUR J. M. (1930) Sur *Glossista infuscata* Meig. et *Anastoechus nitidulus* F., parasites marocains de *Dociostaurus maroccanus* Thunb. *Rev. Path. vég. Ent. agric. Fr.* **17**, 419–430.

LEPPLA N. C., CARLYLE S. L., and CARLYLE T. C. (1974) Effects of surface sterilization and automatic collection on cabbage looper eggs. *J. econ. Ent.* **67**, 33–36.

LERGENMÜLLER E. (1958) Ökologische Untersuchungen am Getreideplattkäfer *Oryzaephilus surinamensis* L. *Z. angew. Zool.* **45**, 31–97.

LEROI B. (1975) Influence d'une plant-hôte des larves (*Apium graveolens* L.) sur la stimulation de la ponte et de la production ovarienne de *Philophylla heraclei* L. (Diptère, Tephritidae). *C. r. hebd. Séanc. Acad. Sci. Paris* (D) **281**, 1015–1018.

LEROY Y. (1969) Rôle de l'ultrastructure tégumentaire lors du cheminement de l'oeuf dans l'oviscapte d'une sauterelle lors de la ponte. *C. r. hebd. Séanc. Acad. Sci. Paris* (D) **269**, 1976–1978.

LESKA W. (1965) Badania nad biologia i szkodliwością kwieciaka malinowca *Anthonomus rubi* Hbst. (Col., Curculionidae). *Polskie Pismo ent.* (B) **1965**, 81–142.

LESKA W. (1967) Badania nad biologia kistnika malinowca—*Byturus tomentosus* F. (Col. Byturidae). *Polskie Pismo ent.* **37**, 357–372 (English summary).

LESPÉRON L. (1937) Recherches cytologiques et expérimentales sur les sécrétions de la soie et sur certains mécanismes excréteurs chez les insectes. *Archs Zool. exp. gén.* **79**, 1–156.

LESSE H. DE (1953) Observations sur la ponte de quelques *Erebia*. *Lambillionea* **53**, 45–47.

LESTAGE J. A. (1921) In Rousseau E., *Les larves et nymphes aquatiques des insectes d'Europe*, Brussels.

LESTAGE J. A. (1922) Note sur la ponte immerge des *Micrasema* (Trichoptera). *Annls biol. lacustre* **11**, 152–162.

LESTON D. (1953) The eggs of Tingitidae (Hem.), especially *Acalypta parvula* (Fallén). *Entomologist's mon. Mag.* **89**, 132–134.

LESTON D. (1954) The eggs of *Anthocoris gallarumulmi* (Deg.) (Hem., Anthocoridae) and *Monanthia humuli* (F.) (Hem., Tingidae), with notes on the eggs of Cimicioidea and Tingiodea. *Entomologist's mon. Mag.* **90**, 99–102.

LESTON D. (1955) Notes on the Ethiopian Pentatomoidea (Hem.): XVIII, the eggs of three

Nigerian shield bugs with a tentative summary of egg forms in Pentatomoidea. *Entomologist's mon. Mag.* **91**, 33–36.

LESTON D. (1961) Egg batches and agglomerates of *Eysarcoris fabricii* (Kirkaldy) (Hem., Pentatomidae). *Entomologist's mon. Mag.* **97**, 27–29.

LESTON D. and SOUTHWOOD T. R. E. (1954) The structure of the egg and the egg-burster of *Sehirus bicolor* (L.) (Hem., Cydnidae). *Entomologist's mon. Mag.* **90**, 291–292.

LEUCKHART R. (1855) Ueber die Mikropyle und den feineren Bau der Schalenhaut bei den Insekteneiern. *Arch. Anat. Physiol. wiss. Med.* **1855**, 90–264.

LEUZINGER H. (1926) Zur Kenntnis der Anatomie und Entwicklungsgeschichte von *Carausius morosus* Br. 1. Eibau und Keimblatterbildung. In Leuzinger H., Wiesmann R., and Lehmann F. E. (1926) *Zur Kenntnis der Anatomie und Entwicklungsgeschichte der Stabheuschrecke* Carausius morosus *Br.*, Gustav Fischer, Jena.

LEVINE E. and CHANDLER L. (1976) Biology of *Bellura gortynoides* (Lepidoptera: Noctuidae), a yellow water lily borer, in Indiana. *Ann. ent. Soc. Am.* **69**, 405–414.

LEVINSON H. Z. and ILAN A. R. B. (1971) Assembling and alerting scents produced by the bedbug *Cimex lectularius* L. *Experientia* **27**, 102–103.

LEWALLEN L. (1954) Biological and toxicological studies of the little housefly. *J. econ. Ent.* **47**, 1137–1141.

LEWIS D. J. (1961) The *Simulium neavei* complex (Diptera, Simuliidae) in Nyasaland. *J. Anim. Ecol.* **30**, 303–310.

LEWIS D. J. (1962) Some recent observations on African Simuliidae. *Proc. 11th Int. Congr. Ent.* **3**, 131–134 (1960).

LEWIS D. J. and DISNEY R. H. L. (1969) A new phoretic *Simulium* from West Cameroon (Diptera: Simuliidae). *Proc. R. ent. Soc. Lond.* (B) **38**, 117–120.

LEWIS D. J., DISNEY R. H. L., and CROSSKEY R. W. (1969) A new phoretic species of *Simulium* (Dipt., Simuliidae) from West Cameroon, with taxonomic notes on allied forms. *Bull. ent. Res.* **59**, 229–239.

LEWIS D. J., LYONS G. R. L., and MARR J. D. M. (1961) Observation on *Simulium damnosum* from the Red Volta in Ghana. *Ann. trop. Med. Parasit.* **55**, 202–210.

LEWIS D. J., REID E. T., CROSSKEY R. W., and DAVIES J. B. (1960) Attachment of immature Simuliidae to other arthropods. *Nature, Lond.* **187**, 618–619.

LEWIS F. B. (1960) Factors affecting assessment of parasitization by *Apanteles fumiferanae* Vier. and *Glypta fumiferanae* (Vier.) on spruce budworm larvae. *Can. Ent.* **92**, 881–891.

LEWIS L. F. and CHRISTENSON D. M. (1973) Influence of grids on mosquito oviposition in steel cemetery vases. *Mosq. News* **33**, 525–528.

LEWIS T. (1973) *Thrips, their Biology, Ecology and Economic Importance*, Academic Press, London.

LEWIS W. J. and JONES R. L. (1971) Substance that stimulates host-seeking by *Microplitis croceipes* (Hymenoptera: Braconidae), a parasite of *Heliothis* species. *Ann. ent. Soc. Am.* **64**, 471–473.

LEWIS W. J., JONES R. L., and SPARKS A. N. (1972) A host-seeking stimulant for the egg parasite *Trichogramma evanescens*: its source and a demonstration of its laboratory and field activity. *Ann. ent. Soc. Am.* **65**, 1087–1089.

LEWIS W. J. and SNOW J. W. (1971) Fecundity, sex ratios, and egg distribution by *Microplitis croceipes*, a parasite of *Heliothis*. *J. econ. Ent.* **64**, 6–8.

LEWIS W. J., SPARKS A. N., and REDLINGER L. M. (1971) Moth odor: a method of host-finding by *Trichogramma evanescens*. *J. econ. Ent.* **64**, 557–558.

LEWIS W. J. and VINSON S. B. (1968) Immunological relationships between the parasite *Cardiochiles nigriceps* Vierick and certain *Heliothis* species. *J. Insect Physiol.* **14**, 613–626.

LEWIS W. J. and VINSON S. B. (1971) Suitability of certain *Heliothis* (Lepidoptera: Noctuidae) as hosts for the parasite *Cardiochiles nigriceps*. *Ann. ent. Soc. Am.* **64**, 970–972.

LEYDIG F. (1867) Der Eierstock und die Samentasche der Insekten. *Nova Acta Acad. Caesar. Leop. Carol.* **33**, 1–88.

LHOSTE J. (1941) Importance relative des soins maternels chez *Forficula auricularia* L. *C. r. hebd. Séanc. Soc. Biol. Paris* **135**, 499–500.

LHOSTE J., RAUCH F., and CAUWER P. (1976) Influence de l'accouplement sur la ponte et la fécondité de *Laspeyresia pomonella* L. *Rev. Zool. agric. Path. vég.* **75**, 93–102.

LIBBEY L. M., MORGAN M. E., PUTNAM T. B., and RUDINSKY J. A. (1974) Pheromones released during inter- and intra-sex response of the scolytid beetle *Dendroctonus brevicomis*. *J. Insect Physiol.* **20**, 1667–1671.

LIBBEY L. M., MORGAN M. E., PUTNAM T. B., and RUDINSKY J. A. (1976) Isomer of antiaggregative pheromone identified from male Douglas-fir beetle: 3-methylcyclohex-3-en-1-one. *J. Insect. Physiol.* **22**, 871–873.

LIEBERMANN J. (1951a) Sobre una nueva forma de oviposición en un Acridio Sudamericano. *Revta Invest. agríc. B. Aires* **5**, 235–280.

LIEBERMANN J. (1951b) Acridios del territorio del Chubut y de la zona militar de Commodoro Rivadavia. *Idia* **4**, 21–32.

LIEBERMANN J. (1958) Sistemática de la oviposición epidáfica en acridoideos (Orth. Caelif. Acrid.). *Rec. Soc. ent. argent.* **20**, 41–44 (1957).

LIECHTI P. M. and BELL W. J. (1975) Brooding behavior of the Cuban burrowing cockroach *Byrsotria fumigata* (Blaberidae, Blattaria). *Insectes soc.* **22**, 35–45.

LIEFTINCK M. A. (1953) Biological and ecological observations on a bark hunting mantid in Java. *Trans. 11th Int. Congr. Ent.* **2** (Symp.) 125–134.

LIKVENTOV A. V. (1960) On the suppressive effect of plantations of oak and lime on the reproduction of the gipsy moth. *Trudȳ vses. Inst. Zashch. Rast.* **15**, 33–40 (in Russian).

LIN-CHOW S. H. and SCHMITT J. B. (1974) Ovipositional behavior of surgically-treated cabbage looper *Trichoplusia ni* (Hubner) (Lepidoptera: Noctuidae) in laboratory. *Pl. Prot. Bull. Taiwan* **16**, 31–34.

LINCOLN D. C. R. (1961) The oxygen and water requirements of the egg of *Ocypus olens* Müller (Staphylinidae, Coleoptera). *J. Insect Physiol.* **7**, 265–272.

LINCOLN D. C. R. (1962) The structure and physiology of some beetle eggs, Part I, PhD thesis, University of Bristol.

LINCOLN D. C. R. (1965) Structure of the egg-shell of *Culex pipiens* and *Mansonia africana* (Culicidae, Diptera). *Proc. zool. Soc. Lond.* **145**, 9–17.

LINDBERG H. (1950) Notes on the biology of Dryinids. *Soc. Scient. Fenn. Comm. Biol. X* **15**, 1–19.

LINDGREN D. L. and VINCENT L. E. (1959) Biology and control of *Trogoderma granarium* Everts. *J. econ. Ent.* **52**, 312–319.

LINDNER E. (1958a) Zur Kenntnis der Eier der Limoniidae (Diptera, Tipuliformia). *Mitt. zool. Mus. Berl.* **34**, 113–133.

LINDNER E. (1958b) Pilzbewohnende Limoniidenlarven unter besonderer Berücksichtigung von *Limonia quadrinotata* Meigen (Diptera). *Tijd. Ent.* **101**, 263–281.

LINDNER E. (1959) Beiträge zur Kenntnis der Larven der Limoniidae (Diptera). *Z. Morph. Ökol. Tiere* **48**, 209–319.

LINDQUIST A. W. (1933) Amounts of dung buried and soil excavated by certain Coprini (Scar.). *J. Kans. ent. Soc.* **6**, 109–123.

LINDQUIST A. W. (1935) Notes on the habits of certain coprophagous beetles and methods of rearing them. *Circ. US Dep. Agric.* **351**, 1–9.

LINDQUIST E. E. (1969) Review of Holarctic tarsonemid mites (Acarina: Prostigmata) parasitizing eggs of Ipine bark beetles. *Mem. ent. Soc. Can.* **60**, 1–111.

LINDQUIST E. E. and BEDARD W. D. (1961) Biology and taxonomy of mites of the genus

Tarsonemoides (Acarina: Tarsonemidae) parasitizing eggs of bark beetles of the genus *Ips*. *Can. Ent.* **93**, 982–999.

LINGREN P. D., RIDGWAY R. L., and JONES S. L. (1968) Consumption by several common arthropod predators of eggs and larvae of two *Heliothis* species that attack cotton. *Ann. ent. Soc. Am.* **61**, 613–618.

LINSLEY E. G. (1936) Studies in the genus *Aulicus spinola* (Coleoptera, Cleridae). *Univ. Calif. Publ. Ent.* **6**, 249–262.

LINSLEY E. G. (1959) Ecology of Cerambycidae. *A. Rev. Ent.* **4**, 99–138.

LINSLEY E. G. and MCSWAIN J. W. (1942a) The parasites, predators, and inquiline associates of *Anthrophora linsleyi*. *Am. Midl. Nat.* **27**, 402–417.

LINSLEY E. G. and MCSWAIN J. W. (1942b) Bionomics of the Meloid genus *Hornia* (Coleoptera). *Univ. Calif. Publ. Ent.* **7**, 189–206.

LINSTOW VON (1911) Die Brennhaare der Spinnerraupen. *Int. ent. Z.* **5**, 241–243.

LINSTOW VON (1914) Zur Biologie und Systematik der Psychiden. *Z. wiss. Insekt Biol.* **10**, 67–71.

LIPA J. J. (1969) Studies on *Arma custos* (Fabr.) (Hemiptera, Pentatomidae). *Pr. nauk. Inst. Ochr. Rośl.* **11**, 197–214 (in Polish; English summary).

LIPA J. J. and WIKLENDT M. (1972) The influence of storage at low temperatures on the viability of eggs and pupae of the spotted cutworm (*Agrotis c-nigrum* L.) (Lepidoptera: Noctuidae). *Pr. nauk. Inst. Ochr. Rośl.* **13**, 163, 168 (1971) (in Polish; English summary).

LIPKOW E. (1968) Zum Eiablage-Verhalten der Staphyliniden. *Pedobiologia* **8**, 208–213.

LISTOV M. V. (1975) Food selectivity and choice of oviposition substrate of *Tribolium destructor* Uytt. (Coleoptera, Tenebrionidae). *Ent. Obozr.* **54**, 515–518 (in Russian; English summary).

LITSINGER J. A. and APPLE J. W. (1973) Oviposition of the alfalfa weevil in Wisconsin. *Ann. ent. Soc. Am.* **66**, 17–20.

LIVINGSTONE D. (1962) On the biology and immature stages of a sap-sucker on *Ziziphus jujuba, Monesteira minutula* Mont., a species new to India (Hemiptera: Tingidae). *Agra Univ. J. Res. (Science)* **11**, 117–129.

LIVINGSTONE D. (1967) On the functional anatomy of the egg of *Tingis buddleiae* Drake (Heteroptera: Tingidae). *J. zool. Soc. India* **19**, 111–119.

LLOYD D. C. (1940) Host selection by hymenopterous parasites of the moth *Plutella maculipennis* Curtis. *Proc. R. Soc.* (B) **128**, 451–484.

LLOYD D. C. (1956) Studies of parasite oviposition behaviour. I. *Mastrus carpocapsae* Cushman (Hymenoptera: Ichneumonidae). *Can. Ent.* **88**, 80–89.

LLOYD D. C. (1958) Studies of parasite oviposition behaviour. II. *Leptomastix dactylopii* Howard (Hymenoptera: Encyrtidae). *Can. Ent.* **90**, 450–461.

LLOYD E. P., MCMEANS J. L., and MERKL M. E. (1961) Preferred feeding and egg laying sites of the boll weevil and the effect of weevil damage on the cotton plant. *J. econ. Ent.* **54**, 979–984.

LOAN C. and HOLDAWAY F. G. (1961) *Pygostolus falcatus* (Nees) (Hymenoptera, Braconidae), a parasite of *Sitona* species (Coleoptera, Curculionidae). *Bull. ent. Res.* **52**, 473–488.

LOBANOV A. M. (1969) Determination of females of Calliohorini and Pollenini (Diptera, Calliphoridae) by oviposition. *Zool. Zh.* **48**, 1189–1196 (in Russian; English summary).

LOBANOV A. M. (1976) Morphology of the ovipositor and classification of flies of the subfamily Muscinae (Diptera, Muscidae). *Zool. Zh.* **55**, 1178–1186.

LOBATÓN MÁRQUEZ M. (1958) Algunas investigaciones sobre el parasitismo de los huevos de *Mescinia peruella* Schaus, en el Valle de Pisco. *Revta per. Ent. agríc.* **1**, 23–24.

LOCKE M. (1964) The structure and formation of the integument in insects. In *The Physiology of Insecta* (M. Rockstein, ed.), **3**, 379–470.

LOGEN D. and HARWOOD R. F. (1965) Oviposition of the mosquito *Culex tarsalis* in response to light cues. *Mosq. News* **25**, 462–465.

LOHER W. and CHANDRASHEKARAN M. K. (1970) Carcadian rhythmicity in the oviposition of the grasshopper *Chorthippus curtipennis. J. Insect Physiol.* **16**, 1555–1566.

LOHER W. and HUBER F. (1964) Experimentelle Untersuchungen am Sexualverhalten des Weibchens der Heuschrecke *Gomphocerus rufus* L. (Acridinae). *J. Insect Physiol.* **10**, 13–36.

LOI G. (1965) I tipulidi (Dipt.) dannosi all'agricoltura. *Circol. Oss. Mal. Piante Pisa (Ent.)* **5**, 1–25.

LONG D. B. (1958) Observations on oviposition in the wheat bulb fly *Leptohylemyia coarctata* (Fall.). *Bull. ent. Res.* **49**, 355–366.

LONG D. B. and ZAHER M. A. (1958) Effect of larval population density on the adult morphology of two species of Lepidoptera, *Plusia gamma* L. and *Pieris brassicae* L. *Entomologia exp. appl.* **1**, 161–173.

LOOR K. A. and DEFOLIART G. R. (1969) An oviposition trap for detecting the presence of *Aedes triseriatus* (Say). *Mosq. News* **29**, 487–488.

LOPES H. DE SOUZA (1937) Contribucão ao conhecimento do gênero *Stylogaster. Archos. Inst. Biol. veg. Rio de J.* **3**, 257–293.

LOPES H. DE SOUZA (1938) Sôbre uma nova espécie do gênero *Stylogaster* Macquart do Brazil. *Mem. Inst. Oswaldo Cruz* **33**, 403–405.

LOPES H. DE SOUZA and MONTEIRO L. (1959) Sôbre algumas espécies brasileiras de *Stylogaster* Macq., com a descricao de quatro espécies novas. *Stud. Ent.* **2**, 1–24.

LOPEZ A. W. (1934) Report of the Entomologist. *Philip. Sugar Ass. Rep. 1931–1932*, 252–279.

LORD F. T. (1947) The influence of spray programs on the fauna of apple orchards in Nova Scotia: II. Oystershell scale. *Can. Ent.* **79**, 196–209.

LOSCHIAVO S. R. (1967) Adult longevity and oviposition of *Trogoderma parabile* Beal (Coleoptera, Dermestidae) at different temperatures. *J. stored Prod. Res.* **3**, 273–282.

LOSCHIAVO S. R. (1968) Effect of oviposition sites on egg production and longevity of *Trogoderma parabile* (Coleoptera; Dermestidae). *Can. Ent.* **100**, 86–89.

LOSCHIAVO S. R. and SINHA R. N. (1966) Feeding, oviposition, and aggregation by the rusty grain beetle *Cryptolestes ferrugineus* (Coleoptera: Cucujidae) on seed-borne fungi. *Ann. ent. Soc. Am.* **59**, 578.

LÖSER S. (1969) Brutfürsorge und Brutpflege bei Laufkäfern der Gattung *Abax. Zool. Anz.* (Suppl.) **33**, 322–326.

LOUGHTON B. G. and WEST A. S. (1962) Serological assessment of spider predation on the spruce budworm *Choristoneura fumiferana* (Clem.) (Lepidoptera: Tortricidae). *Proc. ent. Soc. Ont.* **92**, 176–180.

LOVITT A. E. and SODERSTROM E. L. (1968) Predation on Indian meal moth eggs by *Liposcelis bostrychophilus. J. econ. Ent.* **61**, 1444–1445.

LOWE R. E., FORD H. R., SMITTLE B. J., and WEIDHAAS D. E. (1973) Reproductive behaviour of *Culex pipiens quinquefasciatus* released into a natural population. *Mosq. News* **33**, 221–227.

LOWER H. F. (1957) The acellular covering layers of the immature stages of *Aphodius howitti* (Coleoptera; Scarabaeidae). *J. Morph.* **101**, 149–169.

LOZINSKII V. A. (1961) On the correlation between the weight of the pupae and the number and weight of the eggs of *Lymantria dispar. Zool. Zh.* **40**, 1571–1573 (in Russian; English summary).

LUBBOCK J. (1874) *On the Origin and Metamorphosis of Insects*, 2nd edn, Macmillan, London.

LUBBOCK J. (1882) *Ants, Bees, and Wasps*, London.

LUCA Y. DE (1967) Notes éthologiques sur la ponte et la larve néonate de *Bruchidius ater* (Marsh.) (Col. Burchidae). *Bull. Soc. ent. Fr.* **72**, 16–20.

LUCAS F., SHAW J. T. B., and SMITH S. G. (1957) Amino-acid composition of the silk of

Chrysopa egg-stalks. *Nature, Lond.* **179**, 906–907.
LUCAS W. J. (1920) *A Monograph of the British Orthoptera*, Ray Soc., London.
LUCK R. F. and DAHLSTEN D. L. (1967) Douglas-fir tussock moth (*Hemerocampa pseudotsugata*) egg-mass distribution on white fir in north-eastern California. *Can. Ent.* **99**, 1193–1203.
LUDVIK G. F. (1972) A device for producing artificial squares for boll weevil oviposition and feeding. *J. econ. Ent.* **65**, 1200–1201.
LUDWIG D. (1932) The effect of temperature on the growth curves of the Japanese beetle (*Popillia japonica* Newman). *Physiol. Zoöl.* **5**, 431–447.
LUDWIG D. (1942) The effect of different relative humidities, during the pupal stage, on the reproductive capacity of the luna moth *Tropaea luna* L. *Physiol. Zoöl.* **15**, 48–60.
LUDWIG D. (1943) The effect of different relative humidities, during the pupal stage, on the reproductive capacity of the Cynthia moth, *Samia walkeri* Felder and Felder. *Physiol. Zoöl.* **16**, 381–388.
LUDWIG D. and ANDERSON J. M. (1942) Effects of different humidities, at various temperatures, on the early development of four saturniid moths (*Platysamia cecropia* Linnaeus, *Telea polyphemus* Cramer, *Samia walkeri* Felder and Felder, and *Callosamia promethea* Drury) and on the weights and water contents of their larvae. *Ecology* **23**, 259–274.
LUFF M. L. (1976) Notes on the biology of the developmental stages of *Nebria brevicollis* (F.) (Col., Carabidae) and on their parasites *Phaenoserphys* spp. (Hym., Proctotrupidae). *Entomologist's mon. Mag.* **111**, 249–255.
LÜHMANN M. (1940) Beiträge zur Biologie der Chrysomeliden. 7. Beobachtungen an *Phytodecta rufipes* F. *Ent. Bl.* **36**, 8–11.
LUKASIAK J. (1957) Morphological variability of eggs and of the scale index in *Anopheles maculipennis* Meig., 1818. *Acta parasit. polon.* **5**, 599–612 (in Polish; English summary).
LUKEFAHR M. J. and MARTIN D. F. (1966) Cotton-plant pigments as a source of resistance to the bollworm and tobacco budworm. *J. econ. Ent.* **59**, 176–179.
LUKEFAHR M. J., MARTIN D. F., and MEYER J. R. (1965) Plant resistance to five Lepidoptera attacking cotton. *J. econ. Ent.* **58**, 516–518.
LUM P. T. M. and FLAHERTY B. R. (1969) Effect of mating with males reared in continuous light or in light–dark cycles on fecundity in *Plodia interpunctella* Hübner (Lepidoptera: Phycitidae). *J. stored Prod. Res.* **5**, 89–94.
LUM P. T. M. and FLAHERTY B. R. (1970a) Regulating oviposition by *Plodia interpunctella* in the laboratory by light and dark conditions. *J. econ. Ent.* **63**, 236–239.
LUM P. T. M. and FLAHERTY B. R. (1970b) Effect of continuous light on the potency of *Plodia interpunctella* males (Lepidoptera: Phycitidae). *Ann. ent. Soc. Am.* **63**, 1470–1471.
LUM P. T. M. and FLAHERTY B. R. (1972) Effect of carbon dioxide on production and hatchability of eggs of *Plodia interpunctella* (Lepidoptera: Phycitidae). *Ann. ent. Soc. Am.* **65**, 976–977.
LUM P. T. M. and PHILLIPS R. H. (1972) Combined effects of light and carbon dioxide on egg production of Indian meal moths. *J. econ. Ent.* **65** (5) 1316–1317.
LUMARET J.-P. (1971) Cycle biologique et comportement de ponte de *Percus (Pseudopercus) navaricus* (Col. Carabique). *Entomologiste* **27**, 49–52.
LUMSDEN W. H. R. (1944) *Anopheles hispaniola* Theobald, 1903 (Dipt., Culcid.) from the Emirate of Transjordan. *Bull. ent. Res.* **35**, 1, 3–9.
LUNDBECK W. (1914) Some remarks on the eggs and egg deposition of *Halobates*. *Mindeskr. Steenstrups Føds.* **27**, 13.
LUNDIE A. E. (1940) The small hive beetle *Aethina tumida*. Dep. of Agriculture and Forestry (Ent. Series 3), *Sci. Bull.* 220 *(S. Africa)*, 30 pp.
LYLE G. T. (1908) Ova of *Raphidia notata* (Neuroptera). *Entomologist* **41**, 233.
LYMAN F. E. (1955) Occurrence and description of an embryonic egg burster in the genus

Hexagenia (Ephemeroptera). *Ent. News* **66**, 253–255.

LYNN D. C. and VINSON S. B. (1977) Effects of temperature, host age and hormones upon the encapsulation of *Cardiochiles nigriceps* eggs by *Heliothis* spp. *J. Invert. Path.* **29**, 50–55.

LYON R. L. and FLAKE H. W. (1966) Rearing Douglas-fir tussock moth larvae on synthetic media. *J. econ. Ent.* **59**, 696–698.

LYONET P. (1832) Recherches sur l'anatomie et les métamorphoses de différentes espèces d'insectes. Ouvrage posthume. *Mém. Mus. Hist. nat. Paris* **18**, 120 et seq.

MCCLENDON J. F. (1902) The life history of *Ulula hyalina* Latreille. *Am. Nat.* **36**, 421–429.

MCCLURE H. E. (1932) Incubation of bark-bug eggs (Aradidae). *Ent. News* **43**, 188–189.

MCCLUSKEY E. S. (1967) Circadian rhythms in female ants, and loss after mating flight. *Comp. Biochem. Physiol.* **23**, 665–677.

MCDANIEL I. N., BENTLEY M. D., LEE H. P., and YATAGAI M. (1976) Effects of color and larval-produced oviposition attractants on oviposition of *Aedes triseriatus*. *Envir. Ent.* **5**, 553–556.

MCDONALD J. L. and LU L. C. (1972) Viability of mosquito eggs produced by female mosquitoes denied ovipositing sites. *Mosq. News* **32**, 463–466.

MCFARLANE J. E. (1960a) Structure and function of the egg shell as related to water absorption by the eggs of *Acheta domesticus* (L.). *Can. J. Zool.* **38**, 231–241.

MCFARLANE J. E. (1960b) Permeability of the egg shell of *Acheta domesticus* (L.), and the fate of the vitelline membrane. *Can. J. Zool.* **38**, 1037–1039.

MCFARLANE J. E. (1961) Tyrosinase and the structure of the egg shell of *Acheta domesticus* (L.). *Can. J. Zool.* **39**, 1–9.

MCFARLANE J. E. (1962a) The cuticles of the egg of the house cricket. *Can. J. Zool.* **40**, 13–21.

MCFARLANE J. E. (1962b) The embryonic cuticle of the house cricket, its scales, and their relation to the scales of other cuticles. *Can. J. Zool.* **40**, 23–30.

MCFARLANE J. E. (1962c) The action of pepsin and trypsin on the egg shell of *Acheta domesticus* (L.) (Orthoptera: Gryllidae). *Can. J. Zool.* **40**, 553–557.

MCFARLANE J. E. (1963) Induction *in vitro* by buffer solutions of a structural change occurring normally during embryonic development of the house cricket. *Can. J. Zool.* **41**, 23–28.

MCFARLANE J. E. (1965) The surface structure of various layers of the house cricket eggshell as seen in the scanning electron microscope. *Can. J. Zool.* **43**, 911–913.

MCFARLANE J. E. (1966) The permeability of the cricket eggshell to water. *J. Insect Physiol.* **12**, 1567–1575.

MCFARLANE J. E. (1970) The permeability of the cricket eggshell. *Comp. Biochem. Physiol.* **37**, 133–141.

MCFARLANE J. E. and FURNEAUX P. J. S. (1964) Revised curves for water absorption by the eggs of the house cricket *Acheta domesticus* (L.). *Can. J. Zool.* **42**, 239–243.

MCFARLANE J. E., GHOURI A. S. K., and KENNARD, C. P. (1959) Water absorption by the eggs of crickets. *Can. J. Zool.* **37**, 391–399.

MCFARLANE J. E. and KENNARD C. P. (1960) Further observations on water absorption by the eggs of *Acheta domesticus* (L.). *Can. J. Zool.* **38**, 77–85.

MACFARLANE W. V., MORRIS R. J. H., and HOWARD B. (1958) Heat and water in tropical merino sheep. *Aust. J. agric. Res.* **9**, 217–228.

MCFEELY J. (1971) Some egg laying and larval habits of *Papilio machaon* L. (The swallowtail butterfly). *Entomologist's Rec. J. Var.* **83**, 16–18.

MCGAUGHEY W. H. (1968) Role of salts in oviposition site selection by the black salt-marsh mosquito *Aedes taeniorhynchus* (Wiedmann). *Mosq. News* **28**, 207–217.

MCGOVERN W. L. and CROSS W. H. (1974) Oviposition of a parasite *Bracon mellitor* attacking larvae of the boll weevil inside the cotton square. *Ann. ent. Soc. Am.* **67**, 520–521.

MacGregor M. E. (1916) Resistance of the eggs of *Stegomyia fasciata (Aedes calopus)* to conditions adverse to development. *Bull. ent. Res.* **7**, 81–85.

McGuffin W. C. (1967) Guide to the Lepidoptera of Canada (Lepidoptera). I. Subfamily Sterrhinae. *Mem. ent. Soc. Can.* **50**, 1–67.

McGuffin W. C. (1972) Guide to the Geometridae of Canada (Lepidoptera). II. Subfamily Ennominae. 1. *Mem. ent. Soc. Can.* **86**, 86–159.

McKenzie H. L. (1932) The biology and feeding habits of *Hyperaspis lateralis* Mulsant (Coleoptera, Coccinellidae). *Univ. Calif. Publ. Ent.* **6** (2) 9–20.

McKenzie L. M. and Beirne B. P. (1972) A grape leaf-hopper, *Erythroneura ziczac* (Hom., Cicadellidae), and its mymarid (Hymenoptera) egg-parasite in the Okanagen Valley, British Columbia (Canada). *Can. Ent.* **104**, 1229–1233.

McKittrick F. A. (1964) Evolutionary studies of cockroaches. *Mem. Cornell agric. Exp. Stn* **389**, 1–197.

McKnight M. E. (1971) Biology and habits of *Bracon politiventris* (Hymenoptera: Braconidae). *Ann. ent. Soc. Am.* **64**, 620–624.

McLachlan R. (1870) An unrecorded habit in the life history of certain Trichopterous insects. *Entomologist's mon. Mag.* **16**, 135–136.

McLachlan R. (1892) A *Chrysopa* destructive to coccids in New South Wales. *Entomologist's mon. Mag.* **28**, 50.

MacLellan C. R. (1962) Mortality of codling moth eggs and young larvae in an integrated control orchard. *Can. Ent.* **94**, 655–666.

McLeod J. M. (1972) A comparison of discrimination and of density responses during oviposition by *Exenterus amictorius* and *E. diprionis* (Hymenoptera: Ichneumonidae), parasites of *Neodiprion swainei* (Hymenoptera: Diprionidae). *Can. Ent.* **104**, 1313–1330.

McLintock J. (1951) The continuous laboratory rearing of *Culiseta inornata* (Will.) and a study of the structure and function of the egg shell and the egg raft, Doctoral thesis, McGill University, Quebec.

McLintock J. and Depner K. R. (1954) A review of the life-history and habits of the horn fly *Siphona irritans* (L.) (Diptera: Muscidae). *Can. Ent.* **86**, 20–33.

McMahon E. (1954) Investigations on the hatching of the eggs of the lemon wheat blossom midge. *Sci. Proc. R. Dublin Soc.* (NS) **26**, 339–345.

McMillian W. W. and Wiseman B. R. (1972) Separating egg masses of the fall armyworm. *J. econ. Ent.* **65**, 900–902.

McMullen L. H. (1962) The life-history and habits of *Scolytus unispinosus* Lec. (Col., Scolytidae) in the interior of British Columbia. *Can. Ent.* **94**, 17–25.

McMullen L. H. (1976) Effect of temperature on oviposition and brood development of *Pissodes strobi* (Coleoptera: Curculionidae). *Can. Ent.* **108**, 1167–1172.

McMullen L. H. and Atkins M. D. (1959) Life-history and habits of *Scolytus tsugae* (Swaine) (Coleoptera: Scolytidae) in the interior of British Columbia. *Can. Ent.* **91**, 416–426.

McMullen R. D. (1967) The effects of photoperiod, temperature, and food supply on rate of development and diapause in *Coccinella novemnotata*. *Can. Ent.* **99**, 578–586.

McMurtry J. A., Scriven G. T., and Malone R. S. (1974) Factors affecting oviposition of *Stethorus picipes* (Coleoptera: Coccinellidae), with special reference to photoperiod. *Envir. Ent.* **3**, 123–127.

McNeil J. N. and Rabb R. L. (1973) Life histories and seasonal biology of four hyperparasites of the tobacco hornworm *Manduca sexta* (Lepidoptera: Sphingidae). *Can. Ent.* **105**, 1041–1052.

McNew G. L. (1970) The Boyce Thompson Institute program in forest entomology that led to the discovery of pheromones in bark beetles. *Contr. Boyce Thompson Inst.* **24**, 251–262.

MacPhee A. W. (1953) The influence of spray programs on the fauna of apple orchards in

Nova Scotia. V. The predacious thrips *Haplothrips faurei* Hood. *Can. Ent.* **85**, 33–40.

MacPhee A. W. and Sanford K. H. (1954) The influence of spray programs on the fauna of apple orchards in Nova Scotia. VII. Effects on some beneficial arthropods. *Can. Ent.* **86**, 128–135.

McPherson J. E. (1971) Laboratory rearing of *Euschistus tristigmus tristigmus. J. econ. Ent.* **64** (5) 1339–1340.

McWilliams J. M. and Cook J. M. (1975) Technique for rearing the twolined spittlebug. *J. econ. Ent.* **68**, 421–422.

Macdonald W. W. (1960) On the systematics and ecology of *Armigeres* subgenus *Leicesteria* (Diptera, Culicidae). *Stud. Inst. med. Res. FMS* **29**, 110–153.

Macdonald W. W. and Traub R. (1960) An introduction to the ecology of the mosquitoes of the lowland dipterocarp forest of Selangor, Malaya. *Stud. Inst. med. Res. FMS* **29**, 79–109.

Macdougall R. S. (1930) The warble flies of cattle. *Trans. Highl. agric. Soc. Scot.* **1930**, 1–40.

Machado A. B. M. (1965) Signifição biológica do macho na oviposição de *Neoneura sylvatica* Selys (Odonata—Protoneuridae). *Cienc. Cult. S. Paulo* **17**, 249–250.

Mackerras I. M. (1971) The Tabanidae (Diptera) of Australia. V. Subfamily Tabaninae, tribe Tabanini. *Aust. J. Zool.* (Suppl.) **4**, 1–54.

Mackerass I. M. and Mackerras M. J. (1944) Sheep blowfly investigations. The attractiveness of sheep for *Lucilia cuprina. Bull. Coun. sci. industr. Res. Aust.* **181**, 1–44.

Mackerras M. J. (1933) Observations on the life-histories, nutritional requirements and fecundity of blowflies. *Bull. ent. Res.* **24**, 353–362.

Madden J. L. (1968a) Physiological aspects of host tree favourability for the wood wasp *Sirex noctilio* F. *Proc. ecol. Soc. Aust.* **3**, 147–149.

Madden J. L. (1968b) Behavioural responses of parasites to the symbiotic fungus associated with *Sirex noctilio* F. *Nature, Lond.* **218**, 189–190.

Madden J. L. (1974) Oviposition behaviour of the woodwasp *Sirex noctilio* F. *Aust. J. Zool.* **22**, 341–351.

Madge P. E. (1954) A field study on the underground grass caterpillar *Oncopera fasciculata* (Walker) (Lepidoptera: Hepialide) in South Australia. *Aust. J. Zool.* **2**, 193–204.

Madhavan M. M. (1974) Structure and function of the hydropyle of the egg of the bug *Sphaerodema molestum. J. Insect Physiol.* **20**, 1341–1349.

Madrid F., Vité J. P., and Renwick J. A. A. (1972) Evidence of aggregation pheromones in the ambrosia beetle *Platypus flavicornis* (F.). *Z. angew. Ent.* **72**, 73–79.

Maelzer D. A. (1961) The behaviour of the adult of *Aphodius tasmaniae* Hope (Col., Scarabaeidae) in South Australia. *Bull. ent. Res.* **51**, 643–670.

Maercks H. (1933) Der Einfluss von Temperatur und Luftfeuchtigkeit auf die Embryonalentwicklung der Mehlmottenschlupfwespe *Habrobracon juglandis* Ashmead. *Arb. biol. BundAnst. Land- u. Forstw.* **20**, 347–390.

Magalhães P. S. de (1909) No mundo dos insectos. *Journal de Commercio*, Rio de Janeiro, April 19, p. 4 (reprinted in von Ihering, 1909).

Magalhães Bastos J. A. (1969) Substâncias orgânicas como atraentes para a postura do gorgulho, *Callosobruchus analis* Fabr., no feijão de corda, *Vigna sinensis* Endl. *Pesquisas agropec. bras.* **4**, 127–128.

Maharaj S. (1964) Field studies on the life history of the palm weevil. *J. agric. Soc. Trin. Tob.* **62**, repr. 8 pp.

Mahdihassan S. (1934) Specificity of parasitism by *Eublemma amabilis. Curr. Sci.* **3**, 260–262.

Maillard Y.-P. (1968a) Mise en évidence d'un territoire ovarien modifié en glands de la soie, chez les Coléoptères Hydrophilidae. Données anatomiques comparatives. *C. r. hebd. Séanc. Acad. Sci. Paris* (D) **266**, 500–502.

Maillard Y.-P. (1968b) L'appareil fileur des Coléoptères Hydrophilidae. Données structurales

et fonctionnelles. *Annls Soc. ent. Fr.* (NS) **4**, 503–514.
MAILLARD Y.-P. (1970) Étude comparée de la construction du cocon de ponte chez *Hydrophilus piceus* L. et *Hydrochara caraboides* L. (insecte Coléopt. Hydrophilidae). *Bull. Soc. zool. Fr.* **95**, 71–84.
MAILLARD Y.-P. (1972) Structure fine de la surface de la filière et de la soie du cocon de ponte des Hydrophilidae (Ins. Coléoptères). *C. r. hebd. Séanc. Acad. Sci. Paris* (D) **275**, 75–78.
MAIN H. (1927) A living family of *Anisolabis maritima*. *Proc. ent. Soc. Lond.* **1**, 58.
MAINARDI D., MAINARDI M., and PASQUALI A. (1967) Eliminazione dell'oviposizione preferenziale in tignole (*Ephestia kühniella*) metamorfosate in assenza di stimoli specifici. *Rc. Ist lomb. Sci. Lett.* (B) **101**, 628–630.
MAINARDI D., OTTAVIANI L., and PASQUALI A. (1966) Apprendimento e fattori genetici nel determinismo dell'oviposizione preferenziale in *Ephestia kühniella*. *Atti Accad. naz. Lincei Rc.*, Cl. Sci. fis. mat. nat. **41** (8) 134–138.
MAINARDI M. (1968) Gregarious oviposition and pheromones in *Drosophila melanogaster*. *Boll. Zool.* **35**, 135–136 (Italian summary).
MAKI T. (1966) On the oviposition of tabanid flies. *J. Fac. Agric. Iwate Univ.* **8**, 51–71 (in Japanese; English summary).
MAKINGS P. (1958) The oviposition behaviour of *Achroia grisella* (Fabricius) (Lepidoptera: Galleriidae). *Proc. R. ent. Soc. Lond.* (A) **33**, 136–148.
MALHOTRA C. P. (1958) Bionomics of *Serinetha augur* Fabr. and its association with *Dysdercus cingulatus* Fabr., the red cotton bug. *Indian For.* **84**, 669–671.
MALICKY H. (1973) Trichoptera (Köcherfliegen). *Handb. zool. Berl.* **4** (2) 1–114.
MALLACK J., SMITH L. W., BERRY R. A., and BICKLEY W. E. (1964) Hatching of eggs of three species of aedine mosquitoes in response to temperature and flooding. *Bull. agric. Exp. Stn Univ. Maryland* **A-138**, 1–21.
MALLEA A. R. and SUÁREZ J. H. (1969) Biologia y control de *Cobalus cannae* Herrich-Schäffer, en Mendoza. *Idia* **256**, 6–10.
MALTBY H. L., BURGER T. L., HOLMES M. C., and DEWITT P. R. (1973) The use of an unnatural host, *Lema trilineata trivittata*, for rearing the exotic egg parasite *Anaphes flavipes*. *Ann. ent. Soc. Am.* **66** (2) 298–301.
MANDARON P. (1963) Accouplement, ponte et premier stade larvaire de *Thaumalea testacea* Ruthe (Diptères Nématocères). *Trav. Lab. Piscic. Univ. Grenoble* **54** and **55**, 97–107.
MANGUM C. L., JAMES P. E., and ANDERSON H. V. (1972) A device for treating pink bollworm eggs for suppression of cytoplasmic polyhedrosis virus infection. *J. econ. Ent.* **65**, 289–291.
MANI E. (1968) Biologische Untersuchungen an *Pandemis heparana* (Den. und Schiff.) unter besonderer Berücksichtigung der Faktoren welche die Diapause induzieren und die Eiablage beeinflussen. *Mitt. schweiz ent. Ges.* **40**, 145–203.
MANI M. S. and RAO S. N. (1950) A remarkable example of maternal solicitude in a thrips in India. *Current Sci. Bangalore* **19**, 217.
MANK H. G. (1923) The biology of the Staphylinidae. *Ann. ent. Soc. Am.* **16**, 220–237.
MANNHEIMS B. J. (1935) Beiträge zur Biologie und Morphologie der Blepharoceriden (Dipt.). *Zool. Forschung.* **2**, 1–115.
MANNING A. (1962) A sperm factor affecting the receptivity of *Drosophila melanogaster* females. *Nature, Lond.* **194**, 252–253.
MANSINGH A. and RAWLINS S. C. (1977) Antigonadotrophic action of insect hormone analogues on the cattle tick *Boophilus microplus*. *Naturwissenschaften* **64**, 41.
MANSOUR M. H. (1975) The role of plants as a factor affecting oviposition by *Aphidoletes aphidimyza* (Diptera: Cecidomyiidae). *Entomologia exp. appl.* **18**, 173–179.
MANSOUR M. H. (1976) Some factors influencing egg laying and site of oviposition by *Aphidoletes aphidimyza* (Dipt., Cecidomyiidae). *Entomophaga* **21**, 281–288.

MAPLE J. D. (1937) The biology of *Ooencyrtus johnsoni* (Howard), and the role of the egg shell in the respiration of certain Encyrtid larvae (Hymenoptera). *Ann. ent. Soc. Am.* **30**, 123–154.

MAPLE J. D. (1947) The eggs and first instar larvae of Encyrtidae and their morphological adaptations for respiration. *Univ. Calif. Publ. Ent.* **8**, 25–122.

MARCHAND W. (1920) The early stages of Tabanidae (horse-flies). *Monogr. Rockefeller Inst. med. Res.* **13**, 1–203.

MARKIN G. P. (1970) Foraging behaviour of the Argentine ant in a California *Citrus* grove. *J. econ. Ent.* **63**, 740–744.

MARKKULA M. and MYLLYMÄKI S. (1958a) Investigation into the oviposition on red and Alsike clover and alfalfa of *Apion apricans* Herbst, *A. assimile* Kirby, *A. flavipes* Payk., *A. seniculus* Kirby, and *A. virens* Herbst. (Col., Curculionidae). *Ann. ent. fenn.* **23**, 203–207 (1957).

MARKKULA M. and MYLLYMÄKI S. (1958b) On the size and location of the eggs of *Apion apricans* Herbst, *A. assimile* Kirby, *A. flavipes* Payk., *A. seniculus* Kirby, and *A. virens* Herbst. (Col., Curculionidae). *Ann. ent. fenn.* **24**, 1–11.

MARKKULA M. and MYLLYMÄKI S. (1962a) The distribution, abundance, and biology of the clover head weevil *Phytonomus meles* Fabr. (Col. Curculionidae) in Finland. *Ann. ent. fenn.* **28**, 49–63.

MARKKULA M. and MYLLYMÄKI S. (1962b) The distribution, abundance, and biology of *Apion trifolii* L. (Col., Curculionidae) in Finland. *Ann. ent. fenn.* **28**, 11–24.

MARKKULA M. and ROIVAINEN S. (1961) The effect of temperature, food plant, and starvation on the oviposition of some *Sitona* (Col., Curculionidae) species. *Ann. ent. fenn.* **27**, 30–45.

MARKKULA M. and TINNILÄ A. (1955) Oviposition of the lesser clover leaf weevil *Phytonomus nigrirostris* Fabr. (Col., Curculionidae). *Ann. ent. fenn.* **21**, 26–30.

MARLIÉR G. (1950) Sur deux larves de *Simulium* commensales de nymphes d'Ephémères. *Rev. Zool. Bot. afr.* **43**, 135–144.

MARLIÉR G. (1972) Études sur la productivité des étangs de Haute Belgique: La biologie de *Limnephilus lunatus* Curtis (Trichoptère). *Bull. Inst. R. Sci. nat. Belg.* **47** (40) 1–9 (1971).

MARQUES L. A. DE A. (1933) Tenthredinidae conhecida por mosca de serra, cuja larva ou falsa lagarta é nociva a várias espécies do gênero *Tibouchina* (Biologia de *Bergiana cyanocephala*) (Klug, 1824) Know, 1899. *Inst. Biol. Def. Agric. Min. Agric.* **1933**, 1–11.

MARR J. D. M. (1962) The use of an artificial breeding-site cage in the study of *Simulium damnosum* Theobald. *Bull. Wld Hlth Org.* **27**, 622–629.

MARSHALL J. F. and STALEY J. (1933) Variations in the surface pattern of eggs of *Anopheles maculipennis* (Diptera, Culicidae) obtained in the south of England. *Stylops* **2**, 238–240.

MARSHALL J. F. and STALEY J. (1935) Exhibition of autogenous characteristics by a British strain of *Culex pipiens* L. *Nature, Lond.* **135**, 34.

MARSHALL W. S. (1905) The reproductive organs of the female Maia moth *Hemileuca maia* (Drury). *Trans. Wisconsin Acad. Sci. Arts Lett.* **15**, 1–12.

MARSTON N., CAMPBELL B., and BOLDT P. E. (1975) Mass producing eggs of the greater wax moth *Galleria mellonella* (L.). *Tech. Bull. agric. Res. Serv. US Dep. Agric.* **1510**, 1–15.

MARTELLI G. (1908) Osservazioni fatte sulle cocciniglie dell'olivo e loro parassiti in Puglia ed in Calabria. *Boll. Lab. Zool. gen. agr. R. Scuola Agric. Portici* **2**, 217–296.

MARTELLI G. M. (1964) Notizie su due *Sitona* delle leguminose (*Sitona lineatus* L. e *S. limosus* Rossi) Coleoptera—Curculionidae. *Sci. Tec. agrar.* **4** (3) 15 pp.

MARTIN J. O. (1900) A study of *Hydrometra lineata*. *Can. Ent.* **32**, 70–76.

MARTIN J. T. (1975) Composition of egg-shell and cast larval skin of *Rhodnius prolixus* and pupal case of *Pieris brassicae*. *Insect. Biochem.* **5**, 275–287.

MARTÍNEZ-BERÍNGOLA M. L. (1965) Influencia de la densidad de población larvaria en la duración del desarollo de *Ceratitis capitata* Wied. *Bol. R. Soc. esp. Hist. nat. (Biol.)* **63**, 381–390.

MARTINI E. (1923) *Lehrbuch der medicinischen Entomologie*, Jena.

MARTINOVICH V. (1966) *Acanthiophilus helianthi*, a pest of *Centaurea* seed production in Hungary. *Folia ent. hung.* (SN) **19**, 375–402 (in Hungarian; English summary).

MARTYN E. J. (1965) Studies on the ecology of *Oncopera intricata* Walker (Lepidoptera: Hepialidae). I. Fecundity of the female moths. III. Survival of the eggs under field conditions. *Aust. J. Zool.* **13**, 801–805, 811–815.

MARUCCI P. E. (1955) Notes on the predatory habits and life cycle of two Hawaiian earwigs. *Proc. Hawaii. Ent. Soc.* **15**, 565–569.

MASAKI J. (1959) Studies on rice crane fly (*Tipula aino* Alexander, Tipulidae Diptera) with special reference to the ecology and its protection. *J. Kanto-Tosan agric. Exp. Stn* **13**, 1–195 (in Japanese; English summary).

MASHHOOD ALAM S. (1952) Studies on the structure and working of the "egg-laying mechanism" in *Stenobracon deesae* Cam. (Braconidae, Hymenoptera). *Z. ParasitKde* **15**, 357–368.

MASIRG R. A. (1946a) Experiments on selection for selective egg-deposition in *Drosophila melanogaster*. *C. r. Acad. Sci. URSS* (NS) **51**, 393–396.

MASIRG R. A. (1946b) Inheritance of the ability to select media for egg-laying in *Drosophila melanogaster*. *C. r. Acad. Sci. URSS* (NS) **51**, 545–548.

MASKELL W. M. (1888) On *Henops brunneus*. *Trans. Proc. NZ Inst.* **20**, 106–108.

MATHENY E. L. JR and HEINRICHS E. A. (1972) Chorion characteristics of sod webworm eggs. *Ann. ent. Soc. Am.* **65** (1) 238–246.

MATHESON R. (1912) The Haliplidae of North America, north of Mexico. *Jl NY ent. Soc.* **20**, 156–193.

MATHESON R. and HURLBUT N. S. (1937) Notes on *Anopheles walkeri*. *Am. J. trop. Med.* **17**, 237–242.

MATHIS M. (1935) Sur la nutrition sanguine et à la fécondité de *Stegomyia: Aedes aegypti*. *Bull. Soc. Path. exot.* **28**, 231–234.

MATHUR C. B. (1944) The site of absorption of water by the egg of *Schistocerca gregaria*. *Indian J. Ent.* **5**, 35–40.

MATHUR R. N. (1934) On the biology of the Mantidae (Orthopt.) with notes by C. F. C. Beeson and S. N. Chatterjee. *Indian For. Rec.* **20** (3) 1–26.

MATHUR R. N. (1947) Notes on the biology of some Mantidae. *Ent. Soc. India* **8**, 89–106.

MATSUDA R. (1976) *Morphology and Evolution of the Insect Abdomen*, Pergamon Press, Oxford, 534 pp.

MATSUMOTO Y. and SHONO T. (1973) Ovipositional behaviour of the chestnut gall wasp *Dryocosmus kuriphilus* Yasumatsu (Hymenoptera: Cynipidae). *Jap. J. appl. Ent. Zool.* **17**, 223–225 (Japanese).

MATSUMOTO Y. and THORSTEINSON A. J. (1968) Effect of organic sulfur compounds on oviposition in onion maggot *Hylemya antiqua* Meigen (Diptera: Anthomyiidae). *Appl. Ent. Zool.* **3**, 5–12.

MATSUO K., YOSHIDA Y., and KONOU I. (1972) The scanning electron microscopy of mosquitoes (Diptera: Culicidae). I. The egg surfaces of five species of *Aedes* and *Armigeres subalbatus*. *J. Kyoto Pref. Univ. Med.* **81**, 358–363.

MATSUO K., YOSHIDA Y., and LIEN J. C. (1974) Scanning electron microscopy of mosquitoes. II. The egg surface structure of 13 species of *Aedes* from Taiwan. *J. med. Ent.* **11**, 179–188.

MATTESON J. W. (1966) Flotation technique for extracting eggs of *Diabrotica* spp. and other organisms from soil. *J. econ. Ent.* **59**, 223–224.

MATTHÉE J. J. (1948) Pore canals in the egg membranes of *Locustana pardalina* Walk. *Nature, Lond.* **162**, 226–227.

MATTHÉE J. J. (1951) The structure and physiology of the egg of *Locustana pardalina*. *Sci. Bull. Dep. Agric. S. Afr.* **316**, 1–83.

MATTHEWMAN W. G. and HARCOURT D. G. (1972) Phenology of egg-laying of the cabbage maggot *Hylemya brassicae* (Bouché) on early cabbage in eastern Ontario. *Proc. ent. Soc. Ontario* **102**, 28–35 (1971).
MATTHEWS E. G. (1962) A revision of the genus *Copris* Müller of the Western hemisphere (Coleoptera: Scarabaeidae). *Entomologica am.* (NS) **41**, 1–139.
MATTHYSSE J. G. (1946) Cattle lice, their biology and control. *Bull. Cornell agric. Exp. Stn* **832**, 1–67.
MATTINGLY P. F. (1969a) Mosquito eggs. I. *Mosq. Syst.* **1**, 13–16.
MATTINGLY P. F. (1969b) Mosquito eggs. II. *Mosq. Syst.* **1**, 41.
MATTINGLY P. F. (1969c) Mosquito eggs. III. Tribe Anophelini. *Mosq. Syst.* **1**, 41–50.
MATTINGLY P. F. (1969d) Mosquito eggs. IV. Tribe Sabethini. *Mosq. Syst.* **1**, 74–77.
MATTINGLY P. F. (1969e) Mosquito eggs. V. Genus *Aedes*. Introduction. *Mosq. Syst.* **1**, 78–80.
MATTINGLY P. F. (1970a) Mosquito eggs. VI. Genera *Eretmapodites* and *Culex*. *Mosq. Syst.* **2**, 17–22.
MATTINGLY P. F. (1970b) Mosquito eggs. VII. Genus *Uranotaenia*. *Mosq. Syst.* **2**, 61–67.
MATTINGLY P. F. (1970c) Mosquito eggs. VIII. Genus *Aedes*, subgenus *Mucidus* Theobald. *Mosq. Syst.* **2**, 87–91.
MATTINGLY P. F. (1970d) Mosquito eggs. IX. Genus *Opifex* Hutton. *Mosq. Syst.* **2**, 92–97.
MATTINGLY P. F. (1970e) Mosquito eggs. X. Oviposition in *Neoculex*. *Mosq. Syst.* **2**, 158–159.
MATTINGLY P. F. (1970f) Mosquito eggs. XI. Genera *Orthopodomyia* and *Mimomyia*. *Mosq. Syst.* **2**, 160–164.
MATTINGLY P. F. (1971a) Mosquito eggs. XII. Further notes on genera *Orthopodomyia* and *Mimomyia*. *Mosq. Syst.* **3**, 66–68.
MATTINGLY P. F. (1971b) Mosquito eggs. XIII. Genus *Armigeres* Theobald. *Mosq. Syst.* **3**, 122–129.
MATTINGLY P. F. (1971c) Mosquito eggs. XIV. Genus *Armigeres* Theobald (continued) and *Aedes* subgenus *Alanstonea* Mattingly. *Mosq. Syst.* **3**, 130–137.
MATTINGLY P. F. (1971d) Mosquito eggs. XV. Genera *Heizmannia* Ludlow and *Haemagogus* Williston. *Mosq. Syst.* **3**, 197–201.
MATTINGLY P. F. (1971e) Mosquito eggs. XVI. Genus *Mansonia* (subgenus *Coquillettidia* Dyar) and genus *Ficalbia* Theobald. *Mosq. Syst.* **3**, 202–210.
MATTINGLY P. F. (1971f) Ecological aspects of mosquito evolution. *Parasitologia* **13**, 31–65.
MATTINGLY P. F. (1972a) Mosquito eggs. XVII. Further notes on egg parasitization in genus *Armigeres*. *Mosq. Syst.* **4**, 1–8.
MATTINGLY P. F. (1972b) Mosquito eggs. XVIII. Genus *Mansonia* (subgenera *Rhynchotaenia* Brèthes and *Mansonia* Blanchard) with a further note on genus *Ficalbia* Theobald. *Mosq. Syst.* **4**, 45–49.
MATTINGLY P. F. (1972c) Mosquito eggs. XIX. Genus *Mansonia* (subgenus *Mansonioides* Theobald). *Mosq. Syst.* **4**, 50–59.
MATTINGLY P. F. (1972d) Mosquito eggs. XX. Egg parasitism in *Anopheles* with a further note on *Armigeres*. *Mosq. Syst.* **4**, 84–86.
MATTINGLY P. F. (1972e) Mosquito eggs. XXI. Genus *Culiseta* Felt. *Mosq. Syst.* **4**, 114–127.
MATTINGLY P. F. (1973a) Mosquito eggs. XXII. Eggs of two species of *Haemagogus* Williston (*H.? spegazzinii* Brèth. and *H. lucifer* (H., D. and K.)). *Mosq. Syst.* **5**, 24–26.
MATTINGLY P. F. (1973b) Mosquito eggs. XXIII. Eggs of *Toxorynchites amboinensis* containing two headed monsters. *Mosq. Syst.* **5**, 197–199.
MATTINGLY P. F. (1973c) Mosquito eggs. XXIV. Genus *Deinocerites* Theobald. *Mosq. Syst.* **5**, 221–224.
MATTINGLY P. F. (1974a) Mosquito eggs. XXV. Eggs of some subgenera of *Aedes* with a further note on *Haemagogus*. *Mosq. Syst.* **6**, 41–45.

MATTINGLY P. F. (1974b) Mosquito eggs. XXVI. Further descriptions of sabethine eggs. *Mosq. Syst.* **6**, 231–238.
MATTINGLY P. F. (1975) Mosquito eggs. XXVII. *Mosq. Syst.* **7**, 19–26.
MATTINGLY P. F. (1976) Mosquito eggs. XXVIII. *Culex* subgenera *Melanoconion* and *Mochlostyrax*. *Mosq. Syst.* **8**, 223–231.
MAULIK S. (1938) On the structure of larvae of Hispine beetles. V. (With a revision of the genus *Brontispa* Sharp.) *Proc. zool. Soc. Lond.* **108**, 49–71.
MAULIK S. (1948) Early stages and habits of *Sindia clathrata* (Fabricius) (Cassidinae, Chrysomelidae, Coleoptera). *Ann. Mag. nat. Hist.* **1** (12) 368–371.
MAW M. G. (1960) Notes on the larch sawfly *Pristiphora erichsonii* (Htg.) (Hymenoptera: Tenthredinidae) in Great Britain. *Entomologist's Gaz.* **11**, 43–49.
MAW M. G. (1970) Capric acid as a larvicide and an oviposition stimulant for mosquitoes. *Nature, Lond.* **227**, 1154–1155.
MAX M. G. (1961) Suppression of oviposition rate of *Scambus buolianae* (Htg.) (Hymenoptera: Ichneumonidae) in fluctuating electrical fields. *Can. Ent.* **93**, 602–604.
MAXWELL F. G., JENKINS J. N., and PARROTT W. L. (1967) Influence of constituents of the cotton plant on feeding, oviposition and development of the boll weevil. *J. econ. Ent.* **60**, 1294–1297.
MAXWELL F. G., LAFEVER H. N., and JENKINS J. N. (1966) Influence of the glandless genes in cotton on feeding, oviposition and development of the boll weevil in the laboratory. *J. econ. Ent.* **59**, 585–588.
MAYER K. (1934) Die Metamorphose der Ceratopogonidae (Dipt.). Ein Beitrag zur Morphologie, Systematik, Ökologie und Biologie der Jugenstadien dieser Dipterenfamilie. *Arch. Naturgesch.* **3**, 205–288.
MAYER K. (1961) Untersuchungen über das Wahlverhalten der Fritfliege (*Oscinella frit* L.) beim Anflug von Kulturpflanzen im Feldversuch mit der Fangschalenmethode. *Mitt. biol. BundAnst. Ld- u. Forstw.* **106**, 1–47.
MAYER M. S. and JAMES J. D. (1969) Attraction of *Aedes aegypti* (L.): response to human arms, carbon dioxide, and air currents in a new type of olfactometer. *Bull. ent. Res.* **58**, 629–642.
MAYER M. S. and JAMES J. D. (1970) Attraction of *Aedes aegypti*. II. Velocity of reaction to host with and without additional carbon dioxide. *Entomologia exp. appl.* **13**, 47–53.
MAYNARD SMITH J. (1958) The effects of temperature and of egg-laying on the longevity of *Drosophila subobscura*. *J. exp. Biol.* **35**, 832–842.
MAYO Z. B. and STARKS K. J. (1972) Sexuality of the greenbug *Schizaphis graminum* (Hem., Hom., Aphididae) in Oklahoma. *Ann. ent. Soc. Am.* **65**, 671–678.
MAZZINI M. (1974) Sulla fine struttura del micropilo negli insetti. *Redia* **55**, 343–372.
MAZZINI M. (1975) Sulla fine struttura del micropilo negli insetti II. L'ultrastruttura dell'uovo di *Adalia bipunctata* L. (Coleoptera: Coccinellidae). *Redia* **56**, 185–191.
MAZZINI M. (1976) Fine structure of the insect micropyle. III. Ultrastructure of the egg of *Chrysopa carnea* Steph. (Neuroptera: Chrysopidae). *Int. J. Morph. Embryol.* **5**, 273–278.
MEAD-BRIGGS A. R. and VAUGHAN J. A. (1969) Some requirements for mating in the rabbit flea *Spilopsyllus cuniculi* (Dale). *J. exp. Biol.* **51**, 495–511.
MEATS A. (1967) The relation between soil water tension and rate of development of the eggs of *Tipula oleracea* and *T. paludosa* (Diptera, Nematocera). *Entomologia exp. appl.* **10**, 394–400.
MEATS A. (1968) The effect of exposure to unsaturated air on the survival and development of eggs of *Tipula oleracea* L. and *T. paludosa* Meigen. *Proc. R. ent. Soc. Lond.* (A) **43**, 85–88.
MECZNIKOV E. (1866) Embryologische Studien an Insekten. *Z. wiss. Zool.* **16**, 128–132, 389–500.
MEDEM F. (1951) Biologische Beobachtungen an Psocopteren. *Zool. Jb. (Syst.)* **79**, 591–613.
MEEK C. L. and OLSON J. K. (1977) The importance of cattle hoofprints and tire tracks as

oviposition sites for *Psorophora columbiae* in Texas ricelands. *Envir. Ent.* **6**, 161–166.

MEER MOHR J. C. VAN DER and LIEFTINCK M. A. (1947) Over de biologie van *Antherophagus ludekingi* Grouv. (Col.) in hommelnesten (*Bombus* Latr.) op Sumatra. *Tijd. Ent.* **lxxxviii**, 207–214.

MEGAHED M. M. (1956) A culture method for *Culicoides nubeculosus* (Meigen) (Diptera: Ceratopogonidae) in the laboratory, with notes on the biology. *Bull. ent. Res.* **47**, 107–114.

MEGUŠAR F. (1906) Einfluss abnormer Gravitationseinwirkung auf die Entwicklung bei *Hydrophilus aterrimus* Eschscholtz. *Arch. EntwMech. Organ.* **22**, 141–148.

MEHRA B. P. (1966) Biology of *Chrysopa madestes* Banks (Neuroptera: Chrysopidae). *Indian J. Ent.* **27**, 398–407 (1965).

MEHTA R. C. and SAXENA K. N. (1970) Ovipositional responses of the cotton spotted bollworm *Earias fabia* (Lepidoptera; Noctuidae) in relation to its establishment on various plants. *Entomologia exp. appl.* **13**, 10–20.

MEIJERE J. C. H. DE (1904) Beiträge zur Kenntnis der Biologie und der systematischen Verwandtschaft der Conopiden. *Tijd. Ent.* **46**, 144–225.

MEIJERE J. C. H. DE (1912) Neue Beiträge zur Kenntnis der Conopiden. *Tijd. Ent.* **55**, 184–207.

MEINERT F. (1892) Fortegnelse over Zoologisk Museums Billelarver. *Ent. Medd.* **3**, 168–314.

MEISNER J., ASCHER K. R. S., and LAVIE D. (1974) Factors influencing the attraction to oviposition of the potato tuber moth *Gnorimoschema operculella* Zell. *Z. angew. Ent.* **77**, 179–189.

MEISSNER G. (1855) Beobachtungen über das Eindringen der Samenelemente in den Dotter. *Z. wiss. Zool.* **6**, 272–295.

MELAMED-MADJAR V. (1962) Bionomics of the alfalfa weevil (*Hypera variabilis* Hbst.) in Israel. *Israel J. agric. Res.* **12**, 29–38.

MELAMED-MADJAR V. (1966) Observations on four species of *Sitona* (Coleoptera, Curculionidae) occurring in Israel. *Bull. ent. Res.* **56**, 505–514.

MELDOLA R. (1882) *Proc. ent. Soc. Lond.* **1882**, xxii–xxiii.

MELIS A. (1934) Il grillastro crociato (*Dociostaurus maroccanus* Thunb.) e le sue infestazione in Sardegna. *Atti R. Accad. Georg.* **30** (5) 399–504.

MELLINI E. (1960) Studi sui ditteri larvevoridi. VI. *Bessa selecta* (Meig.) su *Nematus melanaspis* Htg. (Hymenoptera, Tenthredinidae). *Boll. Ist. Ent. Univ. Bologna* **24**, 175–207.

MELLINI E. (1961) Orientamenti e progressi negli studi sul parassitismo degli insetti entomofagi. *Atti Accad. naz. Ital. Ent. Rc.* **8**, 62–85.

MELLINI E. (1965) Studi sui ditteri larvevoridi. XII. *Nemorilla maculosa* Meig. su *Depressaria marcella* Rebel (Lepidoptera, Gelechiidae). *Boll. Ist. Ent. Univ. Bologna* **27**, 145–169 (1964).

MELNIKOV N. (1869) Beiträge zur Embryonalentwicklung der Insekten. *Arch. Naturgesch.* **35**, 136–189.

MEL'NIKOVA N. I. (1962) Observations on the bark beetle *Dendroctonus micans* Kug. near Moscow. *Zool. Zh.* **41**, 234–240 (in Russian; English summary).

MELVILLE A. R. (1946) Annual report of the Entomologist (Coffee Services), 1945. *Rep. Dep. Agric. Kenya 1945*, pp. 51–54.

MENDES FERREIRA A. (1960) Subsidios para o estudo de uma praga do feijão (*Zabrotes subfasciatus* Boh. Coleoptera, Bruchidae) dos climas tropicais. *Garcia de Orta* **8**, 559–581.

MENG H. L., CHANG G. S., and REN S. Z. (1962) Further studies on the cotton bollworm *Heliothis armigera* (Hübner). *Acta ent. sin.* **11**, 71–82 (in Chinese; English summary).

MENON M. A. U. (1938) The egg of *Ficalbia (Mimomyia) hybrida* Leicester. *J. Malar. Inst. India* **1**, 185–186.

MENON M. A. U. and TAMPI M. R. V. (1959) Notes on the feeding and egg-laying habits of *Ficalbia (Mimomyia) chamberlaini*, Ludlow 1904 (Diptera, Culicidae). *Indian J. Malar.* **13**, 13–18.

MER G. (1931) Notes on the bionomics of *Anopheles elutus* Edw. (Dipt., Culic.). *Bull. ent. Res.* **22**, 137–145.

MERCER E. H. and BRUNET P. C. J. (1959) The electron microscopy of the left colleterial gland of the cockroach. *J. biophys. biochem. Cytol.* **5**, 257–262.

MERINO M. G. (1969) Observaciones sobre el comportamiento sexual del escarabajo de la hoja de los cereales, *Oulema melanopus* (L.) (Coleoptera: Chrysomelidae). *Turrialba* **19**, 355–358.

MERLE J. (1968) Fonctionnement ovarien et réceptivité sexuelle de *Drosophila melanogaster* après implantation de fragments de l'appareil génital male. *J. Insect Physiol.* **14**, 1159–1168.

MERLE J. (1970) Role des sécrétions génitales males dans la physiologie de la Drosophile femelle. *Colloques int. Cent. natn. Rech. scient.* **189**, 311–330.

MERLE P. DU (1972) Les prédateurs des Diptères Bombylides associés à la processionaire du pin comme parasites primaires ou sécondaires. *C. r. Acad. Agric. Fr.* **58**, 1006–1012.

MERLE P. DU and DELPECH M. (1973) Dispositifs de terrain destinés à l'étude de la biologie et rôle limitatif de *Villa brunnea* (Dipt., Bombyliidae) vis-à-vis de *Thaumetopoea pityocampa* (Leo., Thaumetopoeidae). *Annls Soc. ent. Fr.* **9**, 471–482.

MERRILL J. H. (1924) Observations on brood-rearing. *Am. Bee J.* **64**, 337–338.

MERTON L. F. H. (1959) Studies in the ecology of the Moroccan locust (*Dociostaurus maroccanus* Thunberg) in Cyprus. *Anti-Locust Bull.* **34**, 1–123.

MERTZ D. B. and CAWTHON D. A. (1973) Sex differences in the cannibalistic roles of adult flour beetles. *Ecology* **54**, 1400–1402.

MERTZ D. B. and ROBERTSON J. R. (1970) Some developmental consequences of handling, egg-eating, and population density for flour beetle larvae. *Ecology* **51**, 989–998.

MERWE C. P. VAN DER (1923) The citrus Psylla (*Trioza merwei*). *J. Dep. Agric. S. Afr.* **7**, 135–141.

METCALF C. L. (1916) Syrphidae of Maine. *Bull. Me agric. Exp. Stn* **253**, 193–264.

METCALF C. L. and FLINT W. P. (1962) *Destructive and Useful Insects. Their Habits and Control*, 4th edn, revd by R. L. Metcalf, McGraw-Hill, New York and London.

MEYER D. (1969) Der Einfluss von Licht und Temperaturschwankungen auf Verhalten und Fekundität des Lärchenwicklers *Zeiraphera diniana* (Gn.) (Lepidoptera: Tortricidae). *Revue suisse Zool.* **76**, 93–141.

MEYER N. F. (1926) Biologie von *Angitia fenestralis* Holmgr. (Hymenoptera, Ichneumonidae) des Parasiten von *Plutella maculipennis* Curt. und einige Worte über Immunität des Insekten. *Z. angew. Ent.* **12**, 139–152.

MEYER O. E. (1970) On adult weight, oviposition preference, and adult longevity in *Anobium punctatum* (Col. Anobiidae). *Z. angew. Ent.* **66**, 103–112.

MIALL L. C. (1895) *The Natural History of Aquatic Insects*, London.

MIALL L. C. and DENNY A. (1886) *The Structure and Life History of the Cockroach (Periplaneta orientalis)*, London.

MIALL L. C. and SHELFORD R. (1897) The structure and life-history of *Phalacrocera replicata*. *Trans. ent. Soc. Lond.* **1897**, 343–361.

MICHAEL R. R. and RUDINSKY J. A. (1972) Sound production in Scolytidae: specificity in *Dendroctonus* beetles. *J. Insect Physiol.* **18**, 2189–2201.

MICHENER C. D. (1947) Mosquitoes of a limited area in southern Mississippi. *Am. Midl. Nat.* **37**, 325.

MICHENER C. D. (1969) Comparative social behaviour of bees. *A. Rev. Ent.* **14**, 299–342.

MICHENER C. D. (1974) *The Social Behaviour of Bees. A Comparative Study*, Harvard University Press, Cambridge, Mass.

MICKOLEIT G. (1973a) Über den Ovipositor der Neuropteroidea und Coleoptera und seine phylogenetische Bedeutung (Insecta, Holometabola). *Z. Morph. Ökol. Tiere* **74**, 37–64.

MICKOLEIT G. (1973b) Anatomie und Funktion des Raphidiopteren-Ovipositors (Insecta,

Neuropteroidea). *Z. Morph. Ökol. Tiere* **76**, 145–171.

MIDDLETON M. I. (1977) The possible discovery of the egg masses of *Nemotaulius punctatolineatus* (Retzius) in Britain (Trichoptera: Limnephilidae). *Entomologist's Gaz.* **28**, 45–50.

MIGER F. (1809) Mémoires sur les larves d'insectes Coléoptères aquatiques. 1er mém. sur le grand Hydrophile. *Annls Mus. Hist. nat.* **14**, 441–459.

MIJUŠKOVIĆ M. and TOMAŠEVIĆ B. (1976) *Pregljevi na agrumima na jugoslovenskom primorju*, Titograd, Yugoslavia; Soc. for Sci. and Art of Montenegro, 204 pp.

MIKA G. (1959) Über das Paarungsverhalten der Wanderheuschrecke *Locusta migratoria* R. und F. und deren Abhängigkeit von Zustand der inneren Geschlechtsorgane. *Zool. Beitr.* (NF) **4**, 153–203.

MILES M. (1958) Studies of British Anthomyiid Flies. IX. Biology of the onion fly *Delia antiqua* (Mg.). *Bull. ent. Res.* **49**, 405–414.

MILES P. W. (1969) Interaction of plant phenols and salivary phenolases in the relationship between plants and Hemiptera. *Entomologia exp. appl.* **12**, 736–744.

MILLER A. (1939) The egg and early development of the stonefly *Pteronarcys proteus* Newman (Plecoptera). *J. Morph.* **64**, 555–609.

MILLER B. S., ROBINSON R. J., JOHNSON J. A., JONES E. T. and PONNAIYA B. W. X. (1960) Studies on the relation between silica in wheat plants and resistance to hessian fly attack. *J. econ. Ent.* **53**, 995–999.

MILLER G. W. (1977) Mortality of *Spodoptera littoralis* (Boisduval) (Lepidoptera: Noctuidae) at non-freezing temperatures. *Bull. ent. Res.* **67**, 143–152.

MILLER M. C., WHITE R., and SMITH C. (1972) Laboratory rearing of *Peridesmia discus* on irradiated alfalfa weevil eggs. *Entomophaga* **17**, 223–229.

MILLER N. C. E. (1934) The developmental stages of some Malayan Rhynchota. *J. Fed. Malay. St. Mus.* **17**, 502–525.

MILLER N. C. E. (1939) Oviposition by *Heteropteryx dilatus* (Parkinson 1798) (Orth., Phasmidae). *Proc. R. ent. Soc. Lond.* (A) **14**, 48.

MILLER N. C. E. (1949) A note on *Palophus tiaratus* St. (Orthoptera, Phasmidae) from Southern Rhodesia. *Proc. R. ent. Soc. Lond.* (A) **24**, 11–13.

MILLER N. C. E. (1956) *The Biology of the Heteroptera*, Hill, London.

MILLER R. L. and HIBBS E. T. (1963) Distribution of eggs of the potato leafhopper *Empoasca fabae* on *Solanum* plants. *Ann. ent. Soc. Am.* **56**, 737–740.

MILLER R. M. and FOOTE B. A. (1976) Biology and immature stages of eight species of Lauxaniidae (Diptera). II. Descriptions of immature stages and discussion of larval feeding habits and morphology. *Proc. ent. Soc. Wash.* **78**, 16–37.

MILLER R. S. (1964) Larval competition in *Drosophila melanogaster* and *D. simulans*. *Ecology* **45**, 132–148.

MILLER T. A., STRYKER R. G., WILKINSON R. N. and ESAH S. (1969) Notes on the use of CO_2 baited CDC miniature light traps for mosquito surveillance in Thailand. *Mosq. News* **29**, 688–689.

MILLOT J. (1938) Le développement et la biologie larvaire des Occodidés (= Cyrtidés), Diptères parasites d'araignées. *Bull. Soc. zool. Fr.* **53**, 162–197.

MILLS R. O. (1924) Some observations and experiments on the irritating properties of the larva of *Parasa hilarata* Staudinger. *Am. J. Hyg.* **5**, 342–363.

MILLS R. R. (1966) A cockroach rearing cage designed for collection of oothecae. *J. econ. Ent.* **59**, 490–491.

MILNE A. (1960) Biology and ecology of the garden chafer *Phyllopertha horticola* (L.). VII. The flight season: male and female behaviour, and concluding discussion. *Bull. ent. Res.* **51**, 353–378.

MILNE A. (1964) Biology and ecology of the garden chafer *Anomala (Phyllopertha) horticola*. IX. Spatial distribution. *Bull. ent. Res.* **54**, 761–795.

MILNE L. J. and MILNE M. (1976) The social behaviour of burying beetles. *Scient. Am.* **235** (August) 84–88.

MILYANOVSKIĬ E. S. and MITROFANOV P. I. (1952) The large Caucasian hepialid—a new pest of grape vines in Abkhazia. *Ent. Obozr.* **32**, 82–85 (in Russian).

MINDER I. F. (1959) Leaf-rollers injurious to fruit crops in the flood-plain of the Oka river. *Rev. Ent. URSS* **38**, 98–110 (in Russian; English summary).

MINEO G. (1970a) L'allevamento del *Prays citri* Mill. (tignola degli agrumi) in laboratorio (I nota). *Boll. Ist. ent. agr. Oss. Fitopath. Palermo* **7**, 135–141.

MINEO G. (1970b) Notizie etologische sul *Prays citri* Mill. (Lep.-Hyponomeutidae). *Boll. Ist. ent. agr. Oss. Fitopath. Palermo* **7**, 277–282.

MINIS D. H. (1965) Parallel peculiarities in the entrainment of a circadian rhythm and photoperiodic induction in the pink boll worm (*Pectinophora gossypiella*). In *Circadian Clocks* (J. Aschoff, ed.), Amsterdam.

MISKIMEN G. W. (1962) Studies of the biological control of *Diatraea saccharalis* F. (Lepidoptera: Crambidae) on St. Croix, US Virgin Islands. *J. Agric. Univ. P. Rico* **46**, 135–139.

MISKIMEN G. W. (1966) Effects of light on mating success and egg-laying activity of the sugarcane borer *Diatraea saccharalis*. *Ann. ent. Soc. Am.* **59**, 280–284.

MISRA B. C. and ISRAEL P. (1968) Anatomical studies of the ovipositional site of plant hoppers and leaf hoppers on rice. *Indian J. Ent.* **30**, 178.

MISRA C. S. (1924) A preliminary account of the tachardiphagous Noctuid moth *Eublemma amabilis*. *Proc. 5th Ent. Meet. Pusa 1923*, pp. 238–247.

MISRA M. P. and GUPTA S. N. (1934) The biology of *Holococera pulverea* Meyr. (Blastobasidae), its predators, parasites, and control. *Indian J. agric. Sci.* **4**, 832–864.

MITCHELL H. C. and CROSS W. H. (1969) Oviposition by the boll weevil in the field. *J. econ. Ent.* **62**, 604–605.

MITCHELL H. C. and CROSS W. H. (1971) Mating of boll weevils in the field. *J. econ. Ent.* **64**, 773–774.

MITCHELL R. (1975) The evolution of oviposition tactics in the bean weevil *Callosobruchus maculatus* (F.). *Ecology* **56**, 696–702.

MITCHELL S., TANAKA N., and STEINER L. F. (1965) Methods of mass culturing melon flies and oriental and Mediterranean fruit flies. ARS 33–104, 22 pp., Washington DC, US Dep. Agric.

MITCHELL W. C. and MAU, R. F. L. (1971) Response of the female southern green stink bug and its parasite *Trichopoda pennipes* to male stink bug pheromones. *J. econ. Ent.* **64**, 856–859.

MITIĆ-MUŽINA N. (1960) The results of studies on the bionomics of the cherry fruit fly near Belgrade. *Zashch. Rast.* **60**, 29–53 (in Serbo-Croat; English summary).

MITROKHIN V. U. (1973) Oviposition of black flies (fam. Simuliidae) in northern Transural. *Parazitologiya* **7**, 87–89 (in Russian; English summary).

MITSUHASHI H. and KOYAMA K. (1975) Oviposition of smaller brown planthopper *Laodelphax striatellus* into various carbohydrate solutions through a Parafilm membrane (Hemiptera: Delphacidae). *Appl. Ent. Zool.* **10**, 123–129.

MITSUHASHI J. (1970) A device for collecting planthopper and leafhopper eggs (Hemiptera: Delphacidae and Deltocephalidae). *Appl. Ent. Zool.* **5**, 47–49.

MITSUHASHI J. and MARAMOROSCH K. (1963) Aseptic cultivation of four virus transmitting species of leafhoppers (Cicadellidae). *Contr. Boyce Thompson Inst.* **22**, 165–173.

MIURA T. (1972) Laboratory and field evaluation of a sonic sifter as a mosquito egg extractor. *Mosq. News* **32**, 432–436.

MIURA T. and TAKAHASHI R. M. (1973) Laboratory and field observations on oviposition preferences of *Aedes nigromaculis* (Diptera: Culicidae). *Ann. ent. Soc. Am.* **66**, 244–251.

MIYAHARA Y., WAKIKADO T., and TANAKA A. (1971) Seasonal changes in the number and size of egg masses of *Prodenia litura* (Lep.). *Jap. J. appl. Ent. Zool.* **15**, 139–143.

MIYAMOTO S. (1953) Biology of *Heloptrephes formosanus* Esaki et Miyamoto, with descriptions of larval stages. *Sieboldia* **1**, 1–10.

MIZUTA K. (1960) Effect of individual number on the development and survival of the larvae of two lymantriid species living in aggregation and in scattering. *Jap. J. appl. Ent. Zool.* **4**, 146–152 (in Japanese; English summary).

MJÖBERG E. (1910) Studien über Mallophagen und Anopluren. *Ark. Zool.* **6** (13) 1–296.

MOCHIDA O. (1964) On oviposition in the brown planthopper *Nilaparvata lugens* (Stål) (Hom., Auchenorrhyncha). II. The number of eggs in an egg group, especially in relation to the fecundity. *Jap. J. appl. Ent. Zool.* **8**, 141–148.

MODÉER A. (1764) Några märkvärdigheter hos Insectet *Cimex* ovatus pallide griseus, abdominis lateribus albo nigroque variis alis albis basi scutelli nigricante. *Vetensk. Akad. Handl.* **25**, 41–57.

MOECK H. A. (1970) Ethanol as the primary attractant for the ambrosia beetle *Trypodendron lineatum* (Coleoptera: Scolytidae). *Can. Ent.* **102**, 985–995.

MOHAMMAD ALI S. (1957) Some bio-ecological studies on *Pseudococcus vastator* Mask. (Coccidae: Hemiptera). *Indian J. Ent.* **19**, 54–58.

MOHYUDDIN A. I. (1972) Distribution, biology and ecology of *Dentichasmias busseolae* Heinr. (Hym., Ichneumonidae), a pupal parasite of graminaceous stemborers (Lep. Pyralidae). *Bull. ent. Res.* **62**, 161–168.

MOISEEVA T. S. (1974a) Weakening of the defence reaction of lepidopterous larvae under simultaneous infestation by two species of parasites. *Zool. Zh.* **53**, 51–57 (in Russian; English summary).

MOISEEVA T. S. (1974b) The defence reaction of larvae of *Pieris rapae* against the endoparasite *Apanteles glomeratus* (Hymenoptera, Braconidae) in different geographical regions. *Zool. Zh.* **53**, 365–367 (in Russian; English summary).

MOLOO S. K. (1971) Some aspects of water absorption by the developing egg of *Schistocerca gregaria*. *J. Insect Physiol.* **17**, 1489–1495.

MONADJEMI N. (1972) Sur les variations de la fécondité d'*Aphelinus asychis* (Hym., Aphelinidae) en fonction des espèces aphidiennes mises à sa disposition à différentes périodes de sa vie imaginale. *Annls Soc. Ent. Fr.* **8**, 451–460.

MONRO J. and BAILEY P. T. (1965) Influence of radiation on ovarian maturation and histolysis of pupal fat body in Diptera. *Nature, Lond.* **207**, 437–438.

MONTEITH L. G. (1955) Host preferences of *Drino bohemica* Mesn. (Diptera: Tachinidae), with particular reference to olfactory responses. *Can. Ent.* **87**, 509–530.

MONTEITH L. G. (1958a) Influence of food plant of host on attractiveness of the host to tachinid parasites with notes on preimaginal conditioning. *Can. Ent.* **90**, 478–482.

MONTEITH L. G. (1958b) Influence of host and its food plant on host-finding by *Drino bohemica* Mesn. (Diptera: Tachinidae) and interaction of other factors. *Proc. 10th Int. Congr. Ent. 1956*, **2**, 603–606.

MONTEITH L. G. (1960) Influence of plants other than the food plants of their host on host-finding by tachinid parasites. *Can. Ent.* **92**, 641–652.

MONTEITH L. G. (1962) Apparent continual changes in the host preferences of *Drion bohemica* Mesn. (Diptera: Tachinidae), and their relation to the concept of host-conditioning. *Anim. Behav.* **10**, 292–299.

MONTEITH L. G. (1963) Daily selection by *Drino bohemica* Mesn. (Diptera: Tachinidae) of four species of hosts. *Can. Ent.* **95**, 162–166.

MONTEITH L. G. (1967) Responses by *Diprion hercyniae* (Hymenoptera: Diprionidae) to its food plant and their influence on its relationship with its parasite *Drino bohemica* (Diptera: Tachinidae). *Can. Ent.* **99**, 682–685.

MONTGOMERY B. E. (1937) Oviposition of *Perithemis* (Odonata, Libellulidae). *Ent. News* **48**, 61–63.

MOORE H. (1900) How long does *Blatta orientalis* Linn. carry its ootheca before deposition? *Ent. Rec. J. Var.* **12**, 79–80.

MOORE H. B. (1970) Incubation time of eggs of *Xyletinus peltatus* (Coleoptera: Anobiidae) under constant temperatures and humidities. *Ann. ent. Soc. Am.* **63**, 617–618.

MOORE I. and NAVON A. (1966) The rearing and some bionomics of the leopard moth *Zeuzera pyrina* L. on an artificial medium. *Entomophaga* **11**, 285–296.

MOORE K. M. (1962) Observations on some Australian forest insects. 8. The biology and occurrence of *Glycaspis baileyi* Moore in New South Wales. 9. A new species of *Glycaspis (Glycapsis)* (Homoptera: Psyllidae). *Proc. Linn. Soc. NSW* **86**, 185–200, 201–202.

MOORE N. W. (1952a) On the so-called "territories" of dragonflies (Odonata–Anisoptera). *Behaviour* **4**, 85–100.

MOORE N. W. (1952b) Notes on the oviposition behaviour of the dragonfly *Sympetrum striolatum* Charpentier. *Behaviour* **4**, 101–103.

MOORE S. T., SCHUSTER M. F., and HARRIS F. A. (1974) Radioisotope technique for estimating lady beetle consumption of tobacco budworm eggs and larvae. *J. econ. Ent.* **67**, 703–705.

MORALES A. E. (1947) Mantidos de la fauna iberica. *Trab. Estac. Fitopatol. agric. Burjasot* **184**, 1–34.

MORAN V. C. and BLOWERS J. R. (1967) On the biology of the South African citrus psylla *Trioza erytreae* (Del Guercio) (Homoptera: Psyllidae). *J. ent. Soc. S. Afr.* **30**, 96–106.

MORAN V. C., BROTHERS D. J., and CASE J. J. (1969) Observations on the biology of *Tetrastichus flavigaster* Brothers and Moran (Hym., Eulophidae) parasitic on psyllid nymphs (Hem., Hom.). *Trans. R. ent. Soc. Lond.* **121**, 41–58.

MORAN V. C. and BUCHAN P. R. (1975) Oviposition by the citrus psylla *Trioza erytreae* (Homoptera: Psyllidae) in relation to leaf hardness. *Entomologia exp. appl.* **18**, 96–104.

MORELAND C. R. and McLEOD W. S. (1956) House fly egg-measuring techniques. *J. econ. Ent.* **49**, 49–51.

MORETTI G. P. (1934) Le uova e la larva di *Anabolia lombarda* Ris (Trichoptera). *Boll. Soc. ent. Ital.* **66**, 21–25.

MORGAN A. H. (1913) A contribution to the biology of mayflies. *Ann. ent. Soc. Am.* **6**, 371–413.

MORGAN F. D. (1966) The biology and behaviour of the beech Buprestid *Nascioides enysi* (Sharp) (Coleoptera: Buprestidae) with notes on its ecology and possibilities for its control. *Trans. R. Soc. NZ (Zool.)* **7**, 159–170.

MORGAN N. C. (1956) The biology of *Leptocerus aterrhimus* Steph. with reference to its availability as food for trout. *J. Anim. Ecol.* **25**, 349–365.

MORI H. (1970) The distribution of the columnar serosa of eggs among the families of Heteroptera in relation to phylogeny and systematics. *Jap. J. Zool.* **16**, 89–98.

MORI H. (1972) Water absorption by the columnar serosa in the eggs of the waterstrider *Gerris paludum insularis*. *J. Insect Physiol.* **18**, 675–681.

MORIARTY F. (1969a) Water uptake and embryonic development in eggs of *Chorthippus brunneus* Thunberg (Saltatoria: Acrididae). *J. exp. Biol.* **50**, 327–333.

MORIARTY F. (1969b) Egg diapause and water absorption in the grasshopper *Chorthippus brunneus*. *J. Insect Physiol.* **15**, 2069–2074.

MORIARTY F. (1970) The significance of water absorption by the developing eggs of five British Acrididae (Saltatoria). *Comp. Biochem. Physiol.* **34**, 657–669.

MORIN J. P. and MARIAU D. (1971) La biologie de *Coelaenomenodera elaeidis* Mlk. III. La reproduction. *Oléagineux* **26**, 373–378.

MORIYAMA T. (1935) On the external observation on the several Lepidopterous eggs. *Ent. World, Tokyo* **3**, 123–135 (in Japanese).

MORRIS H. M. (1921) The larval and pupal stages of the Bibionidae. *Bull. ent. Res.* **12**, 221–232.

MORRIS H. M. (1922) The larval and pupal stages of the Bibionidae. *Bull. ent. Res.* **13**, 189–195.

MORRIS R. F. (1954) A sequential sampling technique for spruce budworm egg surveys. *Can. J. Zool.* **32**, 302–313.

MORRIS R. F. (1967) Influence of parental food quality on the survival of *Hyphantria cunea*. *Can. Ent.* **99**, 24–33.

MORRISON R. K., STINNER R. E., and RIDGWAY R. L. (1976) Mass production of *Trichogramma pretiosum* on eggs of the Angoumois grain moth. *SWest. Entomol.* **1**, 74–80.

MORRISON W. P., PASS B. C., and CRAWFORD C. S. (1972) Effect of humidity on eggs of two populations of the bluegrass webworm. *Envir. Ent.* **1**, 218–221.

MORROW J. A., BATH J. L., and ANDERSON L. D. (1968) Descriptions and key to egg masses of some aquatic midges in southern California (Diptera: Chironomidae). *Calif. Vector Views* **15**, 99–108.

MORTENSON E. W. (1950) The use of sodium hypochlorite to study *Aedes nigromaculis* (Ludlow) embryos (Diptera: Culicidae). *Mosq. News* **10**, 211–212.

MORTENSON E. W., ROTRAMEL G. L., and PRINE J. E. (1972) The use of ovitraps to evaluate *Aedes sierrensis* (Ludlow) populations. *Proc. 40th A. Conf. Calif. Mosquito Control Ass.*, 1972, p. 68.

MORTON K. J. (1890) Notes on the metamorphosis of British Leptoceridae. *Entomologist's mon. Mag.* **26**, 127, 181, 231.

MOSCONA A. (1948) Utilization of mineral constituents of the egg-shell by the developing embryo of the stick insect. *Nature, Lond.* **162**, 62–63.

MOSCONA A. (1950a) Studies of the egg of *Bacillus libanicus* (Orthoptera, Phasmidae). I. The egg envelopes. *Q. Jl microsc. Sci.* **91**, 183–193.

MOSCONA A. (1950b) Studies of the egg of *Bacillus libanicus* (Orthoptera, Phasmidae). II. Moisture, dry material, and minerals in the developing egg. *Q. Jl microsc. Sci.* **91**, 195–203.

MOSEBACH-PUKOWSKI E. (1936) Gibt es einen sozialen Instinkt bei *Necrophorus*? *Forschn u. Fortschr.* **12**, 38–39.

MOSER J. C. (1975) Mite predators of the southern pine beetle. *Ann. ent. Soc. Am.* **68**, 1113–1116.

MOSER J. C., CROSS E. A., and ROTON L. M. (1971) Biology of *Pyemotes parviscolyti* (Acarina: Pyemotidae). *Entomophaga* **16**, 367–379.

MOSER J. C. and ROTON L. M. (1971) Mites associated with southern pine bark beetles in Allen Parish, Louisiana. *Can. Ent.* **103**, 1775–1798.

MOSER J. C., THATCHER R. C., and PICKARD L. S. (1971) Relative abundance of southern pinebeetle associates in east Texas. *Ann. ent. Soc. Am.* **64**, 72–77.

MOTE D. C. (1928) The ox warble flies. *Bull. Ohio agric. Exp. Stn* **428**, 1–45.

MOURIER H. and BANEGAS A. D. (1970) Observations on the oviposition and the ecology of the eggs of *Dermatobia hominis* (Diptera: Cuterebridae). *Vidensk. Meddr dansk naturh. Foren.* **133**, 59–68.

MOUTIA L. A. and MAMET R. (1947a) *Clemora smithi* (Arrow) un ennemi de la canne à sucre à l'Ile Maurice. *Bull. Dep. Agric. Mauritius (Sci. Ser.)* **28**, 7 pp.

MOUTIA L. A. and MAMET R. (1947b) An annotated list of insects and Acarina of economic importance in Mauritius. *Bull. Dep. Agric. Mauritius (Sci. Ser.)* **29**, 1–43.

MRKVA R. (1966) A contribution to the counting and forecasting of the numbers of the winter moth (*Operophtera brumata*). *Lesn. Čas.* **12**, 541–562 (in Czech; English summary).

MUELLER J. F. (1973) Pubic lice from the scalp hair: a report of two cases. *J. Parasit.* **59** (5) 943–944.

MUESEBECK C. F. W. and PARKER D. L. (1933) *Hyposoter disparis* Viereck, an introduced Ichneumonid parasite of the gypsy moth. *J. agric. Res.* **46**, 335–347.

MUIR F. and SHARP D. (1904) On the egg-cases and early stages of some Cassidae. *Trans. R. ent. Soc. Lond.* **1904**, 1–23.

MUIRHEAD THOMSON R. C. (1945) Studies on the breeding places and control *Anopheles gambiae* and *A. gambiae* var. *melas* in coastal districts of Sierra Leone. *Bull. ent. Res.* **36**, 185–252.

MUIRHEAD THOMSON R. C. (1948) Studies on *Anopheles gambiae* and *A. melas* in and around Lagos. *Bull. ent. Res.* **38**, 527–558.

MUKERJI M. K. (1969) Oviposition preference and survival of *Hylemya brassicae* on some cruciferous crops. *Can. Ent.* **101**, 153–158.

MUKHERJEE A. B. and CHOUDHURY A. K. S. (1972) Observations on the feeding habit of *Reduvius* species, a predator of *Corcyra cephalonica* (Stainton). *Indian J. Ent.* **33**, 230–231 (1971).

MULDREW J. A. (1953) The natural immunity of the larch sawfly (*Pristiphora erichsonii* (Htg.)) to the introduced parasite *Mesoleius tenthredinis* Morley in Manitoba and Saskatchewan. *Can. J. Zool.* **31**, 313–332.

MULLA M. S. (1966) Oviposition and emergence period of the eye gnat *Hippelates collusor*. *J. econ. Ent.* **59**, 93–96.

MULLEN G. R. (1977) Acarine parasites of mosquitoes. IV. Taxonomy, life-history and behaviour of *Thyas barbigera* and *Thyasides sphagnorum* (Hydrachnellae Thyasidae). *J. med. Ent.* **13**, 475–485.

MÜLLER A. (1971) Note on the oviposition of *Libellula (Sympetrum) flaveola* Linné. *Entomologist's mon. Mag.* **8**, 127–129.

MÜLLER F. (1888) Die Eier der Haarflügler. *Ent. Nachr.* **17**, 259–261.

MÜLLER H. J. (1951) Über das Schlüpfen der Zikaden (Homoptera, Auchenorrhyncha) aus dem Ei. *Zoologica, Stuttgart* **37**, 1–41.

MÜLLER J. (1969) Untersuchungen zu dem in der Ovipositionszeit von *Hylemyia antiqua* Meig. wirksamen Orientierungsmechanismus. *Wiss. Z. Ernst Moritz Arndt-Univ. Griefswald* **18** (1–2) 73–78.

MÜLLER K. (1938) Histologische Untersuchungen über den Entwicklungsbeginn bei einem Kleinschmetterlung (*Plodia interpunctella*). *Z. wiss. Zool.* **151**, 192–242.

MÜLLER O. (1957) Biologische Studien über den früten Kastanienwickler *Pammene juliana* (Stephens) (Lep. Tortricidae) und seine wirtschaftliche Bedeutung für den Kanton Tessin. *Z. angew. Ent.* **41**, 73–111.

MUMA M. H. (1971) Food habits of Phytoseiidae (Acarina: Mesostigmata) including common species on Florida citrus. *Fla Ent.* **54**, 21–34.

MÜNCHBERG P. (1930) Zur Biologie der Odonatengenera *Brachytron* Evans und *Aeschna* Fbr. Zweite Mitteilung der "Beiträge zur Kenntnis der Biologie der Odonaten Nordostdeutschlands". *Z. Morph. Ökol. Tiere* **20**, 172–232.

MUNDY A. T. (1909) *The Anatomy, Habits, and Psychology of* Chironomus pusio *Meigen (the Early Stages), with Notes on Various other Invertebrates, Chiefly Chironomidae*, Leicester.

MUNGOMERY R. W. (1945) Report of the Division of Entomology and Pathology. *Rep. Bur. Sug. Exp. Stn Qd.* **45**, 20–22.

MUNSTERHJELM G. (1920) Om Chironomidernas agglaggning och aggruper. *Acta Soc. Fauna Flora fenn.* **47**, 1–174.

MURAI S. (1959) Studies on the egg parasites of the rice grasshoppers *Oxya japonica* Willemse and *O. velox* Fabricius. IX. Especially on the distribution and natural enemies of the egg parasites *Scelio muraii* Watanabe and *S. tsuruokensis* Watanabe in Japan. *Bull. Yamagata Univ. (Agric. Sci.)* **3**, 95–100.

MURBACH R. (1961) Observations sur la ponte, la distribution et l'éclosion de *Ceuthorrhynchus picitarsis* Gyll. (Col. Curcul.) sur colza d'automne. *Annu. agric. Suisse* **62**, 203–210.

MURPHY D. H. and GISIN H. (1959) The preservation and microscopic preparation of Anopheline eggs in a lacto-glycerol medium. *Proc. R. ent. Soc. Lond.* (A) **34**, 171–174.

MURPHY H. E. (1919) Observations on the egg-laying of the caddis fly *Brachycentrus nigrosoma* Banks and on the habits of the young larvae. *Jl NY ent. Soc.* **27**, 154–158.

MURPHY H. E. (1922) Notes on the biology of some of our North American species of mayflies. I. The metamorphosis of mayfly mouthparts. II. Notes on the biology of mayflies of the genus *Baëtis*. *Bull. Lloyd Libr.* (Ent. Ser.) **22** (a) 1–46.

MURPHY P. W. (ed.) (1962) *Progress in Soil Zoology*, Butterworths, London.

MURRAY A. (1856) Notice of the leaf-insect (*Phyllium scythe*), lately bred in the Royal Botanic Garden of Edinburgh, with remarks on its metamorphosis and growth. *Edinb. J. Sci. Arts* (NS) **3**, 96–111.

MURRAY M. D. (1955) Oviposition in lice, with reference to *Damalinia ovis*. *Aust. vet. J.* **31**, 320–321.

MURRAY M. D. (1957a) The distribution of the eggs of mammalian lice on their hosts. I. Description of the oviposition behaviour. *Aust. J. Zool.* **5**, 13–18.

MURRAY M. D. (1957b) The distribution of the eggs of mammalian lice on their hosts. II. Analysis of the oviposition behaviour of *Damalinia ovis* (L.). *Aust. J. Zool.* **5**, 19–29.

MURRAY M. D. (1957c) The distribution of the eggs of mammalian lice on their hosts. III. The distribution of the eggs of *Damalinia ovis* (L.) on the sheep. *Aust. J. Zool.* **5**, 173–182.

MURRAY M. D. (1957d) The distribution of the eggs of mammalian lice on their hosts. IV. The distribution of the eggs of *Damalinia equi* (Denny) and *Haematopinus asini* (L.) on the horse. *Aust. J. Zool.* **5**, 183–187.

MURRAY M. D. (1960a) The ecology of lice on sheep. I. The influence of skin temperature on populations of *Linognathus pedalis* (Osborne). *Aust. J. Zool.* **8**, 349–356.

MURRAY M. D. (1960b) The ecology of lice on sheep. II. The influence of temperature and humidity on the development and hatching of the eggs of *Damalinia ovis* (L.). *Aust. J. Zool.* **8**, 357–362.

MURRAY M. D. (1961) The ecology of the louse *Polyplax serrata* (Burm.) on the mouse *Mus musculus* L. *Aust. J. Zool.* **9**, 1–13.

MURRAY M. D. (1963a) The ecology of lice on sheep. III. Differences between the biology of *Linognathus pedalis* (Osborne) and *L. ovillus* (Neumann). *Aust. J. Zool.* **11**, 153–156.

MURRAY M. D. (1963b) The ecology of lice on sheep. IV. The establishment and maintenance of populations of *Linognathus ovillus* (Neumann). *Aust. J. Zool.* **11**, 157–172.

MURRAY M. D. (1963c) The ecology of lice on sheep. V. Influence of heavy rain on populations of *Damalinia ovis* (L.). *Aust. J. Zool.* **11**, 173–182.

MURRAY M. D. (1963d) Influence of temperature on the reproduction of *Damalinia equi* (Denny). *Aust. J. Zool.* **11**, 183–189.

MURRAY M. D. (1968) Ecology of lice on sheep. VI. The influence of shearing and solar radiation on populations and transmission of *Damalinia ovis*. *Aust. J. Zool.* **16**, 725–738.

MURRAY M. D. (1976) Insect parasites of marine birds and mammals. In *Marine Insects* (L. Cheng, ed.), pp. 79–96, North-Holland, Amsterdam.

MURRAY M. D. and GORDON G. (1969) Ecology of lice on sheep. VII. Population dynamics of *Damalinia ovis* (Schrank). *Aust. J. Zool.* **17**, 179–186.

MURRAY M. D. and NICHOLLS D. G. (1965) Studies on the ectoparasites of seals and penguins. I. The ecology of the louse *Lepidophthirus macrorhini* Enderlein on the southern elephant seal *Mirounga leonina* (L.). *Aust. J. Zool.* **13**, 437–454.

MURRAY M. D., SMITH M. S. R., and SOUCEK Z. (1965) Studies on the ectoparasites of seals and penguins. II. The ecology of the louse *Antarctophthirus ogmorhini* Enderlein on the Weddell seal *Leptonychotes weddelli* Lesson. *Aust. J. Zool.* **13**, 761–771.

MURTFELDT M. E. (1887) Traces of maternal affection in *Entylia sinuata* Fabr. *Ent. Am.* **3**, 177–178.
MUSCHINEK G., SZENTESI Á., and JERMY T. (1976) Inhibition of oviposition in the bean weevil (*Acanthoscelides obtectus* Say, Col., Bruchidae). *Acta phytopath. acad. sci. hung.* **11**, 91–98.
MUSPRATT J. (1951) The bionomics of an African *Megarhinus* (Dipt., Culicidae) and its possible use in biological control. *Bull. ent. Res.* **42**, 355–370.
MYERS C. M. (1967) Identification and descriptions of *Aedes* eggs from California and Nevada. *Can. Ent.* **99**, 795–806.
MYERS J. G. (1926) Biological notes on New Zealand Heteroptera. *Trans. NZ Inst.* **56**, 499–511.
MYERS K. (1952) Oviposition and mating behaviour of the Queensland fruit-fly (*Dacus (Strumeta) tryoni* (Frogg.)) and the solanum fruit-fly (*Dacus (Strumeta) cacuminatus* (Hering)). *Aust. J. sci. Res.* (B) **5**, 264–281.

NABROTZKY F. V. and REES D. M. (1968) Biology and behaviour of the gnat *Tendipes utahensis* on the southeast shores of the Great Salt Lake, Utah. *Proc. 21st Annual Meeting Utah Abatement Assoc.*, pp. 45–46 (G. C. Collett and M. F. Benge, eds.), Midvale, Utah.
NAGATOMI A. (1960) Studies in the aquatic snipe flies of Japan. Part II. Descriptions of the eggs (Diptera, Rhagonidae). *Mushi, Fukuoka* **33**, 1–3.
NAGATOMI A. and TANAKA A. (1967) Eggs of *Sepedon sauteri* Hendel. (Diptera, Sciomyzidae). *Kontyû, Tokyo* **35**, 31–33.
NAIR K.-S. S., MCEWEN F. L., and ALEX J. F. (1974) Oviposition and development of *Hylemya brassicae* (Bouché) (Diptera: Anthomyiidae) on cruciferous weeds. *Proc. Ent. Soc. Ontario* **104**, 11–15 (1973).
NAIR K.-S. S., MCEWEN F. L., and SNIECKUS V. (1976) The relationship between glucosinolate content of cruciferous plants and oviposition preferences of *Hylemya brassicae* (Diptera: Anthomyiidae). *Can. Ent.* **108**, 1031–1036.
NAITO A. (1964) Vertical inhabiting distribution of second generation larvae of the rice stem borer *Chilo suppressalis* Walker in the stem of the rice plant. *Jap. J. appl. Ent. Zool.* **8**, 106–110 (in Japanese; English summary).
NAKAMURA H. (1967) Comparative study of adaptability to density in two species of *Callosobruchus*. *Jap. J. Ecol.* **17**, 57–63 (in Japanese; English summary).
NAKAMURA H. A. (1968) A comparative study on the ovipositional behavior of two species of *Callosobruchus* (Coleoptera: Bruchidae). *Jap. J. Ecol.* **18**, 192–197.
NAKASHIMA M. and NOGAMI T. (1975) Oviposition of the oriental tobacco budworm *Helicoverpa asulta asulta* (Guenee) (*Heliothis assulta assulta*) on sweet pepper (*Capsicum*). *Proc. Pl. Prot. Kyushu* **21**, 14–15 (in Japanese).
NAKASHIMA T. (1964) The soil temperature preference of some scarabaeid beetles for egg-laying. *Kontyû, Tokyo* **32**, 28–32 (in Japanese; English summary).
NAKASHIMA Y., and SHIMIZU K. (1976) Feeding and oviposition behaviour of the white-striped longicorn *Batocera lineolata* Chev. *Proc. Pl. Prot. Kyushu* **22**, 61–62.
NAKATA G. (1957) The effect of sodium chloride present in the medium on the oviposition and viability of *Aedes aegypti* Linn. *Botyu-kagaku* **22**, 74–80 (in Japanese; English summary).
NAPPI A. J. (1969) Haemocytic reactions of *Drosophila euronotus* to *Pseudeucoila bochei* and their relationship to encapsulation in *Drosophila melanogaster*. *Diss. Abstr. int.* **30B**, 905.
NAPPI A. J. (1973) The role of melanization in the immune reaction of larvae of *Drosophila algonquin* against *Pseudeucoila bochei*. *Parasitology* **66**, 23–32.
NAPPI A. J. and STOFFOLANO J. G. JR (1971) *Heterotylenchus autumnalis*: hemocytic reactions and capsule formation in the host *Musca domestica*. *Expl Parasit.* **29**, 116–125.
NAPPI A. J. and STOFFOLANO J. G. JR (1972) Haemocytic changes associated with the immune

reaction of nematode-infested larvae of *Orthellia caesarion*. *Parasitology* **65**, 295–302.

NAPPI A. J. and STREAMS F. A. (1969) Haemocytic reactions of *Drosophila melanogaster* to the parasites *Pseudeucoila mellipes* and *P. bochei*. *J. Insect Physiol.* **15**, 1551–1566.

NARAYANAN E. S. and CHAUDHURI R. P. (1954) Studies on *Stenobracon deesae* (Cam.), a parasite of certain lepidopterous borers of graminaceous crops in India. *Bull. ent. Res.* **45**, 647–659.

NARAYANAN E. S., SUBBA RAO B. R., and KAUR R. B. (1959) Host selection and oviposition response in *Apanteles angaleti* Muesebeck (Braconidae: Hymenoptera). *Proc. Indian Acad. Sci.* (B) **49**, 139–147.

NARAYANDAS M. G. (1954) Description of the egg of *Anopheles gigas gigas* Giles, 1901. *Indian J. Malar.* **8**, 19–20.

NASONOVA L. I. (1960) The bionomics of *Cossus cossus*. *Trudy vses. Inst. Zashch. Rast.* **15**, 215–224 (in Russian).

NASR E. S. A. and NASSIF F. (1971) Behaviour of egg-laying and population density of the cotton leaf worm *Spodoptera littoralis* (Boisd.) on different host plants. *Bull. Soc. ent. Égypte* **54**, 541–544 (1970).

NATSKOVA V. (1967) Some studies on the biology and ecology of the buff-tip moth *Phalera bucephala* L. (Lepid.; Notodontidae). *Gradinar. oz. Nauka* **4**, 101–109 (in Bulgarian; English summary).

NAULT L. R., WOOD T. K., and GOFF A. M. (1974) Treehopper (Membracidae) alarm pheromones. *Nature, Lond.* **249**, 387–388.

NAWROT J. (1972) Fertility of the bean weevil (*Acanthoscelides obtectus* Say) (Coleoptera, Bruchidae) in containers of different colours. *Pr. nauk. Inst. Ochr. Rośl.* **13** (1) 145–153.

NEANDER A. (1928) Iakttagesler över parning och äggläggning hos *Lamia (Acanthocinus) aedilis* L. *Ent. Tidskr.* **49**, 202–208.

NEBOISS A. (1957) Note on Australian Triplectidinae (Trichoptera: Leptoceridae). *Beitr. Ent.* **7**, 50–54.

NECHOLS J. R. and TAUBER M. J. (1977) Age-specific interaction between the greenhouse whitefly and *Encarsia formosa*: influence of host on the parasite's oviposition and development. *Envir. Ent.* **6**, 143–149.

NEEDHAM J. G. (1905) Mayflies and midges of New York. *Bull. NY Mus.* **86**, 17–62.

NEEDHAM J. G. (1907) Eggs of *Benacus* and their hatching. *Ent. News* **17**, 113–116.

NEEDHAM J. G. (1924) The male of the parthenogenetic mayfly *Ameletus ludens*. *Psyche, Camb.* **31**, 308–310.

NEEDHAM J. G. and BETTEN C. (1901) Aquatic insects in the Adirondacks. *Bull. NY State Mus.* **47**, 383–612.

NEEDHAM J. G. and HEYWOOD H. B. (1929) *A Handbook of the Dragonflies of North America*, Thomas, Baltimore.

NEEDHAM J. G., TRAVER J. R., and HSU Y. C. (1935) *The Biology of Mayflies*, Comstock, New York.

NEEDHAM J. G. and WESTFALL M. J. (1955) *A Manual of the Dragonflies of North America*, University of California Press.

NEFF S. E. (1964) Snail-killing sciomyzid flies: application in biological control. *Proc. int. Ass. theoret. appl. Limnol.* **15**, 933–939.

NEFF S. E. and BERG C. O. (1961) Observations on the immature stages of *Protodictya hondurana* (Diptera: Sciomyzidae). *Bull. Brooklyn ent. Soc.* **56**, 46–56.

NEFF S. E. and BERG C. O. (1962) Biology and immature stages of *Hoplodictya spinicornis* and *H. setosa* (Diptera: Sciomyzidae). *Trans. Am. ent. Soc.* **88**, 77–93.

NEFF S. E. and BERG C. O. (1966) Biology and immature stages of malacophagous Diptera of the genus *Sepedon* (Sciomyzidae). *Bull. agric. Exp. Stn Virginia Polytech. Inst.* **566**, 1–113.

NEFF S. E. and WALLACE J. B. (1969) Observations on the immature stages of *Cordilura (Achaetella) deceptiva* and *C. (A.) varipes*. *Ann. ent. Soc. Am.* **62**, 775–785.

NEISWANDER R. B. (1958) The distribution and control of bagworms in Ohio. *J. econ. Ent.* **51** (3) 367–368.
NELSON B. C. (1971) Successful rearing of *Colpocephalum turbinatum* (Phthiraptera). *Nature New Biology* **232**, 255.
NELSON D. R., ADAMS T. S., and POMONIS J. G. (1969) Initial studies on the extraction of the active substance inducing monocoitic behaviour in *Musca domestica, Phormia regina,* and *Cochliomyia hominivorax. J. econ. Ent.* **62**, 634–639.
NEMEC S. J. (1969) Use of artificial lighting to reduce *Heliothis* spp. populations in cotton fields. *J. econ. Ent.* **62**, 1138–1140.
NETTLES W. C. JR and BETZ N. L. (1966) Surface sterilization of eggs of the boll weevil with cupric sulfate. *J. econ. Ent.* **59**, 239.
NEUBECKER F. (1967) Beitrag zur Technik der Massenzucht der Getreidemotte *Sitotroga cerealella* (Oliv.). *Anz. Schädlingsk.* **40**, 104–110.
NEUFFER G. (1971) Zur Technik in der Massenzucht der Getreidemotte *Sitotroga cerealella* Oliv. im Insektarium. *Anz. Schädlingsk.* **44**, 19–21.
NEUMANN F. G. (1976) Egg production, adult longevity and mortality of the stick insect *Didymuria violescens* (Leach) (Phasmatodea: Phasmatidae) inhabiting mountain ash forest in Victoria. *J. Aust. ent. Soc.* **15**, 183–190.
NEUMANN K. W. (1943) Die Lebensgeschichte der Käfermilbe *Peocilochirus necrophori* Vitz. nebst Beschreibung aller Entwicklungsstufen. *Zool. Anz.* **142**, 1–21.
NEVILL E. M. and ANDERSON D. (1972) Host preferences of *Culicoides* midges (Diptera: Ceratopogonidae) in South Africa as determined by precipitin tests and light trap catches. *Onderstepoort J. vet. Res.* **39**, 147–152.
NEVILLE A. C. (1960) List of Odonata from Ghana, with notes on their mating, flight, and resting sites. *Proc. R. ent. Soc. Lond.* (A) **35**, 124–128.
NEVILLE A. C. (1975) *Molecular Biology of Arthropod Cuticle*, Springer-Verlag.
NEVILLE A. C. and LUKE B. M. (1969) A two-system model for chitin–protein complexes in insect cuticles. *Tissue and Cell* **1**, 689–707.
NEVILLE A. C. and LUKE B. M. (1971) A biological system producing a self-assembling cholesteric protein liquid crystal. *J. Cell Sci.* **8**, 93–109.
NEW T. R. (1971a) Ovariolar dimorphism and repagula formation in some South American Ascalaphids (Neuroptera). *J. Ent.* (A) **46**, 73–77.
NEW T. R. (1971b) An introduction to the natural history of the British Psocoptera. *Entomologist* **104**, 59–97.
NEW T. R. (1971c) A new species of *Belaphapsocus* Badonnel from Brazil, with notes on its early stages and bionomics (Psocoptera). *Entomologist* **104**, 124–133.
NEWELL I. M. and HARAMOTO F. H. (1968) Biotic factors influencing populations of *Dacus dorsalis* in Hawaii. *Proc. Hawaii. ent. Soc.* **20**, 81–139.
NEWKIRK M. R. (1955) On the eggs of some man-biting mosquitoes. *Ann. ent. Soc. Am.* **48**, 60–66.
NEWKIRK M. R. (1957) On the black-tipped hangingfly (Mecoptera, Bittacidae). *Ann. ent. Soc. Am.* **50**, 302–306.
NEWSTEAD H. and THOMAS H. W. (1910) The mosquitoes of the Amazon region. *Ann. trop. Med. Parasit.* **4**, 141–150.
NICOLAS H. U. (1891) De la ponte de *Leptidea brevipennis* Muls. *Coleopterist* **4**, 56–58.
NIELSEN A. (1936) Das Eierlegen, der Laich und die Larven des I. Stadiums von *Oligoplectrum maculatum* Fourcroy (Trichoptera). *Zool. Anz.* **113**, 255–266.
NIELSEN A. (1942) Über die Entwicklung und Biologie der Trichopteren, mit besonderer Berücksichtigung der Quelltrichopteren Himmerlands. *Arch. Hydrobiol.* (Suppl.) **17**, 255–631.

NIELSEN A. (1943) Postembryonale Entwicklung und Biologie der rheophilen Köcherfliege *Oligoplectrum maculatum* Fourcroy. *Biol. Medd.* **19** (2) 1–87.

NIELSEN A. (1948) Postembryonic development and biology of the Hydroptilidae. A contribution to the phylogeny of the caddis flies and to the question of the origin of the casebuilding instinct. *K. danske Vidensk. Selsk. Biol. Skr.* **5** (1) 1–200.

NIELSEN J. C. (1909) Lagttagelser over entoparasitiske Muscidelarver hos Arthropoder. *Ent. Medd.* **4**, 1–126.

NIELSEN J. C. (1911) Undersogelser over entoparasitiske Muscidenlarver hos Arthropoder. *Vidensk. Meddr. dansk naturh. Foren.* **63**, 1–26.

NIELSEN J. C. (1918) Tachin-Studier. *Vidensk. Meddr dansk naturh. Foren.* **69**, 247–262.

NIELSON M. W. and TOLES S. L. (1968) Observations on the biology of *Acinopterus angulatus* and *Aceratagallia curvata* in Arizona (Homoptera; Cicadellidae). *Ann. ent. Soc. Am.* **61**, 54–56.

NIEMCZYK E. (1967) *Psallus ambiguus* (Fall.) (Heteroptera, Miridae). Cześć I. Morfologia i biologia. *Polskie Pismo ent.* **37**, 797–842 (English summary).

NIEMCZYK E. (1970) Development and fecundity of bark bug—*Anthocoris nemorum* (L.) (Heter., Anthocoridae) reared on *Sitotroga* eggs—*Sitotroga cerealella* Oliv. (Lep, Gelechiidae). *Polskie Pismo ent.* **40**, 857–865 (in Polish; English summary).

NIEMCZYK H. D. and FLESSEL J. K. (1969) Seasonal oviposition habits of the alfalfa weevil in Ohio. *Proc. N. cent. Brch Am. Ass. econ. Ent.* **24**, 32–33.

NIEMCZYK H. D. and FLESSEL J. K. (1970) Population dynamics of alfalfa weevil eggs in Ohio. *J. econ. Ent.* **63**, 242–247.

NIETZKE G. (1939) Aus dem Leben des nebeligen Schildkäfers (*Cassida nebulosa* L.). *Natur Volk* **69**, 355–360.

NIGMANN M. (1908) Anatomie und Biologie von *Acentropus niveus* Oliv. *Zool. Jb. (Syst.)* **26**, 489–560.

NIJHOLT W. W. (1973) The effect of male *Trypodendron lineatum* (Coleoptera: Scolytidae) on the response of field populations to secondary attraction. *Can. Ent.* **105**, 583–590.

NIJVELDT W. (1969) *Gall Midges of Economic Importance . . . Vol. VIII. Gall Midges—Miscellaneous (Zoophagous, Fungivorous, and those that Attack Weeds). Identification of Gall Midges*, Crosby, Lockwood and Sons, London.

NIKOL'SKII V. V. (1911) Migratory or Asiatic locust in Syr-Daria region. *Lyub. Prir.* **3**, 1–13.

NIKOL'SKII V. V. (1925) Parasites of the Asiatic locust and other natural enemies. In "The Asiatic locust *Locusta migratoria* L.". *Trudy Otd. prikl. Ént.* **12**, 215–218 (in Russian).

NINOMIYA E. (1959) Further notes on the immature stages of adphidophagous syrphid flies of Japan. *Sci. Bull. Fac. lib. Arts Educ. Nagasaki Univ.* **10**, 23–52.

NIRULA K. K., ANTONY J., and MENON K. P. V. (1952) Investigations on the pests of the coconut palm. The rhinoceros beetle (*Oryctes rhinoceros* L.). Life history and habits. *Indian Cocon. J.* **5**, 57–70.

NISHIDA T. (1956) An experimental study of the ovipositional behaviour of *Opius fletcheri* Silvestri (Hymenoptera: Braconidae), a parasite of the melon fly. *Proc. Hawaii. ent. Soc.* **16**, 126–134.

NISHIDA T. and NAPOMPETH, B. (1974) Egg distribution on corn plants by the corn earworm moth *Heliothis zea* (Boddie). *Proc. Hawaii. ent. Soc.* **21**, 425–433.

NISHIGAKI J. (1976) Ecological studies of the cupreous chafer *Anomala cuprea* Hope (Coleoptera: Scarabaeidae). VII. Seasonal changes of sex ratio and the egg numbers in the female body of adult chafers, especially of the light trap population and the food-plant population. *Jap. J. Appl. Ent. Zool.* **20**, 164–166.

NISHIJIMA Y. (1960) Host plant preference of the soybean pod borer *Grapholitha glicinivorella* Matsumura (Lep., Eucosmidae). I. Oviposition site. *Entomologia exp. appl.* **3**, 38–47.

NISHIKAWA Y. (1948) The emergence of young Mantids from their egg mass. *Bull. Takarazuka Insectarism* **41**, 9–11 (in Japanese).
NITSCHE H. (1893) Beobachtungen über die Eierdeckschuppen der weiblichen Processionsspinner. *Isis* **1893**, 108–117.
NIXON G. E. J. (1951) *The Association of Ants with Aphids and Coccids*, Commonwealth Inst. Ent., London, 36 pp.
NOLL J. (1963) Über den Einfluss der Temperatur auf die Lebensdauer der Imagines, auf Beginn, Verlauf und Dauer der Eiablage, sowie auf die Eizahlen (Eiproduktion) bei der Kohleule (*Mamestra* (*Barathra*) *brassicae* L.) und seine Bedeutung für den Massenwechsel des Schädlings. *NachrBl. dt. PflSchutz dienst, Berl.* (NF) **17**, 9–24.
NONVEILLER G. (1959) Les prédateurs des pontes de *Lymantria dispar* L. constatés en Yougoslavie au cours de sa gradation de 1945–1950. *Zašt. Bilja* **52–53**, 15–35 (in Serbo-Croat; French summary).
NOORDINK J. P. W. (1965) Enkele toepassingen van radioactieve isotopen bij het oecologisch onderzoek. *Ent. Ber. Amst.* **25**, 130 (English summary).
NOORDINK J. P. W. and MINKS A. K. (1970) Autoradiography: a sensitive method in dispersal studies with *Adoxophyes orana* (Lepidoptera: Tortricidae). *Entomologia exp. appl.* **13**, 448–454.
NORD J. C., GRIMBLE D. G., and KNIGHT F. B. (1972a) Biology of *Oberea schaumii* (Coleoptera: Cerambycidae) in trembling aspen, *Populus tremuloides*. *Ann. ent. Soc. Am.* **65**, 114–119.
NORD J. C., GRIMBLE D. G., and KNIGHT F. B. (1972b) Biology of *Saperda inornata* (Coleoptera: Cerambycidae) in trembling aspen, *Populus tremuloides*. *Ann. ent. Soc. Am.* **65**, 127–135.
NORDLUND D. A., LEWIS W. J., GROSS H. R. JR, and HARRELL E. A. (1974) Description and evaluation of a method for field application of *Heliothis zea* eggs and kairomones for *Trichogramma*. *Envir. Ent.* **3**, 981–984.
NORDLUND D. A., LEWIS W. J., JONES R. L., and GROSS H. R. JR (1976) Kairomones and their use for management of entomophagous insects. IV. Effect of kairomones on productivity and longevity of *Trichogramma pretiosum* Riley (Hymenoptera: Trichogrammatidae). *J. chem. Ecol.* **2**, 67–72.
NORRIS D. M. and CHU, HSIEN-MING (1970) Nutrition of *Xyleborus ferrugineus*. II. A holidic diet for the aposymbiotic insect. *Ann. ent. Soc. Am.* **63**, 1142–1145.
NORRIS K. R. and MURRAY M. D. (1964) Notes on the screw-worm fly *Chrysomya bezziana* (Diptera: Calliphoridae) as a pest of cattle in New Guinea. *CSIRO Aust. Div. Ent. Tech. Pap.* No. **6**, 1–26.
NORRIS M. J. (1936) Experiments on some factors affecting fertility in *Trogoderma versicolor* Creutz. (Coleoptera, Dermestidae). *J. Anim. Ecol.* **5**, 19–22.
NORRIS M. J. (1959) Reproduction in the red locust (*Nomadacris septemfasciata* Serville) in the laboratory. *Anti-Locust Bull.* **36**, 1–46.
NORRIS M. J. (1964) Laboratory experiments on gregarious behaviour in ovipositing females of the desert locust (*Schistocerca gregaria* (Forsk.)). *Entomologia exp. appl.* **6**, 279–303.
NORRIS M. J. (1968) Laboratory experiments on oviposition responses of the desert locust *Schistocerca gregaria* (Forsk.). *Anti-Locust Bull.* **43**, 1–47.
NORRIS M. J. (1970) Aggregation response in ovipositing females of desert locust, with special reference to the chemical factor. *J. Insect Physiol.* **16**, 1493–1515.
NOVÁK K. and SEHNAL F. (1963) The developmental cycle of some species of the genus *Limnephilus* (Trichoptera). *Acta Soc. ent. Bohem.* (*Čsl.*) **60**, 68–80.
NOZATO K. (1969) The oviposition site of the egg of *Rhyacionia duplana simulata* Heinrich on pine shoots. *Jap. J. appl. ent. Zool.* **13**, 22–25 (in Japanese; English summary).
NUNOME J. (1950) On the thickness of the eggshell in the silkworm. *J. seric. Sci. Japan* **19**, 315–322 (in Japanese).

NUORTEVA P. (1970) Histerid beetles as predators of blow-flies (Diptera, Calliphoridae) in Finland. *Ann. zool. fenn.* **7**, 195–198.
NUTTING W. L. (1953a) Observations on the reproduction of the giant cockroach *Blaberus craniifer* Burm. *Psyche, Camb.* **60**, 6–14.
NUTTING W. L. (1953b) A gregarine, *Diplocystis*, in the haemocoele of the roach *Blaberus craniifer* Burm. *Psyche, Camb.* **60**, 126–128.
NUZZACI G. (1969) Nota morfo-biologica sull'*Eulecanium corni* Bouché spp. *apuliae* nov. *Entomologica* **5**, 9–36.
NWANZE K. F. and HORBER E. (1976) Seed coat of cowpeas affect oviposition and larval development of *Callosobruchus maculatus*. *Envir. Ent.* **5**, 213–218.
NWANZE K. F., HORBER E., and PITTS C. W. (1975) Evidence for ovipositional preference of *Callosobruchus maculatus* for cowpea varieties. *Envir. Ent.* **4**, 409–412.
NYIIRA Z. M. (1970) The biology and behavior of *Rhinocoris albopunctatus* (Hemiptera: Reduviidae). *Ann. ent. Soc. Am.* **63**, 1224–1227.

OAKLEY J. N. and UNCLES J. J. (1977) Use of oviposition trays to estimate numbers of wheat bulb fly *Leptohylemia coarctata* eggs. *Ann. appl. Biol.* **85**, 407–409.
OATMAN E. R. and JENKINS L. (1962) The biology of the red-banded leaf roller *Argyrotaenia velutinana* (Wlkr.) in Missouri with notes on its natural control. *Res. Bull. Mo. agric. Exp. Stn* **789**, 1–14.
OATMAN E. R., PLATNER G. R., and GREANY P. D. (1969) The biology of *Orgilus lepidus* (Hymenoptera: Braconidae), a primary parasite of the potato tuberworm. *Ann. ent. Soc. Am.* **62**, 1407–1414.
O'CONNOR B. A., PILLAI J. S., and SINGH S. R. (1955) Notes on the coconut stick insect *Graeffea crouani* Le Guillou. *Agric. J. Fiji* **25**, 89–92 (1954).
ODEBIYI J. A. and OATMAN E. R. (1972) Biology of *Agathis gibbosa* (Hym., Braconidae), a primary parasite of the potato tuberworm (*Phthorimaea operculella*: Lep. Gelechiidae). *Ann. ent. Soc. Am.* **65**, 1104–1114.
ODELL T. M. and ROLLINSON W. D. (1966) A technique for rearing the gipsy moth *Porthetria dispar* (L.) on an artificial diet. *J. econ. Ent.* **59**, 741–742.
ODHIAMBO T. R. (1959) An account of parental care in *Rhinocoris albopilosus* Signoret (Hemiptera–Heteroptera: Reduviidae), with notes on its life history. *Proc. R. ent. Soc. Lond.* (A) **34**, 175–182.
ODHIAMBO T. R. (1960) Parental care in bugs and non-social insects. *New Scient., Lond.* **8**, 449–451.
ODHIAMBO T. R. (1968) The effects of mating on egg production in the cotton stainer *Dysdercus fasciatus*. *Entomologia exp. appl.* **11**, 379–388.
O'DONNELL A. E. and AXTELL R. C. (1965) Predation by *Fuscuropoda vegetans* (Acarina: Uropodidae) on the house fly (*Musca domestica*). *Ann. ent. Soc. Am.* **58**, 403–404.
O'DONNELL A. E. and NELSON E. L. (1967) Predation by *Fuscuropoda vegetans* (Acarina: Uropodidae) and *Macrocheles muscaedomesticae* (Acarina: Macrochelidae) on the eggs of the little house fly *Fannia canicularis*. *J. Kans. ent. Soc.* **40**, 441–443.
OESTER P. T. and RUDINSKY J. A. (1975) Sound production in Scolytidae: stridulation by "silent" *Ips* bark beetles. *Z. angew. Ent.* **79**, 421–427.
O'FARRELL A. F. (1970) Odonata (Dragonflies and damselflies). In *The Insects of Australia. A Textbook for Students and Research Workers*, Melbourne University Press.
O'GOWER A. K. (1958) The oviposition behaviour of *Aëdes australis* (Erickson) (Diptera, Culicidae). *Proc. Linn. Soc. NSW* **83**, 245–250.
O'GOWER A. K. (1963) Environmental stimuli and the oviposition behaviour of *Aedes aegypti*

var. *queenslandis* Theobald (Diptera: Culicidae). *Anim. Behav.* **11**, 189–197.

OHAUS F. (1899–1900) Bericht über eine entomologische Reise nach Zentralbrasilien. *Stett. ent. Ztg.* **60**, 204–245; **61**, 164–191 (1900).

OHAUS F. (1909) Bericht über eine entomologische Studienreise in Sudamerika. *Stett. ent. Ztg.* **70**, 3–139.

OHNO M. (1956) On the Japanese tortoise beetles belonging to the genus *Thlaspida* Weise, with special reference to the morphology and biology of their early stages. (Studies on the tortoise beetles of Japan (part 2).) *J. Tokyo Univ.* **9**, 1–10 (in Japanese).

OKABE T., SUGIYAMA T., and OHI S. (1958) *Jap. J. Zool. tech. Sci.* **29**, 122–128.

OKADA J., MAKI T., and KURODA H. (1934) Studies on the control of rice borers. I. Observations on the liberation of some hymenopterous parasites living in the eggs of the rice borer *Chilo simplex* Butler. *Japan. Min. Agric. Forestry Bull.* **69**, 1–78 (in Japanese).

OKASHA A. Y. K., HASSANEIN A. M. M., and FARAHAT A. Z. (1970) Effects of sublethal high temperature on an insect *Rhodnius prolixus* (Stål). IV. Egg formation, oviposition and sterility. *J. exp. Biol.* **53**, 25–36.

OKELO O. Physiological control of oviposition in the female desert locust *Schistocerca gregaria* Forsk. *Can. J. Zool.* **49**, 969–974.

OKSANEN H., PERTTUNEN V., and KANGAS E. (1970) Studies on the chemical factors involved in the olfactory orientation of *Blastophagus piniperda* (Coleoptera: Scolytidae). *Contr. Boyce Thompson Inst.* **24**, 299–304.

OKSENOV V. (1946) Behaviour of *Deporaus betulae* L. under experimental conditions. *Bull. Acad. Sci. URSS. (Biol. Ser.)* **1946**, 105–116 (in Russian; English summary).

OKU T. and KOBAYASHI T. (1973) Studies on the ecology and control of insects in grasslands. V. Oviposition behaviour of the black cutworm moth *Agrotis ypsilon* Hufnagel, with notes on some larval behaviours. *Bull. Tohoku natn. agric. exp. Stn* **46**, 161–183 (in Japanese; English summary).

OKUMURA G. T. (1966) The dried-fruit moth (*Vitula edmandsae serratilineella* Ragonot) pest of dried fruits and honeycombs. *Bull. Calif. Dep. Agric.* **55**, 180–186.

OKUNI T. (1928) Report of the studies on stored grain pests. Part II. *Rep. Govt. Res. Inst. Formosa* **34**, 1–121 (in Japanese).

OKUNO T. (1961) Feeding tests and some field observations of three aphidophagous Coccinellids. *Publ. ent. Lab. Coll. Agric. Univ. Osaka* **6**, 149–152.

OKUNO T. (1970) Immature stages of two species of the genus *Volucella* Geoffroy (Dipt., Syrphidae). *Kontyû, Tokyo* **38**, 268–270.

OLDBERG G. (1959) *Das Verhalten der solitären Wespen Mitteleuropas (Vespidae, Pompilidae, Sphecidae)*, Berlin.

OLDIGES H. (1959) Der Einflusse der Temperatur auf Stoffwechsel und Eiproduktion von Lepidopteren. *Z. angew. Ent.* **44**, 115–166.

OLDROYD H. (1947) A new species of *Systoechus* (Diptera: Bombyliidae), bred from eggs of the desert locust. *Proc. R. ent. Soc. Lond.* (B) **16**, 105–107.

OLDROYD H. (1964) *The Natural History of Flies*, Weidenfeld and Nicolson, London.

O'LOUGHLIN G. T. (1964) The Queensland fruit fly in Victoria. *J. Agric. Vict. Dep. Agric.* **62**, 391–402.

O'MEARA G. F. and EVANS D. G. (1977) Autogeny in saltmarsh mosquitoes induced by a substance from the male accessory gland. *Nature, Lond.* **267**, 342–344.

OMER S. M. and GILLIES M. T. (1971) Loss of response to carbon dioxide in palpectomized female mosquitoes. *Entomologia exp. appl.* **14**, 251–252.

ONGARO D. (1933) A paraffin in the eggs of *Bombyx mori*. *Annali Chim. Appl.* **23**, 567–572 (in Italian).

ONO T. (1960) Observations on *Culex pipiens* reared through successive generations. First report: On the longevity, the blood-sucking and the oviposition of females after overwintering. *Sanit. injur. Insect* **5**, 19–23 (in Japanese; English summary).

ONSAGER J. A. and MULKERN G. B. (1963) Identification of eggs and egg-pods of North Dakota grasshoppers (Orthoptera: Acrididae). *Tech. Bull. Dep. Ent. agric. Exp. Stn N. Dakota Univ.* **446**, 1–48.

OOMMEN C. N. and NAIR M. R. G. K. (1969) On the biology of *Coccotrypes carpophagus* Horn, a pest of stored arecanut. *Indian J. Ent.* **30**, 314–315 (1968).

OPPONG-MENSAH D. and KUMAR R. (1973) Internal reproductive organs of the cocoa-capsids (Heteroptera, Miridae). *Entomologist's mon. Mag.* **100**, 148–154.

ÖRÖSI-PÁL Z. (1966) Die Bienelaus-Arten. *Angew. Parasitol.* **7**, 139–171.

OROZCO F., ESPEJO M., and CARBONELL E. (1971) Influencia de diversos factores en la puesta del *Tribolium castaneum*. IV. *Influencia del macho. An. Inst. nac. Invest. Agrarias Gen.* **1**, 93–108.

OROZCO F. and GÓMEZ RUANO R. (1970) Influencia de diversos factores en la puesta del *Tribolium castaneum*. III. Influencia del medio alimenticio. *An. Inst. nac. Invest. Agron.* **19**, 403–423.

ORPHANIDES G. M., GONZALEZ D., and BARTLETT B. R. (1971) Identification and evaluation of pink bollworm predators in southern California. *J. econ. Ent.* **64**, 421–424.

ORPHANIDIS P. S., PÉTSIKOU N. A., and PATSAKOS P. G. (1970) Élevage du *Dacus oleae* (Gmel.) sur substrat artificiel. *Annls Inst. phytopath. Benaki* (NS) **9**, 147–169.

OSGOOD C. E. (1971a) An oviposition pheromone associated with the egg rafts of *Culex tarsalis*. *J. econ. Ent.* **64**, 1038–1041.

OSGOOD C. E. (1971b) Wet filter paper as an oviposition substrate for mosquitoes that lay egg-rafts. *Mosq. News* **31**, 32–35.

OSGOOD C. E. and KEMPSTER R. H. (1971) An air-flow olfactometer for distinguishing between oviposition attractants and stimulants of mosquitoes. *J. econ. Ent.* **64**, 1109–1110.

OTSURU M. and MIYAGAWA M. (1950) The eggs of Japanese *Anopheles*. *Jap. J. sanit. Zool.* **1**, 3–4.

OTSURU M., NAGASHIMA Y., NAKAMURA Y., and KISHIMOTO T. (1976) Survey of eggs of the *Anopheles sinensis* sibling species group in Okinawa Is., Japan. *Japan J. sanit. Zool.* **27**, 301–303.

OTSURU M. and OHMORI Y. (1960) Malaria studies in Japan after World War II. Part II. The research for *Anopheles sinensis* sibling species group. *Jap. J. exp. Med.* **30**, 33–65.

OTVOS I. S. (1977) Mortality of overwintering eggs of the eastern hemlock looper in Newfoundland. *Bi-mon. Res. Notes* **33**, 3–5.

OTVOS I. S. and BRYANT D. G. (1972) An extraction method for rapid sampling of easter hemlock looper eggs, *Lambdina fiscellaria fiscellaria* (Lepidoptera: Geometridae). *Can. Ent.* **104**, 1511–1514.

OUCHI M. (1957) Studies on the bionomics of the rice stink-bug *Lagynatomus assimulans* Distant. V. On the hatch and the period of egg and larval stages. VI. The influences of climatic factors on the crawling activity of adults. *Jap. J. Appl. ent. Zool.* **1**, 113–118; **3**, 7–15 (in Japanese; English summary).

OUDEMANS A. C. (1912) Über die Eier von *Lipeurus quadripustulatus*. *Tijd. Ent.* **55**, xxvii–xxx.

OVERMEER W. P. J. (1961) Investigations on species of pear sucker in the Netherlands. *Tijdschr. PlZiekt.* **67**, 281–289 (in Dutch; English summary).

OWUSU-MANU E. (1977) Biology of *Bathycoelia thalassina* (H.-S.) (Heteroptera: Pentatomidae) in Ghana. Published in *Proc. 4th Conf. of W. African Cocoa Entomologists, Ghana, 1974* (R. Kumar, ed.).

PACKARD A. S. (1898) *A Textbook of Entomology Including the Anatomy, Physiology, Embryology and Metamorphoses of Insects for Use in Agricultural and Technical Schools and Colleges as well as by the Working Entomologist*, Macmillan, New York.

PAG H. (1959) *Hyponomeuta*-Arten als Schädlinge im Obstbau. Ein Beitrag zur Biologie, Ökologie und Bekämpfung, unter besonderer Berücksichtigung des Arten- und Rassenproblems. *Z. angew. Zool.* **46**, 129–189.

PAGDEN H. (1934) Notes on hymenopterous parasites of padi insects in Malaya. *Bull. Dep. Agric. FMS* **15**, 1–13.

PAGENSTECHER A. (1864) Die Häutungen der Gespenstheuschrecke *Mantis religiosa*. *Arch. Naturgesch.* **30**, 7–25.

PAILLOT A. (1923) Sur la variabilité du cycle évolutif d'un ichneumonide, parasite nouveau des larves de *Neurotoma nemoralis* L. *C. r. hebd. Séanc. Soc. Biol. Paris* **89**, 1045–1058.

PAILLOT A. (1928) On the natural equilibrium of *Pyrausta nubilalis* Hb. *Sci. Rep. int. Corn Borer Invest.* **1927–1928**, 77–106.

PAINTA F. (1968) Beitrag zur Brutfürsorge des Trichterwicklers *Deporaus betulae*. *Decheniana* **120**, 299–311.

PAINTER R. H. (1951) *Insect Resistance in Crop Plants*, New York.

PAINTER R. H. (1955) Insects on corn and teosinte in Guatemala. *J. econ. Ent.* **48**, 36–42.

PAINTER R. H. (1958) Resistance of plants to insects. *A. Rev. Ent.* **3**, 267–290.

PAINTER R. R. and KILGORE W. W. (1967) Some physical and chemical characteristics of normal eggs, larvae and chorions of the housefly, *Musca domestica*. *Ann. ent. Soc. Am.* **60**, 1163–1166.

PALMEN J. A. (1884) *Über paarige Ausführungsgänge der Geschlechtsorgane bei Insekten. Ein morphologische Untersuchung*, Helsingfors.

PALMER D. F., WINDELS M. B., and CHIANG H. C. (1976) Changes in specific gravity of developing western corn rootworm eggs. *Envir. Ent.* **5**, 621–622.

PALMER M. (1976) Notes on the biology of *Pterombrus piceus* Krombein (Hymenoptera: Tiphiidae). *Proc. ent. Soc. Washington* **78**, 369–375.

PANDE Y. D. (1971) Biology of citrus psylla *Diaphorina citri* Kuw. (Hem. Psyllidae), Hom. *Israel J. Ent.* **6**, 307–311.

PANIS A. (1970) Observations biologiques sur une cochinelle farineuse des plantes ornamentales: *Chorizococcus lounsburyi* (Brain) (Hom., Coccoidea, Pseudococcidae). *Bull. Soc. ent. Fr.* **75**, 75–80.

PANIS A. (1974) Action prédatrice d'*Eublemma scitula* (Lepidoptera, Noctuidae, Erastriinae) dans le sud de la France. *Entomophaga* **19**, 493–500.

PANIZZI A. R. and SMITH J. G. (1977) Biology of *Piezodorus guildinii* oviposition, development time, adult sex ratio, and longevity. *Ann. ent. Soc. Am.* **70**, 35–39.

PANTEL J. (1910) Recherches sur des Diptères à larves entomobies. I. Caractères parasitiques aux points de vue biologique, éthologique et histologique. *Cellule* **26**, 27–216 (1909).

PANTEL J. (1913) Recherches sur les Diptères à larves entomobies. II. Les enveloppes de l'oeuf avec les formations qui en dépendent, les dégâts indirects dur parasitisme. *Cellule* **29**, 7–289 (1912).

PANTEL J. (1919a) Le calcium dans la physiologie normale des Phasmides (Ins. Orth.): oeuf et larve éclosante. *C. r. hebd. Séanc. Acad. Sci. Paris* **168**, 127–129.

PANTEL J. (1919b) Le calcium, forme de reserve dans la femelle des Phasmides; ses formes d'élimination dans les deux sexes. *C. r. hebd. Séanc. Acad. Sci. Paris* **168**, 242–244.

PANTELOURIS E. M. (1957) Size response of developing *Drosophila* to temperature changes. *J. Genet.* **55**, 507–510.

PANTYUKHOV G. A. (1968) A study of the ecology and physiology of the predacious beetle *Chilocorus rubidus* Hope (Coleoptera, Coccinellidae). *Zool. Zh.* **47**, 376–386 (in Russian; English summary).

PAOLI G. (1919) La lotta contro le cavallette in Capitanata nel 1917–18. *Boll. Minist. Agric.* **18**, 1–11.

PAOLI G. (1937a) Osservazioni su alcune particolarità di struttura e funzione dell'apparato genitale famminile di *Dociostaurus maroccanus* Thunb. (Orthopt., Acrididae). *Redia* **23**, 17–26.

PAOLI G. (1937b) Studi sulle cavallette di Foggia (*Dociostaurus maroccanus* Thunb.) e sui loro oofagi (Ditteri Bombiliidi e Coleotteri Meloidi) ed Acari ectofagi (Eritreidi e Trombidiidi). *Redia* **23**, 27–206.

PAOLI G. (1938) Note sulla biologia e sulla filogenesi dei meloidi (Coleoptera). *Mem. Soc. ent. ital.* **16**, 71–96 (1937).

PAPILLON M. (1970a) La compression des oeufs, facteur de diversification des larves chez le criquet pelèrin *Schistocerca gregaria* (Forsk.) en phase grégaire. *C. r. hebd. Séanc. Acad. Sci. Paris* (D) **271**, 1311–1314.

PAPILLON M. (1970b) Influence de la substance spumeuse des oothèques sur la viabilité des oeufs et des larves du criquet pèlerin (*Schistocerca gregaria* Forsk.) en phase grégaire. *C. r. hebd. Séanc. Acad. Sci. Paris* (D) **270**, 1616–1619.

PAPP C. S. (1957) Australian leaf-eating ladybird beetle *Epilachna vigintiopunctata* F. as a possible agricultural pest in the USA (Coleoptera: Coccinellidae). *Bull. S. Calif. Acad. Sci.* **56**, 155–166.

PARADIS R. O. (1957) Observations sur les dégâts causés par le charançon de la prune *Conotrachelus nenuphar* (Hbst.) sur les pommes dans le sudouest de Québec. *Can. Ent.* **89**, 496–502.

PARADIS R. O. and LE ROUX E. J. (1962) A sampling technique for population and mortality factors of the fruit-tree leaf roller *Archips argyrospilus* (Wlk.) (Lepidoptera: Tortricidae) on apple in Quebec. *Can. Ent.* **94**, 561–573.

PARAMONOV S. (1931) Dipterologische Fragmente XXV. *Mém. Cl. Sci. nat. tech. Acad. Sci. Ukr.* **5**, 221–239.

PARFIN S. I. and GURNEY A. B. (1956) The spongilla-flies, with special reference to those of the Western Hemisphere (Sisyridae, Neuroptera). *Proc. US natn. Mus.* **105**, 421–529.

PARIHAR D. R. (1970) Water-balance in developing eggs of the Ak grasshopper *Poekilocerus pictus* (Acridoidea, Pyrgomorphidae). *Z. angew. Ent.* **67**, 9–19.

PARISER K. (1917) Beiträge zur Biologie und Morphologie der einheimischen Chrysopiden. *Arch. Naturgesch.* (A) **83**, 1–57.

PARK T. (1933) Studies in population physiology. II. Factors regulating initial growth of *Tribolium confusum* populations. *J. exp. Zool.* **65**, 17–42.

PARK T. (1934) Observations on the general biology of the flour beetle *Tribolium confusum*. *Q. Rev. Biol.* **9**, 36–54.

PARK T. (1957) Experimental studies of interspecies competition. III. Relation of initial species proportion to competitive outcome in populations of *Tribolium*. *Physiol. Zoöl.* **30**, 22–40.

PARK T. (1962) Beetles, competition, and populations. *Science, Wash.* **138**, 1369–1375.

PARK T., LESLIE P. H., and MERTZ D. B. (1964) Genetic strains and competition in populations of *Tribolium*. *Physiol. Zoöl.* **37**, 97–162.

PARK T., MERTZ D. B., GRODZINSKI W., and PRUS T. (1965) Cannibalistic predation in populations of flour beetles. *Physiol. Zoöl.* **38**, 289–321.

PARK T., MERTZ D. B., and PETRUSEWICZ K. (1961) Genetic strains of *Tribolium*: their primary characteristics. *Physiol. Zoöl.* **34**, 62–80.

PARK T., NATHANSON M., ZIEGLER J. R., and MERTZ D. B. (1970) Cannibalism of pupae by mixed-species populations of adult *Tribolium*. *Physiol. Zoöl.* **43**, 166–184.

PARK T., ZIEGLER J. R., ZIEGLER D. L., and MERTZ D. B. (1974) The cannibalism of eggs by *Tribolium* larvae. *Physiol. Zoöl.* **47**, 37–58.

PARKER A. H. (1950) Studies on the eggs of certain biting midges (*Culicoides* Latreille) occurring in Scotland. *Proc. R. ent. Soc. Lond.* (A) **25**, 43–52.

Parker A. H. (1965) The maternal behaviour of *Pisilus tipuliformis* Fabricius (Hemiptera: Reduviidae). *Entomologia exp. appl.* **8**, 13–19.

Parker F. D. (1970) Seasonal mortality and survival of *Pieris rapae* (Lepidoptera: Pieridae) in Missouri and the effect of introducing an egg parasite, *Trichogramma evanescens*. *Ann. ent. Soc. Am.* **63**, 985–994.

Parker G. A. (1970) The reproductive behaviour and the nature of sexual selection in *Scatophaga stercoraria* L. (Diptera: Scatophagidae). 5. The female's behaviour at the oviposition site. *Behaviour* **37**, 140–168.

Parker H. L. (1926) Note sulla larva del *Polochrum repandum* Spinola (Hym., Sapygidae) parassita della *Xylocopa violacea* L. *Boll. Lab. Zool. gen. agr. Portici* **18**, 268–270.

Parker H. L. (1937) On the oviposition habits of *Stilbula cynipiformis* Rossi. *Proc. ent. Soc. Wash.* **39**, 1–3.

Parker J. B. and Böving A. G. (1924) The blister beetle *Tricrania sanguinipennis*—biology, descriptions of different stages, and systematic relationship. *Proc. US natn. Mus.* **64** (23) 1–40.

Parker J. R. and Wakeland C. (1957) Grasshopper egg pods destroyed by larvae of bee flies, blister beetles, and ground beetles. *Tech. Bull. US Dep. Agric.* **1165**, 1–29.

Parker K. D. and Rudall K. M. (1955) Calcium citrate in an insect. *Biochim. biophys. Acta* **17**, 287.

Parker K. D. and Rudall K. M. (1957) The silk of the egg-stalk of the green lace-wing fly. *Nature, Lond.* **179**, 905–906.

Parkin E. A. (1934) Observations on the biology of the *Lyctus* powder-post beetles, with special reference to oviposition and the egg. *Ann. appl. Biol.* **21**, 495–518.

Parks J. J. and Larsen J. R. (1965) A morphological study of the female reproductive system and follicular development in the mosquito *Aedes aegypti* (L.). *Trans. Am. microsc. Soc.* **84**, 88–98.

Parnell F. R., King H. E., and Ruston D. F. (1949) Jassid resistance and hairiness of the cotton plant. *Bull. ent. Res.* **39**, 539–575.

Parrott W. L., Maxwell F. G., and Jenkins J. N. (1966) Feeding and oviposition of the boll weevil *Anthonomus grandis* (Coleoptera: Curculionidae) on the Rose-of-Sharon, an alternate host. *Ann. ent. Soc. Am.* **59**, 547–550.

Paschke J. D. (1964) Disposable containers for rearing loopers. *J. Insect Pathol.* **6**, 248–251.

Pass B. C. (1967) Observations on oviposition by the alfalfa weevil. *J. econ. Ent.* **60**, 288.

Pass B. C. and Van Meter C. L. (1966) A method for extracting eggs of the alfalfa weevil from stems of Alfalfa. *J. econ. Ent.* **59**, 1294.

Passera L. (1966) La ponte des ouvrières de la fourmi *Plagiolepis pygmaea* Latr. (Hym., Formicidae): Oeufs reproducteurs et oeufs alimentaires. *C. r. hebd. Séanc. Acad. Sci. Paris* **263D**, 1095–1098.

Patel K. K. and Apple J. W. (1967) Ecological studies on the eggs of the northern corn rootworm. *J. econ. Ent.* **60**, 496–500.

Patel R. C. and Kulkarny H. L. (1955) Bionomics of *Urentius echinus* Dist. (Hemiptera–Heteroptera: Tingidae) an important pest of Brinjal (*Solanum melongena* L.) in North Gujarat. *J. Bombay nat. Hist. Soc.* **53**, 86–96.

Patel R. C., Patel J. C., and Patel J. K. (1973) Biology and mass breeding of the tobacco caterpillar *Spodoptera litura* (F.). *Israel J. Ent.* **8**, 131–142.

Patel R. M. (1963) Observations on the life history and control of the sugarcane top borer (*Scirpophaga nivella* F.) in South Gujarat. *Indian J. Sugarcane Res.* **8**, 50–55.

Paterson N. F. (1930) The bionomics and morphology of the early stages of *Paraphaedon tumidulus* Germ. (Coleoptera, Phytophaga, Chrysomelidae). *Proc. zool. Soc. Lond.* **41**, 627–676.

PATTEN W. (1884) The development of Phryganids, with a preliminary note on the development of *Blatta germanica*. *Q. Jl microsc. Sci.* (NS) **24**, 549–602.
PATTERSON H. E. (1962) Status of the East African salt-water-breeding variant of *Anopheles gambiae* Giles. *Nature, Lond.* **195**, 469–470.
PAU R., BRUNET P. C. J., and WILLIAMS M. J. (1971) The isolation and characterization of proteins from the left colleterial gland of the cockroach *Periplaneta americana* (L.). *Proc. R. Soc.* (B) **177**, 565–579.
PAU R. N. and ACHESON R. M. (1968) The identification of 3-hydroxy-4-O-β-D-glucoside of benzyl alcohol in the left colleterial gland of *Blaberus discoidalis*. *Biochim. biophys. Acta* **158**, 206–211.
PAUL C. F., SHUKLA G. N., DAS S. R., and PERTI S. L. (1963) A life-history study of the hide beetle *Dermestes vulpinus* Fab. (Coleoptera: Dermestidae). *Indian J. Ent.* **24**, 167–179 (1962).
PAUL M., GOLDSMITH M. R., HUNSLEY J. R., and KAFATOS F. C. (1972a) Specific protein synthesis in cellular differentiation. Production of eggshell proteins by silkmoth follicular cells. *J. Cell. Biol.* **55**, 653–680.
PAUL M., KAFATOS F. C., and REGIER J. C. (1972b) A comparative study of eggshell proteins in Lepidoptera. *J. supramolecular Structure* **1**, 60–65.
PAULIAN R. (1941) Les premiers états des Staphylinoidea (Coleoptera). Étude de morphologie comparée. *Mém. Mus. Hist. nat. Paris* (NS) **15**, 1–361.
PAVAN M. (1948a) Sulla deposizione della uova in *Morimus asper* Sulz. e *Lamia textor* L. (Col. Cerambycidae). *Atti Soc. ital. Sci. nat.* **87**, 53–60.
PAVAN M. (1948b) Uovo stadio ovulare e primo stadio larvale in *Morimus asper* Sulz. e *Lamia textor* L. (Col. Cerambycidae). *Boll. Zool.* **15**, 49–64.
PAWAN J. L. (1922) On the eggs and oviposition of *Psorophora (Janthinosoma) posticata* Wied. (Culicidae). *Bull. ent. Res.* **12**, 481.
PAWAR A. D. (1975) *Cyrtorhinus lividipennis* Reuter (Miridae: Hemiptera) as a predator of the eggs and nymphs of the brown planthopper and green leafhoppers in Himachal Pradesh, India. *Rice. Ent. Newslett.* **3**, 30–31.
PAWLOWSKY E. N. (1927) *Gifttiere*, Jena.
PAWLOWSKY E. N. and STEIN A. K. (1927) Experimentelle Untersuchungen über die Wirkung der Gifthaare der überwinternden Goldafterraupen (*Euproctis chrysorrhoea*) auf die Menschenhaut. *Z. Morph. Ökol. Tiere* **9**, 616–637.
PAYNE N. M. (1929) Absolute humidity as a factor in insect cold hardiness with a note on the effect of nutrition on cold hardiness. *Ann. ent. Soc. Am.* **22**, 601–620.
PAYNE T. L. (1970) Electrophysiological investigations on response to pheromones in bark beetles. *Contr. Boyce Thompson Inst.* **24**, 275–282.
PAYNE T. L. (1975) Bark beetle olfaction. III. Antennal olfactory responsiveness of *Dendroctonus frontalis* Zimmerman and *D. brevicomis* Le Conte (Coleoptera: Scolytidae) to aggregation pheromones and host tree terpene hydrocarbons. *J. Chem. Ecol.* **1**, 233–242.
PEACOCK J. W., CUTHBERT, R. A., GORE W. E., LANIER G. N., PEARCE G. T., and SILVERSTEIN R. M. (1975) Collection on Porapak Q of the aggregation pheromone of *Scolytus multistriatus* (Coleoptera: Scolytidae). *J. Chem. Ecol.* **1**, 149–160.
PEAKE F. G. G. (1952) On a bruchid seed-borer in *Acacia arabica*. *Bull. ent. Res.* **43**, 317–324.
PEARCE G. T., GORE W. E., SILVERSTEIN R. M., PEACOCK J. W., CUTHBERT R. A., LANIER G. N., and SIMEONE J. B. (1975) Chemical attractants for the smaller European elm bark beetle *Scolytus multistriatus* (Coleoptera: Scolytidae). *J. Chem. Ecol.* **1**, 115–124.
PEARL R. (1932) The influence of density of population upon egg production in *Drosophila melanogaster*. *J. exp. Zool.* **63**, 57–84.

Pearman J. V. (1927) Notes on *Pteroxanium squamosum* Endl. and on the eggs of the Atropidae (Psocoptera). *Entomologist's mon. Mag.* **63**, 107–111.
Pearman J. V. (1928a) Biological observations on British Psocoptera. *Entomologist's mon. Mag.* **64**, 209–218.
Pearman J. V. (1928b) Biological observations on British Psocoptera. II. Hatching and ecdysis. *Entomologist's mon. Mag.* **64**, 239–243.
Pearman J. V. (1929) Additional notes on Psocid biology (oviposition). *Entomologist's mon. Mag.* **65**, 89–90.
Pearman J. V. (1932) Notes on the genus *Psocus*, with special reference to the British species. *Entomologist's mon. Mag.* **68**, 193–204.
Pearman J. V. (1955) The eggs of *Mesopsocus unipunctatus* (Muell.). *Entomologist's mon. Mag.* **91**, 206.
Pedigo L. P. (1971) Ovipositional response of *Plathypena scabra* (Lepidoptera: Noctuidae) to selected surfaces. *Ann. ent. Soc. Am.* **64**, 647–651.
Peet W. B. Jr (1973) Biological studies on *Nidicola marginata* (Hemiptera: Anthocoridae). *Ann. ent. Soc. Am.* **66**, 344–348.
Peleg B. A. and Gothilf S. (1964) Labelling of eggs of the carob moth *Ectomyelois ceratoniae* (Zeller), with P^{32} for ecological studies. *Israel J. agric. Res.* **14**, 75–76.
Peleg B. A. and Rhode R. H. (1967) New methods in mass rearing of the Mediterranean fruit fly in Costa Rica. *J. econ. ent.* **60**, 1460–1461.
Pelerents C. and Van den Brande J. (1961) The effect of γ-rays on eggs of *Ephestia Kuehniella* Zell. *Meded. LandbHoogesch. OpzoekStns Gent.* **26** (3) 9 pp.
Pemberton C. E. and Willard H. F. (1918) A contribution to the biology of fruit-fly parasites in Hawaii. *J. agric. Res.* **15**, 419–466.
Peña, G. L. E. (1968) Natural history notes on *Notiothauma*. *Discovery, New Haven, Conn.* **4**, 43–44.
Pendergrast J. G. (1968) The Aradidae of New Zealand (Hemiptera: Heteroptera). 3. The Aradinae and Calisiinae. *Trans. R. Soc. NZ (Zool.)* **10**, 81–88.
Pener M. P. and Shulov A. (1960) The biology of *Calliptamus palaestinensis* Bdhmr. with special reference to the development of its eggs. *Bull. Res. Council Israel (Zool.)* **9B**, 131–156.
Pengelly D. H. (1964) Oviposition sites and the viability of eggs of *Thymelicus lineola* (Ochs.) (Lepidoptera: Hesperiidae). *Proc. ent. Soc. Ont.* **95**, 102–105.
Pennington K. M. (1940) Notes on early stages and distribution of some rare South African butterflies. *J. ent. Soc. S. Afr.* **3**, 128–130.
Percival E. and Whitehead H. (1928) Observations on the ova and oviposition of certain Ephemeroptera and Plecoptera. *Proc. Leeds Phil. Soc.* **1**, 271–288.
Perez M. Q. (1930) Los parasitos de los pulgones. *Bol. Patol. veg. Ent. Agric.* **4**, 49–64.
Perez Y., Verdier M., and Pener M. P. (1971) The effect of photoperiod on male sexual behaviour in a north Adriatic strain of the migratory locust. *Entomologia exp. appl.* **14**, 245–250.
Perkins P. V. and Watson T. F. (1972) *Nabis alternatus* as a predator of *Lygus hesperus*. *Ann. ent. Soc. Am.* **65**, 625–629.
Perrier E. (1870) Note sur la ponte de la mante religieuse. *Annls Sci. nat. Zool.* **14** (5) 1–2.
Perris E. (1852) Histoire des métamorphoses du *Clambus ensharmensis* Westw., du *Cryptophagus dentatus* Herbst, du *Latridius minutus* Linné, du *Corticaria pubescens* Illig, de l'*Orthopterus piceus* Steph., du *Malachius aeneus* Fabr. et de la *Sapromyza quadripunctata* Fabr. *Annls Soc. ent. Fr.* **10** (2) 571–601.
Perris E. (1853) Histoire des insectes du pin maritime. *Annls Soc. ent. Fr.* **1** (3) 555–644.
Perris E. (1877) *Larves de Coléoptères*, Paris.

Perron J. P., Jasmin J. J., and Lafrance J. (1960) Attractiveness of some onion varieties grown in muck soil to oviposition by the onion maggot (*Hylemya antiqua* (Meig.)) (Anthomyiidae: Diptera). *Can. Ent.* **92**, 765–767.

Perry A. S. and Fay R. W. (1967) Correlation of chemical constitution and physical properties of fatty acid esters with oviposition response of *Aedes aegypti*. *Mosq. News* **27**, 175–183.

Persson B. (1974) Diet distribution of oviposition in *Agrotis ipsilon* (Hfn.), *Agrotis munda* (Walk.) and *Heliothis armigera* (Hbn.) (Lep., Noctuidae), in relation to temperature and moonlight. Experiments in an egg-laying recording apparatus in Queensland, Australia. *Ent. scand.* **5**, 196–208.

Perttunen V., Kangas E., and Oksanen H. (1968) The mechanisms by which *Blastophagus piniperda* L. (Col., Scolytidae) reacts to the odour of an attractant fraction isolated from pine phloem. *Suom. hyönt. Aikak.* **34**, 205–222.

Perttunen V., Oksanen H., and Kangas E. (1970) Aspects of the external and internal factors affecting the olfactory orientation of *Blastophagus piniperda* (Coleoptera: Scolytidae). *Contr. Boyce Thompson Inst.* **24**, 293–297.

Pesce H. and Delgado A. (1971) Poisoning from adult moths and caterpillars. In *Venomous Animals and their Venoms. 3. Venomous Invertebrates* (W. Bücherl and E. Buckley, eds.), pp. 119–156.

Peschken D. (1965) Untersuchungen zur Orientierung aphidophager Schwebfliegen (Diptera: Syrphidae). *Z. angew. Ent.* **55**, 201–235.

Pessôa S. B. and Corrêa C. (1928) Nota sobre a biologia da *Rhyparobia maderae* Fabr. *Rev. Biol. Sci. Med. Rio de J.* **7**, 304–305.

Pessozkaja F. S. (1927) Zur Biologie der Ohrwürmer *Forficula tomis* Kol. und *Labidura riparia* Pall. *Bull. Inst. Sci. Leshaft.* **12**, 51–62.

Peterlík Z. and Štys Z. (1969) A contribution to the bionomics of the lucerne weevil (*O. ligustici*) on hops. *Rostl. Výroba* **15**, 905–914 (in Czech; English summary).

Peters L. L. (1971) Angoumois grain moth egg collection. *J. econ. Ent.* **64**, 1308–1309.

Petersen B. (1954) Egg-laying and habitat selection in some *Pieris* species. *Ent. Tidskr.* **75**, 194–203.

Petersen J. J. (1969) Oviposition response of *Aedes sollicitans*, *Aedes taeniorhynchus* and *Psorophora confinis* to seven inorganic salts. *Mosq. News* **29**, 472–483.

Petersen J. J. and Chapman H. C. (1970) Chemical characteristics of habitats producing larvae of *Aedes sollicitans*, *Aedes taeniorhynchus* and *Psorophora confinis* in Louisiana. *Mosq. News* **30**, 156–161.

Petersen J. J. and Rees D. M. (1966) Selective oviposition response of *Aedes dorsalis* and *Aedes nigromaculis* to soil salinity. *Mosq. News* **26**, 168–174.

Petersen J. J. and Rees D. M. (1967) Comparative oviposition selection preference by *Aedes dorsalis* and *Aedes nigromaculatus* for three inorganic salts in the laboratory. *Mosq. News* **27**, 136–141.

Petersen J. J. and Willis O. R. (1970) Oviposition responses of *Culex pipiens quinquefasciatus* and *Culex salinarius* in the laboratory. *Mosq. News* **30**, 438–444.

Petersen J. J. and Willis O. R. (1971) Effects of salinity on site selection by ovipositing tree hole mosquitoes in Louisiana. *Mosq. News* **31**, 352–355.

Petersen W. (1929) Über die Sphragis und das Spermatophragma der Tagfaltergattung *Parnassius* (Lep.). *D. ent. Z.* **1928**, 407–413.

Peterson A. (1953) *A Manual of Entomological Techniques*, 7th edn, Edwards Bros., Ann Arbor, Mich.

Peterson A. (1960) Photographing eggs of insects. *Fla. Ent.* **43**, 1–7.

Peterson A. (1961) Some types of eggs deposited by moths, Heterocera–Lepidoptera. *Fla. Ent.* **44**, 107–114.

Peterson A. (1962a) Some eggs of moths among the Geometridae–Lepidoptera. *Fla. Ent.* **44**, 109–119.

Peterson A. (1962b) Some eggs of insects that change colour during incubation. *Fla. Ent.* **45**, 81–87.

Peterson A. (1963a) Egg types among moths of the Pyralidae and Phycitidae–Lepidoptera. *Fla. Ent.* suppl. **1**, 1–9.

Peterson A. (1963b) Some eggs of moths among the Amatidae, Arctiidae and Notodontidae–Lepidoptera. *Fla. Ent.* **46**, 169–182.

Peterson A. (1964) Egg types among moths of the Noctuidae (Lepidoptera). *Fla. Ent.* **47**, 71–91.

Peterson A. (1965a) Some eggs of moths among the Olethreutidae and Tortricidae (Lepidoptera). *Fla. Ent.* **48**, 1–8.

Peterson A. (1965b) Some eggs of moths among the Sphingidae, Saturniidae, and Citheroniidae (Lepidoptera). *Fla. Ent.* **48**, 213–219.

Peterson A. (1966) Some eggs of moths among the Liparidae, Lasiocampidae, and Lacosomidae (Lepidoptera). *Fla. Ent.* **49**, 35–42.

Peterson A. (1967a) Some eggs of moths from several families of Microlepidoptera. *Fla. Ent.* **50**, 125–132.

Peterson A. (1967b) Eggs of moths among the Ethmiidae, Acrolophidae and Hepialidae–Microlepidoptera. *Fla. Ent.* **50**, 181–183.

Peterson A. (1968) Eggs of moths from additional species of Geometridae–Lepidoptera. *Fla. Ent.* **51**, 83–94.

Peterson A. (1969) Bagworm photographs: eggs, larvae, pupae and adults of *Thyridopteryx ephemeraeformis* (Psychidae: Lepidoptera). *Fla. Ent.* **52**, 61–65.

Peterson A. (1970) Eggs from miscellaneous species of Rhopalocera–Lepidoptera. *Fla. Ent.* **53**, 65–71.

Peterson B. V. (1959) Observations on mating, feeding, and oviposition of some Utah species of black flies (Diptera: Simuliidae). *Can. Ent.* **91**, 147–155.

Peus F. (1942) Die Stechmücken und ihre Bekämpfung. Teil I. Die Feibermücken des Mittel Meergebietes. *Hyg. Zool.* **8**, 1–150.

Peus F. (1952) Cylindrotominae. In Lindner, *Die Fliegen der palaearktischen Region.* **169**, 1–80.

Peyerimhoff P. de (1901) Le mécanisme de l'éclosion chez les Psoques. *Annls Soc. ent. Fr.* **70**, 149–152.

Peyron J. (1909) *Zur Morphologie der skandinavischen Schmetterlingseier*, pp. 1–304.

Pfaffenberger G. S. and Johnson C. D. (1976) Biosystematics of the first-stage larvae of some North American Bruchidae (Coleoptera). *Tech. Bull. US Dep. Agric.* **1525**, 1–75.

Philippe R. (1970) Rôle de la glande annexe femelle lors de la ponte chez *Chrysopa perla* (L.) (Insectes, Blanipennes). *C. r. hebd. Séanc. Acad. Sci. Paris* (D) **270**, 2448–2450.

Phillips J. R. and Whitcomb W. H. (1962) Field behavior of the adult bollworm *Heliothis zea* (Boddie). *J. Kans. ent. Soc.* **35**, 242–246.

Phillips W. M. (1976) Effects of leaf age on feeding "preference" and egg laying in the chrysomelid beetle *Haltica lythri*. *Physiol. Ent.* **1**, 223–226.

Phipps J. (1949) The structure and maturation of the ovaries in British Acrididae (Orthoptera). *Trans. R. ent. Soc. Lond.* **100**, 233–247.

Phipps J. (1950) The maturation of the ovaries and the relation between weight and maturity in *Locusta migratoria migratoroides* (R. and F.). *Bull. ent. Res.* **40**, 539–557.

Phipps J. (1958) The structure of the ovaries and eggs of some Eumastacidae (Orthoptera, Acridoidea). *Entomologist's mon. Mag.* **94**, 65–66.

Phipps J. (1959) Studies on East African Acridoidea (Orthoptera) with special reference to egg-production, habitats and seasonal cycles. *Trans. R. ent. Soc. Lond.* **111**, 27–56.

Phipps J. (1960) The breeding of *Sphodromantis lineola* Burm. (Dictyoptera Mantidae) in Sierra

Leone. *Entomologist's mon. Mag.* **96**, 192–193.

PHIPPS J. (1970) Observations on the seasonal occurrence and egg production of the grasshopper *Chorthippus curtipennis* Harris (Insecta: Orthoptera) in Newfoundland. *Can. J. Zool.* **48**, 1140–1142.

PICADO C. (1913) Les Broméliacées épiphytes. Considérées comme milieu biologique. *Bull. Sci. Fr. Belg.* **47**, 215–360.

PICK F. (1950) L'inclusion temporaire des oeufs d'*Aedes aegypti* à l'aide de la technique de silico-gel sur lame. *Bull. Soc. Path. exot.* **43**, 364–372.

PICK F. (1962a) Sur le mode de déposition des oeufs par *Triatoma magista* et par *Rhodnius prolixus*. *Annls Parasit. hum. comp.* **37**, 338–347.

PICK F. (1962b) Sur la signification fonctionnelle de l'ornementation des oeufs des Réduvidés hématophages. *Annls Parasit. hum. comp.* **37**, 404–407.

PICKENS L. G. and MORGAN N. O. (1967) A simplified laboratory technique for separating eggs of the face fly from oviposition medium. *J. econ. Ent.* **60**, 1479.

PICKETT A. D. and PATTERSON N. A. (1953) The influence of spray programmes on the fauna of apple orchards in Nova Scotia. IV. A review. *Can. Ent.* **85**, 472–478.

PICKFORD R. (1976) Embryonic growth and hatchability of eggs of the two-striped grasshopper *Melanoplus bivittatus* (Orthoptera: Acrididae) in relation to date of oviposition and weather. *Can. Ent.* **108**, 621–626.

PICKFORD R., EWEN A. B., and GILLOTT C. (1969) Male accessory gland substance: an egg-laying stimulant in *Melanoplus sanguinipes* (F.) (Orthoptera: Acrididae). *Can. J. Zool.* **47**, 1199–1203.

PIENKOWSKI R. L. (1965) The incidence and effect of egg cannibalism in first-instar *Coleomegilla maculata lengi* (Coleoptera: Coccinellidae). *Ann. ent. Soc. Am.* **58**, 150–153.

PIERCE F. N. (1911) Viviparity in Lepidoptera. *Entomologist* **44**, 309–310.

PIERRARD G. and BAURANT R. (1960) La ponte et l'indice sexuel chez *Rhyacionia buoliana* Schiff. (Microl. Tortricidae). *Bull. Inst. agron. Gembloux* **28**, 418–426.

PIERRE, ABBÉ (1904) L'éclosion des oeufs du *Lestes viridis* Van der Lind. *Annls Soc. ent. Fr.* **73**, 477–484.

PIERRE C. (1934) Pariade et ponte chez les Tipulides. *Rev. fr. Ent.* **1**, 29–38.

PIJNACKER L. P. (1971) The origin of the abnormal micropyle apparatus of the eggs of *Carausius morosus* Br. (Cheleutoptera, Phasmidae). *Neth. J. Zool.* **21**, 366–372.

PILLAI J. S. (1962) A celloidin impression technique for recording egg sculpturing in *Aedes* mosquitoes. *Nature, Lond.* **194**, 212–213.

PILLAI M. K. K. and MADHUKAR B. V. R. (1969) Effect of pH on the ovipositional responses of the yellow-fever mosquito *Aedes aegypti* L. *Curr. Sci.* **38** (5) 114–116.

PING C. (1921) The biology of *Ephydra subopaca* Loew. *Mem. Cornell Univ. agric. Exp. Stn* **49**, 561–616.

PISTON J. J. and LANIER G. N. (1974) Pheromones of *Ips pini* (Coleoptera: Scolytidae). Response to interpopulational hybrids and relative attractiveness of males boring in two host species. *Can. Ent.* **106**, 247–251.

PITMAN G. B. (1966) Studies on the pheromone of *Ips confusus* (LeConte). 3. The influence of host material on pheromone production. *Contr. Boyce Thompson Inst.* **23**, 147–157.

PITMAN G. B. (1969) Pheromone response in pine bark beetles: influence of host volatiles. *Science, Wash.* **166**, 905–906.

PITMAN G. B. (1971) *Trans*-Verbenol and alpha-pinene: their utility in manipulation of the mountain pine beetle. *J. econ. Ent.* **64**, 426–430.

PITMAN G. B. (1973) Further observations on Douglure in a *Dendroctonus pseudotsugae* management system. *Envir. Ent.* **2**, 109–112.

PITMAN G. B., HEDDEN R. L., and GARA R. I. (1975) Synergistic effects of ethyl alcohol on the aggregation of *Dendroctonus pseudotsugae* (Col., Scolytidae) in response to pheromones. *Z. angew. Ent.* **78**, 203–208.

PITMAN G. B. and VITÉ J. P. (1969) Aggregation behavior of *Dendroctonus ponderosae* (Coleoptera: Scolytidae) in response to chemical messengers. *Can. Ent.* **101**, 143–149.

PITMAN G. B. and VITÉ J. P. (1970) Field response of *Dendroctonus pseudotsugae* (Coleoptera: Scolytidae) to synthetic frontalin. *Ann. ent. Soc. Am.* **63**, 661–664.

PITMAN G. B. and VITÉ J. P. (1971) Predator–prey response to western pine beetle attractants. *J. econ. Ent.* **64**, 402–404.

PITMAN G. B. and VITÉ J. P. (1974) Biosynthesis of methylcyclohexenone by male Douglas-fir beetle. *Envir. Ent.* **3**, 886–887.

PITMAN G. B., VITÉ J. P., KINZER G. W., and FENTIMAN A. F. (1969) Specificity of population-aggregating pheromones in *Dendroctonus*. *J. Insect. Physiol.* **15**, 363–366.

PLANK H. K. (1948) Biology of the bamboo powder-post beetle in Puerto Rico. *Bull. fed. (agric.) Exp. Stn P. Rico* **44**, 1–29.

PLAUT H. N. (1971) On the biology of the adult of the almond wasp *Eurytoma amygdali* End. (Hym., Eurytomidae), in Israel. *Bull. ent. Res.* **61**, 275–281.

PLAUT H. N. (1973) On the biology of *Paropta paradoxus* (H.-S.) (Lepidoptera: Cossidae) on grapevine (Vitis) in Israel. *Bull. ent. Res.* **63**, 237–245.

PLESKOT G. (1953) Zur Ökologie der Leptophlebiiden (Ins., Ephemeroptera). *Öst. zool. Z.* **4**, 45–107.

PLICHET F. (1957) Le cochenille du noyer en Dordogne. *Rev. Zool. agric.* **56**, 14–18.

PLUGARU S. G. (1965) The oak *Tortrix* in Moldavia. *Vred. polez. Fauna Bespozvon. Moldavii* (2) 15–24 (in Russian).

PLUGARU S. G. (1968) The biology of the oak processionary moth in Moldavia. *Vred. polez. Fauna Bespozvon. Moldavii* (3), 29–41 (in Russian).

PLYATER-PLOKHOTSKAYA V. N. (1963) The accessory genital glands of the female *Blattella germanica* L. (Blattoidea). *Ent. Obozr.* **42**, 550–552 (in Russian).

PO-CHEDLEY D. S. (1969) Radio-sensitivity and water content for yellow mealworm embryos. *J. econ. Ent.* **62**, 1505–1506.

POISSON R. (1923) Accouplement, ponte et éclosion des hémiptères aquatiques. *Bull. biol. Fr. Belg.* **57**, 89–97.

POISSON R. (1926) L'*Anisopus producta* Fieb. Observations sur son anatomie et sa biologie. *Archs Zool. exp. gén.* **65**, 181–208.

POISSON R. (1933) Quelques observations sur la structure de l'oeuf des insectes Hémiptères-Hétéroptères. *Bull. Soc. sci. Bretagne* **10**, 1–38.

POLAK R. A. (1933) Broedzorg bij een Mantidae. *Ent. Ber. Amst.* **8**, 508–509.

POLIVANOVA E. N. (1966) Resistance of embryos of *Eurygaster integriceps* Put. and other Pentatomidae (Hemiptera) to the conditions of various humidity at different temperature. *Zool. Zh.* **45**, 1170–1174 (in Russian; English summary).

POLLARD D. G. (1954) The melon stem-borer *Apomecyna binubila* Pascoe (Coleoptera: Lamiinae) in the Sudan. *Bull. ent. Res.* **45**, 553–561.

POLNIK A. (1960) Effects of some intraspecies processes on competition between two species of flour beetles *Latheticus oryzae* and *Tribolium confusum*. *Physiol. Zoöl.* **33**, 42–57.

POMEROY A. W. J. (1921) The irritating hairs of the wild silk moths of Nigeria. *Bull. Imp. Inst. Lond.* **19**, 311–318.

PONTIN A. J. (1960) Observations on the keeping of aphid eggs by ants of the genus *Lasius* (Hymenoptera: Formicidae). *Entomologist's mon. Mag.* **96**, 198–199.

POOLE A. F. (1967) A note on the oviposition of *Simulium (S.) argyreatum* Meigen. *Entomologist* **100**, 121.

POPE P. (1953a) Studies on the life histories of some Queensland Blattidae (Orthoptera). Part 1.

The domestic species. *Proc. R. Soc. Qd* **63**, 23–46.

POPE P. (1953b) Studies on the life histories of some Queensland Blattidae (Orthoptera). Part 2. Some native species. *Proc. R. Soc. Qd* **63**, 47–59.

POPHAM E. J. (1954) A new and simple method of demonstrating the physical gill of aquatic insects. *Proc. R. ent. Soc. Lond.* (A) **29**, 51–54.

POPOV G. B. (1958a) Ecological studies on oviposition by swarms of the desert locust (*Schistocerca gregaria* Forskål) in eastern Africa. *Anti-Locust Bull.* **31**, 1–70.

POPOV G. B. (1958b) Note on the frequency and the rate of oviposition in swarms of the desert locust (*Schistocerca gregaria* Forskål). *Entomologist's mon. Mag.* **94**, 176–180.

POPOV G. B. (1959) Ecological studies on oviposition by *Locusta migratoria migratorioides* (R. and F.) in its outbreak area in the French Sudan. *Locusta* **6**, 3–63.

POPOV P. A. (1959) Studies of the bionomics of June beetles (*Amphimallon*) in Bulgaria and their control. *Nauchni. Trud. Minist. Zemed. nauch. Inst. Zasht. Rast.* **2**, 33–74 (in Bulgarian; English summary).

POPOV P. A. (1962) Some biological peculiarities of the apple blossom weevil (*Anthonomus pomorum* L.) in Bulgaria. *Izv. nauch. Inst. Zasht. Rast.* **3**, 117–141 (in Bulgarian; English summary).

POPOVA V. (1963) Studies on the bionomics of the mottled lucerne geometrid *Phasiane clathrata* Dup. = *Chiasmia clathrata* L. (Lepidoptera) in the conditions of southern and southwestern Bulgaria. *Rast. Zasht.* **11**, 7–14 (in Bulgarian; English summary).

POPOVA V. (1968) Studies on the bionomics of the brown-spotted lucerne pyralid (*Nomophila noctuella* Hb.) (Lepidoptera, Pyralidae). *RastVŭd. Nauki* **5**, 137–144 (in Bulgarian; English summary).

POPOVICH A. P. (1977) Comparative morphological characteristics of the female genitalia of the tabanids *Tabanus autumnalis autumnalis* and *Hybomitra schineri* Lyn. in Khar'kov and cis-Danube populations. *Vest. Zool.* 1977, **1**, 69–73.

PORTIER P. (1911) Recherches physiologiques sur les insectes aquatiques. *Archs Zool. exp. gén.* **8**, 89–379.

POTGIETER J. T. (1929) A contribution to the biology of the brown swarm locust *Locustana pardalina* (Wlk.) and its natural enemies. *Sci. Bull. Dep. Afric. S. Afr.* **82**, 1–48.

POTTER E. (1938) The internal anatomy of the order Mecoptera. *Trans. R. ent. Soc. Lond.* **87**, 467–502.

POUJADE G. A. (1902) Nouvelle note sur l'*Hydrophilus piceus*. *Bull. Soc. ent. Fr.* **1902**, 206, 219, 238.

POULTON E. B. (1912) The irritating hairs of the moth *Anaphe infracata* Walsingham. *Proc. ent. Soc. Lond.* 1912, lxxviii.

POULTON E. B. (1922) The oviposition of the Mylabrid beetle *M. oculata* Thunb., var. *tricolor* Gerst. *Proc. ent. Soc. Lond.* 1921, xc–xcii.

POUTIERS R. (1922) L'acclimatation de *Cryptolaemus montrouzieri* Muls. dans le midi de la France. *Annls Épiphyt.* **8**, 3–18.

POUZAT J. (1976) Le comportement de ponte de la bruche du haricot en présence d'extrait de plante-hôte. Mise en évidence d'interactions gustatives et tactiles. *C. r. hebd. Séanc. Acad. Sci. Paris* (D) **282**, 1971–1974.

POWELL J. A. and TURNER W. J. (1975) Observations on oviposition behavior and host selection in *Orussus occidentalis* (Hymenoptera : Siricoidea). *J. Kans. ent. Soc.* **48**, 299–307.

POWNING R. F. and IRZYKIEWICZ H. (1962) β-glucosidase in the cockroach (*Periplaneta americana*) and in the puff-ball (*Lycoperdon perlatum*). *Comp. Biochem. Physiol.* **7**, 103–115.

POYARKOFF E. (1914) Essai d'une théorie de la nymphe des insectes holométaboles. *Horae Soc. ent. ross.* **41**, 1–51 (1913).

PRASAD S. K. (1961) The association between the different pests on cabbage. *Indian J. Ent.* **21**,

206–209 (1959).
PRATT H. D. and KIDWELL A. S. (1969) Eggs of mosquitoes found in *Aedes aegypti* oviposition traps. *Mosq. News* **29**, 545–548.
PREDTECHENSKIĬ S. A. (1928) *Locusta migratoria* L. in central Russia. *Rep. Bur. appl. Ent. Leningr.* **3**, 113–199.
PRELL H. (1924) Die biologischen Gruppen der deutschen Rhynchitiden. *Zool. Anz.* **61**, 153–170.
PRELL H. (1930) Zur Kenntnis von Bau und Entstehung einiger Brutbildtypen bei Rindenbrutenden Borkenkäfern. *Z. Morph. Ökol. Tiere* **17**, 625–648.
PRESCOTT H. W. (1955) *Neorhynchocephalus sackenii* and *Trichopsidea clausa*, nemestrinid parasites of grasshoppers. *Ann. ent. Soc. Am.* **48**, 392–402.
PRESCOTT H. W. (1960) Suppression of grasshoppers by nemestrinid parasites (Diptera). *Ann. ent. Soc. Am.* **53**, 513–521.
PRESS J. W., FLAHERTY B. R., DAVIS R., and ARBOGAST R. T. (1973) Development of *Xylocoris flavipes* (Hemiptera: Anthocoridae) on eggs of *Plodia interpunctella* (Lepidoptera: Phycitidae) killed by gamma radiation or by freezing. *Envir. Ent.* **2**, 335–336.
PREVETT P. F. (1953) Notes on the feeding habits and life-history of *Galeruca tanaceti* L. (Col., Chrysomelidae). *Entomologist's mon. Mag.* **89**, 292–293.
PREVETT P. F. (1960) The oviposition and duration of life of a small strain of rice weevil *Calandra oryzae* (L.) in Sierra Leone. *Bull. ent. Res.* **50**, 697–702.
PREVETT P. F. (1966a) Observations on biology in the genus *Caryedon* Schönherr (Coleoptera: Bruchidae) in Northern Nigeria, with a list of associated parasitic Hymenoptera. *Proc. R. ent. Soc. Lond.* (A) **41**, 9–16.
PREVETT P. F. (1966b) The identity of the palm kernel borer in Nigeria, with systematic notes on the genus *Pachymerus* Thunberg (Coleoptera: Bruchidae). *Bull. ent. Res.* **57**, 181–192.
PREVETT P. F. (1968) Notes on the biology, food plants and distribution of Nigerian Bruchidae (Coleoptera), with particular reference to the Northern Region. *Bull. ent. Soc. Nigeria* **1**, 3–6.
PRICE P. W. (1972) Behavior of the parasitoid *Pleolophus basizonus* (Hymenoptera: Ichneumonidae) in response to changes in host and parasitoid density. *Can. Ent.* **104**, 129–140.
PRIGGE M. (1973) Zur Biologie und Entwicklung von *Blaps mucronata* Latr. (Coleoptera: Tenebrionidae). *Z. angew. Ent.* **74**, 130–141.
PRINCE G. J. (1976) Laboratory biology of *Phaenocarpa persimilis* Papp (Braconidae: Alysiinae), a parasitoid of *Drosophila*. *Aust. J. Zool.* **24**, 9–264.
PRINGLE G. (1954) The identification of the adult anopheline mosquitoes of Iraq and neighbouring territories. *Bull. endem. Dis.* **1**, 53–76.
PRIORE R. (1963) Studio morfo-biologico sulla *Rodolia cardinalis* Muls. (Coleoptera Coccinellidae). *Boll. Lab. Ent. agr. Portici* **21**, 63–198.
PRIORE R. (1969) *Dialeurodes citri* (Ashmead) (Homoptera Aleyrodidae) in Campania. (Notes on morphology and biology.) *Boll. Lab. Ent. agr. Portici* **27**, 287–316.
PRITCHARD G. (1967) Laboratory observations on the mating behaviour of the island fruit fly *Rioxa pornia* (Diptera: Tephritidae). *J. Aust. ent. Soc.* **6**, 127–132.
PRITCHARD G. (1969) The ecology of a natural population of Queensland fruit fly *Dacus tryoni*. II. The distribution of eggs and its relation to behaviour. *Aust. J. Zool.* **17**, 293–311.
PROKOPY R. J. (1966) Artificial oviposition devices for apple maggot. *J. econ. Ent.* **59**, 231–232.
PROKOPY R. J. (1967) Factors influencing effectiveness of artificial oviposition devices for apple maggot. *J. econ. Ent.* **60**, 950–955.
PROKOPY R. J. (1968) Visual responses of apple maggot flies *Ragoletis pomonella* (Diptera: Tephritidae). Orchard studies. *Entomologia exp. appl.* **11**, 403–422.

Prokopy R. J. (1972) Evidence for a marking pheromone deterring repeated oviposition in apple maggot flies. *Envir. Ent.* **1**, 326–332.
Prokopy R. J. (1975) Oviposition-deterring fruit marking pheromone in *Rhagoletis fausta*. *Envir. Ent.* **4**, 298–300.
Prokopy R. J. (1976) Feeding, mating and oviposition activities of *Rhagoletis fausta* flies in nature. *Ann. ent. Soc. Am.* **69**, 899–904.
Prokopy R. J., Bennett E. W., and Bush G. L. (1972) Mating behavior in *Rhagoletis pomonella* (Diptera: Tephritidae). II. Temporal organization. *Can. Ent.* **104**, 97–104.
Prokopy R. J. and Boller E. F. (1970) Artificial egging system for the European cherry fruit fly. *J. econ. Ent.* **63**, 1413–1417.
Prokopy R. J. and Boller E. F. (1971) Stimuli eliciting oviposition of European cherry fruit flies *Rhagoletis cerasi* (Diptera: Tephritidae) into inanimate objects. *Entomologia exp. appl.* **14**, 1–14.
Prokopy R. J. and Bush G. L. (1973a) Ovipositional responses to different sizes of artificial fruit by flies of *Rhagoletis pomonella* species group. *Ann. ent. Soc. Am.* **66**, 927–929.
Prokopy R. J. and Bush G. L. (1973b) Oviposition by grouped and isolated apple maggot flies. *Ann. ent. Soc. Am.* **66**, 1197–1200.
Proper A. B. (1931) *Eupteromalus nidulans*, a parasite of the brown-tail and satin moths. *J. agric. Res.* **43**, 37–56.
Protensko A. I. (1955) Materials on the fecundity of *Epicauta erythrocephala* Pall. *Trudȳ Inst. Zool. Parasit.* **3**, 147–151.
Proverbs M. D. and Logan D. M. (1970) A rotating oviposition cage for the codling moth *Carpocapsa pomonella*. *Can. Ent.* **102**, 42–49.
Provine R. R. (1977) Behavioral development of the cockroach (*Periplaneta americana*). *J. Insect Physiol.* **23**, 213–220.
Pruess K. P. (1963) Effects of food, temperature and oviposition site on longevity and fecundity of the army cutworm. *J. econ. Ent.* **56**, 219–221.
Pruess K. P., Weekman G. T., and Somerhalder B. R. (1968) Western corn rootworm egg distribution and adult emergence under two corn tillage systems. *J. econ. Ent.* **61**, 1424–1427.
Prus T. (1968) Some regulatory mechanisms in populations of *Tribolium confusum* and *Tribolium castaneum* Herbst. *Ekol. pol.* (A) **16** (16) 335–374.
Prus T. B. and Park T. (1973) The relation of yeast and flour to the cannibalism of eggs by *Tribolium* larvae. *Ekol. pol.* **21**, 163–171.
Pruszyński S. and Wojnarowska P. (1968) Observations on the hawthorn sawfly *H. crataegi*. *Polskie Pismo ent.* **38**, 189–204 (in Polish; English summary).
Pruthi H. S. (1933) An interesting case of maternal care in an aquatic cockroach *Phlebonotus pallens* Serv. (Epilamprinae). *Curr. Sci.* **1**, 273.
Pryor M. G. M. (1940) On the hardening of the oötheca of *Blatta orientalis*. *Proc. R. Soc.* (B) **128**, 378–393.
Pryor M. G. M., Russell P. B., and Todd A. R. (1946) Protocatechuic acid, the substance responsible for the hardening of the cockroach ootheca. *Biochem. J.* **40**, 627–628.
Przibram H. (1909) Die Lebensgeschichte der Gottesanbeterinnen (Fangheuschrecken). *Z. wiss. InsektBiol.* **3**, 117–123, 147–153.
Puchkov V. G. and Puchkova L. V. (1956) Eggs and larvae of heteroptera—agricultural pests. *Trudȳ vses. ént. Obshch.* **45**, 218–342 (in Russian).
Puchkova L. V. (1955) Eggs of the true Hemiptera–Heteroptera. I. Coreidae. *Ent. Obozr.* **34**, 48–55 (in Russian).
Puchkova L. V. (1956) Eggs of Hemiptera–Heteroptera. II. Lygaeidae. *Ent. Obozr.* **35**, 262–284 (in Russian).

Puchkova L. V. (1957) Eggs of Hemiptera–Heteroptera. III. Coreidae (Supplement). IV. Macrocephalidae. *Ent. Obozr.* **36**, 44–58 (in Russian).
Puchkova L. V. (1959) The eggs of true bugs (Hemiptera–Heteroptera). V. Pentatomoidea, I. *Ent. Obozr.* **38**, 634–648 (in Russian; English summary).
Puchkova L. V. (1961) The eggs of Hemiptera. VI. Pentatomoidea, 2, Pentatomidae and Plataspidae. *Ent. Obozr.* **40**, 131–143 (in Russian).
Puchkova L. V. (1966) The morphology and biology of the egg of the terrestrial bugs (Hemiptera). *Horae Soc. ent. ross.* **51**, 75–132 (in Russian).
Pukowski E. (1933) Ökologische Untersuchungen an *Necrophorus* F. *Z. Morph. Ökol. Tiere* **27**, 518–586.
Pukowski E. (1934) Die Brutpfliege des Totengräbers. *Ent. Bl.* **30**, 109–112.
Putman W. L. (1937) Biological notes on the Chrysopidae. *Can. J. Res.* (D) **15**, 29–37.
Putman W. L. (1942) Notes on the predaceous thrips *Haplothrips subtilissimus* Hal. and *Aeolothrips mekleucus* Hal. *Can. Ent.* **74**, 37–43.
Putman W. L. (1970) Life history and behavior of *Balaustium putmani* (Acarina: Erythraeidae). *Ann. ent. Soc. Am.* **63**, 76–81.
Putnam L. G. (1968) Experiments in the quantitative relations between *Diadegma insularis* (Hymenoptera: Ichneumonidae) and *Microplitis plutellae* (Hymenoptera: Braconidae) with their host *Plutella maculipennis* (Lepidoptera: Plutellidae). *Can. Ent.* **100**, 11–16.
Puttarudriah M. and Channa Basavanna G. P. (1953) Beneficial coccinellids of Mysore. I. *Indian J. Ent.* **15**, 87–96.
Puttler B. (1967) Interrelationship of *Hypera postica* (Coleoptera: Curculionidae) and *Bathyplectes curculionis* (Hymenoptera: Ichneumonidae) in the eastern United States with particular reference to encapsulation of the parasite eggs by the weevil larvae. *Ann. ent. Soc. Am.* **60**, 1031–1038.
Puttler B. (1974) *Hypera postica* and *Bathyplectes curculionis*: encapsulation of parasite eggs by host larvae in Missouri and Arkansas. *Envir. Ent.* **3**, 881–882.
Puttler B. and Bosch R. van den (1959) Partial immunity of *Laphygma exigua* (Hübner) to the parasite *Hyposoter exiguae* (Viereck). *J. econ. Ent.* **52**, 327–329.
Pyenson L. and Sweetman H. L. (1931) The effects of temperature and moisture on the eggs of *Epilachna corrupta* Mulsant (Coccinellidae, Coleoptera). *Bull. Brooklyn ent. Soc.* **26**, 221–226.

Quayle H. J. (1929) The Mediterranean and other fruit flies. *Circ. Calif. agric. Exp. Stn* **315**, 1–19.
Quednau F. W. (1964) Experimental evidence of differential fecundity on red scale (*Aonidiella aurantii* (Mask.)) in six species of *Aphytis* (Hymenoptera, Aphelinidae). *S. Afr. J. agric. Sci.* **7**, 335–340.
Quednau F. W. (1967a) Notes on mating, oviposition, adult longevity, and incubation period of eggs of the larch casebearer *Coleophora laricella* (Lepidoptera: Coleophoridae) in the laboratory. *Can. Ent.* **99**, 397–401.
Quednau F. W. (1967b) Notes on mating behavior and oviposition of *Chrysocharis laricinellae* (Hymenoptera: Eulophidae), a parasite of the larch casebearer (*Coleophora laricella*). *Can. Ent.* **99**, 326–331.
Quednau F. W. (1967c) Ecological observations on *Chrysocharis laricinellae* (Hymenoptera: Eulophidae), a parasite of the larch casebearer (*Coleophora laricella*). *Can. Ent.* **99**, 631–641.
Quednau W. (1956) Die biologischen Kriterien zur Unterscheidung von *Trichogramma*-Arten. *Z. PflKrankh.* **63**, 333–344.
Quednau W. (1957) Über den Einfluss von Temperatur und Luftfeuchtigkeit auf den

Eiparasiten *Trichogramma cacoeciae* Marchal. (Eine biometrische Studie.) *Mitt. biol. BundAnst. Ld- u. Forstw.* **90**, 1–63.

QUEDNAU W. (1959) Über eine Methode zur Messung von Biozönose-Einflüssen unter Verwendung von Eiparasiten der Gattung *Trichogramma* (Hym. Chalcididae). *Z. PflKrankh.* **66**, 77–86.

QUEDNAU W. (1960) Radioaktive Markierung von Schlupfwespen. *Atompraxis* **6**, 427–431.

QUINTANA-MUÑIZ V. and WALKER D. W. (1970) Oviposition preference by gravid sugarcane borer moths in Puerto Rico. *J. econ. Ent.* **63**, 987–988.

QURAISHI M. S. (1965) Hatching pattern of *Aedes aegypti* eggs. *Nature, Lond.* **207**, 882.

QURAISHI M. S., OSMANI M. H., and AHMAD S. H. (1963) Effect of ultrasonic waves on the hatching of *Aedes aegypti* eggs of frequency of 0.5 mc/s. *J. econ. Ent.* **56**, 668–670.

QVRESHI A. H. (1966) A simple technique for rearing larvae of *Tribolium castaneum* (Herbst) in the laboratory. *Rep. Niger. stored Prod. Res. Inst.* **1965**, 105–106.

QURESHI Z. A. and WILBUR D. A. (1966) Effect of sub-lethal gamma on eggs, early, intermediate and last instar larvae of the Angoumois grain moth *Sitotroga cerealella* Oliv. *Proc. agric. Symp. 1966, Dacca*, pp. 112–124.

QURESHI Z. A., WILBUR D. A., and MILLS R. B. (1970) Irradiation of early instars of the Angoumois grain moth. *J. econ. Ent.* **63**, 1241–1247.

RAATIKAINEN M. (1960) The biology of *Calligypona sordidula* (Stål) (Hom., Auchenorrhyncha). *Ann. ent. fenn.* **26**, 229–242.

RABB R. L. and BRADLEY J. R. (1968) The influence of host plants on parasitism of eggs of the tobacco hornworm. *J. econ. Ent.* **61**, 1249–1252.

RABB R. L. and BRADLEY J. R. (1970) Marking host eggs by *Telenomus sphingis*. *Ann. ent. Soc. Am.* **63**, 1053–1056.

RADCLIFFE J. E. and PAYNE E. O. (1969) Feeding preference and oviposition sites of adult grass grub beetles (*Costelytra zealandica* (White)) on pasture plants. *NZ Jl agric. Res.* **12**, 771–776.

RADJABI G. (1971) Étude comparative de la biologie de deux coléoptères Buprestidae nuisibles aux arbres fruitiers en Iran. *Annls Soc. ent. Fr.* (NS) **7**, 201–229.

RAFIQ AHMAD (1973) A note on *Microterys chalcostomus* (Dalm.) predacious on *Eulecanium* eggs in Pakistan. *Tech. Bull. Commonw. Inst. biol. Control* **16**, 1–3.

RAGONOT E. L. (1878) Notes on the eggs and larvae of *Ascalaphus longicornis*. *Ann. Soc. ent. Fr.* **8** (5) 120.

RAGONOT E. L. (1893) Monographie des Phycitinae et des Galleriinae **7**. St Petersburg.

RAGUSA S. (1974) Influence of temperature on the oviposition rate and longevity of *Opius concolor siculus* (Hymenoptera: Braconidae). *Entomophaga* **19**, 61–66.

RAHALKAR G. W. and DOUTT R. L. (1965) A comparison of procedures for making adult endoparasitic wasps with P^{32}. *J. econ. Ent.* **58**, 278–281.

RAHMAN M. (1969) Fluctuation in oviposition by *Pieris rapae* in nature. *Pakist. J. Zool.* **1**, 129–133.

RAHMAN M. (1970) *Exorista flaviceps*, a Tachinid parasite of *Pieris rapae* in South Australia. *J. econ. Ent.* **63**, 836–841.

RAHN R. and PAGANELLI C. V. (1968) Gas exchange in gas gills of diving insects. *Resp. Physiol.* **5**, 145–164.

RAINE J. (1960) Life history and behavior of the bramble leafhopper *Ribautiana tenerrima* (H.-S.) (Homoptera: Cicadellidae). *Can. Ent.* **92**, 10–20.

RAISBECK B. (1972) Pheromone inactivation by the gut of *Periplaneta americana*. *Nature, Lond.* **240**, 107–108.

RAISBECK B. (1975) Mechanisms of pheromone inactivation in the gut of *Periplaneta americana*. *J. Insect Physiol.* **21**, 1141–1149.

RAIT L. (1937) Some observations on the immature stages of *Eusthenia spectabilis* (Westwood). *Trans. R. Soc. S. Aust.* **61**, 74–79.

RAJASEKHARA, K. and CHATTERJI S. (1970) Biology of *Orius indicus* (Hemiptera: Anthocoridae) a predator of *Taeniothrips nigricornis* (Thysanoptera). *Ann. ent. Soc. Am.* **63**, 364–367.

RAJENDRAM G. F. and HAGEN K. S. (1974) *Trichogramma* oviposition into artificial substrates. *Envir. Ent.* **3**, 399–401.

RAJULU G. S. and RENGANATHAN K. (1966) On the stabilization of the ootheca of cockroach *Periplaneta americana*. *Naturwissenschaften* **53**, 136.

RAMMNER W. (1934) Zur Biologie des Schildkäfers *Cassida murraea* L. *Z. wiss. InsektBiol.* **27**, 71–82, 116–123.

RAMSAY G. W. (1965) Development of the ovipositor of *Deinacrida rugosa* Buller (Orthoptera: Gryllacridoidea: Henicidae) and a brief review of the ontogeny and homology of the ovipositor with particular reference to the Orthoptera. *Proc. R. ent. Soc. Lond.* (A) **40**, 41–50.

RANKIN K. (1935) Life history of *Lethocerus americanus* Leidy (Belostomatidae, Hemiptera). *Kans. Univ. Sci. Bull.* **36**, 479–491.

RANKIN M. A., CALDWELL R. L., and DINGLE H. (1972) An analysis of a circadian rhythm of oviposition in *Oncopeltus fasciatus* (Hem., Het., Lygaeidae). *J. exp. Biol.* **56**, 353–359.

RAO B. A., SWEET W. C., and RAO A. M. S. (1938) Ova measurements of *A. stephensi* type and *A. stephensi* var. *mysorensis*. *J. Malar. Inst. India* **1**, 261–266.

RAO V. P. (1947) Short notes and exhibits. *Indian J. Ent.* **8**, 131–132.

RAO Y. R. (1960) *The Desert Locust in India*, New Delhi.

RAROS R. S. and HOLDAWAY F. G. (1968) A simple method for collecting eggs of the northern corn rootworm in the laboratory. *J. econ. Ent.* **61**, 1767–1768.

RASKE A. G. (1974) Mortality of birch casebearer eggs. *Bi-mon. Res. Notes* **30**, 1–2.

RASMUSSEN L. A. (1972) Attraction of mountain pine beetles to small-diameter lodgepole pines baited with *trans*-verbenol and alpha-pinene. *J. econ. Ent.* **65**, 1396–1399.

RATANOV K. N. (1935) Description of the egg-pods of Acrididae. *Bull. W. Siber. Pl. Prot. Stn* **1**, 40–70.

RATTAN L. (1975) Tea leaf weevil (*Systates smei*). *Q. Newsl., Tea Res. Found. Cent. Afr.* **38**, 5.

RAU P. (1918) Maternal care in *Dinocoris tripterus* Fab. *Ent. News* **29**, 75–76.

RAU P. (1943) How the cockroach deposits its egg-case; a study in insect behaviour. *Ann. ent. Soc. Am.* **36**, 221–226.

RAU P. and RAU N. (1913) The biology of *Stagmomantis carolina*. *Trans. Acad. Sci. St Louis* **22**, 1–58.

RAULSTON J. R. (1975) Tobacco budworm: observations on the laboratory adaptation of a wild strain. *Ann. ent. Soc. Am.* **68**, 139–142.

RAUN E. S. (1968) Colored European corn borers and eggs from dye-containing diets. *Proc. N. cent. Brch Am. Ass. econ. Ent.* **22**, 162–163 (1967).

RAWAT B. L. (1939) On the habits, metamorphosis and reproductive organs of *Naucoris cimicoides* L. *Trans. R. ent. Soc. Lond.* **88**, 119–138.

RAWAT R. R. and MODI B. N. (1972) Preliminary study on the biology and natural enemies of tortoise-beetle *Oocassida pudibunda* Boh. (Coleoptera: Chrysomelidae: Cassidinae) in Madhya Pradesh. *Indian agric. Sci.* **42**, 854–856.

RAWLINS W. A. (1940) Biology and control of the wheat wireworm *Agriotes mancus* Say. *Bull. Cornell Univ. agric. Exp. Stn* **738**, 1–30.

RAWLINS W. A. (1967) Oviposition by onion maggot adults fed on a chemically defined diet. *J. econ. Ent.* **60**, 1747–1748.

RAY C. (1960) The application of Bergmann's and Allen's rules to the poikilotherms. *J. Morph.* **106**, 85–108.

RAYBOULD J. N. and YAGUNGA A. S. K. (1969) Studies on the immature stages of the *Simulium neavei* Roubaud complex and their associated crabs in the eastern Usambara Mountains in Tanzania. 2. Investigations in small heavily shaded streams. *Ann. trop. Med. Parasit.* **63**, 289–300.

READ D. C. (1962) Notes on the life history of *Aleochara bilineata* (Gyll.) (Coleoptera: Staphylinidae) and on its potential value as a control agent for the cabbage maggot *Hylemya brassicae* (Bouché) (Diptera: Anthomyiidae). *Can. Ent.* **94**, 417–424.

READ D. P., FEENY P. P., and ROOT R. B. (1970) Habitat selection by the aphid parasite *Diaeretiella rapae* (Hymenoptera: Braconidae) and hyperparasite *Charips brassicae* (Hymenoptera: Cynipidae). *Can. Ent.* **102**, 1567–1578.

READIO P. A. (1927) Studies on the biology of the Reduviidae of America north of Mexico. *Kans. Univ. Sci. Bull.* **17**, 5–291.

READSHAW J. L. (1966) The ecology of the swede midge *Contarinia nasturtii* (Kieff.) (Diptera, Cecidomyiidae). I. Life-history and influence of temperature and moisture on development. *Bull. ent. Res.* **56**, 685–700.

READSHAW J. L. and BEDFORD G. O. (1971) Development of the egg of the stick insect *Didymuria violescens* with particular reference to diapause. *Aus. J. Zool.* **19**, 141–158.

RÉAUMUR R. A. F. DE (1738) *Mémoires pour servir à l'histoire des insectes* **4**, 377–379.

RÉAUMUR R. A. F. DE (1740) *Mémoires pour servir à l'histoire des insectes* **5**.

REED D. K., HART W. G., and INGLE S. J. (1968) Laboratory rearing of brown soft scale and its hymenopterous parasites. *Ann. ent. Soc. Am.* **61**, 1443–1446.

REED W. (1965) *Heliothis armigera* (Hb.) (Noctuidae) in western Tanganyika. I. Biology, with special reference to the pupal stage. *Bull. ent. Res.* **56**, 117–125.

REEKS W. A. (1954) An outbreak of the larch sawfly (*Pristiphora erichsonii* (Htg.)) in the maritime provinces (Hymenoptera: Tenthredinidae) and the role of parasites in its control. *Can. Ent.* **86**, 471–480.

RÉGNIER P. R. (1931) Les invasions d'acridiens au Maroc de 1927 à 1931. *Direc. gén. Agric. Déf. Cult.* **3**, 1–139.

REICHARDT, H. (1973) A critical study of the suborder Myxophaga, with a taxonomic revision of the Brazilian Torridincolidae and Hydroscaphidae (Coleoptera). *Archos Zool. S. Paulo* **24**, 73–162.

REICHART G. (1967) New data to the biology of *Capnodis tenebrionis* L. (Coleoptera). *Acta zool. hung.* **13**, 395–408.

REICHSTEIN T., EUW J. VON, PARSONS J., and ROTHSCHILD M. (1968) Heart poisons in the Monarch butterfly. *Science, Wash.* **161**, 861–866.

REID J. A. (1953) The *Anopheles hyrcanus* group in south-east Asia (Diptera: Culicidae). *Bull. ent. Res.* **44**, 5–76.

REID J. A. (1962) The *Anopheles barbirostris* group (Diptera, Culicidae). *Bull. ent. Res.* **53**, 1–57.

REID J. A. (1965) A revision of the *Anopheles-aitkenii* group in Malaya and Borneo. *Ann. trop. Med. Parasit.* **59**, 106–125.

REID R. W. (1957) The bark beetle complex associated with lodgepole pine slash in Alberta. Part III. Notes on the biologies of several predators with special reference to *Enoclerus sphegeus* Fab. (Coleoptera: Cleridae) and two species of mites. *Can. Ent.* **89**, 111–120.

REID R. W. (1962a) Biology of the mountain pine beetle *Dendroctonus monticolae* Hopkins in the East Kootenay region of British Columbia. I. Life cycle, brood development, and flight periods. *Can. Ent.* **94**, 531–538.

REID R. W. (1962b) Biology of the mountain pine beetle *Dendroctonus monticolae* Hopkins, in the East Kootenay region of British Columbia. II. Behaviour in the host, fecundity, and internal changes in the female. *Can. Ent.* **94**, 605–613.

REID R. W. (1963) Biology of the mountain pine beetle *Dendroctonus monticolae* Hopkins in the East Kootenay region of British Columbia. III. Interaction between the beetle and its host, with emphasis on brood mortality and survival. *Can. Ent.* **95**, 225–238.

REID R. W. (1969) The influence of humidity on incubating bark beetle eggs. *Can. Ent.* **101**, 182–183.

REIKHARDT A. N. (1923) Work of the expedition of the Ministry of Agriculture for the study of the parasites of Acrididae in Siberia in 1922. *Izv. sibirsk. ent. Byu.* **2**, 38–45 (in Russian).

REINERT J. F. (1972a) Description of the egg of *Aedes (Aedimorphus) domesticus* (Theobald) (Diptera: Culicidae). *Mosq. Syst.* **4**, 60–62.

REINERT J. F. (1972b) Description of the egg of *Aedes (Diceromyia) furcifer* (Edwards) (Diptera: Culicidae). *Mosq. Syst.* **4**, 87–89.

REINHARD H. J. (1938) The sorghum webworm (*Celama sorghiella* Riley). *Bull. Texas agric. Exp. Stn* **559**, 1–35.

REISEN W. K. and BASIO R. G. (1972) Oviposition trap surveys conducted on four USAF installations in the western Pacific. *Mosq. News* **32**, 107–108.

REMAUDIÈRE G. (1947) Sur les principaux parasites du Criquet Migrateur (*Locusta migratoria* L.) dans ses foyers des Landes de Gascogne. I. Ennemis des oeufs et des oothèques. *Bull. Soc. ent. Fr.* **52**, 63–64.

REMAUDIÈRE G. and SAFAVI M. (1963) Sur l'origine et le mode de formation de la membrane interne des oeufs de Pentatomidae (Het.) parasites par *Asolcus* spp. (Hym., Scelionidae). *Rev. Path. vég. Ent. agric. Fr.* **42**, 227–231.

REMMERT H. (1960) Über die Eiablage von *Trichocladius vitripennis* (Meigen) (Diptera, Chironomidae). *Kieler Meeresforsch.* **16**, 236–237.

REMPEL J. G. and CHURCH N. S. (1965) The embryology of *Lytta viridana* Le Conte (Coleoptera: Meliodae). I. Maturation, fertilization and cleavage. *Can. J. Zool.* **43**, 915–925.

RENNER M. (1952) Analyse der Kopulationsbereitschaft des Weibchens der Feldheuschrecke *Euthystira brachyptera* Ocsk. in ihrer Abhängigkeit von Zustand des Geschlechtsapparates. *Z. Tierpsychol.* **9**, 122–154.

RENSING L. and HARDELAND R. (1967) Zur Wirkung der circadianen Rhythmik auf die Entwicklung von *Drosophila. J. Insect Physiol.* **13**, 1547–1568.

RENWICK J. A. A. (1967) Identification of two oxygenated terpenes from the bark beetles *Dendroctonus frontalis* and *D. brevicomis. Contr. Boyce Thompson Inst.* **23**, 355–360.

RENWICK J. A. A. (1970) Chemical aspects of bark beetle aggregation. *Contr. Boyce Thompson Inst.* **24**, 337–341.

RENWICK J. A. A., HUGHES P. R., PITMAN G. B., and VITÉ J. P. (1976) Oxidation products of terpenes identified from *Dendroctonus* and *Ips* bark beetles. *J. Insect Physiol.* **22**, 725–727.

RENWICK J. A. A., HUGHES P. R., and TANLETIN DEJ. TY (1973) Oxidation products of pinene in the bark beetle *Dendroctonus frontalis. J. Insect Physiol.* **19**, 1735–1740.

RENWICK J. A. A., HUGHES P. R., and VITÉ J. P. (1975) The aggregation pheromone system of a *Dendroctonus* bark beetle in Guatemala. *J. Insect Physiol.* **21**, 1095–1098.

RENWICK J. A. A., PITMAN G. B., and VITÉ J. P. (1966) Detection of a volatile compound in hindguts of male *Ips confusus* (LeConte) (Coleoptera: Scolytidae). *Naturwissenschaften* **53**, 83–84.

RENWICK J. A. A. and VITÉ J. P. (1969) Bark beetle attractants: mechanism of colonization by *Dendroctonus frontalis. Nature, Lond.* **224**, 1222–1223.

RENWICK J. A. A. and VITÉ J. P. (1970) Systems of chemical communication in *Dendroctonus. Contr. Boyce Thompson Inst.* **24**, 283–292.

RENWICK J. A. A. and VITÉ J. P. (1972) Pheromones and host volatiles that govern aggregation

of the six-spined engraver beetle *Ips calligraphus*. *J. Insect Physiol.* **18**, 1215–1219.
RETHFELDT C. (1924) Die Viviparität bei *Chrysomela varians* Schaller. *Zool. Jb. (Anat.)* **46**, 245–302.
RETNAKARAN A. and FRENCH J. (1971) A method for separating and surface sterilizing the eggs of the spruce budworm *Choristoneura fumiferana* (Lepidoptera: Tortricidae). *Can. Ent.* **103**, 712–716.
RETTENMEYER C. W. (1961) Observations on the biology and taxonomy of flies found over swarm raids of army ants (Diptera: Tachinidae, Conopidae). *Kans. Univ. Sci. Bull.* **42**, 993–1066.
REYES A. V. (1969) Biology and host range of *Plodia interpunctella* Hubner (Pyralididae: Lepidoptera). *Philipp. Ent.* **1**, 301–311.
REYES P. V., GOMA P. C., EBORA P., ESCANDOR N. B., and MONTANA A. C. (1970) The water bugs, *Diplonychus rusticus* and *Diplonychus rusticus* var. *marginicollis* as predators of the snail hosts of *Fasciola gigantica* in the Philippines. *Philipp. Anim. Ind.* **27**, 41–64.
RHODE R. H. (1957) A diet for Mexican fruit flies. *J. econ. Ent.* **50**, 215.
RICE L. A. (1954) Observations on the biology of ten notonectid species found in the Douglas Lake, Michigan Region. *Am. Midl. Nat.* **51**, 105–132.
RICE M. J. and McRAE T. M. (1977) Contact chemoreceptors on the ovipositor of *Locusta migratoria* L. *J. Aust. ent. Soc.* **15**, 364.
RICE R. E. (1969a) Bionomics of *Enoclerus barri* (Coleoptera: Cleridae). *Can. Ent.* **101**, 382–386.
RICE R. E. (1969b) Response of some predators and parasites of *Ips confusus* (LeC.) (Coleoptera: Scolytidae) to olfactory attractants. *Contr. Boyce Thompson Inst.* **24**, 189–194.
RICE R. E. (1976) A comparison of monitoring techniques for the navel orangeworm. *J. econ. Ent.* **69**, 25–28.
RICH E. R. (1956) Egg cannibalism and fecundity in *Tribolium*. *Ecology* **37**, 233–255.
RICHARDS J. G. and KING P. E. (1967) Chorion and vitelline membranes and their role in re-absorbing eggs of the Hymenoptera. *Nature, Lond.* **214**, 601–602.
RICHARDS O. W. (1940) The biology of the small white butterfly (*Pieris rapae*), with special reference to the factors controlling its abundance. *J. Anim. Ecol.* **9**, 243–288.
RICHARDS O. W. (1947) Observations on grain weevils *Calandra* (Col., Curculionidae). I. General biology and oviposition. *Proc. zool. Soc. Lond.* **117**, 1–43.
RICHARDS P. G. and MORRISON F. O. (1972) The egg and chorion of *Pollenia rudis* (Fabricius) (Diptera: Calliphoridae). *Can. J. Zool.* **50**, 1676–1678.
RICHARDSON C. H. (1925) *The Oviposition Response of Insects.* Bull. 1324, US Dep. Agric., Washington.
RICHMOND C. A. and MARTIN D. F. (1966) Technique for mass rearing of the pink bollworm by infesting diet medium with eggs. *J. econ. Ent.* **59**, 762–763.
RICHMOND E. A. (1920) Studies on the biology of the aquatic Hydrophilidae. *Bull. Am. Mus. nat. Hist.* **42**, 1–94.
RICHTER G. (1964) Untersuchungen zur Prognose des Ausflugs und der Eireife von *Melolontha*. *NachrBl. dt. PflSchutzdienst, Berl.* **18**, 178–183.
RICHTER H. C. (1870) Eggs of bird parasites. *Hardwicke's Science-gossip*, June, pp. 132–133.
RICHTER P. O. (1945) Coprinae of eastern North America with descriptions of the larvae and keys to genera and species (Coleoptera: Scarabaeidae). *Bull. Ky agric. Exp. Stn* **477**, 1–23.
RIDDIFORD L. M. and ASHENHURST J. B. (1973) The switchover from virgin to mated behaviour in female *Cecropia* moths: the role of the bursa copulathrix (*Hyalophora cecropia*: Lep. Saturniidae). *Biol. Bull. Woods Hole* **144**, 162–171.
RIEK E. F. (1970) Neuroptera (Lacewings). In the *Insects of Australia. A Textbook for Students and Research Workers*, Melbourne University Press.
RIEMANN J. G. (1973) Ultrastructure of the ejaculatory duct region producing the male

housefly accessory material. *J. Insect Physiol.* **19**, 213–223.

RIEMANN J. G., MOEN D. J., and THORSON B. J. (1967) Female monogamy and its control in house flies. *J. Insect Physiol.* **13**, 407–418.

RIEMANN J. G. and THORSON B. J. (1969) Effect of male accessory material on oviposition and mating by female houseflies. *Ann. ent. Soc. Am.* **62**, 828–834.

RIEMANN J. G., THORSON B. J., and RUUD R. L. (1974) Daily cycle of release of sperm from the testes of the Mediterranean flour moth. *J. Insect Physiol.* **20**, 195–208.

RIES E. (1932) Die Progresse der Eibildung und des Eiwachstums bei Pediculiden und Mallophagen. *Z. Zellforsch.* **16**, 314–388.

RILEY C. V. (1874) Katydids. *6th Ann. Rep. Insects Missouri*, pp. 150–169.

RILEY C. V. (1876) On the curious egg mass of *Corydalus cornutus* (Linn.) and on the eggs that have hitherto been referred to that species. *Proc. Am. Ass. Adv. Sci.* **25**, 275–279.

RILEY C. V. (1877) The hellgrammite fly *Corydalus cornutus* (Linn.). *A. Rep. Noxious, Beneficial and Other Insects of Missouri* **9**, 125–129.

RILEY C. V. (1878) Invertebrate enemies. In *First Annual Report of the United States Entomological Commission for the Year 1877, Relating to the Rocky Mountain Locust*, pp. 284–334, Washington.

RILEY C. V. (1880) Further facts about the natural enemies of locusts. In *Second Report of the United States Entomological Commission for the Years 1878 and 1879, Relating to the Rocky Mountain Locust and the Western Cricket*, pp. 259–271, Washington.

RILEY C. V. (1884a) Remarks on the bag-worm *Thyridopteryx ephemeraeformis*. *Proc. biol. Soc. Wash.* **2**, 80–83.

RILEY C. V. (1884b) Annual address of the President. *Proc. ent. Soc. Wash.* **1**, 23–24.

RILEY C. V. (1892) The yucca moth and yucca pollination. *Ann. Rep. Missouri bot. Garden* **3**, 99–158.

RILEY R. C. and FORGASH A. J. (1967) *Drosophila melanogaster* eggshell adhesive. *J. Insect Physiol.* **11**, 509–518.

RIMANDO L. C., COREY R. A., and SUN (YUN-PEI) (1966) Mass rearing of the western spotted cucumber beetle. *J. econ. Ent.* **59**, 230–231.

RIMES G. D. (1951) Some new and little-known shore-bugs (Heteroptera: Saldidae) from the Australian region. *Trans. R. Soc. S. Aust.* **74**, 135–145.

RINGOLD G. B., GRAVELLE P. J., MILLER D., FURNISS M. M., and MCGREGOR M. D. (1975) Characteristics of Douglas-fir beetle infestation in northern Idaho resulting from treatment with Douglure. *Res. Note, Intermount. For. Range Exp. Stn For. Serv. USDA* **189**, 1–10.

RINGS R. W. (1969) Contributions to the bionomics of the green fruitworms: the life history of *Lithophane laticinerea*. *J. econ. Ent.* **62**, 1388–1393.

RINGS R. W. (1970) Contributions to the bionomics of the green fruitworms: the life history of *Orthosia hibisci*. *J. econ. Ent.* **63**, 1562–1568.

RIORDAN D. F. and PESCHKEN D. P. (1970) A method for obtaining P^{32}-labelled eggs of the flea beetle *Altica carduorum* (Coleoptera: Chrysomelidae). *Can. Ent.* **102**, 1613–1616.

RISBEC J. (1946) Action de prédateurs et de parasites sur *Schistocerca gregaria* au Sénégal. *Bull. Off. anti-acrid.* No. 2, 5–16.

RISBEC J. (1951) Les Diptères nuisibles au riz de Camargue au début de son développement. *Rev. Path. vég. Ent. agric. Fr.* **30**, 211–227.

RIVERA J. and HILL R. B. (1935) Persistencia de los caracteres diferenciales de los huevos, larvas y adultos, en diferentes generaciones de *Anopheles maculipennis (atroparvus)*. *Medicina Paises Calidos* **8** (7) 1–7.

RIVIÈRE J. L. (1976) Fluctuations de la ponte chez *Pales pavida* Meigen (Dipt. Tachinidae). *Bull. Soc. ent. Fr.* **80**, 181–183.

RIZKI T. M. (1968) Hemocyte encapsulation of streptococci in *Drosophila. J. Invert. Path.* **12**, 339–343.
ROBERT P. A. (1958) *Les Libellules (Odonates)*, Delachawx and Niestle, Neuchatel and Paris.
ROBERTS F. H. S. (1950) The tail-switch louse of cattle *Haematopinus quadripertusus* Fahrenholz. *Aust. vet. J.* **26**, 136.
ROBERTS F. H. S. and O'SULLIVAN P. J. (1948) Studies on the behaviour of adult Australasian anophelines. *Bull. ent. Res.* **39**, 159–178.
ROBERTS R. A. (1937a) Biology of the bordered mantid *Stagmomantis limbata* Hahn (Orthoptera, Mantidae). *Ann. ent. Soc. Am.* **30**, 96–110.
ROBERTS R. A. (1937b) Biology of the minor mantid *Litaneutria minor* Scudder (Orthoptera, Mantidae). *Ann. ent. Soc. Am.* **30**, 111–121.
ROBERTS R. H. (1971) Effect of amount of CO_2 on collection of Tabanidae in Malaise traps. *Mosq. News* **31**, 551–558.
ROBERTSON F. W. (1960) The ecological genetics of growth in *Drosophila*. Body size and developmental time on different diets. *Genet. Res. Camb.* **1**, 288–304.
ROBERTSON I. A. D. (1954) The numbers of eggs in pods of the red locust *Nomadacris septemfasciata* (Serville) (Orth., Acrididae). *Entomologist's mon. Mag.* **90**, 254–255.
ROBERTSON I. A. D. (1958) The reproduction of the red locust *Nomadacris septemfasciata* (Serv.) (Orthoptera, Acrididae) in an outbreak area. *Bull. ent. Res.* **49**, 479–496.
ROBINSON C. H. (1925) The oviposition response of insects. *Bull. US Dep. Agric.* **1324**, 1–17.
ROBINSON D. M. (1956) A study of cranefly eggs (Diptera: Tipulidae) with particular reference to the micropyle and the terminal filament. *Proc. Univ. Durham phil. Soc.* **12**, 175–182.
ROBINSON I. (1953) The postembryonic stages in the life cycle of *Aulacigaster leucopeza* (Meigen) (Diptera Cyclorrhapha: Aulacigasteridae). *Proc. R. ent. Soc. Lond.* (A) **28**, 77–84.
ROBINSON W. H. and FOOTE B. A. (1968) Biology and immature stages of *Megaselia aequalis*, a phorid predator of slug eggs. *Ann. ent. Soc. Am.* **61**, 1587–1594.
ROBLES-CHILLIDA E. M., MUÑIZ M., and BLANCO-MARCO E. (1970) Influencia del CINa sobre la ultrastructura del corión de *Dacus oleae* Gmel. (Dipt. Trypetidae). *Graellsia* **25**, 325–333.
ROBREDO F., CONDE M., and ALONSO DE MEDINA F. J. (1974) Contribución al conocimiento de la bioecologia de *Rhyacionia duplana* Hb. (Lep. Tortricidae): estudio de su puesta. *Bol. Est. Central Ecol.* **3**, 57–69.
ROCK G. C. (1968) Growth of *Argyrotaenia velutinana* by egg cannibalism. *Ann. ent. Soc. Am.* **61**, 1034.
RODRIGUEZ J. G., SINGH P., and TAYLOR B. (1970) Manure mites and their role in fly control. *J. med. Ent.* **7**, 335–341.
RODRIGUEZ J. G. and WADE C. F. (1961) The nutrition of *Macrocheles muscaedomesticae* (Acarina: Macrochelidae) in relation to its predatory action on the house fly egg. *Ann. ent. Soc. Am.* **54**, 782–788.
RODRIGUEZ J. G., WADE C. F., and WELLS C. N. (1962) Nematodes as a natural food for *Macrocheles muscaedomesticae* (Acarina: Macrochelidae), a predator of the house fly egg. *Ann. ent. Soc. Am.* **55**, 507–511.
ROEPKE W. (1919) Zwei neue javanische Embiiden: *Oligotoma maerens* und *O. nana* m.; zugleich ein Beitrag zur Naturgeschichte der Embiiden. *Treubia* **1**, 5–22.
ROFFEY J. (1958) Observations on the biology of *Trox procerus* Har. (Coleoptera, Trogidae), a predator of eggs of the desert locust, *Schistocerca gregaria* (Forsk.). *Bull. ent. Res.* **49**, 449–465.
ROGERS C. E. (1974) Bionomics of the carrot beetle in the Texas Rolling Plains. *Envir. Ent.* **3**, 969–974.
ROGERS C. E., JACKSON H. B., and EIKENBARY R. D. (1972) Voracity and survival of *Propylea 14-punctata* preying upon greenbugs. *J. econ. Ent.* **65**, 1313–1316.

Rogers D. (1972) The Ichneumon wasp *Venturia canescens*: oviposition and avoidance of superparasitism. *Entomologia exp. appl.* **15**, 190–194.

Rogers J. S. (1949) The life history of *Megistocera longipennis* (Macquart) (Tipulidae, Diptera), a member of the Neuston fauna. *Occ. Pap. Mus. Zool. Univ. Mich.* **521**, 1–14.

Rolston L. H., Mayes R., Edwards P., and Wingfield M. (1965) Biology of the eggplant tortoise beetle (Coleoptera: Chrysomelidae). *J. Kans. ent. Soc.* **38**, 362–366.

Rommel E. (1961) Nutritional biology and brood-care-conduct of the Spanish moon-horned beetle *Copris hispanus*. *Biol. Zbl.* **80**, 327–346.

Rommel E. (1967) Ernährungsbiologie und Brutpflegeverhatten des kleinen Mondhornhäfers *Copris lunaris* (L.) (Col., Scarab). Eine Vergleichsstudie zu den Arbeiten über den spanischen Mondhornkäfer *Copris hispanis* (L.). *NachrBl. bayer. Ent.* **16**, 8–13, 20–28.

Roonwal M. L. (1936) The growth-changes and structure of the egg of *Locusta migratoria migratorioides* R. and F. (Orthoptera: Acrididae). *Bull. ent. Res.* **27**, 1–14.

Roonwal M. L. (1954a) The eggwall of the African migratory locust *Locusta migratoria migratorioides* Reich. and Frm. (Orthoptera, Acrididae). *Proc. natn. Inst. Sci. India* **20**, 361–370.

Roonwal M. L. (1954b) Size, sculpturing, weight and moisture content of the developing eggs of the desert locust, *Schistocerca gregaria* (Forskål) (Orthoptera, Acrididae). *Proc. natn. Inst. Sci. India* **20**, 388–398.

Roonwal M. L. (1954c) Structure of the egg-masses and their hairs in some species of *Lymantria* of importance to forestry (Insecta: Lepidoptera: Lymantriidae). *Indian For. Rec. Ent.* (NS) **8**, 265–276.

Roonwal M. L. (1976) Ecology and biology of the grasshopper *Hieroglyphus nigrorepletus* Bolivar (Orthoptera: Acrididae). I. Egg-pods, diapause, prolonged viability and annual hatching rhythm. *Z. angew. Zool.* **63**, 171–185.

Rosay B. (1959a) Gross external morphology of embryos of *Culex tarsalis* Coquillet (Diptera: Culicidae). *Ann. ent. Soc. Am.* **52**, 481–484.

Rosay B. (1959b) Expansion of eggs of *Culex tarsalis* Coquillet and *Aedes nigromaculis* (Ludlow) (Diptera: Culicidae). *Mosq. News* **19**, 270–273.

Rose A. H. (1958) Some notes on the biology of *Monochamus scutellatus* (Say) (Coleoptera: Cerambycidae). *Can. Ent.* **89**, 547–553 (1957).

Rose R. I. and McCabe J. M. (1973) Laboratory rearing techniques for the southern corn rootworm. *J. econ. Ent.* **66**, 398–400.

Rosel A. (1969) Oviposition, egg development and other features of the biology of 5 species of Lyctidae (Coleoptera). *J. Aust. ent. Soc.* **8**, 145–152.

Ross E. S. (1944) A revision of the Embioptera, or web-spinners, of the New World. *Proc. US natn. Mus.* **94**, 401–504.

Ross H. H. (1929) The life history of the German cockroach. *Trans. Ill. Acad. Sci.* **21**, 84–93.

Ross R. H. Jr, Monroe R. E., and Butcher J. W. (1971) Studies on techniques for the xenic and aseptic rearing of the European pine shoot moth *Rhyacionia buoliana* (Lepidoptera: Olethreutidae). *Can. Ent.* **103**, 1449–1454.

Rossetto C. J., Arruda H. V., and Silva W. J. da (1977) Localização dos ovos *Sitophilus zeamais* Motschulsky (Coleoptera: Curculionidae) em milho em palha e debulhado. *Anais Soc. Ent. Brasil* **4**, 21–27.

Rosskothen P. (1949) Der Brutfürsorgeinstinkt des Trichterwicklers (*Deporaus betulae* L.). *Ent. Bl.* **41–44**, 67–76.

Rosskothen P. (1952) Der Brutfürsorgeinstinkt des Eiehenblattrollers (*Attelabus nitens*, Scop. Curculionidae). *Ent. Bl.* **47**, 54–57.

Roth L. H. (1916) Observations on the growth and habits of the stick insect *Carausius morosus* Br.; intended as a contribution towards a knowledge of variation in an organism which reproduces itself by the parthenogenetic method. *Trans. ent. Soc. Lond.* **1916**, 345–386.

ROTH L. M. (1948) A study of mosquito behaviour. An experimental laboratory study of the sexual behaviour of *Aedes aegypti* (Linnaeus). *Am. Midl. Nat.* **40**, 265–352.

ROTH L. M. (1967a) Water changes in cockroach oöthecae in relation to the evolution of ovoviviparity and viviparity. *Ann. ent. Soc. Am.* **60**, 928–946.

ROTH L. M. (1967b) The evolutionary significance of rotation of the oöthecae in the Blattaria. *Psyche, Camb.* **74**, 85–103.

ROTH L. M. (1968a) Oviposition behavior and water changes in the oöthecae of *Laphoblatta brevis* (Blattaria: Blattellidae: Plectopterinae). *Psyche, Camb.* **75**, 99–106.

ROTH L. M. (1968b) Oothecae of the Blattaria. *Ann. ent. Soc. Am.* **61**, 83–111.

ROTH L. M. (1968c) Ovarioles of the Blattaria. *Ann. ent. Soc. Am.* **61**, 132–140.

ROTH L. M. (1970) Evolution and taxonomic significance of reproduction in Blattaria. *A. rev. Ent.* **15**, 75–96.

ROTH L. M. (1971) Additions to the oöthecae, uricose glands, ovarioles and tergal glands of Blattaria. *Ann. ent. Soc. Am.* **64**, 127–141.

ROTH L. M. (1974) Control of oöthecae formation and oviposition in Blattaria. *J. Insect Physiol.* **20**, 821–844.

ROTH L. M. and STAY B. (1962) Oöcyte development in *Blattella germanica* and *Blattella vaga* (Blattaria). *Ann. ent. Soc. Am.* **55**, 633–642.

ROTH L. M. and WILLIS E. R. (1950) The oviposition of *Dermestes ater* Degeer, with notes on bionomics under laboratory conditions. *Am. Midl. Nat.* **44**, 427–447.

ROTH L. M. and WILLIS E. R. (1952) A study of cockroach behaviour. *Am. Midl. Nat.* **47**, 66–129.

ROTH L. M. and WILLIS E. R. (1954a) *Anastatus floridanus* (Hymenoptera: Eupelmidae) a new parasite on the eggs of the cockroach *Eurycotis floridana*. *Trans. Am. ent. Soc.* **80**, 29–41.

ROTH L. M. and WILLIS E. R. (1954b) The biology of the cockroach egg parasite, *Tetrastichus hagenowii* (Hymenoptera: Eulophidae). *Trans. Am. ent. Soc.* **80**, 53–72.

ROTH L. M. and WILLIS E. R. (1954c) The reproduction of cockroaches. *Smithson, misc. Coll.* **122** (12) 1–49.

ROTH L. M. and WILLIS E. R. (1955a) Relation of water loss to hatching of eggs from detached oöthecae of *Blattella germanica* (L.). *J. econ. Ent.* **48**, 57–70.

ROTH L. M. and WILLIS E. R. (1955b) The water content of cockroach eggs during embryogenesis in relation to oviposition behavior. *J. exp. Zool.* **128**, 489–509.

ROTH L. M. and WILLIS E. R. (1955c) Intra-uterine nutrition of the "beetle-roach" *Diploptera dytiscoides* (Serv.) during embryogenesis, with notes on its biology in the laboratory (Blattaria: Diplopteridae). *Psyche, Camb.* **62**, 55–68.

ROTH L. M. and WILLIS E. R. (1958a) The biology of *Panchlora nivea* with observations on the eggs of other Blattaria. *Trans. Am. ent. Soc.* **83**, 195–207.

ROTH L. M. and WILLIS E. R. (1958b) An analysis of oviparity and viviparity in the Blattaria. *Trans. Am. ent. Soc.* **83**, 221–238.

ROTH L. M. and WILLIS E. R. (1961) A study of bisexual and parthenogenetic strains of *Pycnoscelus surinamensis* (Blattaria: Epilamprinae). *Ann. ent. Soc. Am.* **54**, 12–25.

ROTH M. R. and STAY B. (1962) A comparative study of oöcyte development in false ovoviviparous cockroaches. *Psyche, Camb.* **68**, 165–208.

ROTHERAM S. (1967) Immune surface of eggs of a parasitic insect. *Nature, Lond.* **214**, 700.

ROTHERAM S. (1973a) The surface of the egg of a parasitic insect. I. The surface of the egg and first-instar larva of *Nemeritis*. *Proc. R. Soc.* (B) **183**, 179–194.

ROTHERAM S. (1973b) The surface of the egg of a parasitic insect. II. The ultrastructure of the particulate coat on the egg of *Nemeritis*. *Proc. R. Soc.* (B) **183**, 195–204.

ROTHSCHILD G. H. L. (1971) The biology and ecology of rice-stem borers in Sarawak (Malaysian Borneo). *J. appl. Ecol.* **8**, 287–322.

Rothschild M., Reichstein T., von Euw J., Aplin R., and Harman R. R. M. (1970) Toxic Lepidoptera. *Toxicon* **8**, 293–299.

Rothschild M. and Schoonhoven L. M. (1977) Assessment of egg load by *Pieris brassicae* (Lepidoptera: Pieridae). *Nature, Lond.* **266**, 352–355.

Roubaud E. (1927) L'éclosion de l'oeuf et les stimulants d'éclosion chez le *Stegomyia* de la fièvre jaune. Application à la lutte antilarvaire. *C. r. hebd. Séanc. Acad. Sci. Paris* **184**, 1491–1492.

Roubaud E. (1935) La microstructure du flotteur de l'oeuf dan les races biologiques de *Culex pipiens* L. *Bull. Soc. Path. exot.* **28**, 443–445.

Roubaud E. and Colas-Belcour J. (1927) Action des diastases dans le déterminisme d'éclosion de l'oeuf chez le moustique de la fièvre jaune (*Stegomyia fasciata*). *C. r. hebd. Séanc. Acad. Sci. Paris* **184**, 248–249.

Rowley W. A. and Peters D. C. (1972) Scanning electron microscopy of the egg-shell of four species of Diabrotica (Coleoptera: Chrysomelidae). *Ann. ent. Soc. Am.* **65**, 1188–1191.

Roy D. N. and Siddons L. B. (1939) Egg of *A. philippinensis* Ludl. *J. Malar. Inst. India* **2**, 159–164.

Roy P. and Dasgupta B. (1971) Behaviour of *Chrysomya megacephala* (Fabr.) and *Hemipyrellia ligurriens* (Wied.), as parasites of living animals under experimental conditions. *S. Afr. J. med. Sci.* **36** (4) 85–91.

Rozeboom L. E. (1937a) On *Anopheles albitarsis* Lynch Arribalzaga in Panama. *Southern med. J.* **30**, 950–951.

Rozeboom L. E. (1937b) The egg of *Anopheles pseudopunctipennis* in Panama. *J. Parasit.* **23**, 538–539.

Rozeboom L. E. (1938) The eggs of the *Nysorrhyncus* group of *Anopheles* (Culicidae) in Panama. *Am. J. Hyg.* **27**, 95–107.

Rozeboom L. E. (1942) Subspecific variations among neotropical *Anopheles* mosquitoes, and their importance in the transmission of malaria. *Am. J. trop. Med.* **22**, 235–255.

Rozeboom L. E., Rosen L., and Ikeda J. (1973) Observations on oviposition by *Aedes (S.) albopictus* Skuse and *A. (S.) polynesiensis* Marks in nature. *J. med. Ent.* **10**, 397–399.

Rozhošný R. and Knutson L. V. (1970) Taxonomy, biology, and immature stages of Palearctic *Pteromicra*, snail-killing Diptera (Sciomyzidae). *Ann. ent. Soc. Am.* **63**, 1434–1459.

Rubtsov I. A. (1972) Phoresy in blackflies (Diptera, Simuliidae) and new phoretic species from nymphs of Ephemeroptera. *Ent. Obozr.* **51**, 403–411 (in Russian; English summary).

Rubtsova N. N. (1961) On the development of the eggs of the oak tortrix (*Tortrix viridana* L.) in the Voronezh region. *Zool. Zh.* **40**, 1665–1676 (in Russian; English summary).

Rudall K. M. (1956) Protein ribbons and sheets. *Lectures on the Scientific Basis of Medicine* **5**, 217–230.

Rudall K. M. and Kenchington W. (1971) Arthropod silks: the problem of fibrous proteins in animal tissues. *A. Rev. Ent.* **16**, 73–96.

Rudinsky J. A. (1966) Host selection and invasion by the Douglas-fir beetle *Dendroctonus pseudotsugae* Hopkins in coastal Douglas-fir forests. *Can. Ent.* **98**, 98–111.

Rudinsky J. A. (1968) A pheromone-mask by the female *Dendroctonus pseudotsugae* Hopk., an attraction regulator (Coleoptera: Scolytidae). *Pan-Pacif. Ent.* **44**, 248–250.

Rudinsky J. A. (1969) Masking of the aggregation pheromone in *Dendroctonus pseudotsugae* Hopk. *Science, Wash.* **166**, 884–885.

Rudinsky J. A. (1970) Sequence of Douglas-fir beetle attraction and its ecological significance. *Contr. Boyce Thompson Inst.* **24**, 311–314.

Rudinsky J. A. (1973a) Multiple functions of the southern pine beetle pheromone verbenone. *Envir. Ent.* **2**, 511–514.

Rudinsky J. A. (1973b) Multiple functions of the Douglas fir beetle pheromone 3-methyl-2-

cyclohexen-1-one. *Envir. Ent.* **2**, 579–585.
RUDINSKY J. A. and DATERMAN G. E. (1964) Response of the ambrosia beetle *Trypodendron lineatum* (Oliv.) to a female-produced pheromone. *Z. angew. Ent.* **54**, 300–303.
RUDINSKY J. A., FURNISS M. M., KLINE L. N., and SCHMITZ R. F. (1972) Attraction and repression of *Dendroctonus pseudotsugae* (Coleoptera: Scolytidae) by three synthetic pheromones in traps in Oregon and Idaho. *Can. Ent.* **104**, 815–822.
RUDINSKY J. A., KINZER G. W., FENTIMAN A. F. JR, and FOLTZ R. L. (1972) *Trans*-verbenol isolated from Douglas-fir beetle: laboratory and field bioassays in Oregon. *Envir. Ent.* **1**, 485–488.
RUDINSKY J. A., KLINE L. N., and DIEKMAN J. D. (1975) Response-inhibition by four analogues of MCH an antiaggregative pheromone of the Douglas-fir beetle. *J. econ. Ent.* **68**, 527–528.
RUDINSKY J. A. and MICHAEL R. R. (1972) Sound production in Scolytidae: chemostimulus of sonic signal by the Douglas-fir beetle. *Science, Wash.* **175**, 1386–1390.
RUDINSKY J. A. and MICHAEL R. R. (1973) Sound production in Scolytidae: stridulation by female *Dendroctonus* beetles. *J. Insect Physiol.* **19**, 689–705.
RUDINSKY J. A. and MICHAEL R. R. (1974) Sound production in Scolytidae: "rivalry" behaviour of male *Dendroctonus* beetles. *J. Insect Physiol.* **20**, 1219–1230.
RUDINSKY J. A., MORGAN M., LIBBEY L. M., and MICHAEL R. R. (1973) Sound production in Scolytidae: 3-methyl-2-cyclohexen-1-one released by the female Douglas fir beetle in response to male sonic signal. *Envir. Ent.* **2**, 505–509.
RUDINSKY J. A., MORGAN M. E., LIBBEY L. M., and PUTNAM T. B. (1974a) Antiaggregative-rivalry pheromone of the mountain pine beetle, and a new arrestant of the southern pine beetle. *Envir. Ent.* **3**, 90–98.
RUDINSKY J. A., MORGAN M. E., LIBBEY L. M., and PUTNAM T. B. (1974b) Additional components of the Douglas fir beetle (Col., Scolytidae) aggregative pheromone and their possible utility in pest control. *Z. angew. Ent.* **76**, 65–77.
RUDINSKY J. A., MORGAN M. E., LIBBEY L. M., and PUTNAM T. B. (1976) Release of frontalin by male Douglas fir beetle. *Z. angew. Ent.* **81**, 267–269.
RUDINSKY J. A., NOVÁK V., and ŠVIHRA P. (1971a) Pheromone and terpene attraction in the bark beetle *Ips typographus* L. *Experentia* **27**, 161–162.
RUDINSKY J. A., NOVÁK V., and ŠVIHRA P. (1971b) Attraction of the bark beetle *Ips typographus* L. to terpenes and a male-produced pheromone. *Z. angew. Ent.* **67**, 179–188.
RUDINSKY J. A. and RYKER L. C. (1976) Sound production in Scolytidae: rivalry and premating stridulation of male Douglas-fir beetle. *J. Insect Physiol.* **22**, 997–1003.
RUDINSKY J. A., SARTWELL C. JR, GRAVES T. M., and MORGAN M. E. (1974) Granular formulation of methylcyclohexenone: an antiaggregative pheromone of the Douglas fir and spruce bark beetles (Col., Scolytidae). *Z. angew. Ent.* **75**, 254–263.
RUDINSKY J. A. and ZETHNER-MØLLER O. (1967) Olfactory responses of *Hylastes nigrinis* (Coleoptera: Scolytidae) to various host materials. *Can. Ent.* **99**, 911–916.
RUDNEV D. F. (1951) Determination of the egg production of the Gypsy moth by the pupae. *Zool. Zh.* **30**, 224–228 (in Russian).
RÜHM W. (1969) Zur Populationsdynamik der Kriebelmücken, insbesondere von *Boophthora erythrocephala* der Geer und des *Odagmia ornata*-Komplexes. *Z. angew. Ent.* **63**, 212–227.
RÜHM W. (1971) Eiablagen einiger Simuliidenarten. *Angew. Parasit.* **12**, 68–78.
RÜHM W. (1972) Zur Populationsdynamik von *Boophthora erythrocephala* de Geer. *Z. angew. Ent.* **71**, 35–44.
RÜHM W. (1975) Freilandbeobachtungen zum Funktionskreis der Eiablage verschiedener Simuliidenarten unter besonderer Berücksichtigung von *Simulium argyreatum* Meig. (Dipt. Simuliidae). *Z. angew. Ent.* **78**, 321–334.

Rummel R. W. and Turner E. C. Jr (1970) A refined technique for counting face fly eggs. *J. econ. Ent.* **63**, 1378–1379.

Rupérez A. (1960) Localización del huevo del *Balaninus elephas* Gyll. con relación al daño denominado "melazo" de la bellota de encina. (*Q. ilex* Oerst.) *Boln Serv. Plagas for.* **3**, 133–145.

Russ K. (1955) Eine neue Methode zur Erzielung massierter Eiablage von *Cheimatobia brumata* L. (Kleiner, Frostspanner) und einige Beobachtungen über die Biologie der Falter. *PflSchBer.* **16**, 163–172.

Russell M. P. (1962) Effects of sorghum varieties on the lesser rice weevil *Sitophilus oryzae* (L.). I. Oviposition, immature mortality, and size of adults. *Ann. ent. Soc. Am.* **55**, 678–685.

Russell P. F. and Rao T. R. (1942) On the swarming, mating and ovipositing behaviour of *Anopheles culicifacies*. *Am. J. trop. Med.* **22**, 417–427.

Russev B. (1959) "Vol de compensation pour la ponte" de *Palingenia longicauda* Oliv. (Ephem.) contre le courant de Danube. *C. r. Acad. Bulg. Sci.* **12**, 165–168.

Rust R. W. and Thorn R. W. (1973) The biology of *Stelis chlorocyanea*, a parasite of *Osmia nigrifrons* (Hymenoptera: Megachilidae). *J. Kans. Ent. Soc.* **46**, 548–562.

Ruzaev K. S. (1958) Materials on the bioecology of *Otiorrhynchus turca* Boh. *Zool. Zh.* **37**, 855–865 (in Russian; English summary).

Ryan M. F. (1973a) The natural mortality of wheat-bulb fly eggs in bare fallow soils. *J. appl. Ecol.* **10**, 869–874.

Ryan M. F. (1973b) The natural mortality of wheat-bulb fly larvae. *J. appl. Ecol.* **10**, 875–879.

Ryan M. F. and Ryan J. (1973) The natural mortality of cabbage root fly eggs in peatland. *Sci. Proc. R. Dublin Soc.* (B) **3**, 195–199.

Ryan R. B. (1974) Reduced oviposition by *Ephialtes ontario* and *Itoplectis quadricingulatus* in a humid environment. *Ann. ent. Soc. Am.* **67**, 928–930.

Rygg T. and Sömme L. (1972) Oviposition and larval development of *Hylemya floralis* (Fallén) (Dipt., Anthomyiidae) on varieties of swedes and turnips. *Norsk ent. Tidsskr.* **19**, 81–90.

Ryker L. C. and Rudinsky J. A. (1976) Sound production in Scolytidae: aggressive and mating behaviour of the mountain pine beetle. *Ann. ent. Soc. Am.* **69**, 677–680.

Saakyan-Baranova A. A. and Muzafarov S. S. (1971) *Kermococcus quercus* and its parasites. *Zashch. Rast.* **16** (9) 48–49.

Saakyan-Baranova A. A. and Muzafarov S. S. (1972) The structure, biology and interrelations of *Kermococcus quercus* L. (Homoptera, Kermococcidae) and its parasites (Hymenoptera, Chalcidoidea). II. Hyperparasites and relations between *Kermococcus quercus* L. and its parasites. *Ent. Obozr.* **51**, 697–715 (in Russian; English summary).

Saba F. (1970) Parasites, predators, and diseases in a rearing culture of *Diabrotica balteata*. *J. econ. Ent.* **63**, 1674.

Sabrosky C. W. (1958) A new genus and two new species of Chamaemyiidae (Diptera) feeding on *Orthezia* scale insects. *Bull. Brooklyn ent. Soc.* **52**, 114–117.

Saccuman G. (1963) Contributo alla conoscenza della *Euproctis chrysorrhoea* L. Cenni sulla morfologia, biologia e mezzi di lotta. *Boll. Lab. Ent. agr. Portici* **21**, 271–322.

Sailer R. I. (1950) A thermophobic insect. *Science, Lancaster, Pa* **112**, 743.

St. Quentin D. (1934) Beobachtungen und Versuche an Libellen in ihren Jagdrevieren. *Konowia* **13**, 275–282.

St. Quentin D. (1962) Der Eilegeapparat der Odanaten. *Z. Morph. Ökol. Tiere* **51**, 165–189.

St. Quentin D. and Beier M. (1968) Odonata (Libellen). In *Handbuch der Zoologie* **4** (2) (2/6) 1–39.

Sakagami S. F., Montenegro M. J., and Kerr W. E. (1965) Behaviour studies of the stingless bees, with special reference to the oviposition process. V. *Melipona quadrifasciata anthidioides* Lepeletier. *J. Fac. Sci. Hokkaido Univ. (Zool.)* **15**, 578–607.

SAKAGAMI S. F. and ONIKI Y. (1963) Behaviour studies of the stingless bees, with special reference to the oviposition process. 1. *Melipona compressipes manaosensis* Schwarz. *J. Fac. Sci. Hokkaido Univ. (Zool.)* **15**, 300–318.

SAKAGAMI S. F. and ZUCCHI R. (1967) Behaviour studies of the stingless bees, with special reference to the oviposition process. 6. *Trigona (Tetragona) clavipes. J. Fac. Sci. Hokkaido Univ. (Zool.)* **16**, 292–313.

SAKAGAMI S. F. and ZUCCHI R. (1968) Oviposition behavior of an Amazonic stingless bee *Trigona (Duckeola) ghilianii. J. Fac. Sci. Hokkaido Univ. (Zool.)* **16**, 564–581.

SAKANOSHITA A. and YANAGITA Y. (1972) Fundamental studies on the reproduction of diamond back moth *Plutella maculipennis* Curtis. I. Effects of environmental factors on emergence, copulation and oviposition. *Proc. Ass. Pl. Prot. Kyushu* **18**, 11–12 (in Japanese; English summary).

SALAMA H. S. and ATA M. A. (1972) Reactions of mosquitoes to chemicals in their oviposition sites. *Z. angew. Ent.* **71**, 53–57.

SALAVIN R. J. (1958) Notas biológicas sobre la mosca *Servaisia (Protodexia) arteagai* (Blanch.) Rob. (Diptera Sarcophagidae) parásito de la tucura. *Revta Invest. agríc. B. Aires* **12**, 299–310.

SALIBA L. J. (1974) The adult behaviour of *Cerambyx dux* Faldermann. *Ann. ent. Soc. Am.* **67**, 47–50.

SALITERNIK Z. (1942) The macroscopic differentiation of anopheline eggs according to their pattern on the surface of the water. *Bull. ent. Res.* **33**, 221.

SALITERNIK Z. (1957) *Anopheles sacharovi* Favr in Israel. *Riv. Parassit.* **18**, 248–265.

SALKELD E. H. (1972) The chorionic architecture of *Zelus exsanguis* (Hem., Reduviidae) (Het.). *Can. Ent.* **104**, 433–442.

SALKELD E. H. (1973) The chorionic architecture and shell structure of *Amathes c-nigrum* (Lepidoptera: Noctuidae). *Can. Ent.* **105**, 1–10.

SALKELD E. H. (1975) Biosystematics of the genus *Euxoa* (Lepidoptera: Noctuidae). IV. Eggs of the subgenus *Euxoa* Hbn. *Can. Ent.* **107**, 1137–1152.

SALKELD E. H. (1976) Biosystematics of the genus *Euxoa* (Lepidoptera: Noctuidae). VII. Eggs of the subgenera *Chorizagrotis, Crassivesica, Longivesica, Orosagrotis* and *Pleonectopoda. Can. Ent.* **108**, 1371–1395.

SALKELD E. H. and POTTER C. (1953) The effect of age and stage of development of insect eggs on their resistance to insecticides. *Bull. ent. Res.* **44**, 527–580.

SALKELD E. H. and WILKES A. (1968) The ultrastructure on insect surfaces by scanning electron microscopy. *Can. Ent.* **100**, 1–4.

SALT G. (1936) Experimental studies in insect parasitism. IV. The effects of superparasitism on populations of *Trichogramma evanescens. J. exp. Biol.* **13**, 363–375.

SALT G. (1955) Experimental studies in insect parasitism. VIII. Host reactions following artificial parasitization. *Proc. R. Soc. (B)* **144**, 380–398.

SALT G. (1956) Experimental studies in insect parasitism. IX. The reactions of a stick insect to an alien parasite. *Proc. R. Soc. (B)* **146**, 93–108.

SALT G. (1957) Experimental studies in insect parasitism. X. The reactions of some endopterygote insects to an alien parasite. *Proc. R. Soc. (B)* **147**, 167–184.

SALT G. (1960) Experimental studies in insect parasitism. XI. The haemocytic reaction of a caterpillar under varied conditions. *Proc. R. Soc. (B)* **151**, 446–467.

SALT G. (1961) The haemocytic reaction of insects to foreign bodies. *The Cell and the Organism*, pp. 175–192, Cambridge.

SALT G. (1963a) Experimental studies in insect parasitism. XII. The reactions of six

exopterygote insects to an alien parasite. *J. Insect Physiol.* **9**, 647–669.
SALT G. (1963b) The defense reactions of insects to metazoan parasites. *Parasitology* **53**, 527–642.
SALT G. (1964) The ichneumonid parasite *Nemeritis canescens* (Gravenhorst) in relation to the wax moth *Galleria mellonella* (L.). *Trans. R. Ent. Soc. Lond.* **116**, 1–14.
SALT G. (1965) Experimental studies in insect parasitism. XIII. The haemocytic reaction of a caterpillar to eggs of its habitual parasite. *Proc. R. Soc.* (B) **162**, 303–318.
SALT G. (1968) The resistance of insect parasitoids to the defence reactions of their hosts. *Biol. Rev.* **43**, 200–232.
SALT G. (1970) *The Cellular Defence Reactions of Insects*, Cambridge Monogr. Exp. Biol.
SALT G. (1975) The fate of an internal parasitoid *Nemeritis canescens* in a variety of insects. *Trans. R. ent. Soc. Lond.* **127**, 141–161.
SALT G. and BOSCH R. VAN DEN (1967) The defense reactions of three species of *Hypera* (Coleoptera, Curculionidae) to an ichneumon wasp. *J. Invert. Path.* **9**, 164–177.
SALT R. W. (1949) Water uptake in eggs of *Melanoplus bivittatus* (Say). *Can. J. Res.* (D) **27**, 236–242.
SALT R. W. (1952) Some aspects of moisture absorption and loss in eggs of *Melanoplus bivittatus* (Say). *Can. J. Zool.* **30**, 55–82.
SALZEN A. E. (1960) The growth of the locust embryo. *J. Embryol. exp. Morph.* **8**, 139–162.
SAMANIEGO A. and STERRINGA J. T. (1973) Estudios sobre el barrenador *Hypsipyla grandella* (Zeller) (Lepidoptera, Pyralidae). XXI. Un nuevo método para obtener oviposición en cautividad. *Turrialba* **23**, 367–370.
SANCHEZ F. F. and LAIGO F. M. (1968) Notes on the cacao tussock moth *Orgyia australis postica* Walker (Lymantridae, Lepidoptera). *Philipp. Ent.* **1**, 67–71.
SANDER K. and VOLLMAR H. (1967) Vital staining of insect eggs by incorporation of trypan blue. *Nature, Lond.* **216**, 174–175.
SANDERS W. (1962) The behaviour of the Mediterranean fruit-fly *Ceratitis capitata* Wied. during oviposition. *Z. Tierpsychol.* **19**, 1–28.
SANDERS W. (1969a) Die Eiablagehandlung der Mittelmeerfruchtfliege *Ceratitis capitata* Wied. Ihre Abhängigkeit von der Oberflachen- und Innenfeuchte der Fruchte. *Z. Tierpsychol.* **26**, 236–242.
SANDERS W. (1969b) Die Eiablagehandlung der Mittelmeerfruchtfliege *Ceratitis capitata* Wied. Der Einfluss der Zwischenflüge auf die Wahl des Eiablageortes. *Z. Tierpsychol.* **26**, 853–865.
SANDNER H. (1960) Untersuchungen über den Einfluss einiger biotischer Faktoren auf die Populationsentwicklung von *Calandra granaria* L. und *Sitophilus oryzae* L. *Ontogeny of Insects. Acta Symp. Evol. Praha 1959*, pp. 321–324.
SANDNER H. (1962) Studies on the rôle of the density factor and the manner of its action on populations of *A. obtectus*. *Ekol. pol.* (B) **8**, 179–186 (in Polish; English summary).
SANDNER H. and KOT J. (1962) Studies on the effect of the population density of some species of the genus *Trichogramma* on the formation of host–parasite relations. *Zesz. probl. Postep. Nauk roln.* **35**, 143–150 (in Polish; German summary).
SANDNER H. and PANKANIN M. (1973) Effect of the presence of food on egg laying by *Acanthoscelides obtectus* (Say) (Coleoptera, Bruchidae). *Polskie Pismo Ent.* **43**, 811–817.
SANFORD J. W. (1976) Sugarcane borer: seasonal distribution and fate of eggs deposited on sugarcane in Louisiana. *J. Ga Ent. Soc.* **11**, 332–334.
SANFORD K. H. (1964) Eggs and oviposition sites of some predacious Mirids on apple trees (Miridae: Hemiptera). *Can. Ent.* **96**, 1185–1189.
SANG J. H. (1949) The ecological determinants of population growth in a *Drosophila* culture. III. Larval and pupal survival. *Physiol. Zoöl.* **22**, 183–202.
SANG J. H. (1950) Population growth in *Drosophila* cultures. *Biol. Rev.* **25**, 188–219.

SANKARAN T. (1955) The natural enemies of *Ceroplastes pseudoceriferus* Green (Hemiptera, Coccidae). *J. sci. Res. Banaras Hindu Univ.* **5**, 100–119.

SANKARAN T., SRINATH D., and KRISHNA K. (1966) Studies on *Gesonula punctifrons* Stål (Orthoptera: Acrididae: Cyrtacanthacridinae) attacking water-hyacinth in India. *Entomophaga* **11**, 433–440.

SANTIS L. DE (1945) El bicho de cesto (*Oiketicus kirbyi* Guild.). *Boln Univ. nat. La Plata* **8**, 1–15.

SANTORO DE CROUZEL I. and SALAVIN R. G. (1943) Contribución al estudio de los *Neorhynchocephalus argentinos* (Diptera: Nemestrinidae). *An. Soc. cient. argent.* **136**, 145–177.

SAPUNOV V. B. and KAÏDANOV L. Z. (1976) The effect of transplantation of paragonia on the oviposition of females of *Drosophila melanogaster. Vest. leningr. Univ. Biologiya* no. 15, 138–143.

SARDESAI J. B. (1961) Effects of the density of the adult population on the oviposition of *Caryedon gonagra* F. *Garcia de Orta* **9**, 227–233 (in Portuguese; English summary).

SÁRINGER G. (1957) A repcedarázs (*Athalia rosae* L. (*colibri* Christ.) Tenthredinidae, Hym.). *Ann. Inst. Prot. Pl. hung.* **7** (1952–6), 125–183 (in Hungarian; German summary).

SARLET L. (1949–1968) Descriptions and figures of eggs of European Lepidoptera. *Lambillionea* **49**, 59–60, 62–67, 97–100, 108; **50**, 29–33; **53**, 26–32, 54–63, 71–75; **54**, 17–23, 45–49, 70–74, 86–89; **55**, 4–7, 59–64, 67–77, 94–97; **56**, 6–7, 35–36, 43–48, 68–76, 93–101; **57**, 106–109; **58**, 11–16, 31–34, 49–52, 67–69, 86–88, 95–96, 107–110; **59**, 15–20, 36, 40–42, 45–49, 63–67, 72–80, 96–98; **60**, 14–17, 36–38, 47–52, 82–85; **61**, 71–73, 97–98; **62**, 70–71; **63**, 12–14, 49–57; **65**, 13–16; **66**, 28–32, 45–48, 77–80, 99–102; **67**, 7–9.

SARLET L. (1964) Iconographie des oeufs de Lépidoptères belges (Rhopalocera, Heterocera). Deuxième partie. Heterocera: Bombycides—Sphingides. *Mém. Soc. r. ent. Belg.* **29**, 1–172.

SARLET L. G. (1967) Iconographie des oeufs de Lépidoptères belges. Deuxième partie. Heterocera, Bombycidae, Sphingidae. *Bull. ann. Soc. r. ent. Belg.* **130**, 294–301.

SARLET L. (1969) Iconographie des oeufs de Lépidoptères belges. 4. Geometridae (Suite). *Rev. verviét. Hist. nat.* **26**, 53–54, 66–92, 103–111.

SARLET L. G. (1970a) Iconographie des oeufs de Lépidoptères belges. 4. Geometridae (à suivre). *Rev. verviét. Hist. nat.* **27**, 23–32.

SARLET L. G. (1970b) Iconographie des oeufs de Lépidoptères belges. 4. Geometridae (à suivre). *Rev. verviét Hist. nat.* **27**, 72–80.

SARLET L. G. (1970c) Iconographie des oeufs de Lépidoptères belges. 4. Geometridae (à suivre). *Rev. verviét. Hist. nat.* **27**, 96–103.

SAUNDERS D. S. and DODD C. W. H. (1972) Mating, insemination, and ovulation in the tsetse fly *Glossina morsitans. J. Insect Physiol.* **18**, 187–197.

SAUNDERS R. C. and HSIAO T. H. (1970) Biology and laboratory propagation of *Amblymerus bruchophagi* (Hymenoptera: Pteromalidae), a parasite of the alfalfa seed chalcid. *Ann. ent. Soc. Am.* **63**, 744–749.

SAUNDERS S. S. (1881) Exhibit of *Callostoma fascipennis* and Bombyliidae. *Proc. ent. Soc. Lond.* **5**, xiv–xix.

SAUNT J. W. (1921) Oviposition of *Pachyrrhina crocata* L. *Entomologist's mon. Mag.* **57**, 186.

SAUPE R. (1928) Zur Kenntnis der Lebensweise der Riesenschabe *Blabera fusca* Brunner und der Gewachschausschabe *Pycnoscelus surinamensis* L. *Z. angew. Ent.* **14**, 461–500.

SAUSSURE H. DE (1872) *Mission scientifique au Mexique. Recherches zoologiques.* Pt. V. Orthoptères. Famille des Mantides, pp. 202–295.

SAVILOV A. I. (1967) Oceanic insects of the genus *Halobates* (Hemiptera, Gerridae) in the Pacific. *Okeanologia* **7**, 252–260 (in Russian).

SAXOD R. (1964) L'oeuf, l'éclosion, la cuticule embryonnaire, et la larve néonate de *Gyrinus substriatus* Steph. *Trav. Lab. Piscic. Univ. Grenoble* **56**, 17–28.

SBORSHCHIKOVA M. P. (1964) Pest of the shoots of rice. *Zashch. Rast. Vredit. Bolez.* 1964, pt. 5, 33–34 (in Russian).
SCHAEFER C. H. (1962) Life history of *Conophthorus radiatae* (Coleoptera: Scolytidae) and its principal parasite *Cephalonomia utahensis* (Hymenoptera: Bethylidae). *Ann. ent. Soc. Am.* **55**, 569–577.
SCHAEFER H. A. (1949) Biologische und ökologische Beobachtungen an Psylliden (Hemiptera). *Verh. naturf. Ges. Basle* **60**, 25–41.
SCHALLER F. (1956) Die Endosymbiose und Brutpflege der Erdwanze *Brachypelta aterrima* (Heteropt., Cydnidae). *Verh. dt. zool. Ges.* **1956**, 118–123.
SCHÄTZ W. (1954) Die Eier von *Orth. pistacina* F., *Iota* Cl. und *circellaris* Hufn. und ihre Ablage (Lep., Noct.). *NachrBl. Bayer Ent.* **3**, 1–3.
SCHEDL K. E. (1936) Der Schwammspinner (*Portheria dispar* L.) in Eurasien, Afrika und Neuengland. *Monogr. angew. Ent.* **12**, 1–242.
SCHEDL K. E. (1958) Breeding habits of arboricole insects in Central Africa. *Proc. 10th Int. Congr. Ent.* **1**, 183–197 (1956).
SCHEDL W. (1964) Biologie des gehöckerten Eichenbolzbohrers, *Xyleborus monographus* Fab. (Scolytidae, Coleoptera). *Z. angew. Ent.* **53**, 411–428.
SCHEIDTER F. (1923) Über einen bisher wenig beachteten Blattroller, *Rhynchites* (*Deporaus tristis* F.). *Z. angew. Ent.* **9**, 390–394.
SCHELTEMA R. S. (1968) Ocean insects. *Oceanus* **14**, 9–12.
SCHENK J. A. and BENJAMIN D. M. (1969) Notes on the biology of *Ips pini* in central Wisconsin jack pine forests. *Ann. ent. Soc. Am.* **62**, 480–485.
SCHERF H. (1956) Zum feineren Bau der Eigelege von *Galeruca tanaceti* L. (Coleopt., Chrysom.). *Zool. Anz.* **157**, 124–130.
SCHERF H. (1963) Ein Beitrag zur Kenntnis zweier Pinnipedierlause (*Antarctophthirius trichechi* Boheman und *Echinophthirius horridus* Olfers). *Z. ParasitKde* **23**, 16–44.
SCHERF H. (1966) Beobachtungen an Ei und Gelege von *Galeruca tanaceti* L. (Coleoptera, Chrysomelidae). *Biol. Zbl.* **85**, 7–17.
SCHIEFERDECKER H. (1968) Zur Biologie und Massenzucht der Getreidemotte *Sitotroga cerealella* Olivier. 1. Beitrag: Zur Eiablage von *Sitotroga cerealella* Olivier (Lepidoptera: Gelechiidae). *Beitr. Ent.* **18**, 329–345.
SCHILDER F. A. and SCHILDER M. (1928) Die Nahrung der Coccinelliden und ihre Beziehung zur Verwandtschaft der Arten. *Arb. biol. BundAnst. Land- u. Forstw.* **16**, 213–282.
SCHIÖDTE J. C. (1861–83) De metamorphosi eleutheratorum observationes. *Nat. Tidsskr.* (3) **1–13**.
SCHLAEGER D. A. and FUCHS M. S. (1974a) Effects of dopa-decarboxylase inhibition on *Aedes aegypti* eggs: evidence for sclerotization. *J. Insect Physiol.* **20**, 349–357.
SCHLAEGER D. A. and FUCHS M. S. (1974b) Dopa decarboxylase activity in *Aedes aegypti*: a preadult profile and its subsequent correlation with ovarian development. *Develop. Biol.* **38**, 209–219.
SCHLAEGER D. A. and FUCHS M. S. (1974c) Localization of dopa decarboxylase in adult *Aedes aegypti* females. *J. exp. Zool.* **187**, 217–221.
SCHLICK W. (1887) Yngleforhold hos *Sperchus emarginatus*. *Ent. Medd.* **1**, 26–27.
SCHLINGER E. I., BOSCH R. VAN DEN, and DIETRICK E. J. (1959) Biological notes on the predaceous earwig *Labidura riparia* (Pallas), a recent immigrant to California (Dermaptera: Labiduridae). *J. econ. Ent.* **52**, 247–249.
SCHLINGER E. I. and HALL J. C. (1961) The biology, behavior, and morphology of *Trioxys* (*Trioxys*) *utilis*, an internal parasite of the spotted alfalfa weevil aphid *Therioaphis maculata* (Hymenoptera: Braconidae, Aphidiinae). *Ann. ent. Soc. Am.* **54**, 34–45.
SCHMID J. M. (1969) *Laphria gilva* (Diptera: Asilidae), a predator of *Dendroctonus ponderosae* in

the Black Hills of South Dakota. *Ann. ent. Soc. Am.* **62**, 1237–1241.

SCHMID J. M. (1970) *Medetera aldrichii* (Diptera: Dolichopodidae) in the Black Hills. I. Emergence and behavior of adults. *Can. Ent.* **102**, 705–713.

SCHMID J. M. (1971) *Medetera aldrichii* (Diptera: Dolichopodidae) in the Black Hills. II. Biology and densities of the immature stages. *Can. ent.* **103**, 848–853.

SCHMIDT C. T. (1935) Biological studies on the nitidulid beetles found in pineapple fields. *Ann. ent. Soc. Am.* **28**, 475–511.

SCHMIDT G. T. (1974) Host-acceptance behavior of *Campoletis sonorensis* toward *Heliothis zea*. *Ann. ent. Soc. Am.* **67**, 835–844.

SCHMIEGE D. C. (1965) The fecundity of the black-headed budworm *Acleris variana* (Fern.) (Lepidoptera: Tortricidae) in coastal Alaska. *Can. Ent.* **97**, 1226–1230.

SCHMITZ G. (1958) *Helopeltis* du contonnier en Afrique centrale. *Publs Inst. natn. Étude agron. Congo belge (Sci.)* **71**, 1–178.

SCHMITZ H. (1930) Phoriden aus Eipaketen von *Locusta migratoria* in Daghestan. *Natuurh. Maandbl.* **19**, 67–69.

SCHNEIDER F. (1941) Entwicklung und Eiresorption in den Ovarien des Puppenparasite *Brachymeria euploeae* West. (Chalcididae). *Z. angew. Ent.* **28**, 211–288.

SCHNEIDER F. (1950) Die Abwehrreaktion des Insektenblutes und ihre Beeinflussung durch die Parasiten. *Vjschr. naturf. Ges. Zürich* **95**, 22–44.

SCHNEIDER H. (1955) Vergleichende Untersuchungen über Parthenogenese und Entwicklungsrythmen bei einheimischen Psocopteren. *Biol. Zbl.* **74**, 273–310.

SCHNEIDER I. and FARRIER M. H. (1969) New hosts, distribution and biological notes on an imported ambrosia beetle *Xylosandrus germanus* (Coleoptera: Scolytidae). *Can. Ent.* **101**, 412–415.

SCHNEIDER I. and RUDINSKY J. A. (1969) The site of pheromone production in *Trypodendron lineatum* (Coleoptera: Scolytidae): bio-assay and histological studies of the hindgut. *Can. Ent.* **101**, 1181–1186.

SCHÖNHERR J. (1970) Evidence of an aggregating pheromone in the ash-bark beetle *Leperisinus fraxini* (Coleoptera: Scolytidae). *Contr. Boyce Thompson Inst.* **24**, 305–307.

SCHÖNHERR J. (1972) Pheromon beim Kiefern-Borkenkäfer "Waldgärtner", *Myelophilus piniperda* L. (Coleopt., Scolytidae). *Z. angew. Ent.* **71**, 410–413.

SCHORR H. (1957) Zur Verhaltensbiologie und Symbiose von *Brachypelta aterrima* Först (Cydnidae, Heteroptera). *Z. Morph. Ökol. Tiere* **45**, 561–602.

SCHOUTEDEN H. (1903) La sollicitude maternelle chez les Hemiptères. *Rev. Univ. Brux.* **8**, 771–777.

SCHRADER F. (1926) Notes on the English and American races of the greenhouse whitefly (*Trialeurodes vaporarium*). *Ann. appl. Biol.* **13**, 183–195.

SCHREMMER F. (1959) Freilandbeobachtungen zur Eiablage von *Mantispa pagana* Fbr. (Neuroptera, Planipennia). *Z. Morph. Ökol. Tiere* **48**, 412–423.

SCHROEDER W. J. (1969) Stimulation of mating and oviposition of hickory shuckworm moths by pecan nuts. *J. ecọn. Ent.* **62**, 1244–1245.

SCHUBERT W. (1927) Biologische Untersuchungen über die Rubenblattwanze *Piesma quadrata* Fieb., im schlesischen Befallgebiet. *Z. angew. Ent.* **13**, 128–155.

SCHUCH K. (1938) Über den Einfluss der Feuchtigkeit auf das Eistadium des Maikäfers (*Melolontha melolontha* L.). *Arb. physiol. angew. Ent.* **5**, 220–225.

SCHUDER D. L. (1967) A technique for recovering sod webworm eggs from sod. *Proc. N. cent. Brch Am. Ass. econ. Ent.* **21**, 34 (1966).

SCHULTZ V. G. M. (1935) Lepidopterologische Beiträge. Einige Fälle von oophagen und chrysalidophagen Kannibalismus bei Groszschmetterlingsraupen. *Int. ent. Z.* **28**, 501–504.

SCHUMACHER F. (1917) Brutpflege bei der Wanze *Clinocoris griseus* L. *Ent. Mitt.* **6**, 243–249.

Schurr K. and Holdaway F. G. (1966) Periodicity in oviposition of *Ostrinia nubilalis* (Hbn) (Lepidoptera: Pyraustidae). *Ohio J. Sci.* **66**, 76–80.

Schurr K. and Holdaway F. G. (1970) Olfactory responses of female *Ostrinia nubilalis* (Lepidoptera: Pyraustinae). *Entomologia exp. appl.* **13**, 455–461.

Schütte F. (1957) Untersuchungen über die Populationsdynamik des Eichenwicklers (*Tortrix viridana* L.). Teil I. *Z. angew. Ent.* **40**, 1–36.

Schvester D. (1957a) Contribution à l'étude écologique des Coléoptères Scolytides. Essai d'analyse des facteurs de fluctuation des populations chez *Ruguloscolytus rugulosus* Muller 1818. *Annls Épiphyt.* **8** (num. hors seríe), 1–162.

Schvester D. (1957b) Sur le comportement de ponte et le développement de la bruche (*Acanthoscelides obsoletus* Say) chez diverses variétés de haricots en stock. *Rev. Zool. agric. Path. vég.* **56**, 120–125.

Schwarz E. (1923) The reason why *Catacola* eggs are occasionally deposited on plants upon which the larva cannot survive; and a new variation (Lepid., Noctuidae). *Ent. News* **34**, 272–273.

Scott A. W. (1863) Description of an ovo-viparous moth belonging to the genus *Tinea*. *Trans. ent. Soc. NSW* **1**, 33–36.

Scott E. I. (1939) An account of the developmental stages of some aphidophagous Syrphidae (Dipt.) and their parasites (Нуменорt.). *Ann. appl. Biol.* **26**, 509–532.

Scott M. T. S. (1952) Observations on the bionomics of the sheep body louse *Damalina ovis*. *Aust. J. agric. Res.* **3**, 60–67.

Scudder G. G. E. (1957) The systematic position of *Dicranocephalus* Hahn. 1826 and its allies (Hemiptera: Heteroptera). *Proc. R. ent. Soc. Lond.* (A) **32**, 147–158.

Scudder G. G. E. (1961) The comparative morphology of the insect ovipositor. *Trans. R. ent. Soc. Lond.* **113**, 25–40.

Sechriest R. E. and Treece R. E. (1963) The biology of the lesser clover leaf weevil *Hypera nigrirostris* (Fab.) (Coleoptera: Curculionidae) in Ohio. *Res. Bull. Ohio agric. Exp. Stn* **956**, 1–28.

Seddiqi P. M. (1972) Studies on longevity, oviposition, fecundity and development of *Callosobruchus chinensis* L. (Coleoptera: Bruchidae). *Z. angew. Ent.* **72**, 66–72.

Séguy E. (1944) Insectes ectoparasites (Mallophaga, Anoploures, Siphonaptera). *Faune Fr.* **43**, 1–684.

Séguy E. (1946) Trois chloropides nouveaux parasites d'oothèques d'Orthoptères. *Encycl. ent.* (B) **10**, 5–7.

Séguy E. and Baudot E. (1922) Note sur les premiers états du *Bombylius fugax* Wied. *Bull. Soc. ent. Fr.* **1922**, 139–141.

Sein F. Jr (1923) Cucarachas. *Circ. P. Rico agric. Exp. Stn Insular Stn Rio Piedras* **64**, 1–12.

Seitz A. (1889) Ueber Schmetterlingseier. *Zool. Jb. (Syst.)* **4**, 485–492.

Sekido S. and Sogawa K. (1976) Effects of salicylic acid on probing and oviposition of the rice plant and leafhoppers (Homoptera: Deltocephalidae). *Appl. Ent. Zool.* **11**, 75–81.

Sekul A. A., Sparks A. N., Beroza M., and Bierl B. A. (1975) A natural inhibitor of the corn earworm moth sex attractant. *J. econ. Ent.* **68**, 603–604.

Selander J., Kangas E., Perttunen V., and Oksanen H. (1973) Olfactory responses of *Hylobius abietis* L. (Col., Curculionidae) to substances naturally present in pine phloem or their synthetic counterparts. *Ann. ent. fenn.* **39**, 40–45.

Selhime A. G. (1956) *Brumus suturalis*, a beneficial lady beetle. *Fla Ent.* **39**, 65–68.

Senevet G. (1933) A propos des races d'*Anopheles maculipennis*. *Archs Inst. Pasteur Algér.* **11**, 12–14.

Senevet G. and Andarelli L. (1955) Races et variétés de l'*Anopheles claviger* Meigen, 1804. *Archs Inst. Pasteur Algér.* **33**, 128–137.

SENEVET G. and ANDARELLI L. (1956) Les *Anopheles* de l'Afrique du Nord et du Bassin Méditerranéen. *Encycl. ent.* (A) **33**, 1–280.

SENGALEWITSCH G. (1966) Schädliche Cossidae an Obst- und Forstgehölzen und ihre Bekämpfung in Bulgarien. *Beitr. Ent.* **16**, 693–706.

SENGEL P. and BULLIÈRE D. (1966) Ponte naturelle et provoquée chez *Blabera craniifer* Burm. (Insectes Dictyoptères). *C. r. hebd. Séanc. Acad. Sci. Paris* **262**, 1286–1288.

SENGUPTA G. C. and BEHURA B. K. (1957) On the biology of *Lema praeusta* Fab. *J. econ. Ent.* **50**, 471–474.

SEPULVEDA R. (1955) Biologia del *Mecistorhinus tripterus* F. (Hem., Pentatomidae) y su possible influencia en la transmissión de la moniliasis de cacao. *Cacao Colomb.* **4**, 15–42.

SERGEEVA T. K. (1974) Serological detection of predators of *Neodiprion sertifer* (Hymenoptera, Diprionidae). *Zool. Zh.* **53**, 710–719 (in Russian; English summary).

SERGENT E. (1936) Note sur les oeufs d'*Anopheles* d'eaux saumâtres du Littoral Algérien. *Archs Inst. Pasteur Algér.* **14**, 109–118.

SERGENT E. (1937) Oeufs d'*Anopheles maculipennis* de France et d'Algérie. *Archs Inst. Pasteur Algér.* **15**, 212–216.

SERGENT E. (1938) Détail anatomique des oeufs d'*Anopheles maculipennis*. *Archs Inst. Pasteur Algér.* **16**, 318–319.

SERGENT E. (1939) Sur l'oeuf d'*Anopheles maculipennis melanoon* Hackett du littoral algérois. *Archs Inst. Pasteur Algér.* **17**, 59–61.

SERGENT E. and TRENSZ F. (1933) Note préliminaire sur la morphologie des oeufs d'*Anopheles maculipennis* de France et d'Algérie. *Archs Inst. Pasteur Algér.* **11**, 9–11.

SERRE P. (1904) Observation biologique. *Bull. Soc. ent. Fr.* **1904**, 254.

SERVICE M. W. (1960) Effect of blood fractions upon oviposition in *Culex (Culex) fatigans* Wiedemann (Dipt., Culicidae). *Entomologist's mon. Mag.* **96**, 98–100.

SERVICE M. W. (1968) A method for extracting mosquito eggs from soil samples taken from oviposition sites. *Ann. trop. Med. Parasit.* **62**, 478–480.

SERVICE M. W. (1973a) Mortalities of the larvae of the *Anopheles gambiae* Giles complex and detection of predators by the precipitin test. *Bull. ent. Res.* **62**, 359–369.

SERVICE M. W. (1973b) Study of the natural predators of *Aedes cantans* (Meigen) using the precipitin test. *J. med. Ent.* **10**, 503–510.

SERVICE M. W. and LYLE P. T. W. (1975) Detection of the predators of *Simulium damnosum* by the precipitin test. *Ann. trop. Med. Parasit.* **69**, 105–108.

SESHAGIRI RAO D. (1954) Notes on rice moth *Corcyra cephalonica* (Stainton) (family: Galleriidae; Lepidoptera). *Indian J. Ent.* **16**, 95–114.

SETTY L. R. (1940) Biology and morphology of some North American Bittacidae (order Mecoptera). *Am. Midl. Nat.* **23**, 257–353.

SÉVERIN H. C. (1946) Grasshoppers ovipositing in a pile of coal. *Bull. Brooklyn ent. Soc.* **40**, 158–159 (1945).

SÉVERIN H. H. P. (1910) A study of the structure of the egg of the walking-stick *Diapheromera femorata* Say; and the biological significance of the resemblance of phasmid eggs to seeds. *Ann. ent. Soc. Am.* **3**, 83–92.

SÉVERIN H. H. P. and SÉVERIN H. C. (1910) The effect of moisture and dryness on the emergence from the egg of the walking-stick *Diapheromera femorata* Say. *J. econ. Ent.* **3**, 479.

SÉVERIN H. H. P., SÉVERIN H. C., and HARTUNG W. J. (1916) The stimuli which cause the eggs of leaf-ovipositing Tachinidae to hatch. *Psyche, Camb.* **22**, 132–137.

SHAAYA E. and BODENSTEIN D. (1969) The function of the accessory sex glands in *Periplaneta americana* (L.). II. The rôle of the juvenile hormone in the synthesis of protein and protocatechuic acid glucoside. *J. exp. Zool.* **170**, 281–292.

Shaaya E. and Sekeris C. E. (1970) The formation of protocatechuic acid-4-0,β-glucoside in *Periplaneta americana* and the possible rôle of the juvenile hormone. *J. Insect Physiol.* **16**, 323–330.

Shade R. E., Hansen H. L., and Wilson M. C. (1970) A partial life table of the cereal leaf beetle *Oulema melanopus* in northern Indiana. *Ann. ent. Soc. Am.* **63**, 52–59.

Shade R. E. and Wilson M. C. (1969) Oviposition of the cereal leaf beetle. *Proc. N. cent. Brch Am. Ass. econ. Ent.* **24**, 22.

Shagov E. M. (1968) The effect of temperature on the predacious bug *Perillus bioculatus* Fabr. (Heteroptera, Pentatomidae). *Zool. Zh.* **47**, 563–570 (in Russian; English summary).

Shands W. A., Holmes R. L., and Simpson G. W. (1970) Improved laboratory production of eggs of *Coccinella septempunctata*. *J. econ. Ent.* **63**, 315–317.

Shands W. A. and Simpson G. W. (1972) Insect predators for controlling aphids on potatoes. 4. Spatial distribution of introduced eggs of two species of predators in small fields. *J. econ. Ent.* **65**, 805–809.

Shands W. A., Simpson G. W., and Gordon C. C. (1972) Insect predators for controlling aphids on potatoes. 5. Numbers of eggs and schedules for introducing them in large field cages. *J. econ. Ent.* **65**, 810–817.

Shannon R. C. and Putman P. (1934) The biology of *Stegomyia* under laboratory conditions. I. The analysis of factors which influence larval development. *Proc. ent. Soc. Wash.* **36**, 185–216.

Shapinskiĭ D. V. (1923a) Biological observations on the Acrididae of the Chelyabinsk district, Orenburg, in 1916. *Bull. Soc. ent. Moscou* **2** (2), xxvi–xxix (in Russian).

Shapinskiĭ D. V. (1923b) Notes. Orthoptérologiques. 1. Sur la biologie du *Chrysochraon dispar* Germ. 2. Nouveaux parasites des acridiens. *Bull. Soc. ent. Moscou* **2**, 57–69 (in Russian).

Shapiro I. D. and Vilkova N. A. (1963) On the oviposition sites of *Oscinella frit* L. (Diptera, Chloropidae). *Rev. Ent. URSS* **42**, 138–150 (in Russian).

Shapiro V. A. (1956) The principal parasites of *Porthetria dispar* L. and the prospects of using them. *Zool. Zh.* **35**, 251–265 (in Russian).

Shapiro V. A. (1976) *Apanteles*—a parasite of the cabbage white butterfly. *Zashch. Rast.* no. 10, 17–18.

Shapiro V. A., Topolovskii V. A., and Makarenko G. N. (1971) Freezing preservation of eggs of the bug *Graphosoma lineatum* L. and their suitability to the mass rearing of *Trissolcus grandis* Thoms., an oophage of *Eurygaster integriceps* Put. *Bull. vses. nauchno-issled. Inst. Zashch. Rast.* **18**, 21–26 (in Russian; English summary).

Shapovalov A. A. (1956) Biology and ecology of *Aegeria apiformis* Clerk. in forest shelter belts. *Zool. Zh.* **35**, 583–587 (in Russian; English summary).

Sharan R. K. and Sahni S. L. (1960) Provisional embryonic cuticles of *Dysdercus cingulatus* (Hem.; Pyrrchoridae). *Ann. ent. Soc. Am.* **53**, 538–541.

Sharifi S. (1972) Oviposition site and egg plus staining as related to development of two species of *Sitophilus* in wheat kernels. *Z. angew. Ent.* **71**, 428–431.

Sharifi S. and Zarea N. (1970) Biology of the citrus butterfly *Papilio demoleus demoleus* (Lepidoptera: Papilionidae). *Ann. ent. Soc. Am.* **63**, 1211–1213.

Sharma M. L. and Laviolette R. (1968) Comportement de ponte chez le puceron du pin rouge *Schizolachnus piniradiatae* (Davidson) (Aphididae: Homoptera). *Phytoprotection* **49**, 87–89.

Sharma V. K. and Chatterji S. M. (1972) Preferential oviposition and antibiosis in different maize germplasms against *Chilozonellus* (Swin.) under cage conditions. *Indian J. Ent.* **33**, 299–311 (1971).

Sharp D. (1898) Account of the Phasmidae with notes on the eggs. In Willey A. (1898), *Zoological Results Based on Material Collected in New Britain, New Guinea, Loyalty Islands*

and Elsewhere, Part I, Cambridge.
SHAW E. (1925) New genera and species (mostly Australian) of Blattidae, with notes, and some remarks on Tepper's types. *Proc. Linn. Soc. NSW* **1**, 171–213.
SHAW J. G. and STARR D. F. (1946) Development of the immature stages of *Anastrepa serpentina* in relation to temperature. *J. agric. Res.* **72**, 265–276.
SHAW J. T., ELLIS R. O., and LUCKMANN W. H. (1976) Apparatus and procedure for extracting corn rootworm eggs from soil. *Biol. Notes, nat. Hist. Surv. Div. State Ill.* **96**.
SHAW M. W. (1971) Egg-laying by some dipterous pests of cultivated Cruciferae in north-east Scotland. *J. appl. Ecol.* **8**, 353–365.
SHAW M. W. (1972) The separation of eggs of *Erioischia brassicae* and *E. floralis* in swede crops. *Plant Path.* **21**, 10–15.
SHELFORD R. (1901) Notes on some Bornean insects. *Rep. Br. Ass.* (D), 689–691.
SHELFORD R. (1906) Studies of the Blattidae. VI. Viviparity amongst the Blattidae. *Trans. ent. Soc. Lond.* **1906**, 509–514.
SHELFORD R. (1907) The larva of *Collyris emarginatus* Dej. *Trans. ent. Soc. Lond.* **1907**, 83–90.
SHELFORD R. (1909) Two remarkable forms of mantid oothecae. *Trans. ent. Soc. Lond.* **1909**, 509–514.
SHELFORD R. (1912) Oothecae of Blattidae. *Ent. Rec.* **24**, 283–287.
SHELFORD R. (1916) *A Naturalist in Borneo*, London.
SHELFORD V. E. (1909) Life-histories and larval habits of the tiger beetles (Cicindelidae). *J. Linn. Soc. (Zool.)* **30**, 157–184.
SHEPARD M., WADDILL V., and KLOFT W. (1973) Biology of the predaceous earwig *Labidura riparia* (Dermaptera: Labiduridae). *Ann. ent. Soc. Am.* **66**, 837–841.
SHEPHERD R. F. (1966) Factors influencing the orientation and rates of activity of *Dendroctonus ponderosae* Hopkins (Coleoptera: Scolytidae). *Can. Ent.* **98**, 507–518.
SHEPHERD R. F. and BROWN C. E. (1971) Sequential egg-band sampling and probability methods of predicting defoliation by *Malacosoma disstria* (Lasiocampidae: Lepidoptera). *Can. Ent.* **103**, 1371–1379.
SHEPHERD R. F. and GRAY T. G. (1972) Solution separation and maximum likelihood density estimates of hemlock looper (Lepidoptera: Geometridae) eggs in moss. *Can. Ent.* **104**, 751–754.
SHERMAN M. and TAMASHIRO M. (1956) Biology and control of *Araecerus levipennis* Jordan (Coleoptera: Anthribidae). *Proc. Hawaii ent. Soc.* **16**, 138–148.
SHERWOOD R. C. and POND D. D. (1954) A simple method of rearing *Hylemyia brassicae* (Bouché) (Diptera: Anthomyiidae). *Can. Ent.* **86**, 178–179.
SHIELDS K. S. (1976) The development of *Blepharipa pratensis* and its histopathological effects on the gypsy moth *Lymantria dispar*. *Ann. ent. Soc. Am.* **69**, 667–670.
SHIGA M. and NAKANISHI A. (1968) Variation in the sex ratio of *Gregopimpla himalayensis* Cameron (Hymenoptera: Ichneumonidae) parasitic on *Malacosoma neustria testacea* Motschulsky (Lepidoptera: Lasiocampidae), with considerations on the mechanism. *Kontyû, Tokyo* **36**, 369–376.
SHIMIZU J. T. and HAGEN K. S. (1967) An artificial ovipositional site for some Heteroptera that insert their eggs into plant tissue. *Ann. ent. Soc. Am.* **60**, 1115–1116.
SHIPP E. (1964) Rates of development in eggs from three populations of *Didymuria violescens* (Leach) (Phasmatodea). *Proc. Linn. Soc. NSW* **88**, 287–294 (1963).
SHOREY H. H. (1964) The biology of *Trichoplusia ni* (Lepidoptera: Noctuidae). III. Response to the oviposition substrate. *Ann. ent. Soc. Am.* **57**, 165–170.
SHOREY H. H., ANDRES L. A., and HALE R. L. (1962) The biology of *Trichoplusia ni* (Lepidoptera: Noctuidae). I. Life history and behavior. *Ann. ent. Soc. Am.* **55**, 591–597.
SHOWERS W. B., REED G. L., and OLOUMI-SADEGHI H. (1974) Mating studies of female

European corn borers: relationship between deposition of egg masses on corn and captures in light traps. *J. econ. Ent.* **67**, 616–619.

SHRIVER D. and BICKLEY W. E. (1964) The effect of temperature of hatching of eggs of the mosquito *Culex pipiens quinquefasciatus* Say. *Mosq. News* **24**, 137–140.

SHTEĬNBERG D. M. (1961) Host–parasite relations in entomophagous insects. The possibility of the development of larvae of certain parasitic Hymenoptera in hosts new for them. *Dokl. Akad. Nauk SSSR* **138**, 1477–1480 (in Russian).

SHULOV A. (1952a) The development of eggs of *Schistocerca gregaria* (Forskål) in relation to water. *Bull. ent. Res.* **43**, 469–476.

SHULOV A. (1952b) Observations on the behaviour and the egg development of *Tmethis pulchripennis asiaticus* Uv. *Bull. Res. Coun. Israel* **2**, 249–254.

SHULOV A. S. (1970) The development of eggs of the red locust *Nomadacris septemfasciata* (Serv.) and the African migratory locust *Locusta migratoria migratorioides* (R. & F.), and its interruption under particular conditions of humidity. *Anti-Locust Bull.* **48**, 1–22.

SHULOV A. and PENER M. P. (1959) A contribution to knowledge of the development of the egg of *Locusta migratoria migratorioides* (R. & F.). *Locusta* **6**, 73–88.

SHULOV A. and PENER M. P. (1963) Studies on the development of eggs of the desert locust *Schistocerca gregaria* Forskål and its interruption under particular conditions of humidity. *Anti-Locust Bull.* **41**, 1–59.

SHUTE P. G. (1936) A study of laboratory-bred *Anopheles maculipennis* var. *atroparvus*, with special reference to egg laying. *Ann. trop. Med. Parasit.* **30**, 11–16.

SIDDIQI Z. A., SHRI RAM, and ROHATGI K. K. (1966) Studies on life and seasonal history of stalk borer of sugarcane *Chilo auricilia* Dudg. in Bihar. *Indian J. Ent.* **27**, 442–449 (1965).

SIEBOLD C. T. E. VON (1837) Fernere Beobachtungen über die Spermatozoen der wirbellosen Tiere. *Arch. Anat. Physiol. wiss. Med.* **3**, 381–439.

SIJAZOV M. (1913) Contribution à la biologie des coprophages (Coléoptèra: Scarabaeidae). *Russk. ent. Obozr.* **13**, 113–131.

SIKES E. K. and WIGGLESWORTH V. B. (1931) The hatching of insects from the egg and the appearance of air in the tracheal system. *Q. Jl microsc. Sci.* **74**, 165–192.

SIKKA S. M., SAHNI V. M., and BUTANI D. K. (1966) Studies on jassid resistance in relation to hairiness of cotton leaves. *Euphytica* **15**, 383–388.

SIKORA H. (1915) Beiträge zur Biologie von *Pediculus vestimenti*. *Zentbl. Bakt.* **76** (1) 523–537.

SILFVENIUS A. J. [SILTALA A. J.] (1906) Trichopterologische Untersuchungen. I. Über den Laich der Trichopteren. *Acta Soc. Fauna Flora fenn.* **28** (4) 1–128.

SILTALA A. J. (1907) Zusätze zu meinem Aufsatze über den Laich der Trichopteren. *Arch. Hydrobiol.* **2**, 527–533.

SILVA G. DE M. (1973) The use of parafilm as an ovipositioning device. A contribution to the mass rearing of the olive fly *Dacus oleae* Gmel. *Agron. Lusitana* **35**, 85–90.

SILVER G. T. (1959) A method for sampling eggs of the black-headed budworm *Acleris variana* (Fern.) on *Tsuga heterophylla* in British Columbia. *J. For.* **57**, 203–205.

SILVERSTEIN R. M., BROWNLEE R. G., BELLAS T. E., WOOD D. L., and BROWNE L. (1968) Brevicomin: principal sex attractant in the frass of the female western pine beetle. *Science, Wash.* **159**, 889–891.

SILVERSTEIN R. M., RODIN J. O., and WOOD D. L. (1966) Sex attractants in frass produced by male *Ips confusus* in ponderosa pine. *Science, Wash.* **154**, 509–510.

SILVESTRI F. (1918) Descrizione e notizie biologiche di alcuni *Imenotteri* Chalcididi parassiti di uova di cicale. *Boll. Lab. Zool. gen. agric. Portici* **12**, 252–265.

SILVESTRI F. (1919) Contribuzioni alla conoscenza degli insetti dannosi e dei loro simbionti. IV. La cocciniglia del nocciuola (*Eulecanium coryli* L.). *Boll. Lab. Zool. gen. agric. Portici* **13**, 70–192.

SILVESTRI F. (1920) Contribuzione alla conoscenza dei parassiti delle ova del grilletto canterino (*Oecanthus pellucens* Scop.). *Boll. Lab. Zool. gen. agric. Portici* **14**, 219–250.

SILVESTRI F. (1921) Contribuzioni alla conoscenza biologica degli *Imenotteri* parassiti. V. Sviluppo del *Platygaster dryomyiae* Silv. (Fam. Proctotrupidae). *Boll. Lab. Zool. gen. agric. Portici* **11**, 299–326 (1916).

SILVESTRI F. (1943) *Compendio di entomologia applicata (agraria-forestale-medica-veterinaria).* Portici. Parte speciale **2**, 1–512.

SILVESTRI F. (1946) Descrizione di due specie neotropicali di *Zorotypus* (Ins., Zoraptera). *Boll. Lab. Ent. agr. Portici* **7**, 1–12.

SILVESTRI F., MARTELLI G., and MASI L. (1908) Sugli *Imenotteri* parassiti ectofagi della mosca delle olivo fino ad ora osservati nell'Italia meridionale e sulla loro importanza del combattere la mosca stessa. *Boll. Lab. Zool. gen. Agric. Portici* **2**, 18–82.

SIMANTON F. L. (1916) The terrapin scale, an important insect enemy of peach orchards. *Bull. US Dep. Agric.* **35**, 1–96.

SIMMONDS F. J. (1956a) Superparasitism by *Spalangia drosophilae* Ashm. *Bull. ent. Res.* **47**, 361–376.

SIMMONDS F. J. (1956b) An investigation of the possibilities of biological control of *Melittomma insulare* Fairm. (Coleoptera, Lymexylonidae), a serious pest of coconut in the Seychelles. *Bull. ent. Res.* **47**, 685–702.

SIMMONDS H. W. (1958) The housefly problem in Fifi and Samoa. *S. Pacif. Com. Q. Bull.* **8**, 29–30.

SIMMONS P. and ELLINGTON G. W. (1925) The ham beetle *Necrobia rufipes* de Geer. *J. agric. Res.* **30**, 845–863.

SIMMONS P., REED W. D., and MCGREGOR E. A. (1931) Fig insects in California. *Circ. US Dep. Agric.* **157**, 1–71.

SINADSKIĬ YU. V. and SULAĬMANOV KH. A. (1976) The tamarisk cicada. *Zashch. Rast.* no. 5, 46–47.

SINGH B. (1952) Observation on the biology of the cricket *Gymnogryllus humeralis* Walker (Insecta, Orthoptera, Gryllidae) in the Dera Dun Insectary. *J. zool. Soc. India* **4**, 47–60.

SINGH H. N. and BYAS (1973) Host selection in the tobacco-caterpillar *Spodoptera litura* (Fabricius) (Lepidoptera: Noctuidae). *Indian J. agric. Sci.* **43**, 357–360.

SINGH K., AGRAWAL N. S., and GIRISH G. K. (1974) The oviposition and development of *Sitophilus oryzae* (L.) in different high-yielding varieties of wheat. *J. stored Prod. Res.* **10**, 105–111.

SINGH K. R. P. and BROWN A. W. A. (1957) Nutritional requirements of *Aedes aegypti* L. *J. Insect Physiol.* **1**, 199–220.

SINGH O. P. and TEOTIA T. P. S. (1970) A simple method of mass culturing melon fruit-fly *Dacus cucurbitae* Coquillet. *Indian J. Ent.* **32**, 28–31.

SINGH R. P. (1969) A simple technique for rendering host eggs inviable for the laboratory rearing of *Trichogramma* spp. *Indian J. Ent.* **31**, 83–84.

SINGH S. R. and SODERSTROM E. L. (1963) Sexual maturity of the rice weevil *Sitophilus oryzae* (L.) as indicated by sperm transfer and viable eggs. *J. Kans. ent. Soc.* **36**, 32–34.

SINGH Z., WHITE C. E., and LUCKMANN W. H. (1973) Notes on *Amyotea malabarica*, a predator of *Nezara viridula* in India. *J. econ. Ent.* **66**, 551–552.

SIVERLY R. E. (1975) Blood meal identifications of *Culex pipiens pipiens* (northern house mosquito) collected during 1972 and 1973 in Delaware and Henry Counties, Indiana. (Abstract.) *Proc. Indiana Acad. Sci.* **84**, 284 (1974).

SKUF'IN K. V. (1954) The ecology of *Chrysops relictus* Mg., Tabanidae, Diptera. Note 2. Ecology of oviposition. *Zool. Zh.* **33**, 1289–1292 (in Russian).

SLATER F. W. (1899) The egg carrying habit of *Zaitha*. *Am. Nat.* **33**, 931–933.

SLATER J. A. (1976) The immature stages of Lygaeidae (Hemiptera: Heteroptera) of southwest Australia. *J. Aus. ent. Soc.* **15**, 101–126.

SLIFER E. H. (1937) The origin and fate of the membranes surrounding the grasshopper egg together with some experiments on the source of the hatching enzyme. *Q. Jl microsc. Sci.* **79**, 493–506.

SLIFER E. H. (1938) The formation and structure of a special water-absorbing area in the membranes covering the grasshopper egg. *Q. Jl microsc. Sci.* **80**, 437–458.

SLIFER E. H. (1946) The effects of xylol and other solvents on diapause in the grasshopper egg, together with a possible explanation for the action of these agents. *J. exp. Zool.* **102**, 333–356.

SLIFER E. H. (1949) Changes in certain of the grasshopper egg coverings during development as indicated by fast green and other dyes. *J. exp. Zool.* **110**, 183–203.

SLIFER E. H. (1950) A microscopical study of the hydropyle and hydropyle cells in the developing egg of the grasshopper *Melanoplus differentialis*. *J. Morph.* **87**, 239–274.

SLIFER E. H. (1958) Diapause in the eggs of *Melanoplus differentialis* (Orthoptera, Acrididae). *J. exp. Zool.* **138**, 259–282.

SLIFER E. H. and SEKHON S. S. (1963) The fine structure of the membranes which cover the egg of the grasshopper *Melanoplus differentialis*, with special reference to the hydropyle. *Q. Jl microsc. Sci.* **104**, 321–334.

SLOBODCHIKOFF C. N. (1973) Behavioural studies of three morphotypes of *Therion circumflexum* (Hymenoptera: Ichneumonidae). *Pan-Pacif. Ent.* **49**, 197–206.

SMEE C. (1937) Report of the entomologist. *Rep. Agric. Nysald.* **1936**, 20–24.

SMEREKA E. P. (1965) The life history and habits of *Chrysomela crotchi* Brown (Coleoptera: Chrysomelidae) in northwestern Ontario. *Can. Ent.* **97**, 541–549.

SMIRNOFF W. (1953) *Chrysopa vulgaris* Schneider prédateur important de *Parlatoria blanchardi* Targ. dans les palmeraies de l'Afrique du Nord (Planip., Chrysopidae). *Bull. Soc. ent. Fr.* **58**, 146–152.

SMIT B. (1934) A study of the most important insect pests that confront the *Citrus* grower in the Eastern Province. *S. Afr. J. Sci.* **31**, 439–441.

SMIT C. J. B. (1939) Field observations on the brown locust in an outbreak centre. *Sci. Bull. Dep. Agric. S. Afr.* **190**, 1–143.

SMITH B. C., STARRATT A. N., and BODNARYK R. P. (1973) Oviposition responses of *Coleomegilla maculata lengi* (Coleoptera: Coccinellidae) to the wood and extracts of *Juniperus virginiana* and to various chemicals. *Ann. ent. Soc. Am.* **66**, 452–456.

SMITH B. C. and WILLIAMS R. R. (1976) Temperature relations of adult *Coleomegilla maculata lengi* and *C. m. medialis* (Coleoptera: Coccinellidae) and responses to ovipositional stimulants. *Can. Ent.* **108**, 925–930.

SMITH C. N. (ed.) (1966) *Insect Colonization and Mass Production*, Academic Press, London.

SMITH D. S. (1959) Note on destruction of grasshopper eggs by the field cricket *Acheta assimilis luctuosus* (Serville) (Orthoptera: Gryllidae). *Can. Ent.* **91**, 127.

SMITH D. S. (1966) Fecundity and oviposition in the grasshoppers *Melanoplus sanguinipes* (F.) and *Melanoplus bivittatus* (Say). *Can. Ent.* **98**, 617–621.

SMITH D. S. (1969) Oviposition, development of eggs, and chromosome complement of eggs of virgin females of the grasshopper *Melanoplus sanguinipes*. *Can. Ent.* **101**, 23–27.

SMITH D. S. (1970) Crowding in grasshoppers. I. Effect of crowding within one generation on *Melanoplus sanguinipes*. *Ann. ent. Soc. Am.* **63**, 1775–1776.

SMITH D. S., TELFER W. H., and NEVILLE A. C. (1971) Fine structure of the chorion of a moth *Hyalophora cecropia*. *Tissue & Cell* **3**, 477–498.

SMITH E. H. (1965) Laboratory rearing of the peach tree borer and notes on its biology. *J. econ. Ent.* **58**, 228–236.

SMITH F. F., WEBB R. E., BOSWELL A. L., and COMBS G. F. (1970) A circular rotating cage for obtaining uniform oviposition by *Liriomyza munda* in exposed plants. *J. econ. Ent.* **63**, 655–656.
SMITH G. J. C. (1969) Host selection and oviposition behavior of *Nasonia vitripennis* (Hymenoptera: Pteromalidae) on two host species. *Can. Ent.* **101**, 533–538.
SMITH H. S. and ARMITAGE H. M. (1931) The biological control of mealybugs attacking *Citrus*. *Bull. Calif. agric. Exp. Stn* **509**, 1–74.
SMITH H. S. and COMPERE H. (1928) A preliminary report on the insect parasites of the black scale, *Saissetia oleae* (Bern.). *Univ. Calif. Publ. Ent.* **4**, 231–334.
SMITH H. S. and COMPERE H. (1931) Notes on *Ophelosia crawfordi*. *J. econ. Ent.* **24**, 1109–1110.
SMITH K. G. V. (1955) Notes on the egg and first instar larva of *Volucella bombylans* L. (Dipt., Syrphidae). *Entomologist's mon. Mag.* **91**, 52–54.
SMITH K. G. V. (1966) The larva of *Thecophora occidensis*, with comments upon the biology of Conopidae (Diptera). *J. Zool. Lond.* **149**, 263–276.
SMITH K. G. V. (1967) The biology and taxonomy of the genus *Stylogaster* Macquart, 1935 (Diptera: Conopidae, Stylogasterinae) in the Ethiopian and Malagasy regions. *Trans. R. ent. Soc. Lond.* **119**, 47–69.
SMITH L. B. (1962) Observations on the oviposition rate of the rusty grain beetle *Cryptolestes ferrugineus* (Steph.) (Coleoptera: Cucujidae). *Ann. ent. Soc. Am.* **55**, 77–82.
SMITH L. M. (1930) *Macrorileya oecanthi* Ashm., a hymenopterous egg parasite of the tree crickets. *Univ. Calif. Publ. Ent.* **5**, 165–172.
SMITH L. M. (1932) The shot hole borer. *Circ. Cal. agric. Ext. Serv.* **64**, 1–13.
SMITH L. W. JR (1969) The relationship of mosquitoes to oxidation lagoons in Columbia, Missouri. *Mosq. News* **29**, 556–563.
SMITH O. J., DUNN P. H., and ROSENBERGER J. H. (1955) Morphology and biology of *Sturmia harrisinae* Coquillett (Diptera), a parasite of the western grape leaf skeletonizer. *Univ. Calif. Publ. Ent.* **10**, 321–357.
SMITH R. C. (1921) A study of the biology of the Chrysopidae. *Ann. ent. Soc. Am.* **14**, 27–35.
SMITH R. C. (1922a) Hatching in three species of Neuroptera. *Ann. ent. Soc. Am.* **15**, 169–176.
SMITH R. C. (1922b) The biology of the Chrysopidae. *Mem. Cornell Univ. agric. Exp. Stn* **58**, 1291–1372.
SMITH R. C. (1923) The life histories and stages of some hemerobiids and allied species (Neuroptera). *Ann. ent. Soc. Am.* **16**, 129–151.
SMITH R. C. (1934) Notes on the Neuroptera and Mecoptera of Kansas, with keys for the identification of species. *J. Kans. ent. Soc.* **7**, 120–145.
SMITH R. H. (1956) The rearing of *Lyctus planicollis* and the preparation of wood for control tests. *J. econ. Ent.* **49**, 127–129.
SMITH W. W. and JONES D. W. JR (1972) Use of artificial pools for determining presence, abundance, and oviposition preferences of *Culex nigripalpus* Theobald in the field. *Mosq. News* **32**, 244–245.
SMITHERS C. N. (1965) A bibliography of the Psocoptera (Insecta). *Aust. Zool.* **13**, 137–209.
SMITTLE B. J., LOWE R. E., PATTERSON R. S., and CAMERON A. L. (1975) Winter survival and oviposition of ^{14}C-labeled *Culex pipiens quinquefasciatus* Say in northern Florida. *Mosq. News* **35**, 54–56.
SMITTLE B. J. and PATTERSON R. S. (1970) Transfer of radioactivity to eggs and larvae by female *Culex pipiens quinquefasciatus* Say treated as larvae with ^{32}P. *Mosq. News* **30**, 93–94.
SNAPP O. I. (1930) Life history and habits of the plum curculio in the Georgia peach belt. *Tech. Bull. US Dep. Agric.* **188**, 1–90.
SNODGRASS R. E. (1923) The fall webworm. *Smithson. Rept.* **1921**, 395–414.

SNOW W. F. (1971) The spectral sensitivity of *Aedes aegypti* (L.) at oviposition. *Bull. ent. Res.* **60**, 683–696.

SNYDER T. E. (1916) Egg and manner of oviposition of *Lyctus planicollis*. *J. agric. Res.* **6**, 273–276.

SOANS J. S. and SOANS A. B. (1972) Cannibalism in the coffee bean weevil *Araecerus fasciculatus* De Geer (Coleoptera, Anthribidae). *J. Bombay nat. Hist. Soc.* **69**, 210–211.

SOBOLEVA-DOKUCHAEVA I. I. and PODOPLELOV I. I. (1972) Antisera specific to soluble proteins in items of the diet of Carabidae. *Zool. Zh.* **51**, 280–286 (in Russian; English summary).

SÖFNER L. (1941) Zur Entwicklungsbiologie und Ökologie der einheimischen Psocopterenarten *Ectopsocus meridionalis* (Ribaga) 104, und *Ectopsocus briggsi* McLachlan 1899. *Zool. Jb. (Syst.)* **74**, 323–360.

SOKOLOFF A. (1955) Competition between sibling species of the *pseudoobscura* subgroup of Drosophila. *Ecol. Monogr.* **25**, 387–409.

SOKOLOFF A. and LERNER I. M. (1967) Laboratory ecology and mutual predation of *Tribolium* species. *Am. Nat.* **101**, 261–276.

SOL R. (1964) Zum Einfluss des Lichtes bei der Eiablage der Brachfliege (*Phorbia coarctata* Fall.). *Z. PflKrankh.* **71**, 177–179.

SOL R. (1971) Beitrag zur Frage der Eiablage der Brachfliege (*Phorbia coarctata* Fall.) auf Kulturflächen, künstlicher Brache und im Laboratorium sowie ihres Fluges im Freien. *Z. angew. Ent.* **67**, 397–411.

SOLAR E. DEL (1968) Selection for and against gregariousness in the choice of oviposition sites by *Drosophila pseudoobscura*. *Genetics* **58**, 275–282.

SOLAR E. DEL and PALOMINO H. (1966) Tendencia gregaria en *Drosophila melanogaster*. *Biológica, Santiago* **39**, 27–31.

SOLIMAN M. H. (1971) Selection of site of oviposition by *Drosophila melanogaster* and *D. simulans*. *Am. Midl. Nat.* **86**, 487–493.

SOLINAS M. (1965) Studi sui ditteri cecidomiidi. II. *Jaapiella medicaginis* Rübsaamen e *Anabremia inquilina* sp.n. *Entomologica* **1**, 211–281.

SOLOMON J. D. (1967) Carpenterworm oviposition. *J. econ. Ent.* **60**, 309.

SOLOMON J. D. and NEEL W. W. (1974) Fecundity and oviposition behavior in the carpenterworm *Prionoxystus robiniae*. *Ann. ent. Soc. Am.* **67**, 238–240.

SOMAN R. S. and REUBEN R. (1970) Studies on the preference shown by ovipositing females of *Aedes aegypti* for water containing immature stages of the same species. *J. med. Ent.* **7**, 485–489.

SOMCHOUDHURY A. K. and MUKHERJEE A. B. (1972) Bionomics of *Acaropsis docta* (Berlese) (Acarina: Cheyletidae), a predator on the eggs of some pests of stored grains. *Indian J. Ent.* **33**, 79–81 (1971).

SÖMME L. and RYGG T. (1972) The effect of physical and chemical stimuli on oviposition in *Hylemya floralis* (Fallén) (Dipt., Anthomyiidae). *Norsk ent. Tidsskr.* **19**, 19–24.

SOMMERMAN K. M. (1943) Description and bionomics of *Caecilius manteri* n. sp. (Corrodentia). *Proc. ent. Soc. Wash.* **45**, 29–39.

SONLEITNER F. J. (1961) Factors affecting egg cannibalism and fecundity in populations of adult *Tribolium castaneum* Herbst. *Physiol. Zoöl.* **34**, 233–255.

SOPER R. S., SHEWELL G. E., and TYRRELL D. (1976) *Colcondamyia auditrix* nov. sp. (Diptera: Sarcophagidae), a parasite which is attracted by the mating song of its host (*Okanagana rimosa*) (Homoptera: Cicadidae). *Can. Ent.* **108**, 61–68.

SOTO P. E. (1974) Ovipositional preference and antibiosis in relation to resistance to a sorghum shoot fly. *J. econ. Ent.* **67**, 265–267.

SOUTHWOOD T. R. E. (1956) The structure of the eggs of the terrestrial Heteroptera and its relationship to the classification of the group. *Trans. R. ent. Soc. Lond.* **108**, 163–221.

SOUTHWOOD T. R. E. and FEWKES D. W. (1961) The immature stages of the commoner British Nabidae (Heteroptera). *Trans. Soc. Br. Ent.* **14**, 147–166.

SOUTHWOOD T. R. E. and LESTON D. (1959) *Land and Water Bugs of the British Isles*, London.

SOUTHWOOD T. R. E. and SCUDDER G. G. E. (1956) The bionomics and immature stages of the thistle lace bugs (*Tingis ampliata* H.-S. and *T. cardui* L., Hem., Tingidae). *Trans. Soc. Br. Ent.* **12**, 93–112.

SPARKS M. R. (1970) A surrogate leaf for oviposition by the tobacco hornworm. *J. econ. Ent.* **63**, 537–540.

SPARKS M. R. (1973) Physical and chemical stimuli affecting oviposition preference of *Manduca sexta* (Lepidoptera: Sphingidae). *Ann. ent. Soc. Am.* **66**, 571–573.

SPARKS M. R. and CHEATHAM J. S. (1970) Responses of a laboratory strain of the tobacco hornworm *Manduca sexta* to artificial oviposition sites. *Ann. ent. Soc. Am.* **63**, 428–431.

SPENCER G. J. (1931) The oviposition habits of *Rhynchocephalus sackeni* Will. *Proc. ent. Soc. Br. Columb.* **28**, 21–24.

SPENCER G. J. (1932) Further notes on *Rhynchocephalus sackeni* Will. *Proc. ent. Soc. Br. Columb.* **29**, 25–27.

SPENCER G. J. (1958) On the Nemestrinidae of British Columbia dry range lands. *Proc. 10th Int. Congr. Ent.* **4**, 503–509 (1956).

SPEYER E. R. (1922) Shot-hole borer of tea: damage caused to the tea bush. *Bull. Dep. Agric. Ceylon* **60**, 1–16.

SPEYER E. R. (1923) Notes upon the habits of Ceylonese ambrosia-beetles. *Bull. ent. Res.* **14**, 11–23.

SPEYER W. (1929a) Der Apfelblattsauger *Psylla mali* Schmidberger. *Monogr. Pflanzenschutz.* **1**, 1–127.

SPEYER W. (1929b) Die Embryonalentwicklung und das Ausschlüpfen der Junglarven von *Psylla mali* Schm. *Z. wiss. InsektBiol.* **24**, 215–220.

SPIELMAN A., LEAHY M. G., and SKAFF V. (1967) Seminal loss in repeatedly mated female *Aedes aegypti*. *Biol. Bull. Woods Hole* **132**, 404–412.

SPILLER D. (1948) Effect of humidity on hatching of eggs of the common house borer *Anobium punctatum* De Geer. *NZ Jl Sci. Tech.* **30** (B) 163–165.

SPILLER D. (1960) Artificial egg-laying sites for the common house-borer *Anobium punctatum* De Geer. *NZ Ent.* **2** (5) 19–23.

SPILLER D. (1964) Numbers of eggs laid by *Anobium punctatum* (De Geer). *Bull. ent. Res.* **55**, 305–311.

SPITTLER H. (1969) Beiträge zur Morphologie, Biologie und Ökologie des Sattelmückenparasiten *Platygaster equestris* nov. spec. (Hymenoptera, Proctotrupoidea, Scelionidae) unter besonderer Berücksichtigung seines abundanzdynamischen Einflusses auf *Haplodiplosis equestris* Wagner (Diptera, Cecidomyiidae). Teile I–II. *Z. angew. Ent.* **63**, 353–381; **64**, 1–34.

SPOONER J. D. (1967) Bush katydid oviposition behaviour. *Am. Zool.* **7**, 798.

SPRADBERY J. P. (1968) A technique for artificially culturing parasites of woodwasps (Hymenoptera: Siricidae). *Entomologia exp. appl.* **11**, 257–260.

SPRADBERY J. P. (1969) The biology of *Pseudorhyssa sternata* Merrill (Hym., Ichneumonidae), a cleptoparasite of Siricid woodwasps. *Bull. ent. Res.* **59**, 291–297 (1968).

SPRADBERY J. P. (1970a) Host finding by *Rhyssa persuasoria* (L.), an Ichneumonid parasite of Siricid woodwasps. *Anim. Behav.* **18**, 103–114.

SPRADBERY J. P. (1970b) The biology of *Ibalia drewseni* Borries (Hymenoptera: Ibaliidae), a parasite of Siricid woodwasps. *Proc. R. ent. Soc. Lond.* (A) **45**, 104–113.

SPRADBERY J. P. (1973) The responses of *Ibalia* species (Hymenoptera: Ibaliidae) to the fungal symbionts of siricid woodwasp hosts. *J. Ent.* (A) **48** (2) 217–222.

SPRADBERY J. P. and SANDS D. P. A. (1976) Reproductive system and terminalia of the old world screw-worm fly *Chrysomya bezziana* Villeneuve (Diptera: Calliphoridae). *International J. insect Morph. Embryol.* **5**, 409–421.

SPRAGUE I. B. (1956) The biology and morphology of *Hydrometra martini* Kirkaldy. *Kans. Univ. Sci. Bull.* **38**, 579–693.

SPRINGETT B. P. (1968) Aspects of the relationship between burying beetles *Necrophorus* spp. and the mite *Poecilochirus necrophori* Vitz. *J. Anim. Ecol.* **37**, 417–424.

SRIVASTAVA A. S. and GUPTA B. P. (1967) Studies on the bionomics of dhaincha stem borer *Azygophleps scalaris* Fabr. *Labdev J. Sci. Technol.* **5**, 264–265.

SRIVASTAVA B. K. and BHATIA S. K. (1959) The effect of host species on the oviposition of *Callosobruchus chinensis* Linn. (Coleoptera, Bruchidae). *Ann. Zool. Agra* **3**, 37–42.

SRIVASTAVA P. D. (1956) Studies on the copulatory and egg-laying habits of *Hieroglyphus nigrorepletus* Bol. (Orthoptera, Acrididae). *Ann. ent. Soc. Am.* **49**, 167–170.

SRIVASTAVA R. P. and BOGAWAT J. K. (1969) Descriptions of the immature stages of a fruit-sucking moth *Othreis materna* (L.) (Lep., Noctuidae), with notes on its bionomics. *Bull. ent. Res.* **59**, 275–280 (1968).

SRIVASTAVA R. P. and PANDEY Y. D. (1968) Body weight of castor semi-looper *Achaea janata* Linn. (Lepidoptera: Noctuidae) in relation to its host plants. *Labdev J. Sci. Technol.* **6-B**, 56–57.

ŠROT M. (1962) Contribution to knowledge of the bionomics of the poplar borer *Saperda carcharias* L. *Pr. výzk. Úst. lesn.* **25**, 85–114 (in Czech.; English summary).

ŠROT M. (1963) Škodlivé rozšíreni nesytky ovádové (*Paranthrene tabaniformis* Rott.) a nesytky včelové (*Aegeria apiforme* Cl.) no topolech v ČSSR. *Lesn. čas.* **9**, 145–158 (in Czech.; German summary).

STADELBACHER E. A. (1969) Chambers for incubating individual insect eggs and egg parasites. *J. econ. Ent.* **62**, 253–254.

STADELBACHER E. A. and SCALES A. L. (1973) Technique for determining oviposition preference of the bollworm and tobacco budworm for varieties and experimental stocks of cotton. *J. econ. Ent.* **66**, 418–421.

STÄDLER E. (1974) Host plant stimuli affecting oviposition behavior of the eastern spruce budworm. *Entomologia exp. appl.* **17**, 176–188.

STÄDLER E. (1975) Täglicher Aktivitätsrhythmus der Eiablage bei der Möhrenfliege *Psila rosae* Fab. (Diptera: Psilidae). *Mitt. schweiz. ent. Ges.* **48**, 133–139.

STÄGER R. (1930) Einige Notizen über *Anechura bipunctata*. *Ent. Z.* **43**, 271–273.

STAHLER N. and SEELEY D. C. JR (1971) Effect of age and host on oviposition of *Anopheles stephensi* in the laboratory. *J. econ. Ent.* **64**, 561–562.

STANDIFER L. N., POOLE H. K., and DOULL K. M. (1971) Egg production, oviposition and survival of isolated queen honey bees fed experimental diets. *Ann. ent. Soc. Am.* **64**, 228–232.

STANILAND L. N. (1922) Hoverflies: their habits and economic importance. *Fruit Grow.* **53**, 143–144.

STANLEY J. (1942) A mathematical theory of the growth of populations of the flour beetle *Tribolium confusum* Duv. V. The relation between the limiting value of egg populations in the absence of hatching and the sex ratio of the group of adult beetles used in a culture. *Ecology* **23**, 24–31.

STANLEY J. (1964) Washing the eggs of *Tribolium* for gregarine control. *Can. J. Zool.* **42**, 920.

STARK B. P. and GAUFIN A. R. (1976) The nearctic genera of Perlidae (Plecoptera). *Misc. Publ. ent. Soc. Am.* **10**, 1–80.

STARK B. P. and SZCZYTKO S. W. (1976) The genus *Beloneuria* (Plecoptera: Perlidae). *Ann. ent. Soc. Am.* **69**, 1120–1124.

STARK R. W. and DAHLSTEN D. L. (1965) Notes on the distribution of eggs of a species in the *Neodiprion fulviceps* complex (Hymenoptera: Diprionidae). *Can. Ent.* **97**, 550–552.

STARRATT A. N. and OSGOOD C. E. (1972) An oviposition pheromone of the mosquito *Culex tarsalis*: diglyceride composition of the active fraction. *Biochim. biophys. Acta* **280**, 187–193.

STARRATT A. N. and OSGOOD C. E. (1973) 1,3-Diglycerides from eggs of *Culex pipiens quinquefasciatus* and *Culex pipiens pipiens*. *Comp. Biochem. Physiol.* **46B**, 857–859.

STAVRAKI H. G. (1976) Effects of diet and temperature on development, fecundity and longevity of a *Trichogramma* sp., parasite of olive moth (Praysoleae). *Z. angew. Ent.* **81**, 381–386.

STAVRAKI-PAULOPOULOU H. G. (1967) Contribution à l'étude de la capacité reproductrice et de la fécondité réelle d'*Opius concolor* Szepl. (Hymenoptera, Braconidae). *Annls Épiphyt.* **17**, 391–435 (1966).

STAY B. and GELPERIN A. (1966) Physiological basis of ovipositional behaviour in the false ovoviviparous cockroach *Pycnoscelus surinamensis* (L.). *J. Insect Physiol.* **12**, 1217–1226.

STAY B., KING A., and ROTH L. M. (1960) Calcium oxalate in the oothecae of cockroaches. *Ann. ent. Soc. Am.* **53**, 79–86.

STAY B. and ROTH L. M. (1962) The collateral glands of cockroaches. *Ann. ent. Soc. Am.* **55**, 124–130.

STEBBING E. P. (1910) A note on the lac insect (*Tachardia lacca*), its life history, propagation and collection. *Ind. For. Mem.* **1**, 1–82.

STEELE C. W. and SAWYER W. H. (1944) The brown tail moth. *J. Maine med. Ass.* **35**, 137.

STEELE R. W. (1970) Copulation and oviposition behaviour of *Ephestia cautella* (Walker) (Lepidoptera: Phycitidae). *J. stored Prod. Res.* **6**, 229–245.

STEER W. (1929) The eggs of some Hemiptera–Heteroptera. *Entomologist's mon. Mag.* **65**, 34–38.

STEFANI R. (1956) Il problema della parthenogenesi in *Haploembia soleri* Ramb. *Mem. Acad. Lincei* **5** (8) 127–200.

STEFANOV D. and KEREMIDCHIEV M. (1961) The possibility of using some predators and parasitic insects (entomophagous insects) in the biological control of the gypsy moth (*Lymantria dispar* L.) in Bulgaria. *Nauchni Trud. vissh. lesotekh. Inst.* **9**, 157–168 (in Bulgarian; German summary).

STEHR F. W. and COOK E. F. (1968) A revision of the genus *Malacosoma* Hübner in North America (Lepidoptera: Lasiocampidae): systematics, biology, immatures, and parasites. *Bull. US natn. Mus.* **276**, 1–321.

STEIN E. (1963) Zur Biologie von *Drosophila flava* Fall. (Diptera, Drosophilidae), einer Minierfliege an Kulturcruciferen. *Z. angew. Ent.* **52**, 39–56.

STEIN J. D. (1974) Unusual oviposition sites of *Paleacrita vernata*. *J. Kans. ent. Soc.* **47**, 483–485.

STEIN W. (1972) Der Einfluss unterschiedlicher Laboratoriumshaltung auf Eiablage und Lebensdauer von *Apion virens* Herbst (Col., Curculionidae). *Z. angew. Zool.* **59**, 353–360.

STEINER P. (1937) Hausbockuntersuchungen. I. Über den Einfluss von Temperatur und Feuchtigkeit auf das Eistadium und Bemerkungen zur Biologie der Imago. *Z. angew. Ent.* **23**, 531–546.

STEINHAUS E. A. (1946) *Insect Microbiology. An Account of the Microbes Associated with Insects and Ticks with Special Reference to the Biologic Relationships Involved,* Comstock Publ. Co. Inc., Ithaca, NY.

STEINHAUS E. A. (1949) *Principles of Insect Pathology,* New York and London.

STEINHAUS E. A. (1957) Microbial diseases of insects. *A. Rev. Microbiol.* **11**, 165–182.

STEINHAUSEN W. (1950) Vergleichende Morphologie, Biologie und Ökologie der Entwicklungstadien der in Niedersachsen heimischen Schildkäfer (Cassidinae Chrysomelidae

Coleoptera) und deren Bedeutung für die Landwirtschaft. Diss. Tech. Hochch. Carolo-Wilh. Braunschweig, pp. 1–69.
STELLWAG F. (1928) *Die Weinbauinsekten der Kulturländer*, Berlin.
STENSETH C. (1970) Studies on aphids of plum in Norway (Hemiptera: Homoptera: Aphididae). *Meld. Norg. LandbrHöisk.* **49**, 1–21.
STENSETH C. (1971) Morphology and life-cycle of *Ampullosiphon stachydis* Heikinheimo (Hemiptera: Homoptera: Aphididae). *Norsk ent. Tidsskr.* **18**, 9–13.
STEPANOV P. T. (1881) Metamorphosis of Diptera, family Bombyliidae. *Trudȳ Obshch. Ispyt. Prir. imp. Khar'kov* **15**, 9 pp. (in Russian).
STEPHEN F. M. and DAHLSTEN D. L. (1976) The arrival sequence of the arthropod complex following attack by *Dendroctonus brevicomis* (Coleoptera: Scolytidae) in ponderosa pine. *Can. Ent.* **108**, 283–304.
STEPHENS C. S. (1962) *Oiketicus kirbyi* (Lepidoptera: Psychidae) a pest of bananas in Costa Rica. *J. econ. Ent.* **55**, 381–386.
STEPHENS G. S. (1976) Transportation and culture of the egg predator *Tytthus mundulus* (Breddin). *Ent. Newsl. S. Afr.* no. 3.
STEPHENS S. G. (1959) Laboratory studies of feeding and oviposition preferences of *Anthonomus grandis* Boh. *J. econ. Ent.* **52**, 390–396.
STEPHENS S. G. and LEE HONG SUK (1961) Further studies on the feeding and oviposition preferences of the boll weevil (*Anthonomus grandis*). *J. econ. Ent.* **54**, 1085–1090.
STERN V. M. and SMITH R. F. (1960) Factors affecting egg production and oviposition in populations of *Colias philodice eurytheme* Boisduval (Lepidoptera: Pieridae). *Hilgardia* **29**, 411–454.
STERNLICHT M. (1973) Parasitic wasps attracted by the sex pheromone of their Coccid host. *Entomophaga* **18**, 339–342.
STEVANOVIC D. (1961) Ecology and population dynamics of the *Aeropus sibiricus* L. on Kopaonik. *Posebna Izdanja, Belgrade* **8**, 1–87.
STEVENSON J. H. (1955) Onderzoek naar de wijze, waarop de koolzaadgalmug (*Dasyneura brassicae* Winn.) haar eieren legt op koolzaad (*Brassica napus* L.). *Tijdschr. PlZiekt.* **61**, 81–87.
STEWART K. W. and WALTON R. R. (1964) Oviposition and establishment of the southwestern corn borer on corn. *J. econ. Ent.* **57**, 628–631.
STINNER R. E. (1976) Ovipositional response of *Venturia canescens* (Grav.) (Hymenoptera: Ichneumonidae) to various host and parasite densities. *Res. Popul. Ecol.* **18**, 57–73.
STITZ H. (1900) Der Genitalapparat der Mikrolepidopteren. *Zool. Jb. (Anat.)* **14**, 135–176.
STITZ H. (1931) Planipennia. *Biologie Tiere Dtl.* **35**, 67–304.
STOCKARD C. R. (1908) Habits, reactions, and mating instincts of the "walking stick" *Aplopus mayeri*. *Publ. Carnegie Inst.* **2**, 43–59.
STOCKEL J. (1968) Premiers résultats sur la mise en évidence d'un caractère de résistance du maïs aux infestations de *Sitotroga cerealella* Oliv. en culture. *Rev. Zool. Agric. appl.* **67**, 14–20.
STOCKEL J. (1969) Influence de la présence de grains de maïs comme stimulus de la ponte de *Sitotroga cerealella* Oliv. (Lép., Gelechiidae). *C. r. hebd. Séanc. Acad. Sci. Paris* (D) **268**, 2941–2943.
STOCKEL J. (1973) The functioning of the reproductive system in the female of *Sitotroga cerealella* (Lep., Gelechiidae). *Annls Soc. Ent. Fr.* **9**, 627–645.
STOCKEL J. and TURTAUT P. (1972) Utilisation d'un "oviposimètre" pour l'étude du rythme de ponte de *Sitotroga cerealella* Oliv. *Rev. Comp. anim.* **6**, 59–66.
STONE M. W. (1965) Biology and control of the lima-bean pod borer in southern California. *Tech. Bull. US Dep. Agric.* **1321**, 1–46.
STONE M. W. and HOWLAND A. F. (1944) Life history of the wireworm *Melanotus longulus* (Lec.) in Southern California. *Tech. Bull. US Dep. Agric.* **858**, 1–30.

STONE W. S. and REYNOLDS F. H. K. (1939) Hibernation of anopheline eggs in the tropics. *Science, NY* **90**, 371–372.

STOWER W. J., POPOV G. B., and GREATHEAD D. J. (1958) Oviposition behaviour and egg mortality of the desert locust *Schistocerca gregaria* Forskål on the coast of Eritrea. *Anti-Locust Bull.* **30**, 1–33.

STRAATMAN R. (1962) Notes on certain Lepidoptera ovipositing on plants which are toxic to their larvae. *J. Lepid. Soc.* **16**, 99–103.

STRASSEN R. (1957) Zur Ökologie des *Velleius dilatatus* Fabricius, eines als Raumgast bei *Vespa crabro* Linnaeus lebenden Staphyliniden (Ins. Co.). *Z. Morph. Ökol. Tiere* **46**, 244–292.

STRAUS-DURCKHEIM H. (1828) *Considérations générales sur l'anatomie comparée des animaux articules*, Paris.

STRAWINSKI D. (1935) Historja naturalna Korowea sosnowgo *Aradus cinnamomeus* Pnz. (Hemiptera–Heteroptera). *Roczn. Nauk rol.* **13**, 644–693 (1925).

STREAMS F. A. (1968a) Factors affecting the susceptibility of *Pseudeucoila bochei* eggs to encapsulation by *Drosophila melanogaster. J. Invert. Path.* **12**, 379–387.

STREAMS F. A. (1968b) Defense reactions of *Drosophila* species (Diptera: Drosophilidae) to the parasite *Pseudeucoila bochei* (Hymenoptera: Cynipidae). *Ann. ent. Soc. Am.* **61**, 158–164.

STREAMS F. A. (1971) Encapsulation of insect parasites in superparasitized hosts. *Entomologia exp. appl.* **14**, 484–490.

STREAMS F. A. and GREENBERG L. (1969) Inhibition of the defense reaction of *Drosophila melanogaster* parasitised simultaneously by the wasps *Pseudeucoila bochei* and *Pseudeucoila mellipes. J. Invert. Path.* **13**, 371–377.

STRETTON G. B. (1943) Some observations on the leaf-rolling habits of *Byctiscus populi* L. (Col., Curculionidae). *Entomologist's mon. Mag.* **79**, 252–255.

STRICKLAND E. H. (1923) Biological notes on parasites of prairie cutworms. *Tech. Bull. Dep. Agric. Can. Stat.* **22**, 1–40.

STRIDE G. O. (1953) On the nutrition of *Carpophilus hemipterus* L. (Coleoptera: Nitidulidae). *Trans. R. ent. Soc. Lond.* **104**, 171–194.

STRIDE G. O. and STRAATMAN R. (1963) On the biology of *Mecas saturnina* and *Nupserha antennata*, cerambycid beetles associated with *Xanthium* species. *Aust. J. Zool.* **11**, 446–469.

STRIDE G. O. and WARWICK E. P. (1960) On two species of *Epilachna* (Coleoptera: Coccinellidae) from Australia. *Proc. Linn. Soc. NSW* **85**, 208–214.

STRIDE G. O. and WARWICK E. P. (1962) Ovipositional girdling by a North American Cerambycid beetle, *Mecas saturnina. Anim. Behav.* **10**, 112–117.

STRIEBEL H. (1960) Zur Embryonalentwicklung der Termiten. *Acta trop.* **17**, 193–260.

STRONG F. E. and SHELDAHL J. A. (1970) The influence of temperature on longevity and fecundity in the bug *Lygus hesperus* (Hemiptera: Miridae). *Ann. ent. Soc. Am.* **63**, 1509–1515.

STRONG F. E., SHELDAHL J. A., HUGHES P. R., and HUSSEIN E. M. K. (1970) Reproductive biology of *Lygus hesperus* Knight. *Hilgardia* **40**, 105–147.

STRÜBING H. (1956) Über Beziehungen zwischen Oviduct, Eiablage, und natürlicher Verwandtschaft einheimischer Delphaciden. *Zool. Beitr.* **2**, 331–357.

STRÜBING H. (1957) Die Oviducktrüsen der Delphaciden (Hom. Auchenorrhyncha) und ihre Bedeutung für die Eiablage. *Verh. dt. zool. Ges.* **1956**, 361–366.

STUARDO C. (1935) Algunas observaciones sobre las costumbres y metamorfosis de *Hirmoneura articulata* Ph. *Revta chil. Hist. nat.* **1934**, 197–202.

STUART A. M. (1957) *Ephialtes brevicornis* (Grav.) as an external parasite of the diamond-back moth *Plutella maculipennis* (Curt.). *Bull. ent. Res.* **48**, 477–488.

STUBBS A. E. (1970) Observations on the oviposition of *Leptarthus brevirostris* (Mg.) (Diptera: Asilidae). *Entomologist* **103**, 289–293.

STUCKENBERG B. R. (1960) Diptera (Brachycera): Rhagionidae. In *South African Animal Life: Results of the Lund University Expedition in 1950–51* **7**, 216–308.
STUCKENBERG B. R. (1963) A study on the biology of the genus *Stylogaster*, with the description of a new species from Madagascar (Diptera, Conopidae). *Rev. Zool. Bot. afr.* **28**, 251–275.
STULTZ H. T. (1955) The influence of spray programs on the fauna of apple orchards in Nova Scotia. VIII. Natural enemies of the eye-spotted bud moth *Spilonota ocellana* (D. & S.) (Lepidoptera: Olethreutidae). *Can. Ent.* **87**, 79–85.
STURM H. (1961) Brutpflege bei dem Käfer *Omaspides specularis* Er. (Cassidinae, Chrysomelidae). *Zool. Anz.* **166**, 8–26.
ŠTUSÁK J. M. (1957) Beitrag zur Kenntnis der Eier der Tingiden (Heteroptera: Tingidae). *Beitr. Ent.* **7**, 20–28.
ŠTUSÁK J. M. (1958) Zweiter Beitrag zur Kenntnis der Eier der Tingiden (Hemiptera–Heteroptera, Tingidae). *Acta Soc. ent. Čsl.* **55**, 361–371.
ŠTUSÁK J. M. (1961) Dritter Beitrag zur Kenntnis der Eier der Tingiden (Heteroptera, Tingidae). *Acta Soc. ent. Čsl.* **58**, 71–88 (Czech summary).
ŠTUSÁK J. M. (1962) Immature stages of *Elasmotropis testacea* (H.–S.) and notes on the bionomics of the species (Heteroptera: Tingidae). *Acta Soc. ent. Čsl.* **59**, 19–27.
SUBBA RAO B. R. and GOPINATH K. (1961) The effects of temperature and humidity on the reproductive potential of *Apanteles angaleti* Muesebeck (Braconidae: Hymenoptera). *Entomologia exp. appl.* **4**, 119–122.
SUBRA R. (1971) Études écologiques sur *Culex pipiens fatigans* Wiedemann, 1828 (Diptera, Culicidae) dans une zone urbaine de savane soudanienne ouest-africaine. Rhythme de ponte et facteurs conditionnant l'oviposition. *Cahiers ORSTOM (Ent. med. Parasit.)* **9**, 317–332.
SUBRAMANIAM V. K. (1955) Control of the fluted scale in peninsular India. *Indian J. Ent.* **16**, 391–415 (1954); **17**, 103–120.
SUBRAMANIAN T. R. (1957) Life-history of the *Calotropis* or "Ak" weevil *Paramecops farinosa* Wiedemann. *Indian J. Ent.* **19**, 31–36.
SUBRAMANIAN T. R. (1958) Description and life-history of a new weevil of the genus *Protocylas* from Coimbatore. *Indian J. Ent.* **19**, 204–213 (1957).
SUBRAMANIAN T. R. (1959a) The biology of the weevil *Alcidodes bubo* (Fabricius) (Coleoptera: Curculionidae). *J. Bombay nat. Hist. Soc.* **56**, 82–94.
SUBRAMANIAN T. R. (1959b) Biology of *Alcidodes affaber* Aurivillius. *Indian J. agric. Sci.* **29**, 81–89.
SU-FANG M. (1964) Studies on the different types of *Anopheles (A.) hyrcanus sinensis* within the *Anopheles (A.) hyrcanus* groups. I. Comparative morphological study of eggs. *Acta ent. sin.* **13** (6) 862–871.
SUGIMOTO T. (1966) Relation of decrease of water-surface level and survival of egg mass of *Tendipes dorsalis* Meigen. *Sci. Rept. Kyoto Pref. Univ. (Agric.)* **18**, 63–69 (in Japanese; English summary).
SULLIVAN C. R. (1960) The effect of physical factors on the activity and development of adults and larvae of the white pine weevil *Pissodes strobi* (Peck). *Can. Ent.* **92**, 732–745.
SULLIVAN C. R. (1961) The effect of weather and the physical attributes of white pine leaders on the behaviour and survival of the white pine weevil *Pissodes strobi* Peck in mixed stands. *Can. Ent.* **93**, 721–741.
SUMAROKA A. F. (1967) Factors affecting the sex ratio in *Aphytis proclia* Wlk. (Hymenoptera, Aphelinidae) an external parasite of the San José scale. *Ent. Obozr.* **46**, 299–310 (English translation: *Ent. Rev.* **46**, 179–185).
SUMIMOTO M., KONDO T., and KAMIYAMA Y. (1974) Attractants for the scolytid beetle *Cryphalus fulvus*. *J. Insect Physiol.* **20**, 2071–2077.

Sumimoto M., Suzuki T., Shiraga M., and Kondo T. (1975) Further attractants for the scolytid beetle *Taenoglyptes fulvus*. *J. Insect Physiol.* **21**, 1803–1806.

Sundby R. A. (1968) Some factors influencing the reproduction and longevity of *Coccinella septempunctata* Linnaeus (Coleoptera: Coccinellidae). *Entomophaga* **13**, 197–202.

Surtees G. (1958) Laboratory studies on the survival of the eggs of *Aedes (Stegomyia) aegypti* under adverse conditions. *W. Afr. med. J.* (NS) **7**, 52–53.

Surtees G. (1964) Observations on some effects of temperature and isolation on fecundity of female weevils *Sitophilus granarius* (L.) (Coleoptera, Curculionidae). *Entomologia exp. appl.* **7**, 249–252.

Suzuki K. and Hirata S. (1924) Genetic studies on the adhesive character of the silkworm egg. *Jap. J. Zool.* **1**, 113.

Svec H. J. (1964) Laboratory rearing of the tomato hornworm *Protoparce quinquemaculata* (Haworth). *Can. J. Zool.* **42**, 717.

Svensson S. A. (1966) Studier över vissa vinteraktiva inselters biologi. *Norsk. ent. Tidsskr.* **13**, 335–338.

Swaby J. A. and Rudinsky J. A. (1976) Acoustic and olfactory behaviour of *Ips pini* (Say) (Coleoptera: Scolytidae) during host invasion and colonisation. *Z. angew. Ent.* **81**, 421–432.

Swadener S. O. and Yonke T. R. (1973) Immature stages and biology of *Apiomerus crassipes* (Hemiptera: Reduviidae). *Ann. ent. Soc. Am.* **66**, 188–196.

Swailes G. E. (1961) Laboratory studies on mating and oviposition of *Hylemya brassicae* (Bouché) (Diptera: Anthomyiidae). *Can. Ent.* **93**, 940–943.

Swailes G. E. (1967) A substrate for oviposition by the cabbage maggot in laboratory cultures. *J. econ. Ent.* **60**, 619–620.

Swailes G. E. (1971) Reproductive behaviour and effects of the male accessory gland substance in the cabbage maggot *Hylemya brassicae*. *Ann. ent. Soc. Am.* **64**, 176–179.

Swain R. B., Green W., and Portman R. (1938) Notes on oviposition and sex ratio in *Hyposoter pilosulus* Prov. (Hym., Ichneumonidae). *J. Kans. ent. Soc.* **11**, 7–9.

Swaine J. M. (1918) Canadian bark beetles. Part II. A preliminary classification with an account of the habits and means of control. *Tech. Bull. Dep. Agric. Can. (Ent. Br.)* **14**, 1–143.

Sweeney P. R., Church N. S., Rempel J. G., and Gerrity R. G. (1968) The embryology of *Lytta viridana* Le Conte (Coleoptera: Meloidae). III. The structure of the chorion and micropyles. *Can. J. zool.* **46**, 213–217.

Sweet M. H. (1964) The biology and ecology of the Rhyparochrominae of New England (Heteroptera: Lygaeidae). Part 1. *Ent. Am.* **43**, 1–124.

Sweet W. C. and Rao B. A. (1937) Races of *A. stephensi* Liston, 1901. *Indian Med. Gaz.* **72**, 1–12.

Sweet W. C. and Rao B. A. (1938) Measurements of *A. culicifacies* ova. *J. Malar. Inst. India* **1**, 33–35.

Sweetman H. L. (1938) Physical ecology of the firebrat *Thermobia domestica* (Packard). *Ecol. Monogr.* **8**, 285–311.

Sweetman H. L. (1958) *The Principles of Biological Control*, Iowa.

Swynnerton C. F. M. (1915a) Experiments on some carnivorous insects, especially the driver ant *Dorylus*; and with butterflies' eggs as prey. *Trans. ent. Soc. Lond.* **1915**, 317–350.

Swynnerton C. F. M. (1915b) Further notes on the eggs of butterflies. *Trans. ent. Soc. Lond.* **1915**, 428–430.

Sychevskaya V. I. (1974) The biology of *Eucoila trichopsila* Hartig (Hymenoptera, Cynipoidea), parasitoid of larvae of synanthropic flies of the family Sarcophagidae (Diptera). *Ent. Obozr.* **53**, 54–65 (in Russian; English summary).

Syed R. A. (1971) Studies on trypetids and their natural enemies in West Pakistan. V. *Dacus*

(Strumeta) cucurbitae Coquillett. *Tech. Bull. Common. Inst. biol. Control* **14**, 63–75.
SYME P. D. (1969) Interaction between *Pristomerus* sp. and *Orgilus obscurator*, two parasites of the European pine shoot moth. *Bi-Mon. Res. Notes* **25** (4) 30–31.
SYMMONS P. and CARNEGIE A. J. M. (1959) Some factors affecting breeding and oviposition of the red locust *Nomadacris septemfasciata* (Serv.). *Bull. ent. Res.* **50**, 333–353.
SYMS E. E. (1933) Notes on British Mecoptera. *Proc. S. Lond. nat. Hist. Soc.* **1933–34**, 84–88.
SZENTESI Á. (1972) Studies on the mass rearing of *Acanthoscelides obtectus* Say (Col., Bruchidae). *Acta phytopath. acad. Sci. hung.* **7**, 453–463.
SZENTESI Á. (1976) Effect of oviposition stimuli and subsequent matings on the viability of eggs of *Acanthoscelides obtectus* Say (Coleoptera, Bruchidae). *Acta phytopath. acad. Sci. hung.* **10**, 417–424 (1975).
SZMIDT A. (1960) Beiträge zur Biologie von *Dirhicnus alboannulatus* (Ratz.) (Hym., Chalcididae) als Grundlage einer Massenzucht für die biologische Bekämpfung. *Entomophaga* **5**, 155–163.
SZUJECKI A. (1966) Notes on the appearance and biology of eggs of several Staphylinidae (Coleoptera) species. *Bull. Acad. Polon. Sci.* (Cl. II) **14**, 169–175.
SZUMKOWSKI W. (1953) Observations on Coccinellidae. I. Coccinellids as predators of lepidopterous eggs and larvae in Venezuela. *Trans. 9th Int. Congr. Ent.* **1**, 778–781.

TAFT H. M., HOPKINS A. R., and JAMES W. (1963) Differences in reproductive potential, feeding rate, and longevity of boll weevils mated in the fall and in the fall and spring. *J. econ. Ent.* **56**, 180–181.
TAHER EL-S. M. and ABD EL-RAHMAN H. A. (1960) On the biology and life history of the pink bollworm *Pectinophora gossypiella* (Saunders) (Lepidoptera: Gelechiidae). *Bull. Soc. ent. Égypte* **44**, 71–90.
TAKADA M. and KITANO H. (1971) Studies on the larval haemocytes in the cabbage white butterfly *Pieris rapae crucivora* Boisduval, with special reference to haemocyte classification, phagocytic activity and encapsulation capacity. *Kontyû, Tokyo* **39**, 385–394 (in Japanese; English summary).
TAKAHASHI F. (1956) On the effect of population density on the power of reproduction of the almond moth *Ephestia cautella*. I. On the relationship between the body size of the moth and its fecundity and longevity. *Jap. J. appl. Zool.* **21**, 78–82 (in Japanese; English summary).
TAKAHASHI F. (1961) On the effect of population density on the power of the reproduction of the almond moth *Ephestia cautella*. VII. The effect of larval density on the number of larval molts and the duration of each larval instar. *Jap. J. appl. Ent. Zool.* **5**, 185–190 (in Japanese; English summary).
TAKAHASHI R. (1921) Parental care of *Canthao ocellatus*. *Trans. nat. Hist. Soc. Formosa* **11** (54) 6 (in Japanese).
TAKAHASHI Y. (1959) Structure and permeability of the chorion in the silkworm egg. *Jap. J. appl. Ent. Zool.* **3**, 80–85 (in Japanese; English summary).
TAKAIE H. (1967) Oviposition habit of *Micoterys clauseni* Compere. *Kontyû, Tokyo* **35**, 336 (in Japanese).
TAKIZAWA Y. and KATSUNO S. (1954) A cyto-histological study on the mucous gland in the female Eri-silkworm moth *Philosamia cynthia ricini*, with special reference to the mitochondria. *Mem. Fac. Agric. Hokkaido Univ.* **2**, 79–84 (in Japanese; English summary).
TAKSDAL G. (1963) Ecology of plant resistance to the tarnished plant bug *Lygus lineolaris*. *Ann. ent. Soc. Am.* **56**, 69–74.
TAKSDAL G. (1965) Resistens hos planter mot åtak av skadedyr. *Naturen* **6**, 370–383.

TAMAKI Y. (1967) Examination of several conditions in the mass rearing of the smaller tea tortrix on the artificial diet for successive generations. *Bull. agric. Chem. Inspect. Stn* **7**, 56–60 (in Japanese; English summary).

TAMANINI L. (1956) Osservazioni biologiche e morphologiche sugli *Aradus betulinus* Fall., *A. corticalis* L., *A. pictus* Bär. (Hemiptera: Aradidae). *Stud. Trent. Sci. Nat.* **33**, 3–53.

TAMURA I., KISHINO K., and IIZIMA N. (1957) Studies on the bionomics of *Notiphila sekiyai* Koizumi (Ephydridae) and its control (preliminary report). *Jap. J. appl. Ent. Zool.* **1**, 125–130 (in Japanese; English summary).

TANAKA A., WAKIKADO T., and OUCHI Y. (1971) Ecology of the army worm *Mythimna (Leucania) separata* (Walker). II. Field survey of oviposition by sorghum dry-leaf trap. *Proc. Ass. Pl. Prot. Kyushu* **17**, 86–88 (in Japanese).

TANAKA N. (1965) Artificial egging receptacles for three species of tephritid flies. *J. econ. Ent.* **58**, 177–178.

TANAKA Y. (1950) Studies on hibernation with special reference to photoperiodicity and breeding of the Chinese Tussar-silkworm. *J. seric. Sci. Japan* **19**, 580–590.

TANDON S. K. (1967) The effect of infested leaves on the oviposition of *Phytomyza atricornis* Meigen (Diptera: Agromyzidae). *Proc. Indian Sci. Congr.* **54**, 463.

TANG C. C. (1976) Ecological studies on green asparagus thrip *Frankliniella intonsa* Trybom in Taiwan. *J. Agric. Res. China* **25**, 299–309.

TANGEL F. and HAMMERSCHLAG G. (1908) Untersuchungen über die Beteilung der Eischale am Stoffwechsel des Eiinhaltes während der Bebrutung. *Arch. ges. Physiol.* **121**, 423–436.

TANNER G. D. (1969) Oviposition traps and population sampling for the distribution of *Aedes aegypti* (L.). *Mosq. News* **29**, 116–121.

TANTAWY A. O. (1961) Effects of temperature on productivity and genetic variance of body size in populations of *Drosophila pseudoobscura*. *Genetics* **46**, 227–238.

TARANUKHA M. D. (1967) The effects of species and varieties of cereal crops on the fecundity and viability of the noxious Pentatomid. *Zool. Zh.* **46**, 701–709 (in Russian; English summary).

TARDIF R. and SECREST J. P. (1970) Devices for cleaning and counting eggs of the gypsy moth. *J. econ. Ent.* **63**, 678–679.

TASHIRO H. (1976) Biology of the grass webworm *Herpetogramma licarsisalis* (Lepidoptera: Pyraustidae) in Hawaii. *Ann. ent. Soc. Am.* **69**, 797–803.

TAWFIK M. F. S. (1970) Phagocytosis in insects against endoparasitic invasion. *Bull. Soc. ent. Égypte* **53**, 199–203.

TAWFIK M. F. S., ABUL-NASR S., and EL-HUSSEINI M. M. (1973) The biology of *Labidura riparia* Pallas (Dermaptera: Labiduridae). *Bull. Soc. ent. Égypte* **56**, 75–92.

TAWFIK M. F. S. and ATA A. M. (1973) Comparative description of the immature forms of *Orius albidipennis* (Reut.) and *O. laevigatus* (Fieb.) (Hem.: Anthocoridae) (Het.). *Bull. Soc. ent. Égypte* **57**, 73–77.

TAWFIK M. F. S. and ATA A. M. (1974a) The life-history of *Orius albidipennis* (Reut.) (Hemiptera–Heteroptera: Anthocoridae). *Bull. Soc. ent. Égypte* **57**, 117–126 (1973).

TAWFIK M. F. S. and ATA A. M. (1974b) The life-history of *Orius laevigatus* (Fieber) (Hemiptera: Anthocoridae). *Bull. Soc. ent. Égypte* **57**, 145–151.

TAWFIK M. F. S. and EL-HUSSEINI M. M. (1972) The life-history of the anthocorid predator *Blaptostethus piceus* Fieber var. *pallescens* Poppius (Hemiptera: Anthocoridae). *Bull. Soc. ent. Égypte* **55**, 239–252.

TAWFIK M. F. S. and NAGUI A. (1966) The biology of *Montandoniella moraguesi* Puton, a predator of *Gynaikothrips ficorum* Marchal, in Egypt (Hemiptera–Heteroptera: Anthocoridae). *Bull. Soc. ent. Égypte* **49**, 181–200 (1965).

TAYLOR J. S. (1930) Notes on the solanum tortoise beetle *Aspidomorpha hybrida* Boh., Cassidae, with especial reference to the eastern Transvaal. *S. Afr. J. Nat. Hist.* **6**, 382–385.

TAYLOR J. S. (1945) Notes on the olive beetle *Argopistes sexvittatus* Bryant. *J. ent. Soc. S. Afr.* **8**, 49–52.

TAYLOR N. and MILLS R. R. (1976) Evidence for the presence of mandelic acid dehydrogenase and benzoyl formate decarbonoxylase in the American cockroach. *Insect Biochem.* **6**, 85–87.

TAYLOR R. L. (1929) The biology of the white pine weevil *Pissodes strobi* (Peck), and a study of its insect parasites from an economic viewpoint. *Ent. Am.* **9**, 166–246; **10**, 1–86.

TAYLOR T. A. (1969) Observations on oviposition behaviour and parasitisation in *Trichogramma semifumatum* (Perkins) (Hymenoptera, Trichogrammatidae). *Ghana J. Sci.* **9**, 30–34.

TAYLOR T. H. C. (1935) The campaign against *Aspidiotus destructor* Sign. in Fiji. *Bull. ent. Res.* **26**, 1–100.

TAYLOR T. H. C. (1937) *The Biological Control of an Insect in Fiji. An Account of the Coconut Leaf-mining Beetle and its Parasite Complex*, Imp. Inst. Ent., London.

TEDDERS W. L. and OSBURN M. (1970) Morphology of the reproductive systems of *Gretchena bolliana*, the pecan bud moth (Lepidoptera, Olethreutidae). *Ann. ent. Soc. Am.* **63**, 786–789.

TEICHERT M. (1961) Biologie und Brutfursorgenverhalten von *Copris lunaris* L. *Proc. 11th Int. Congr. Ent.* **1**, 621–625 (1961).

TEJADA A. (1968) Lepidopterismo y erucismo en la hoya del Huallaga. *Revta Soc. Peruana Derm.* **2** (2) 143–148.

TELFORD A. D. (1957) The pasture *Aedes* of central and northern California. The egg stage: gross embryology and resistance to desiccation. *Ann. ent. Soc. Am.* **50**, 537–543.

TEMPELIS C. H. (1975) Host-feeding patterns of mosquitoes with a review of advances in analysis of blood meals by serology. *J. med. Ent.* **11**, 635–653.

TEOTIA T. P. S. and PANDEY R. C. (1968) The effect of different natural foods on the oviposition, fecundity and development of *Cadra (Ephestia) cautella* Walker. *Labdev J. Sci. Technol.* **6-B** (3) 145–150.

TEOTIA T. P. S. and SINGH V. S. (1966) The effect of host species on the oviposition fecundity and development of *Callosobruchus chinensis* Linn. (Bruchidae: Coleoptera). *Bull. Grain Technol.* **4**, 3–10.

TEOTIA T. P. S. and SINGH V. S. (1968) On the oviposition behaviour and development of *Sitophilus oryzae* Linn. in various natural foods. *Indian J. Ent.* **30**, 119–124.

TER-GRIGORYAN M. A. (1976) The biology of the Ararat cochineal insect *Porphyrophora hamelii* Brandt (Homoptera, Coccoidea, Margarodidae). *Ent. Obozr.* **55**, 300–307.

TERRANOVA A. C. and LEOPOLD R. A. (1971) A bioassay technique for investigating mating refusal in female house flies. *Ann. ent. Soc. Am.* **64**, 263–266.

TERRANOVA A. C., LEOPOLD R. A., DEGRUGILLIER M. E., and JOHNSON J. R. (1972) Electrophoresis of the male accessory secretion and its fate in the mated female. *J. Insect Physiol.* **18**, 1573–1591.

TERRY R. J. (1952) Some observations on *Scatella silacea* Loew. (Ephydridae) in sewage filter beds. *Proc. Leeds phil. lit. Soc. (Sci. Sect.)* **6**, 104–111.

TERRY T. W. (1905) Leaf hoppers and their natural enemies. V. Forficulidae. *Bull. Hawaii. Sug. Ass. ent. Ser.* **1**, 163–174.

TESH R. B., CHANIOTIS B. N., ARONSON M. D., and JOHNSON K. M. (1971) Natural host preferences of Panamanian Phlebotomine sandflies as determined by precipitin test. *Am. J. trop. Med. Hyg.* **20**, 150–156.

TEYROVSKÝ V. (1924) Das Ei und Eiablage von *Cymatia coleoptrata* F. *Jubilejni sbornik Čsl. ent. spol.* **1924**, 100–101.

THAGGARD C. W. and ELIASON D. A. (1969) Field evaluation of components for an *Aedes aegypti* (L.) oviposition trap. *Mosq. News* **29**, 608–612.

THALENHORST W. (1968) Zur Kenntnis der Fichtenblattwespen. VIII. Eizahl und Eiablage. *Z. PflKrankh. PflPath. PflSchutz.* **75**, 338–350.

THALMANN J. (1968) Note sur la ponte de *Meloe majalis* (Col., Meloidae). *C. r. Soc. Sci. nat. phys. Maroc.* **34**, 16–17.

THANH-XUAN N. (1969) Étude de la résistance au froid et de la capacité d'acclimation de *Psylla buxi* L. (Homoptera, Psyllidae). *C. r. hebd. Séanc. Acad. Sci. Paris* (D) **268**, 1410–1413.

THANH-XUAN N. (1970a) Recherches sur la morphologie et la biologie de *Psyllopsis fraxini* (Hom., Psyllidae). *Annls Soc. ent. Fr.* (NS) **6**, 757–773.

THANH-XUAN N. (1970b) Influence de la température et de la photopériode sur la reproduction d'un Psylle du Poire, *Psylla pyri* L. (Insécte Homoptera, Psyllidae). *C. r. hebd. Séanc. Acad. Sci. Paris* (D) **271**, 2336–2338.

THANH-XUAN N. (1971) Effet du groupement sur la reproduction et la longévité de *Psylla pyri* L. *C. r. hebd. Séanc. Acad. Sci. Paris* (D) **272**, 1782–1784.

THEOBALD F. V. (1901) *A Monograph of the Culicidae or Mosquitoes*, **1**, British Museum (Nat. Hist.), London.

THEODOR O. (1925) Observations on the Palestinian *Anopheles*. *Bull. ent. Res.* **15**, 377–382.

THEOWALD B. (1967) Familie Tipulidae (Diptera, Nematocera), Larven und Puppen. *BestimmBüch. Bodenfauna Eur.* **7**, 1–100.

THERON P. P. A. (1945) The artificial conditioning of lepidopterous larvae for attack by ectoparasitic ichneumonid larvae. *J. ent. Soc. S. Afr.* **8**, 111–116.

THEWKE S. E. and PUTTLER B. (1970) Aerosol application of lepidopterous eggs and their susceptibility to parasitism by *Trichogramma*. *J. econ. Ent.* **63**, 1033–1034.

THIAGARAJAN K. B. (1939) The habits of the common earwig of Annamalainagar *Euborellia stali* (Dohrn.). *J. Bombay nat. Hist. Soc.* **40**, 721–723.

THIENEMANN A. (1923) Hydrobiologische Untersuchungen an Quellen. V. Die Trichopteren-Fauna der Quellen Holsteins (Mit einem Anhang über die Metamorphose der Beraeinen). *Z. wiss. InsektBiol.* **18**, 126–134, 179–186.

THIENEMANN A. (1954) *Chironomus. Leben, Verbreitung und wirtschaftliche Bedeutung der Chironomiden*, Stuttgart.

THOMAS A. (1969) Étude du déterminisme de l'oviposition chez *Carausius morosus* Br. (Phasmides, Chéleutoptères). *C. r. hebd. Séanc. Acad. Sci. Paris* **269**, 2424–2427.

THOMAS D. C. (1938) A list of the Hemiptera–Heteroptera in Hertfordshire with notes on their biology. *Trans. Herts. nat. Hist. Soc. Fld. Cl.* **20**, 314–327.

THOMAS D. C. (1954) Notes on the biology of some Hemiptera–Heteroptera. *Entomologist* **87**, 25–30.

THOMAS H. D. (1943) Preliminary studies on the physiology of *Aedes aegypti* (Diptera: Culicidae). I. The hatching of the eggs under sterile conditions. *J. Parasit.* **29**, 324–327.

THOMAS R. T. S. (1964) Some aspects of life history, genetics, distribution, and taxonomy of *Aspidomorpha adhaerens* (Weber, 1801) (Cassidinae, Coleoptera). *Tijd. Ent.* **107**, 167–264.

THOMAS V. (1966) Studies on life-history and biology of *Dysdercus cingulatus* F. (Hemiptera: Pyrrhocoridae), a pest of malvaceous crops in Malaysia. *Malaysian agric. J.* **45**, 417–428.

THOMPSON C. B. (1919) The development of the castes of nine genera and thirteen species of termites. *Biol. Bull. Woods Hole* **36**, 379–398.

THOMPSON V. and BODINE J. H. (1936) Oxygen consumption and rates of dehydration of grasshopper eggs. *Physiol. Zoöl.* **9**, 455–470.

THOMPSON W. R. (1915a) Contribution à la connaissance de la larve planidium. *Bull. Biol. Fr. Belg.* **48**, 319–349.

THOMPSON W. R. (1915b) Sur le cycle évolutif de *Fortisia foeda*, Diptère parasite d'un *Lithobius*.

C. r. Soc. Biol. Paris **78**, 413–416.

THOMPSON W. R. (1920) Note sur *Rhacodineura antiqua* Fall., tachinaire parasite des forficules. *Bull. Soc. ent. Fr.* **1920**, 199–201.

THOMPSON W. R. (1923a) Recherches sur les Diptères parasites. Les larves primaires des Tachinidae du groupe des Echinomyiinae. *Annls Épiphyt.* **9**, 137–201.

THOMPSON W. R. (1923b) Recherches sur la biologie des Diptères parasites. *Bull. Biol. Fr. Belg.* **57**, 174–237.

THOMPSON W. R. (1928) A contribution to the study of the dipterous parasites of the European earwig *Forficula auricularia* L. *Parasitology* **20**, 123–156.

THOMPSON W. R. (1963) The tachinids of Trinidad. III. The goniines with microtype eggs (Diptera, Tachinidae). *Stud. Ent. Petrópolis* (NS) **6**, 257–404.

THOMSEN L. C. (1937) Aquatic Diptera. Part V. Ceratopogonidae. *Mem. Cornell agric. Exp. Stn* **210**, 57–80.

THOMSON M. G. (1958) Egg sampling for the western hemlock looper. *For. Chron.* **34**, 248–256.

THOMSON V. (1966) The biology of the lesser grain borer *Rhizopertha dominica* (Fab.). *Bull. Grain Technol.* **4**, 163–168.

THONTADARYA T. S. and CHANNA BASAVANNA G. P. (1960) Mode of egg-laying in Tingidae (Hemiptera). *Mysore agric. J.* **35**, 13–14.

THONTADARYA T. S. and CHANNA-BASAVANNA G. P. (1968) Mode of egg-laying in *Perigrinus maidis* (Ashmead) (Araeopidae: Homoptera). *Mysore J. agric. Sci.* **2**, 338–339.

THORPE W. H. (1930) The biology of the petroleum fly *Psilopa petrolei* Coq. *Trans. ent. Soc. Lond.* **78**, 331–334.

THORPE W. H. (1936) A new type of respiratory interrelation between an insect (Chalcid) parasite and its host (Coccidae). *Parasitology* **28**, 517–540.

THORPE W. H. (1938) Further experiments on olfactory conditioning in a parasitic insect. The nature of the conditioning response. *Proc. R. Soc.* (B) **126**, 370–397.

THORPE W. H. (1942) Observations on *Stomoxys ochrosoma* Speiser (Diptera: Muscidae) as an associate of army ants (Dorylinae) in East Africa. *Proc. R. ent. Soc. Lond.* (A) **17**, 38–41.

THORPE W. H. (1950) Plastron respiration in aquatic insects. *Biol. Rev.* **25**, 344–390.

THORPE W. H. and CRISP D. J. (1947a) Studies on plastron respiration. I. The biology of *Aphelocheirus* (Hemiptera, Aphelocheiridae (Naucoridae)) and the mechanism of plastron retention. *J. exp. Biol.* **24**, 227–269.

THORPE W. H. and CRISP D. J. (1947b) Studies on plastron respiration. II. The respiratory efficiency of the plastron in *Aphelocheirus*. *J. exp. Biol.* **24**, 270–303.

THORPE W. H. and CRISP D. J. (1947c) Studies on plastron respiration. III. The orientation responses of *Aphelocheirus* (Hemiptera, Aphelocheiridae (Naucoridae)) in relation to plastron respiration, together with an account of specialized pressure receptors in aquatic insects. *J. exp. Biol.* **24**, 310–328.

THORPE W. H. and CRISP D. J. (1949) Studies on plastron respiration. IV. Plastron respiration in the Coleoptera. *J. exp. Biol.* **26**, 219–260.

THORPE W. H. and JONES F. G. W. (1937) Olfactory conditioning in a parasitic insect and its relation to the problem of host selection. *Proc. R. Soc.* (B) **124**, 56–81.

THROCKMORTON L. H. (1962) The problem of phylogeny in the genus *Drosophila*. In Wheeler, *Studies in genetics*, II, Univ. Tex. Publ. 6205, 207–343.

THROCKMORTON L. H. (1966) The relationships of the endemic Hawaiian Drosophilidae. In *Studies in genetics*, III, Univ. Tex. Publ. 6615, 335–396.

THURSTON R., SMITH W. T., and COOPER B. P. (1966) Alkaloid secretion by trichomes of *Nicotiana* species and resistance to aphids. *Entomologia exp. appl.* **9**, 428–432.

THYGESEN T. (1962) Om muligheden for en mere effektiv bekaempelse af rapsens alvorligste

skadedyr, skulpegalmyggen (*Dasyneura brassicae*). *Ugeskr. Landm.* **15**, 7 pp.
TICHOMIROFF A. (1885) Chemische Studien über die Entwicklung der Insecteneier. *Hoppe-Seyler's Z. physiol. Chem.* **9**, 518–532.
TICHÝ V. and KUDLER J. (1962) A contribution to the understanding of the effect of birds on the course of outbreaks of *B. piniarius. Lesnictví* **8**, 151–166 (in Czech.; English summary).
TIDWELL M. A. and HAYS K. L. (1971) Oviposition preferences of some Tabanidae (Diptera). *Ann. ent. Soc. Am.* **64**, 547–549.
TIEGS O. W. (1941) The "dorsal organ" of Collembolan embryos. *Q. Jl microsc. Sci.* (NS) **83**, 153–169.
TIEGS O. W. (1944) The "dorsal organ" of the embryo of *Campodea. Q. Jl microsc. Sci.* **84**, 33–47.
TILDEN, J. W. (1949) Oviposition of *Cryptocephalus confluens* Say (Coleoptera, Chrysomelidae). *Ent. News* **60**, 151–154.
TILLYARD R. J. (1916) Studies on the Australian Neuroptera. IV. The families Ithonidae, Hemerobiidae, Sisyridae, Berothidae, and the new family Trichomatidae; with a discussion of their characters and relationships, and descriptions of new and little-known genera and species. *Proc. Linn. Soc. NSW* **41**, 269–332.
TILLYARD R. J. (1917) *The Biology of Dragonflies*, Cambridge University Press.
TILLYARD R. J. (1918) Studies in Australian Neuroptera. 7. The life history of *Psychopsis elegans* (Guér.). *Proc. Linn. Soc. NSW* **43**, 787–818.
TILLYARD R. J. (1922) The life-history of the Australian moth lace-wing, *Ithone fusca* Newman (Order Neuroptera Planipennia). *Bull. ent. Res.* **13**, 205–223.
TILLYARD R. J. (1926) *The Insects of Australia and New Zealand*, Sydney.
TINGEY W. M. and LEIGH T. F. (1975) Height preference of lygus bugs for oviposition on caged cotton plants. *Envir. Ent.* **3** (2) 350–351 (1974).
TINGEY W. M., LEIGH T. F., and HYER A. H. (1973a) Lygus bug resistant cotton. *Calif. Agric.* **27** (11) 8–9.
TINGEY W. M., LEIGH T. F., and HYER A. H. (1973b) Three methods of screening cotton for ovipositional nonpreference by *Lygus* bugs. *J. econ. Ent.* **66**, 1312–1314.
TISSEUIL J. (1935) Contribution à l'étude de la papillonite guyanaise. *Bull. Soc. Pat. exot.* **28**, 719–721.
TITSCHACK E. (1922) Beiträge zu einer Monographie der Kleidermotte *Tineola biselliella. Z. techn. Biol.* **10**, 168 pp.
TITSCHACK E. (1930) Untersuchungen über das Wachstum, den Nahrungsverbrauch und die Eierzeugung. *Z. Morph. Ökol. Tiere* **17**, 471–551.
TIWARI N. K. (1967) Notes on survey of *Citrus* infestation at Nagpur and adjoining areas in January–February 1966. *Cecidologia indica* **2**, 53–56.
TOBIAS V. I. (1976) Origin and development of the ability to paralyse the host in parasitic Hymenoptera and wasps. *Ent. Obozr.* **55**, 308–310.
TOD M. E. (1973) Notes on beetle predators of molluscs. *Entomologist* **106**, 196–201.
TODD D. H. (1956) A preliminary account of *Dasyneura mali* Kieffer (Cecidomyidae: Dipt.) and an associated Hymenopterous parasite in New Zealand. *NZ Jl Sci. Tech.* (A) **37**, 462–464.
TODD J. W. and LEWIS W. J. (1976) Incidence and oviposition patterns of *Trichopoda pennipes* (F.), a parasite of the southern green stink bug, *Nezara viridula* (L.). *J. Geo. ent. Soc.* **11**, 50–54.
TOKMAKOĞLU C., SOYLU O. Z., and DEVECIOĞLU H. (1967) The bionomics of *Ectomyelois ceratoniae* and investigations on control methods. *Bitki Koruma Bült.* **7**, 91–106 (in Turkish; English summary).
TOLL S. (1941) Larviparie bei *Coleophora leucapennella* Hb. *Mitt. dt. ent. Ges.* **10**, 55–56.

Tomita M. and Karashina J. (1929) Beiträge zur Embryochemie der Reptilien. VI. Über das Verhalten der Fette bei der Bebrütung von Meerschild-kröteneiern. *J. Biochem. Tokyo* **10**, 375–377.

Tonapi G. T. (1959) A note on the eggs of *Gerris fluviorum* F. with a brief description of its neanide (Hem., Gerridae). *Entomologist's mon. Mag.* **95**, 29–31.

Tonkes P. R. (1933) Recherches sur les poils urticants des chenilles. *Bull. biol. Fr. Belg.* **67**, 44–99.

Torre-Bueno J. R. de la (1943) Maternal solicitude in *Gargaphia iridescens* Champion. *Bull. Brooklyn ent. Soc.* **37**, 131 (1942).

Tostowaryk W. and McLeod J. M. (1972) Sequential sampling for egg clusters of the Swaine jack pine sawfly *Neodiprion swainei* (Hymenoptera: Diprionidae). *Can. Ent.* **104**, 1343–1347.

Tothill J. D. (1918) Some notes on the natural control of the oyster-shell scale *Lepidosaphes ulmi* L. *Bull. ent. Res.* **9**, 183–196.

Tothill J. D. (1922) The natural control of the fall webworm (*Hyphantria cunea* Drury) in Canada. *Tech. Bull. Dep. Agric. Can.* **3**, 1–107.

Touzeau J. and Vonderheyden F. (1968) L'élevage semi-industriel des tordeuses de la grappe destinées au piègeage sexuel. *Phytoma* **20** (197) 25–30.

Townsend C. H. T. (1908) A record of results from rearings and dissections of Tachinidae. *Tech. Ser. US Bur. Ent.* **12**, 95–118.

Townsend C. H. T. (1938) *Manual of Myiology in Twelve Parts. Part VII. Oesteroid Generic Diagnosis and Data. Gymnosomatini to Senostomatini*, São Paulo.

Townsend C. H. T. (1942) *Manual of Myiology in Twelve Parts. Part XII. General Consideration of the Oesteromuscaria. Geologic History and Geographic Distribution—Environment and Responses—Relations to Man—Hosts and Flowers—Bibliography and Plates*, São Paulo.

Toye S. A. (1970) The structure of the egg of *Zonocerus variegatus* (L.) (Orthoptera, Acridoidea). *Rev. Zool. Bot. afr.* **81**, 19–28.

Trägårdh I. (1930) Some aspects in the biology of longicorn beetles. *Bull. ent. Res.* **21**, 1–8.

Trager W. (1935) The culture of mosquito larvae free from living micro-organisms. *Am. J. Hyg.* **22**, 18–25.

Travassos filho L. (1945) Sôbre a familia Acanthopidae Burmeister, 1838, emend. (Mantodea). *Archos Zool. S. Paulo* **4**, 157–232.

Travis V. B. (1953) Laboratory studies on the hatching of marsh-mosquito eggs. *Mosq. News* **13**, 190–198.

Traynier R. M. M. (1965) Chemostimulation of oviposition by the cabbage root fly *Erioischia brassicae* (Bouché). *Nature, Lond.* **207**, 218–219.

Traynier R. M. M. (1967) Stimulation of oviposition by the cabbage root fly *Erioischia brassicae*. *Entomologia exp. appl.* **10**, 401–412.

Traynier R. M. M. (1975) Field and laboratory experiments on the site of oviposition by the potato moth *Phthorimaea operculella* (Zell.) (Lepidoptera, Gelechiidae). *Bull. ent. Res.* **65**, 391–398.

Trehan K. N. (1940) Studies on the British whiteflies (Homoptera–Aleyrodidae). *Trans. R. ent. Soc. Lond.* **90**, 575–616.

Treherne R. C. and Buckell E. R. (1924) The grasshoppers of British Columbia, with particular reference to the influence of injurious species on the range lands of the province. *Bull. Dep. Agric. Can.* **39**, 1–40.

Trelka D. G. and Foote B. A. (1970) Biology of slug-killing *Tetanocera* (Diptera: Sciomyzidae). *Ann. ent. Soc. Am.* **63**, 877–895.

Trembley H. L. (1955) Mosquito culture techniques and experimental procedures. *Bull. Am. Mosq. Control Ass.* **3**.

TRENSZ F. (1933) Étude expérimentale sur la fonction des chambres à air de l'oeuf d'*Anopheles maculipennis*. *Archs. Inst. Pasteur Algér.* **11**, 192–197.

TRIGGIANI O. (1973) Note biologiche sulla *Deraecoris flavilinea* Costa (Rhynchota–Heteroptera). *Entomologica* **9**, 137–145.

TRPIŠ M. (1972) Dry season survival of *Aedes aegypti* eggs in various breeding sites in the Dar es Salaam area, Tanzania. *Bull. Wld Hlth Org.* **47**, 433–437.

TRPIŠ M. (1974) An improved apparatus for separation of mosquito eggs from soil. *Zentbl. Bakt. ParasitKde Infekt. Hyg. Erste Abteilung Originale* (A) **226**, 418–423.

TRPIŠ M. and HORSFALL W. R. (1967) Eggs of floodwater mosquitoes (Diptera: Culicidae). XI. Effect of medium on hatching of *Aedes stictus*. *Ann. ent. Soc. Am.* **60**, 1150–1152.

TRUCKENBRODT W. (1964) Zytologische und entwicklungs-physiologische Untersuchungen am besamten und am parthenogenetischen Ei von *Kalotermes flavicollis* Fabr. Reifung, Furchungsablauf und Bildung der Keimanlage. *Zool. Jb. (Anat.).* **81**, 359–434.

TSACAS L. (1959) La ponte et l'oeuf d'*Empis (Polyblepharis) livida* L. (Dipt., Empididae). *Bull. Soc. zool. Fr.* **84**, 483–485.

TSAI PAN-HUA and LI YA-TSE (1960) A preliminary investigation on the elm Chrysomelid (*Ambrostoma quadriimpressum* Motsch.). *Acta ent. sin.* **10**, 143–170 (in Chinese; Russian summary).

TSAI PAN-HUA, LIU YU-CH'IAO, HOW T'AU-CH'IEN, LI CHIH-YIN, and HO CHUNG (1958) Preliminary study on the development of the pine caterpillar, *Dendrolimus punctatus* Walker, under different conditions of injury of the host plant. *Acta ent. sin.* **8**, 327–334 (in Chinese; English summary).

TSALEV M. (1966) A contribution to the morphology, bionomics and control of the spherical rose scale in Bulgaria. *RastVŭd. Nauki* **3** (7) 3–14 (in Bulgarian; English summary).

TSIROPOULOS G. J. (1972) Storage temperatures for eggs and pupae of the olive fruit fly. *J. econ. Ent.* **65**, 100–102.

TSIROPOULOS G. J. and TZANAKAKIS M. E. (1970) Mating frequency and inseminating capacity of radiation-sterilized and normal males of the olive fruit fly. *Ann. ent. Soc. Am.* **63**, 1007–1010.

TSUJITA M. (1948) A cytological study on the mucous glands in the female silkworm-moth, with special reference to the roles played by the nucleus, the mitochondria and the Golgi bodies in secretion. *Bull. agric. Exp. Stn Tokyo* **12**, 633–648 (in Japanese; English summary).

TUCK J. B. and SMITH R. C. (1939) Identification of the eggs of mid-western grasshoppers by the chorionic sculpturing. *Tech. Bull. Kans. agric. Exp. Stn* **48**, 1–39.

TUFT P. H. (1950) The structure of the insect egg-shell in relation to the respiration of the embryo. *J. exp. Biol.* **26**, 327–334.

TULI S., MOOKHERJEE P. B., and SHARMA G. C. (1966) Effect of temperature and humidity on the fecundity and development of *Cadra cautella* Walk. in wheat. *Indian J. Ent.* **28**, 305–317.

TURNER D. A. (1971) Olfactory perception of live hosts and carbon dioxide by the tsetse fly *Glossina morsitans orientalis* Vanderplank. *Bull. ent. Res.* **61**, 75–96.

TURNHOUT H. M. T. VAN and LAAN P. A. VAN DER (1958) Control of *Lygus campestris* on carrot seed crops in North Holland. *Tijdschr. PlZiekt.* **64**, 301–306.

TURNIPSEED S. G. and RABB R. L. (1963) Some factors influencing oviposition by the tobacco wireworm *Conoderus vespertinus* (Coleoptera: Elateridae). *Ann. ent. Soc. Am.* **56**, 751–755.

TURNIPSEED S. G. and RABB R. L. (1965) Effects of temperature and moisture on survival of eggs of the tobacco wireworm. *J. econ. Ent.* **58**, 1155–1156.

TUTT W. J. (1899) *A Natural History of the British Lepidoptera. A Textbook for Students and Collectors*, **1**, London.

Tutt W. J. (1900) *A Natural History of the British Lepidoptera. A Textbook for Students and Collectors*, **2**, London.
Tutt W. J. (1907–8) *A Natural History of the British Lepidoptera, their World-wide Variation and Geographical Distribution. A Textbook for Students and Collectors*, **9**, London.
Tuxen S. L. (1944) The hot springs, their animal communities and their zoogeographical significance. *The Zoology of Iceland* **1** (2) 1–207, Copenhagen.
Tyagi A. K. and Girish G. K. (1975) Studies on the oviposition site of *Sitophilus oryzae* on wheat and effect of size and outer surface. *Bull. Grain Tech.* **13**, 144–150.
Tychen P. H. and Vincent J. F. V. (1976) Correlated changes in mechanical properties of the intersegmental membrane and bonding between proteins in the female adult locust. *J. Insect Physiol.* **22**, 115–125.
Tyndale-Biscoe M. (1971) Protein-feeding by the males of the Australian bushfly *Musca vetustissima* Wlk. in relation to mating performance. *Bull. ent. Res.* **60**, 607–614.
Tynegar M. O. T. (1969) Eggs of two species of *Tripteroides* Giles from New Caledonia (Dipt., Culicidae). *J. Aust. ent. Soc.* **8**, 214–216.
Tyzzer E. E. (1907) The pathology of the brown-tail moth dermatitis. *J. med. Res.* **16** (11) 43–64.

Ubelaker J. E., Cooper N. B., and Allison V. F. (1970) Possible defensive mechanism of *Hymenolepis diminuta* cysticercoids to hemocytes of the beetle *Tribolium confusum*. *J. Invert. Path.* **16**, 310–312.
Ubelaker J. E. and Keller B. L. (1964) Observations of eggs and larvae of *Cuterebra emasculator* (Fitch, 1856). *Trans. Kans. Acad. Sci.* **67**, 713–715.
Ubelaker J. E., Payne E., Allison V. F., and Moore D. V. (1973) Scanning electron microscopy of the human pubic louse *Pthirus pubis* (Linnaeus, 1758). *J. Parasit.* **59**, 913–919.
Uematsu H. (1972) Studies on *Marietta carnesi* (Hymenoptera: Aphelinidae), a hyperparasite of Diaspine scales (Homoptera: Diaspididae). I. Host relation. *Jap. J. appl. Ent. Zool.* **16**, 187–192 (in Japanese; English summary).
Uematsu H. (1976) Studies on *Marietta carnesi* (Hymenoptera: Aphelinidae), a hyperparasite of diaspine scales (Homoptera: Diaspididae). III. Reproductive capacity. *Jap. J. appl. Ent. Zool.* **20**, 115–119.
Ueno H. (1960) On the bionomics and control of the wood-boring beetles (Ipidae, Coleoptera) attacking persimmons in Japan. *Jap. J. appl. Ent. Zool.* **4**, 166–172 (in Japanese; English summary).
Uhler L. D. (1951) Biology and ecology of the goldenrod gall fly *Eurosta solidaginis* (Fitch). *Mem. Cornell Univ. agric. Exp. Stn* **300**, 1–51.
Ullyett G. C. (1945) Oviposition by *Ephestia kühniella* Zell. *J. ent. Soc. S. Afr.* **8**, 53–59.
Unzicker J. D. (1968) The comparative morphology and evolution of the internal female reproductive system of Trichoptera. *Ill. biol. Monogr.* **40**, 1–72.
Urbino C. M. (1936) The eggs of some Philippine *Anopheles*. *Mon. Bull. Bur. Hlth Philipp.* **16**, 261–273.
Urquhart F. A. (1937) Some notes on the sand cricket *Tridactylus apicalis* Say. *Can. Fld Nat.* **51**, 28–29.
Urquhart F. A. (1938) The oviposition and cannibalistic habits of the narrow-winged katydid *Phaneroptera pistillata* Brunner. *Can. Fld Nat.* **52**, 51–53.
Usinger R. L. (1946) Notes and descriptions of *Ambrysus* Stål with an account of the life history of *Ambrysus mormon* Montd. (Hemiptera, Naucoridae). *Kans. Univ. Sci. Bull.* **31**, 185–209.
Usinger R. L. (1956) Aquatic Hemiptera. In *Aquatic Insects of California* (R. L. Usinger, ed.), Berkeley.

USINGER R. L. and MATSUDA R. (1959) *Classification of the Aradidae (Hemiptera–Heteroptera)*, Brit. Mus. (Nat. Hist.), pp. 1–410.

USMAN S. (1967) Relation between the weights of the rice moths on emergence and their fecundity and longevity. *Mysore J. agric. Sci.* **1**, 123–127.

UTIDA S. (1967) Collective oviposition and larval aggregation in *Zabrotes subfasciatus* (Boh.) (Coleoptera: Bruchidae). *J. stored Prod. Res.* **2**, 315–322.

UTIDA S. (1971) Influence of temperature on the number of eggs, mortality and development of several species of Bruchid infesting stored beans. *Jap. J. appl. Ent. Zool.* **15**, 23–30 (in Japanese; English summary).

UVAROV B. (1966) *Grasshoppers and Locusts. A Handbook of General Acridology.* Volume I. *Anatomy, Physiology, Development, Phase Polymorphism, Introduction to Taxonomy*, Cambridge University Press.

VAIDYA V. G. (1956) On the phenomenon of drumming in egg-laying female butterflies. *J. Bombay nat. Hist. Soc.* **54**, 216–217.

VAIDYA V. G. (1969) Investigations of the role of visual stimuli in the egg-laying and resting behaviour of *Papilio domoleus* L. (Papilionidae, Lepidoptera). *Anim. Behav.* **17**, 350–355.

VALADARÈS DA COSTA M. (1955) Influence du milieu nutritif sur la ponte de *Drosophila melanogaster*. *C. r. hebd. Séanc. Acad. Sci. Paris* **241**, 1857–1860.

VALCOVIC L. R. and GROSCH D. S. (1968) Apholate-induced sterility in *Bracon hebetor*. *J. econ. Ent.* **61**, 1514–1517.

VALDIMOROVA M. A. and SMIRNOV E. S. (1938) Intraspecific competition and interspecific competition between *Musca domestica* L. and *Phormia groenlandica* Zeit. *Med. Parazit. Moscow* **7**, 755–777 (in Russian).

VALEK D. A. and COPPEL H. C. (1972) Ovipositional site preferences of the oak defoliating grasshopper *Dendrotettix quercus* in Wisconsin. *Trans. Wis. Acad. Sci. Arts Lett.* **60**, 225–230.

VALERY-MAYET M. (1875) Mémoire sur les moeurs et métamorphoses d'une nouvelle espèce de coléoptère de la famille des vesicants. Le *Sitaris colletis*. *Annls Soc. ent. Fr.* **5**, 62–95.

VALLE K. J. (1926) Zur Eiablage einiger Odonaten. *Not. Ent.* **6**, 106–109.

VANCE A. M. (1931) *Apanteles thompsoni* Lyle. A braconid parasite of the European corn borer. *Tech. Bull. US Dep. Agric.* **233**.

VANCE A. M. (1932) The biology and morphology of the braconid *Chelonus annulipes* Wesm., a parasite of the European corn borer. *Tech. Bull. US Dep. Agric.* **294**, 1–48.

VANDERZANT E. S. and DAVICH T. B. (1958) Laboratory rearing of the boll weevil: a satisfactory larval diet and oviposition studies. *J. econ. Ent.* **51**, 288–291.

VANDERZANT E. S. and DAVICH T. B. (1961) Artificial diets for the adult boll weevil and techniques for obtaining eggs. *J. econ. Ent.* **54**, 923–928.

VANDERZANT E. S., RICHARDSON C. D., and DAVICH T. B. (1960) Feeding and oviposition by the boll weevil on artificial diets. *J. econ. Ent.* **52**, 1138–1143 (1959).

VAN LEEUWEN E. R. (1947) Increasing production of codling moth eggs in an oviposition chamber. *J. econ. Ent.* **40**, 744–745.

VARGAS L. (1941) Nota sobre los huevecillos de *Anopheles* mexicanos. *Gaceta Med. Mex.* **1941**, 107–123.

VARGAS L. (1942) El huevo de *Anopheles barberi* Coquillet, 1903. *Revta Inst. Salubr. Enferm. trop. Méx.* **3**, 329–331.

VARMA B. K. (1954) Notes on *Cassida circumdata* Hbst., *Cassida indicola* Duv. and *Glyphocassis trilineata* Hope (Coleoptera: Chrysomelidae: Cassidinae) as pests of sweet potato (*Ipomoea batatas*) at Kanpur. *Indian J. agric. Sci.* **24**, 261–263.

Varma B. K. (1963) A study of the development and structure of the female genitalia and reproductive organs of *Galerucella birmanica* Jac. (Chrysomelidae, Coleoptera). *Indian J. Ent.* **25**, 224–232.

Vasev A. (1968) Studies on the bionomics and means of control of the stone-fruit leaf sawfly (*Neurotoma nemoralis* L., Tenthredinidae, Hymenoptera). *Gradinarska Lozarska Nauka* **5** (4) 15–26 (in Bulgarian; English summary).

Vasiljević L. and Injac M. (1967) Neke posledice zračenja gubarevih jaja radioaktivnim kobaltom 60. *Zašt. Bilja* **18** (93–95) 55–65 (English summary).

Vasilyan V. V. (1960) The effect of radiation on the development of *Pectinophora malvella*. *Rev. Ent. URSS* **39**, 599–604 (in Russian; English summary).

Vedy J. (1971) Is phthiriasis of the eyelid harmless for the eyeball? *Méd. trop.* **31**, 463–465.

Veeresh G. K. and Puttarudraiah M. (1968) A study of the bionomics of *Lymantria ampla* Wlk. (Lymantriidae: Lepidoptera). *Mysore J. agric. Sci.* **2**, 61–69.

Velasquez C. C. (1968) *Phthirus pubis* from human eyelashes in the Philippines. *J. Parasit.* **54**, 1140.

Venkatesh M. V., Chandra H., and Ahluwalia P. J. S. (1974) Reaction of gravid females of the desert locust *Schistocerca gregaria* Forsk. to some soil types. *Prog. Rep. Field Res. Sta. (United Nations Develop. Prog. Desert Locust Project) Tech. Ser.*, No. AGP/DL/TS/**14**, 53–60.

Venkatraman T. V. and Subba Rao B. R. (1954) The mechanism of oviposition in *Stenobracon deesae* (Cam.) (Hymenoptera: Braconidae). *Proc. R. ent. Soc. Lond.* (A) **29**, 1–8.

Vepsalainen K. (1973) Developmental rates of some Finnish *Gerris* Fabr. species (Het., Gerridae) in laboratory cultures. *Ent. scand.* **4** (3) 206–216.

Vereecke A. and Pelerents C. (1969) Sensitivity to gamma radiation of *Tribolium confusum* eggs at various developmental stages. *Entomologia exp. appl.* **12**, 62–66.

Verhoeff C. (1892) Zur Kenntnis des biologischen Verhaltnisses zwischen wirth- und parasiten-Bienenlarven. *Zool. Anz.* **15**, 41–43.

Verhoeff C. (1912) Über Dermapteren. Zur Kenntnis der Brutpflege unserer Ohrwürmer. *Z. wiss. InsektBiol.* **8**, 381–385; **9**, 21–24, 55–58.

Verma J. P., Mathur Y. K., Sharma S. K., and Batra R. C. (1969) New record of a staphylinid beetle *Paederus* sp. (Coleoptera: Staphylinidae) as a predator of *Euxoa segetum* Schiffermüller (Lepidoptera: Noctuidae) in India. *Indian J. Ent.* **31**, 77–78.

Verma J. S. (1955a) Comparative toxicity of some insecticides to *Peregrinus maidis* (Ashm.) and its egg-predator. *J. econ. Ent.* **48**, 205–206.

Verma J. S. (1955b) Biological studies to explain the failure of *Cyrtorhinus mundulus* (Breddin) as an egg-predator of *Peregrinus maidis* (Ashmead) in Hawaii. *Proc. Hawaii. ent. Soc.* **15**, 623–634.

Verma J. S. (1956) Effects of demeton and schradan on *Peregrinus maidis* (Ashm.) and its egg-predator *Cyrtorhinus mundulus* (Breddin). *J. econ. Ent.* **49**, 58–63.

Verson E. (1875) Il micropilo delle uova del buco da seta. *Boll. Backic. Padova* **1875**, 37–41.

Vey A. (1971) Études des réactions cellulaires anticryptogamiques chez *Galleria mellonella* L.: structure et ultrastructure des granulomes à *Aspergillus niger* V. Tiegh. *Annls Zool. Écon. anim.* **3**, 17–30.

Vey A. and Gotz P. (1975) Humoral encapsulation in Diptera (Insecta): comparative studies *in vitro*. *Parasitology* **70**, 77–86.

Vick K. W., Burkholder W. E., and Smittle B. J. (1972) Duration of mating refractory period and frequency of second matings in female *Trogoderma inclusum* (Coleoptera: Dermestidae). *Ann. ent. Soc. Am.* **65**, 790–793.

Vieira R. M. S. (1952) *A mosca da fruta (*Ceratitis capitata *(Wied.)) na Ilha da Madeira*, 219 pp.

Villacorta A., Bell R. A., and Callenbach J. A. (1971) An artificial plant stem as an oviposition site for the wheat stem sawfly. *J. econ. Ent.* **64**, 752–753.

VILLIERS A. (1946) *Faune Emp. Fr. V. Coléoptères cérambycides de l'Afrique du Nord*, 152 pp.

VILLIERS A. (1948) *Faune Emp. Fr. IX. Hémiptères reduviides de l'Afrique Noire*, Paris, Office de la Recherche Scientifique Coloniale, 488 pp.

VINCENT J. F. V. (1975a) Locust oviposition: stress softening of the extensible intersegmental membranes. *Proc. R. Soc. Lond.* **188**, 189–204.

VINCENT J. F. V. (1975b) How does the female locust dig her oviposition hole? *J. Ent.* (A) **50**, 175–181.

VINCENT J. F. V. (1976) Design for living—the elastic-sided locust. In *The Insect Integument* (H. R. Hepburn, ed.), pp. 401–419, Elsevier.

VINCENT J. F. V. and PRENTICE J. H. (1973) Rheological properties of the extensible intersegmental membrane of the adult female locust. *J. mater. Sci.* **8**, 624–630.

VINCENT J. F. V. and WOOD S. D. E. (1972) Mechanism of abdominal extension during oviposition in *Locusta*. *Nature, Lond.* **235**, 167–168.

VINCKE I. and LELEUP N. (1949) Description de la pupe, de la larve et de l'oeuf de *Anopheles concolor* Edwards. *Rev. Zool. Bot. afr.* **42**, 245–247.

VINSON S. B. (1971) Defense reaction and hemocytic changes in *Heliothis virescens* in response to its habitual parasitoid *Cardiochiles nigriceps*. *J. Invert. Path.* **18**, 94–100.

VINSON S. B. (1972) Factors involved in successful attack on *Heliothis virescens* by the parasitoid *Cardiochiles nigriceps*. *J. Invert. Path.* **20**, 118–123.

VINSON S. B. (1974) The role of the foreign surface and female parasitoid secretions on the immune response of an insect. *Parasitology* **68**, 27–33.

VINSON S. B. and BARRAS D. J. (1970) Effects of the parasitoid, *Cardiochiles nigriceps*, on the growth, development, and tissues of *Heliothis virescens*. *J. Insect Physiol.* **16**, 1329–1338.

VINSON S. B., HENSON R. D., and BARFIELD C. S. (1976) Ovipositional behavior of *Bracon mellitor* Say (Hymenoptera: Braconidae), a parasitoid of boll weevil (*Anthonomus grandis* Boh.). I. Isolation and identification of a synthetic releaser of ovipositor probing. *J. Chem. Ecol.* **2**, 431–440.

VINSON S. B. and LEWIS W. J. (1973) Teretocytes: growth and numbers in the hemocoel of *Heliothis virescens* attacked by *Microplitis croceipes*. *J. Invert. Path.* **22**, 351–355.

VINSON S. B. and SCOTT J. R. (1975) Particles containing DNA associated with the oocyte of an insect parasitoid. *J. Invert. Path.* **25** (3) 375–378.

VISWANATHAN T. R. and ANANTHAKRISHNAN T. N. (1973) On partial ovoviviparity in *Tiarothrips subramanii* (Ramakrishna) (Thysanoptera: Insecta). *Curr. Sci.* **42**, 649–650.

VITÉ J. P. (1961) The influence of water supply on oleoresin exudation pressure and resistance to bark beetle attack in *Pinus ponderosa*. *Contr. Boyce Thompson Inst.* **21**, 37–66.

VITÉ J. P. (1970) Pest management systems using synthetic pheromones. *Contr. Boyce Thompson Inst.* **24**, 343–350.

VITÉ J. P. (1975) Möglichkeiten und Grenzen der Pheromonanwendung in der Borkenkäfernbekämpfung. *Z. angew. Ent.* **77**, 325–329.

VITÉ J. P., BAKKE A., and RENWICK J. A. A. (1972) Pheromones in *Ips* (Coleoptera: Scolytidae): Occurrence and production. *Can. Ent.* **104**, 1967–1975.

VITÉ J. P., ISLAS S. F., RENWICK J. A. A., HUGHES P. R., and KLIEFOTH R. A. (1974) Biochemical and biological variation of southern pine beetle populations in North and Central America. *Z. angew. Ent.* **75**, 422–435.

VITÉ J. P. and PITMAN G. B. (1968) Bark beetle aggregation: effects of feeding on the release of pheromones in *Dendroctonus* and *Ips*. *Nature, Lond.* **218**, 169–170.

VITÉ J. P. and PITMAN G. B. (1969a) Aggregation behaviour of *Dendroctonus brevicomis* in response to synthetic pheromones. *J. Insect. Physiol.* **15**, 1617–1622.

VITÉ J. P. and PITMAN G. B. (1969b) Insect and host odors in the aggregation of the western pine beetle. *Can. Ent.* **101**, 113–117.

VITÉ J. P. and PITMAN G. B. (1970) Management of western pine beetle populations; use of chemical messengers. *J. econ. Ent.* **63**, 1132–1135.

VITÉ J. P., PITMAN G. B., FENTIMAN A. F. JR, and KINZER G. W. (1972) 3-Methyl-2-cyclohexen-1-ol isolated from *Dendroctonus*. *Naturwissenschaften* **59**, 469.

VITÉ J. P. and RENWICK J. A. A. (1968) Insect and host factors in the aggregation of the southern pine beetle. *Contr. Boyce Thompson Inst.* **24**, 61–63.

VITÉ J. P. and RENWICK J. A. A. (1971a) Population aggregating pheromone in the bark beetle *Ips grandicollis*. *J. Insect Physiol.* **17**, 1699–1704.

VITÉ J. P. and RENWICK J. A. A. (1971b) Inhibition of *Dendroctonus frontalis* response to frontalin by isomers of brevicomin. *Naturwissenschaften* **58**, 418.

VITÉ J. P. and WILLIAMSON D. L. (1970) *Thanasimus dubius*: prey perception. *J. Insect Physiol.* **16**, 233–239.

VITÉ J. P. and WOOD D. L. (1961) A study on the applicability of the measurement of oleoresin exudation pressure in determining susceptibility of second growth ponderosa pine to bark beetle infestation. *Contr. Boyce Thompson Inst.* **21**, 67–78.

VLASBLOM A. G. and WOLVEKAMP H. P. (1957) On the function of the "funnel" on the nest of the water beetle *Hydrous piceus* L. *Physiol. comp. Oecol.* **4**, 240–246.

VOEGELE J. (1968) Bionomics and morphology of *Graphosoma semipunctata* Fabricius (Heteroptera, Pentatomidae). *Awamia* **20**, 43–102 (1966).

VOELKEL H. (1924) Zur Biologie und Bekämpfung des Khapräkafers, *Trogoderma granarium* Everts. *Arb. biol. BundAnst. Land- u. Forstw.* **13** (2) 129–171.

VOGEL R. (1925) Bemerkungen zum weiblichen Geschlechtsapparat bei *Periplaneta orientalis*. *Zool. Anz.* **64**, 56–62.

VOIGT E. (1952) Ein Haareinschluss mit Phthirapteren-Eiern in Bernstein. *Mitt. miner.-geol. StInst. Hamb.* **21**, 59–74.

VOLK S. (1964) Untersuchungen zur Eiablage von *Syrphus corollae* Fabr. (Diptera: Syrphidae). *Z. angew. Ent.* **54**, 365–386.

VORHIES C. T. (1909) Studies on the Trichoptera of Wisconsin. *Trans. Wis. Acad. Sci. Arts Lett.* **16**, 647–738.

VORIS R. (1939a) The immature stages of the genera *Ontholestes*, *Creophilus* and *Staphylinus*. Staphylinidae, Coleoptera. *Ann. ent. Soc. Am.* **32**, 288–303.

VORIS R. (1939b) Immature staphylinids of the genus *Quedius* (Col., Staph.). *Ent. News* **6**, 151–156.

VOSS F. (1921) Embryonalmechanismen. *Verh. dt. zool. Ges.* **26**, 38–39.

VOSSELER J. (1905) Die Wanderheuschrecken in Usambara im Jahre 1903/1904, zugleich ein Beitrag zu ihrer Biologie. *Ber. Land- u. Forstw. Dt.-Ostafr.* **2**, 291–374.

VOSSELER J. (1919) Some work of the insectary division, etc. *Mon. Bull. Calif. Commn Hort.* **8**, 231–239.

VOUKASSOVITCH P. (1939) Contribution à l'étude de la fonction des ovaires chez un Coléoptère: *Phytodecta fornicata* Brügg. *Bull. Acad. Sci. Belgrade* (B) **5**, 113–125.

VOY A. (1947a) Sur la structure des organes génitaux accessoires femelles de la Blatte *Blatta orientalis* L. *C. r. hebd. Séanc. Acad. Sci. Paris* **225**, 767–769.

VOY A. (1947b) Sur les organes accessoires femelles des Orthoptères. *C. r. hebd. Séanc. Acad. Sci. Paris* **225**, 1382–1384.

VOY A. (1949) Contribution à l'étude anatomique et histologique des organes accessoires de l'appareil génital femelle chez quelques espèces d'Orthoptéroïdes. *Annls Sci. nat.* **11** (11) 269–345.

VOY A. (1954a) Sur l'existence de deux catégories d'oeufs dans la ponte globale du phasme (*Clonopis gallica* Charp.). *C. r. hebd. Séanc. Acad. Sci. Paris* **238**, 625–627.

Voy A. (1954b) Sur la répartition des deux catégories d'oeufs et la variation de la durée apparante du développement embryonnaire, au cours de la ponte estivale, chez le phasme (*Clonopsis gallica* Charp.). *C. r. hebd. Séanc. Acad. Sci. Paris* **239**, 196–198.

Voy A. (1956) Sur la biologie du développement embryonnaire chez une espèce de Chéleutoptères (Orthoptéroide) français: *Clonopsis gallica* Charp. (Bacillinae). *C. r. Ass. fr. Avanc. Sci. Paris* **73**, 462–464 (1955).

VUKASOVIĆ P. and GLUMAC S. (1967) Uticaj hrane na plodnost i dužina života pasuljevog žiška (*Acanthoscelides obtectus* Say). *Zašt. Bilja* **18** (93–95) 11–20 (French summary).

WACHTER S. (1925) The hatching of the eggs of *Peripsocus californicus* Banks. *Pan-Pacif. Ent.* **2**, 87–89.

WADDILL V. and SHEPARD M. (1974) Potential of *Geocoris punctipes* (Hemiptera: Lygaeidae) and *Nabis* spp. (Hemiptera: Nabidae) as predators of *Epilachna varivestis* (Coleoptera: Coccinellidae). *Entomophaga* **19**, 421–426.

WADE F. W. and RODRIGUEZ J. G. (1961) Life history of *Macrocheles muscaedomesticae* (Acarina: Macrochelidae), a predator of the house fly. *Ann. ent. Soc. Am.* **54**, 776–781.

WADSWORTH J. T. (1915) On the life-history of *Aleochara bilineata* Gyll., a staphylinid parasite of *Chortophila brassicae* Bouché. *J. econ. Biol.* **10**, 1–27.

WAGNER W. (1913) Ueber die Biologie von *Conomelus limbatus* Fabr. *Z. wiss. InsektBiol.* **9**, 120–122.

WALCH E. W. and WALCH-SORGDRAGER G. B. (1935) The eggs of some Netherlands-Indian Anophelines. *Trans. Far East Ass. trop. Med. 9th Congr.*, 65–81 (in Dutch in *Geneesk. Tijd. Ned.-Ind.* **75**, 1700–1730).

WALCH E. W. and WALCH-SORGDRAGER G. B. (1936) Over de morphologische eigenschappen van verschillende *subpictus* eieren. *Geneesk. Tijd. Ned.-Ind.* **76**, 394–422.

WALDBAUER G. P. and KOGAN M. (1975) Position of bean leaf beetle eggs in soil near soybeans determined by a refined sampling procedure. *Envir. Ent.* **4**, 375–380.

WALDBAUER G. P., YAMAMOTO R. T., and BOWERS W. S. (1964) Laboratory rearing of the tobacco hornworm *Protoparce sexta* (Lepidoptera: Sphingidae). *J. econ. Ent.* **57**, 93–95.

WALKER D. W. and FIGUEROA M. (1964) Biology of the sugarcane borer *Diatraea saccharalis* (Lepidoptera: Crambidae) in Puerto Rico. III. Oviposition rate. *Ann. ent. Soc. Am.* **57**, 515–516.

WALKER D. W. and QUINTANA V. (1969) Mating and oviposition behaviour of the coffee leaf miner *Leucoptera coffeella* (Lepidoptera: Lyonetiidae). *Proc. ent. Soc. Wash.* **71**, 88–90.

WALKER E. M. (1919) The terminal abdominal structures of orthopteroid insects: a phylogenetic study. *Ann. ent. Soc. Am.* **12**, 267–316.

WALKER J. K. (1957) A biological study of *Collops balteatus* Lec. and *Collops vittatus* (Say). *J. econ. Ent.* **50**, 395–399.

WALKER J. K. (1959) Some observations on the development of the boll weevil on the wine cup, *Callirhoe involucrata* (Nutt.) A. Gray. *J. econ. Ent.* **52**, 755–756.

WALKER W. F. and MENZER R. E. (1969) Chorionic melanization in the eggs of *Aedes aegypti*. *Ann. ent. Soc. Am.* **62**, 7–11.

WALLACE D. R. and CAMPBELL I. M. (1965) Method for characterizing sawfly egg pigmentation. *Nature, Lond.* **207**, 1363–1364.

WALLER J. B. (1967) Handling techniques for eggs and larvae of *Wiseana* sp. (Hepialidae). *NZ Ent.* **4**, 57–60.

WALLIS R. C. (1954a) A study of oviposition activity of mosquitoes. *Am. J. Hyg.* **60**, 135–168.

WALLIS R. C. (1954b) The effect of population density and of NaCl concentrations in test series in laboratory experiments with ovipositing *Aedes aegypti*. *Mosq. News* **14**, 200–204.

WALLIS R. C. (1954c) Observations on oviposition of two *Aedes* mosquitoes (Diptera, Culicidae). *Ann. ent. Soc. Am.* **47**, 393–396.
WALLIS R. C. and LANG C. A. (1956) Egg formation and oviposition in blood-fed *Aedes aegypti* L. *Mosq. News* **16**, 283–286.
WALLIS R. C. and LITE S. W. (1970) Axenic rearing of *Culex salinarius*. *Mosq. News* **30**, 427–429.
WALLIS R. C. and WHITMAN L. (1968) Oviposition of *Culiseta morsitans* (Theobald) and comments on the life cycle of the American form. *Mosq. News* **28**, 198–200.
WALLNER W. E. and ELLIS T. L. (1976) Olfactory detection of gypsy moth pheromone and egg masses by domestic canines. *Envir. Ent.* **5**, 183–186.
WALLWORK J. H. and RODRIQUEZ J. G. (1963) The effect of ammonia on the predation rate of *Macrocheles muscaedomesticae* (Acarina: Macrochelidae) on house fly eggs. *Adv. Acarol.* **1**, 60–69.
WALOFF N. (1950) The egg pods of British short-horned grasshoppers (Acrididae). *Proc. R. ent. Soc. Lond.* (A) **25**, 115–126.
WALOFF N. (1954) The number and development of ovarioles of some Acridoidea (Orthoptera) in relation to climate. *Physiol. comp. Oecol.* **3**, 370–390.
WALOFF N. (1974) Biology and behaviour of some species of Dryinidae (Hymenoptera). *J. Ent.* (A) **49**, 97–109.
WALOFF N., NORRIS M. J., and BROADHEAD E. C. (1948) Fecundity and longevity of *Ephestia elutella* Hübner (Lep., Phycitidae). *Trans. R. ent. Soc. Lond.* **99**, 245–268.
WALOFF N. and RICHARDS O. W. (1958) The biology of the chrysomelid beetle *Phytodecta olivacea* (Forster) (Coleoptera: Chrysomelidae). *Trans. R. ent. Soc. Lond.* **110**, 99–116.
WALSINGHAM, LORD (1907) Descriptions of new North American Tineid moths, with a generic table of the family Blastobasidae. *Proc. US nat. Mus.* **33**, 197–228.
WALT J. F. and MCPHERSON J. E. (1973) Descriptions of immature stages of *Stethaulax marmorata* (Hemiptera: Scutelleridae) (Het.) with notes on its life history. *Ann. ent. Soc. Am.* **66**, 1103–1108.
WALTERS J. and MCMULLEN L. H. (1956) Life history and habits of *Pseudohylesinus nebulosus* (Leconte) (Coleoptera: Scolytidae) in the interior of British Columbia. *Can. Ent.* **88**, 197–202.
WALTON G. A. (1936) Oviposition in the British species of *Notonecta* (Hemipt.). *Trans. Soc. Br. Ent.* **3**, 49–57.
WALTON G. A. (1962) The egg of *Agraptocorixa gestroi* Kirkaldy (Hemiptera–Heteroptera: Corixidae). *Proc. R. ent. Soc. Lond.* (A) **37**, 104–106.
WALTON W. R. (1914) A new tachinid parasite of *Diabroctica vittata*. *Proc. ent. Soc. Wash.* **16**, 11–14.
WANDOLLECK B. (1923) Über die Eier der *Lipeurus*-Arten. *Z. SchädlBekämpf.* **1**, 20–23.
WARD R. H. and KOK L. T. (1975) Oviposition response of *Ceuthorhynchidius horridus*, an introduced thistle-weevil, to different photophases. *Envir. Ent.* **4**, 658–660.
WARD R. D. and READY P. A. (1975) Chorionic sculpturing in some sandfly eggs (Diptera, Psychodidae). *J. Ent.* (A) **50**, 127–134.
WATANABE M. (1968) Studies on bionomics of deer fly *Chrysops suavis* Loew (Tabanidae, Diptera). I. Observations on oviposition and hatch. *Jap. J. sanit. Zool.* **19**, 87–92 (in Japanese; English summary).
WATERHOUSE C. O. (1912) Mantid oöthecae. *Proc. ent. Soc. Lond.* **1912**, cxxv–cxxvii.
WATERHOUSE F. L., ONYEARU A. K., and AMOS T. G. (1971) Oviposition of *Tribolium* in static environments incorporating controlled temperature and humidity gradients. *Oikos* **22**, 131–135.
WATERSTON A. R. (1951) Observations on the Moroccan locust (*Dociostaurus maroccanus*

Thunberg) in Cyprus, 1950. 4. Observations on adult locusts. *Anti-Locust Bull.* **10**, 36–52.
WATERSTON J. (1922) Observations on the life-history of a liotheid, parasite of the curlew (*Numenius arquata* Linn.). *Entomologist's mon. Mag.* **58**, 243–247.
WATT M. N. (1914) Descriptions of the ova of some of the Lepidoptera of New Zealand. *Trans. NZ Inst.* **46**, 65–95.
WATTAL B. L. and MAMMEN M. L. (1959) A preliminary note on colour preference for oviposition of *Musca domestica nebulo* Fab. *Indian J. Malar.* **13**, 185–187.
WAY M. J. and BANKS C. J. (1964) Natural mortality of eggs of the black bean aphid, *Aphis fabae* Scop., on the spindle tree, *Euonymus europaeus* L. *Ann. appl. Biol.* **54**, 255–267.
WEARING C. H., CONNOR P. J., and AMBLER K. D. (1973) Olfactory stimulation of oviposition and flight activity of the codling moth *Laspeyresia pomonella*, using apples in an automated olfactometer. *NZ Jl Sci.* **16**, 697–710.
WEARING C. H. and HUTCHINGS R. F. N. (1973) α-farnesene, a naturally occurring oviposition stimulant for the codling moth *Laspeyresia pomonella*. *J. Insect Physiol.* **19**, 1251–1256.
WEBB D. W., PENNY N. D., and MARLIN J. C. (1975) The Mecoptera, or scorpion flies of Illinois. *Bull. Ill. nat. Hist. Surv.* **31**, 251–316.
WEBB R. E. and SMITH F. F. (1970) Survival of eggs of *Liriomyza munda* in chrysanthemum during cold storage. *J. econ. Ent.* **63**, 1359–1361.
WEBBER L. G. (1955) The relationship between larval and adult size of the Australian sheep blowfly *Lucilia cuprina* (Wied.). *Aust. J. Zool.* **3**, 346–353.
WEBER H. (1930) *Biologie der Hemipteren. Eine Naturgeschichte der Schnabelkerfe*, J. Springer, Berlin.
WEBER H. (1931) Lebensweise und Umweltbeziehungen von *Trialeurodes vaporariorum* (Westwood) (Homoptera–Aleurodina). Erster Beitrag zu einer Monographie dieser Art. *Z. Morph. Öekol. Tiere* **23**, 575–753.
WEBER H. (1939) Zur Eiablage und Entwicklung der Elefantenlaus *Haematomyzus elephantis* Piaget. *Biol. Zbl.* **59**, 98–109.
WEBER R. G. and THOMPSON H. E. (1976) Oviposition-site characteristics of the elm leaf beetle *Pyrrhalta luteola* (Mueller) in north-central Kansas (Coleoptera: Chrysomelidae). *J. Kansas ent. Soc.* **49**, 171–176.
WEBSTER F. M. (1893) Methods of oviposition in the Tipulidae. *Bull. Ohio agric. Exp. Stn* (Tech. Ser.) **1**, 151–154.
WEBSTER F. M. and PHILLIPS W. J. (1912) The spring grain-aphis or green bug. *Bull. US Bur. Ent.* **110**, 94–103.
WEFELSCHEID H. (1912) Über die Biologie und Anatomie von *Plea* Leach. *Zool. Jb.* **32**, 387–474.
WEGNER A. M. R. (1955) Biological notes on *Megacrania wegneri* Willemse and *M. alpheus* Westwood. *Treubia* **23**, 47–52.
WEGOREK W. and KRZYMÁNSKA J. (1968) The biochemical causes of resistance in some lupin varieties to the pea aphid *Acyrthosiphon pisum* Harris. *Pr. nauk. Inst. Ochr. Rośl.* **10** (1) 7–30 (in Polish; English summary).
WEIDEMANN G. (1971) Zur Biologie von *Pterostichus metallicus* F. (Coleoptera, Carabidae). *Fauna Ökol. Mitt.* **4**, 30–36.
WEIDNER H. (1936) Beiträge zu einer Monographie der Raupen mit Giftbaaren. *Z. angew. Ent.* **23**, 432–484.
WEIDNER H. (1937) Die Gradflügler der Nordmark und Nordwestdeutschlands. *Verh. naturw. Ver. Hamb.* **26**, 25–64.
WEIDNER H. (1970) Isoptera. In *Handbuch der Zoologie*, Berlin.
WEIDNER H. (1972) Copeognatha (Psocodea). *Handbuch der Zoologie* **4** (2) 2/16, 1–94.
WEIMAN H. L. (1910) A study of germ cells of *Leptinotarsa signaticollis*. *J. Morph.* **21**, 135–216.

WEIR J. S. (1959) Egg masses and early larval growth in *Myrmica*. *Insectes soc.* **6**, 187–201.
WEISER J. (1969) *An Atlas of Insect Diseases*, Shannon, Irish Univ., Prague, Academia.
WEISS A. (1890) Bemerkungen über die Lebensweise eines befruchteten *Hydrophilus piceus*. *Stett. ent. Z.* **50**, 343.
WEISS H. B. (1914) Some facts about the egg-nest of *Paratenodera sinensis*. *Ent. News* **25**, 279–282.
WEISS H. B. (1919) Notes on *Gargaphia tiliae* Walsh, the linden lace-bug. *Proc. biol. Soc. Wash.* **32**, 165–168.
WEISSMAN-STRUM A. and KINDLER S. H. (1963) Hatching of *Aedes aegypti* (L.) eggs, a two-stage mechanism. *J. Insect Physiol.* **9**, 839–847.
WELCH J. (1964) Culturing: *Gnathoceros cornutus*. *Tribolium Infor. Bull.* **7**, 82.
WELCH P. S. (1916) Contribution to the biology of certain aquatic Lepidoptera. *Ann. ent. Soc. Am.* **9**, 159–187.
WELLSO S. G., CONNIN R. V., and HOXIE R. P. (1973) Oviposition and orientation of the cereal leaf beetle. *Ann. ent. Soc. Am.* **66**, 78–83.
WENK P. and RAYBOULD J. N. (1972) Mating, blood feeding, and oviposition of *Simulium damnosum* Theobald in the laboratory. *Bull. Wld Hlth Org.* **47**, 627–634.
WERNER R. A. (1972a) Aggregation behaviour of the beetle *Ips grandicollis* in response to host-produced attractants. *J. Insect Physiol.* **18**, 423–437.
WERNER R. A. (1972b) Aggregation behaviour of the beetle *Ips grandicollis* in response to insect-produced attractants. *J. Insect Physiol.* **18**, 1001–1013.
WERNER R. A. (1972c) Response of the beetle *Ips grandicollis* to combinations of host and insect produced attractants. *J. Insect Physiol.* **18**, 1403–1412.
WESELOH R. M. (1969) Biology of *Cheiloneurus noxius* with emphasis on host relationships and oviposition behavior. *Ann. ent. Soc. Am.* **62**, 299–305.
WESELOH R. M. (1971a) Influence of host deprivation and physical host characteristics on host selection behavior of the hyperparasite *Cheiloneurus noxius* (Hymenoptera: Encyrtidae). *Ann. ent. Soc. Am.* **64**, 580–586.
WESELOH R. M. (1971b) Influence of primary (parasite) hosts on host selection of the hyperparasite *Cheiloneurus noxius* (Hymenoptera: Encyrtidae). *Ann. ent. Soc. Am.* **64**, 1233–1236.
WESELOH R. M. (1972a) Sense organs of the hyperparasite *Cheiloneurus noxius* (Hymenoptera: Encyrtidae) important in host selection processes. *Ann. ent. Soc. Am.* **65**, 41–46.
WESELOH R. M. (1972b) Influence of gypsy moth egg mass dimensions and microhabitat distribution on parasitization by *Ooencyrtus kuwanai*. *Ann. ent. Soc. Am.* **65**, 64–69.
WESELOH R. M. (1974) Host recognition by the gypsy moth larval parasitoid *Apanteles melanoscelus*. *Ann. ent. Soc. Am.* **67**, 583–587.
WESENBERG-LUND C. (1908) Über tropfende Laichmassen. *Int. Rev. ges. Hydrobiol. Hydrogr.* **1**, 869–871.
WESENBERG-LUND C. (1910) Über die Biologie von *Glyphotaelius punctatolineatus* Retz. nebst Bemerkungen über das freilebende Puppenstadium der Wasserinsekten. *Int. Rev. ges. Hydrobiol. Hydrogr.* **3**, 93–114.
WESENBERG-LUND C. (1911) Über die Biologie der *Phryganea grandis* und über die Mechanik ihres Gehäusebaues. *Int. Rev. ges. Hydrobiol. Hydrogr.* **4**, 65–90.
WESENBERG-LUND C. (1913) Fortpflanzungsverhältnisse: Paarung und Eiablage der Süsswasserinsekten. *Fortschr. naturw. Forsch.* **8**, 161–286.
WESENBERG-LUND C. (1943) *Biologie der Süsswasserinsekten*, Copenhagen.
WEST M. J. and ALEXANDER R. D. (1963) Subsocial behaviour in a burrowing cricket *Anurogryllus muticus* (De Geer). Orthoptera: Gryllidae. *Ohio J. Sci.* **63**, 19–24.
WESTDAL P. H. and BARRETT C. F. (1960) Life-history and habits of the sunflower maggot

Strauzia longipennis (Wied.) (Diptera: Trypetidae) in Manitoba. *Can. Ent.* **92**, 481–488.
WESTIGARD P. H., GENTNER L., and BUTT B. A. (1976) Codling moth: egg and first instar mortality on pear with special reference to varietal susceptibility. *Envir. Ent.* **5**, 51–54.
WESTWOOD J. O. (1888) Notes on the life-history of various species of neuropterous genus *Ascalaphus*. *Trans. ent. Soc. Lond.* **1888**, 1–12.
WETMORE A. (1916) Birds of Porto Rico. *Bull. US Dep. Agric.* **326**, 1–140.
WETZEL T. and MENDE F. (1972) Untersuchungen zur Eiablage der Brachfliege (*Leptohylemyia coarctata* Fallén) (Diptera: Anthomyiidae). *Arch. PflSchutz* **8**, 125–132.
WETZEL T., MENDE F., and LUTZE G. (1972) Kontrolle der Eiablage der Brachfliege (*Leptohylemyia coarctata* Fallén) als Voraussetzung für die Befallsprognose und Bekämpfung. *Arch. PflSchutz* **8**, 471–476.
WEYER F. (1928) Untersuchungen über die Keimdrüsen bei Hymenopterenarbeiterinnen. *Z. wiss. Zool.* **131**, 345–501.
WEYER F. (1939) Eistruktur und Rassen bei *Anopheles maculipennis*. *Proc. 7th Int. Congr. Ent.* **3**, 1715–1722.
WEYRAUCH W. K. (1929) Experimentelle Analyse der Brutpflege des Ohrwurmes *Forficula auricularia* L. *Biol. Zbl.* **49**, 543–558.
WHARTON R. H. (1953) The habits of adult mosquitoes in Malaya. IV. Swarming of anophelines in nature. *Ann. trop. Med. Parasit.* **47**, 285–290.
WHEELER R. E. and JONES J. C. (1960) The mechanics of copulation in *Aedes aegypti* (L.) mosquitoes. *Anat. Rec.* **138**, 388.
WHEELER R. M. (1949) The subgenus *Pholadoris (Drosophila)* with descriptions of two new species. *Univ. Texas Publ.* **4920**, 143–156.
WHEELER W. M. (1889) The embryology of *Blatta germanica* and *Doryphora decimlineata*. *J. Morph.* **3**, 291–386.
WHEELER W. M. (1923) *Social Life Among Insects*, New York.
WHEELER W. M. (1928) *The Social Insects. Their Origin and Evolution*, New York.
WHEELER W. M. (1930) *Demons of the Dust*, New York.
WHITCOMB W. D. (1965) The carrot weevil in Massachusetts. Biology and control. *Bull. Mass. agric. Exp. Stn* **550**, 1–30.
WHITCOMB W. H. (1967) Bollworm predators in northeast Arkansas. *Ark. Fm Res.* **16** (3) 2.
WHITCOMB W. H. and BELL R. (1960) Ground beetles on cotton foliage. *Fla Ent.* **43**, 103–104.
WHITE E. B. and LEGNER E. F. (1966) Notes on the life history of *Aleochara taeniata*, a staphylinid parasite of the house fly *Musca domestica*. *Ann. ent. Soc. Am.* **59**, 573–577.
WHITE E. G. and HUFFAKER C. B. (1969) Regulatory processes and population cyclicity in laboratory populations of *Anagasta kühniella* (Zeller) (Lepidoptera: Phycitidae). I. Competition for food and predation. *Res. Popul. Ecol. Kyoto Univ.* **11**, 57–83.
WHITE T. C. R. (1968) Uptake of water by eggs of *Cardiaspina densitexta* (Homoptera: Psyllidae) from leaf of host plant. *J. Insect Physiol.* **14**, 1669–1683.
WHITE T. C. R. (1970) Some aspects of the life history, host selection, dispersal, and oviposition of adult *Cardiaspina densitexta* (Homoptera: Psyllidae). *Aust. J. Zool.* **18**, 105–117.
WHITE W. B. (1970) Radiography to facilitate bagworm egg counts. *J. econ. Ent.* **63**, 910–911.
WHITEHEAD D. L., BRUNET P. C. J., and KENT P. W. (1960) Specificity *in vitro* of a phenoloxidase system from *Periplaneta americana* (L.). *Nature, Lond.* **185**, 610.
WHITEHEAD D. L., BRUNET P. C. J., and KENT P. W. (1965a) Observations on the nature of the phenoloxidase system in the secretion of the left colleterial gland in *Periplaneta americana* (L.). I. The specificity. *Proc. Cent. Afr. Sci. Med. Congr.*, pp. 351–364, Pergamon.
WHITEHEAD D. L., BRUNET P. C. J., and KENT P. W. (1965b) Observations on the nature of the phenoloxidase system in the secretion of the left colleterial gland of *Periplaneta americana*

(L.). II. *Proc. Cent. Afr. Sci. Med. Congr.*, pp. 365–383, Pergamon.

WHITSEL R. B. and SCHOEPPNER R. F. (1970) Observations on follicle development and egg production in *Leptoconops torrens* (Diptera: Ceratopogonidae) with method for obtaining viable eggs. *Ann. ent. Soc. Am.* **63**, 1498–1502.

WHITTEN J. M. (1976) Definition of insect instars in terms of "apolysis" or "ecdysis". *Ann. ent. Soc. Am.* **69**, 556–559.

WIĄCKOWSKI S. K. and KOT J. (1962) Preliminary tests on the biological control of *Cydia funebrana* by the introduction of the egg-parasite *T. cacoeciae*. *Zesz. probl. Postep. Nauk. roln.* **35**, 167–175 (in Polish; English summary).

WICHMANN H. E. (1953) Untersuchungen über *Ips typographus* L. und seine Umwelt. Die Ameisen. *Z. angew. Ent.* **35**, 201–206.

WICHMANN H. E. (1955) Zur derzeitigen Verbreitung des Japanischen Nutzholzborkenkäfers *Xylosandrus germanus* Blandf. im Bundesgebiete. *Z. angew. Ent.* **37**, 250–258.

WICHMANN H. E. (1957) Untersuchungen an *Ips typographus* L. und seiner Umwelt. *Z. angew. Ent.* **40**, 433–440.

WICHT M. C. JR, RODRIGUEZ J. G., SMITH W. T. JR, and JALIL M. (1971) Attractant to *Macrocheles muscaedomesticae* (Acarina) present in the housefly *Musca domestica*. *J. Insect Physiol.* **17**, 63–67.

WIDSTROM N. W. and BURTON R. L. (1970) Artificial infestation of corn with suspensions of corn earworm eggs. *J. econ. Ent.* **63**, 443–446.

WIGGINS G. B. (1973) A contribution to the biology of caddisflies (Trichoptera) in temporary pools. *Life Sci. Contr. R. Ont. Mus.* **88**, 1–28.

WIGGLESWORTH V. B. (1932) The hatching organ of *Lipeurus columbae* Linn. (Mallophaga), with a note on its phylogenetic significance. *Parasitology* **24**, 365–367.

WIGGLESWORTH V. B. (1956) Formation and involution of striated muscle fibres during the growth and moulting cycles of *Rhodnius prolixus* Hemiptera. *Q. Jl microsc. Sci.* **97**, 465–480.

WIGGLESWORTH V. B. (1965) *The Principles of Insect Physiology*, 6th edn, revised, London.

WIGGLESWORTH V. B. (1973) The significance of "apolysis" in the moulting of insects. *J. Ent.* (A) **47**, 141–149.

WIGGLESWORTH V. B. and BEAMENT J. W. L. (1950) The respiratory mechanisms of some insect eggs. *Q. Jl microsc. Sci.* **91**, 429–452.

WIGGLESWORTH V. B. and BEAMENT J. W. L. (1960) The respiratory structures in the eggs of higher Diptera. *J. Insect Physiol.* **4**, 184–189.

WIGGLESWORTH V. B. and SALPETER M. M. (1962) The aeroscopic chorion of the egg of *Calliphora erythrocephala* Meig. (Diptera) studied with the electron microscope. *J. Insect Physiol.* **8**, 635–641.

WIGHTMAN J. A. (1968) A study of oviposition site, mortality and migration in the first (overwintering) generation of *Lygocoris pabulinus* (Fallen) (Heteroptera: Miridae) on blackcurrant shoots (1966–7). *Entomologist* **101**, 269–275.

WIGHTMAN J. A. (1972) The egg of *Lygocoris pabulinus* (Het., Miridae) (Hem.). *NZ Jl Sci.* **15**, 88–89.

WIGHTMAN J. A. and FARRELL J. A. K. (1973) Rearing *Costelytra zealandica* (Coleoptera: Scarabaeidae). 3. Obtaining eggs from field-collected adults. *NZ Jl Sci.* **16**, 21–25.

WIGHTMAN J. A. and FOWLER M. (1974) Rearing *Costelytra zealandica* (Coleoptera: Scarabaeidae). 5. Decontamination of eggs. *NZ Jl Zool.* **1**, 225–230.

WIKLUND C. (1974) Oviposition preferences in *Papilio machaon* in relation to the host plants of the larvae. *Entomologia exp. appl.* **17**, 189–198.

WILBERT H. (1959) Der Einfluss des Superparasitismus auf den Massenwechsel der Insekten. *Beitr. Ent.* **9**, 93–139.

WILBERT H. (1967) Mechanische und physiologische Abwehrreaktionen einiger Blattlausarten (Aphididae) gegen Schlupfwespen (Hymenoptera). *Entomophaga* **12**, 127–137.

WILCKE J. L. (1941) Biologie en Morphologie van *Psylla buxi* L. *Tijdschr. PlZiekt.* **47**, 41–89.

WILDBOLZ T. (1958) Über die Orientierung des Apfelwicklers bei der Eiablage. *Mitt. schweiz. ent. Ges.* **31**, 25–34.

WILDE G. and SCHOONHOVEN A. VAN (1976) Mechanism of resistance to *Empoasca kraemeri* in *Phaseolus vulgaris*. *Envir. Ent.* **5**, 251–255.

WILDE J. DE, SLOOFF R., and BONGERS W. (1960) A comparative study of feeding and oviposition preferences in the Colorado beetle (*Leptinotarsa decemlineata* Say). *Meded. LandbHoogesch. OpzoekStns Gent* **25**, 1340–1346.

WILDE W. H. A. (1970) Common plantain as a host of pear *Psylla* (*Psylla pyricola*: Hem., Hom., Psyllidae). *Can. Ent.* **102**, 384.

WILDERMUTH V. L. and CAFFREY D. J. (1916) The New Mexico range caterpillar and its control. *Bull. US Dep. Agric.* **443**, 1–12.

WILEY G. O. (1922) Life history notes on two species of Saldidae (Hemiptera) found in Kansas. *Kans. Univ. Sci. Bull.* **14**, 299–310.

WILKINSON D. S. (1926) The Cyprus processionary caterpillar (*Thaumetapoea wilkinsoni* Tanis). *Bull. ent. Res.* **17**, 163–182.

WILKINSON J. D. and DAUGHERTY D. M. (1970) The biology and immature stages of *Bradysia impatiens* (Diptera: Sciaridae). *Ann. ent. Soc. Am.* **63**, 656–660.

WILKINSON R. C. (1971) Slash-pine sawfly *Neodiprion merkeli*. 1. Oviposition pattern and descriptions of egg, female larva, pupa and cocoon. *Ann. ent. Soc. Am.* **64**, 241–247.

WILLE H. P. (1950) Untersuchungen über *Psylla pyri* L. und andere Birnblattsaugeranten im Wallis. *Promotionsarbeit Edig. Tech. Hochsch. Zürich*, 113.

WILLE H. and WILDBOLZ T. (1953) Beobachtungen über die Eiablage des Maikäfers und die Entwicklung des Engerlings im Laboratorium. *Mitt. schweiz. ent. Ges. (Bull. Soc. ent. Suisse)* **26**, 219–224.

WILLIAMS C. B. (1913) Some biological notes on *Raphidia maculicollis* Steph. *Entomologist* **46**, 6–8.

WILLIAMS C. B. (1914) *Phytodecta viminalis*, a viviparous British beetle. *Entomologist* **47**, 249–250.

WILLIAMS C. B. and BUXTON P. A. (1916) Oötheca formation in *Sphodromantis guttata* Thunb. *Trans. ent. Soc. Lond.* **1916**, 86–100.

WILLIAMS C. E. (1904) Notes on the life history of *Gongylus gongyloides*, a mantis of the tribe Empusides and a floral simulator. *Trans. ent. Soc. Lond.* **1904**, 125–137.

WILLIAMS C. M., ADKISSON P. L., and WALCOTT C. (1965) Physiology of insect diapause. XV. The transmission of photoperiod signals to the brain of the oak silkworm *Antheraea pernyi*. *Biol. Bull. Woods Hole* **128**, 497–507.

WILLIAMS F. X. (1919) Philippine wasp studies. *Bull. Hawaii. Sug. Ass. Ent. Ser.* **14**, 1–186.

WILLIAMS F. X. (1936) Two water beetles that lay their eggs in the frothy egg masses of a frog or tree toad. *Pan-Pacif. Ent.* **12**, 6–7.

WILLIAMS J. L. (1940) The anatomy of the internal genitalia and the mating behaviour of some Lasiocampid moths. *J. Morph.* **67**, 411–438.

WILLIAMS J. L. (1941a) The internal genitalia of yucca moths and their connection with the alimentary canal. *J. Morph.* **69**, 217–224.

WILLIAMS J. L. (1941b) The internal genitalia of the evergreen bagworm and the relation of the female genital ducts to the alimentary canal. *Proc. Penn. Acad. Sci.* **15**, 53–58.

WILLIAMS J. L. (1945) The anatomy of the internal genitalia of some Coleoptera. *Proc. ent. Soc. Wash.* **47**, 73–91.

WILLIAMS J. R. (1957) The sugarcane Delphacidae and their natural enemies in Mauritius.

Trans. R. ent. Soc. Lond. **109**, 65–110.

WILLIAMS J. R. (1962) The reproduction and fecundity of the sugar cane stalk borer *Proceras sacchariphagus* Bojer (Lep., Crambidae). *Proc. 11th Int. Congr. Cane Sugar Tech.*, pp. 611–625.

WILLIAMS J. R. (1970) Studies on the biology, ecology and economic importance of the sugarcane scale insect *Aulacaspis tegalensis* (Zhnt.) (Diaspididae) in Mauritius. *Bull. ent. Res.* **60**, 61–95.

WILLIAMS L. H. (1972) Anobiid beetle eggs consumed by a psocid (Psocoptera: Liposcelidae). *Ann. ent. Soc. Am.* **65**, 533–536.

WILLIAMS L. H. and MAULDIN J. K. (1974) Anobiid beetle *Xyletinus peltatus* (Coleoptera: Anobiidae), oviposition on various woods. *Can. Ent.* **106**, 949–955.

WILLIAMS R. T. (1970a) *In vitro* studies on the environmental biology of *Goniodes colchici* (Denny) (Mallophaga: Ischnocera). 1. The effects of temperature and humidity on the bionomics of *G. colchici*. *Aust. J. Zool.* **18**, 379–389.

WILLIAMS R. T. (1970b) *In vitro* studies on the environmental biology of *Goniodes colchici* (Denny) (Mallophaga: Ischnocera). 2. The effects of temperature and humidity on water loss. *Aust. J. Zool.* **18**, 391–398.

WILLIAMS T. R. (1968) The taxonomy of the East African river-crabs and their association with the *Simulium neavi* complex. *Trans. R. Soc. trop. Med. Hyg.* **62**, 29–34.

WILLIAMS T. R. (1974) Egg membranes of Simuliidae. *Trans. R. Soc. trop. Med. Hyg.* **68**, 15–16.

WILLIAMSON D. L. (1971) Olfactory discernment of prey by *Medetera bistriata* (Diptera: Dolichopodidae). *Ann. ent. Soc. Am.* **64**, 586–589.

WILLIAMSON E. B. (1901) Manner of oviposition of *Trachopteryx thoreyi*. *Ent. News.* **12**, 1–3.

WILLIS E. R., RISER G. R., and ROTH L. M. (1958) Observations on reproduction and development in cockroaches. *Ann. ent. Soc. Am.* **51**, 53–69.

WILLIS J. H. and BRUNET P. C. J. (1966) The hormonal control of colleterial gland secretion. *J. exp. Biol.* **44**, 363–378.

WILSON B. R. (1960) Some chemical components of the egg-shell of *Drosophila melanogaster*. I. Amino acids. II. Amino sugars and elements. *Ann. ent. Soc. Am.* **53**, 170–173, 732–735.

WILSON C. B. (1923a) Life history of the scavenger water beetle *Hydrous (Hydrophilus) triangularis* and its economic relation to fish breeding. *Bull. US Bur. Fish.* **39**, 9–38.

WILSON C. B. (1923b) Water beetles in relation to pond fish culture with life-histories of those found in fishponds at Fairpont, Iowa. *Bull. US Bur. Fish.* **39**, 231–345.

WILSON D. D., RIDGWAY R. L., and VINSON S. B. (1974) Host acceptance and oviposition behavior of the parasitoid *Campoletis sonorensis* (Hymenoptera: Ichneumonidae). *Ann. ent. Soc. Am.* **67**, 271–274.

WILSON F. (1962) Adult reproductive behaviour in *Asolus basalis* (Hymenoptera: Scelionidae). *Anim. Behav.* **10**, 385.

WILSON F. (1963) *Australia as a Source of Beneficial Insects for Biological Control*, Tech. Comm. no. 3, Commonwealth Inst. Biol. Control, pp. 1–28.

WILSON G. R. and HORSFALL W. R. (1970) Eggs of floodwater mosquitoes. XII. Installment hatching of *Aedes vexans* (Diptera: Culicidae). *Ann. ent. Soc. Am.* **63**, 1644–1647.

WILSON H. F. and MILUM V. G. (1927) Winter protection for the honeybee colony. *Res. Bull. Wis. agric. Exp. Stn* **75**, 1–47.

WILSON L. F. (1959) Branch "tip" sampling for determining abundance of spruce budworm egg masses. *J. econ. Ent.* **52**, 618–621.

WILSON L. F. (1963) Host preference for oviposition by the spruce budworm in the Lake States. *J. econ. Ent.* **56**, 285–288.

WILSON L. F. (1964a) Oviposition site of the spruce budworm *Choristoneura fumiferana* modified by light. *Ann. ent. Soc. Am.* **57**, 643–645.

WILSON L. F. (1964b) Observations on geo-orientation of sprucebudworm, *Choristoneura fumiferana* adults. *Ann. ent. Soc. Am.* **57**, 645–648.
WILSON L. F. (1975) Spatial and seasonal distribution of pine root collar weevil eggs in young red pine plantations. *Great Lakes Ent.* **8**, 115–121.
WILSON M. E. (1962) Laboratory studies on the life history of the palm weevil. *J. agric. Soc. Trin. Tob.* **62**, repr. 9 pp.
WILSON R. L. (1973) Rearing lygus bugs on green beans: a comparison of two oviposition cages. *J. econ. Ent.* **66**, 810–811.
WILTON D. P. (1968) Oviposition site selection by the tree-hole mosquito *Aedes triseriatus* (Say). *J. med. Ent.* **5**, 189–194.
WINBURN T. F. and PAINTER R. H. (1932) Insect enemies of the corn earworm (*Heliothis obsoleta* Fabr.). *J. Kans. ent. Soc.* **5**, 1–28.
WINDELS M. B. and CHIANG H. C. (1975) Distribution of second-brood European corn borer egg masses on field and sweet corn plants. *J. econ. Ent.* **68**, 133.
WINGET R. N., REES D. M., and COLLETT G. C. (1964) Preliminary investigation of the brine flies in the Great Salt Lake, Utah. *Proc. 22nd Ann. Mtg Utah Mosq. Abatem. Ass.*, pp. 16–18.
WINGO C. W., THOMAS G. D., CLARK G. N., and MORGAN C. E. (1974) Succession and abundance of insects in pasture manure: relationship to face fly survival. *Ann. ent. Soc. Am.* **67**, 386–390.
WIRTH W. W. (1971) The brine flies of the genus *Ephydra* in North America (Diptera: Ephydridae). *Ann. ent. Soc. Am.* **64**, 357–377.
WISEMAN B. R., WIDSTROM N. W., and McMILLIAN W. W. (1974) Methods of application and numbers of eggs of the corn earworm required to infest ears of corn artificially. *J. econ. Ent.* **67**, 74–76.
WISHART G., DOANE J. F., and MAYBEE G. E. (1957) Notes on beetles as predators of eggs of *Hylemya brassicae* (Bouché) (Diptera: Anthomyiidae). *Can. Ent.* **88**, 634–639 (1956).
WISHART G. and MONTEITH E. (1954) *Trybliographa rapae* (Westw.) (Hymenoptera: Cynipidae). A parasite of *Hylemya* spp. (Diptera: Anthomyiidae). *Can. Ent.* **86**, 145–154.
WITHYCOMBE C. L. (1923) Notes on the biology of some British Neuroptera (Planipennia). *Trans. ent. Soc. Lond.* **1922**, 501–594.
WITHYCOMBE C. L. (1925) Some aspects of the biology and morphology of the Neuroptera, with special reference to the immature stages and their possible phylogenetic significance. *Trans. ent. Soc. Lond.* **1924**, 303–411.
WITTER J. A. and KULMAN H. M. (1969) Estimating the number of eggs per egg mass of the foresttent caterpillar, *Malacosoma disstria* (Lepidoptera: Lasiocampidae). *Michigan Ent.* **2**, 63–71.
WITTER J. A., KULMAN H. M., and HODSON A. C. (1972) Life tables for the forest tent caterpillar. *Ann. ent. Soc. Am.* **65**, 25–31.
WOGLUM R. S. and MACGREGOR E. A. (1958) Observations of the life history and morphology of *Agulla bractea* Carpenter (Neuroptera: Raphidiodea: Raphidiidae). *Ann. ent. Soc. Am.* **51**, 129–141.
WOGLUM R. S. and MACGREGOR E. A. (1959) Observation on the life history and morphology of *Agulla astuta* (Banks) (Neuroptera: Raphidiodea: Raphidiidae). *Ann. ent. Soc. Am.* **52**, 489–502.
WOHLGEMUTH R. (1967) Über die Ei- und Larvalentwicklung von *Trogoderma angustum* Sol. (Dermestidae). *Anz. Schädlingsk.* **40**, 83–91.
WOKE P. A. (1955) Deferred oviposition in *Aedes aegypti* (Linnaeus) (Diptera: Culicidae). *Ann. ent. Soc. Am.* **48**, 39–46.
WOLCOTT G. N. (1950) The insects of Puerto Rico. *J. agric. Univ. P. Rico* **32**, 1–224 (1948).
WOLFE L. S. (1953) A study of the genus *Uropetala* Selys (order Odonata) from New Zealand.

Trans. R. Soc. NZ **80**, 245–275.
WOLFF M. and KRAUSSE A. (1922) *Die forstlichen Lepidopteren*, Jena.
WOLLASTON T. V. (1862) On the *Euphorbia*-infesting Coleoptera of the Canary Islands. *Trans. ent. Soc. Lond.* Series 3, **1**, 136–189.
WOLLBERG Z. (1966) Attack and oviposition behaviour of *Agrothereutes tunetanus* Haber. (Hymenoptera, Ichneumonidae). *Israel J. Zool.* **15**, 28.
WONG T. T. Y., CLEVELAND M. L., and DAVIS D. G. (1969) Sex attraction and mating of lesser peach tree borer moths. *J. econ. Ent.* **62**, 789–792.
WOO WEI-CHÜN, YIEN YÜ-HUA, and TSAI NIEN-HUA (1965) Distribution pattern of egg masses of European corn borer in corn fields at whorl stage and its practical implications. *Acta ent. sin.* **14**, 515–522 (in Chinese; English summary).
WOOD D. L. (1962) Experiments on the interrelationship between oleoresin exudation pressure in *Pinus ponderosa* and attack by *Ips confusus* (Lec.) (Coleoptera: Scolytidae). *Can. Ent.* **94**, 473–477.
WOOD D. L., BROWNE L. E., SILVERSTEIN R. M., and RODIN J. O. (1966) Sex pheromones of bark beetles. I. Mass production, bio-assay, source, and isolation of the sex pheromone of *Ips confusus* (Lec.). *J. Insect Physiol.* **12**, 523–536.
WOOD G. W. and NEILSON W. T. A. (1956) Notes on the black army cutworm *Actebia fennica* (Tausch.) (Lepidoptera: Phalaenidae), a pest of the low-bush blueberry in New Brunswick. *Can. Ent.* **88**, 93–96.
WOOD T. K. (1974) Aggregating behavior of *Umbonia crassicornis* (Homoptera: Membracidae). *Can. Ent.* **106**, 169–173.
WOOD T. K. (1975) Defense in two pre-social membracids (Homoptera: Membracidae). *Can. Ent.* **107**, 1227–1231.
WOOD T. K. (1976a) Alarm behaviour of brooding female *Umbonia crassicornis* (Homoptera: Membracidae). *Ann. ent. Soc. Am.* **69**, 340–344.
WOOD T. K. (1976b) Biology and presocial behaviour of *Platycotis vittata* (Homoptera: Membracidae). *Ann. ent. Soc. Am.* **69**, 807–811.
WOOD T. K. and PATTON R. L. (1971) Egg froth distribution and deposition by *Enchenopa binotata* (Homoptera: Membracidae). *Ann. ent. Soc. Am.* **64**, 1190–1191.
WOOD-MASON J. (1890) On a viviparous caddis-fly. *Ann. Mag. nat. Hist.* **6** (6) 139–141.
WOODRING J. P. and MOSER J. C. (1970) Six new species of anoetid mites associated with North American Scolytidae. *Can. Ent.* **102**, 1237–1257.
WOODROW D. F. (1965a) The responses of the African migratory locust *Locusta migratoria migratorioides* R. & F. to the chemical composition of the soil at oviposition. *Anim. Behav.* **13**, 348–356.
WOODROW D. F. (1965b) Laboratory analysis of oviposition behaviour in the red locust *Nomadacris septemfasciata* (Serv.). *Bull. ent. Res.* **55**, 733–745.
WOODS W. C. (1917) The biology of the alder flea-beetle. *Bull. Me agric. Exp. Stn* **265**, 250–284.
WOODSIDE A. M., BISHOP J. L., and PIENKOWSKI R. L. (1968) Winter oviposition by the alfalfa weevil in Virginia. *J. econ. Ent.* **61**, 1230–1232.
WOOL D. (1969) Depth distribution of adult and immatures of two *Tribolium castaneum* strains in pure and mixed cultures. *Res. Popul. Ecol. Kyoto Univ.* **11** (2) 137–149.
WRIGHT D. W. and HUGHES R. D. (1959) Controlling cabbage root fly. *New Scient.* **6**, 74–75.
WRIGHT D. W., HUGHES R. D., and WORRALL J. (1960) The effect of certain predators on the numbers of cabbage root fly *Erioischia brassicae* (Bouché) and on the subsequent damage caused by the pest. *Ann. appl. Biol.* **48**, 756–763.
WROBLEWSKI A. (1958) The Polish species of the genus *Micronecta* Kirk. (Heteroptera, Corixidae). *Ann. zool. Warsaw* **17**, 247–381.

Wu Wei-chün, Yen Yü-hua, and Tsai Ning-hua (1970) The distribution pattern of the egg masses of the European corn borer in corn fields at the whorl stage and its practical implications. *Acta ent. sin.* no. **4**, 25–33 (1965, 1968).

Wulker G. (1928) Zur Kenntnis der Stachelbeerblattwespen. *Z. angew. Ent.* **13**, 419–450.

Wygodzinsky P. (1950) Schizopterinae from Angola (Cryptostemmatidae, Hemiptera). *Publ. cult. Mus. Dundo, Comp. Diam. Ang.* **7**, 9–47.

Wygodzinsky P. (1966) A monograph of the Emesinae (Reduviidae, Hemiptera). *Bull. Am. Mus. nat. Hist.* **133**, 614 pp.

Wyk L. E. van (1952) The morphology and histology of the genital organs of *Leucophaea maderae* (Fabr.) (Blattidae, Orthoptera). *J. ent. Soc. S. Afr.* **15**, 3–62.

Wyl E. von (1976) Paragonial proteins of *Drosophila melanogaster* adult male: electrophoretic separation and molecular weight estimation. *Insect Biochem.* **6**, 193–199.

Wyl E. von and Steiner E. (1977) Paragonial proteins of *Drosophila melanogaster* adult male: in vitro synthesis. *Insect Biochem.* **7**, 15–20.

Wylie H. G. (1958) Factors that affect host finding by *Nasonia vitripennis* (Walk.) (Hymenoptera: Pteromalidae). *Can. Ent.* **90**, 597–608.

Wylie H. G. (1960) Some factors that affect the annual cycle of the winter moth *Operophtera brumata* (L.) (Lepidoptera: Geometridae) in Western Europe. *Entomologia exp. appl.* **3**, 93–102.

Wylie H. G. (1963) Some effects of host age on parasitism by *Nasonia vitripennis* (Walk.) (Hymenoptera: Pteromalidae). *Can. Ent.* **95**, 881–886.

Wylie H. G. (1966) Some mechanisms that affect the sex ratio of *Nasonia vitripennis* (Walk.) (Hymenoptera: Pteromalidae) reared from superparasitized housefly pupae. *Can. Ent.* **98**, 645–653.

Wylie H. G. (1967) Some effects of host size on *Nasonia vitripennis* and *Muscidifurax raptor* (Hymenoptera: Pteromalidae). *Can. Ent.* **99**, 742–748.

Wylie H. G. (1970) Oviposition restraint of *Nasonia vitripennis* (Hymenoptera: Pteromalidae) on hosts parasitized by other hymenopterous species. *Can. Ent.* **102**, 886–894.

Wylie H. G. (1971) Oviposition restraint of *Muscidifurax raptor* (Hymenoptera: Pteromalidae) on parasitized housefly pupae. *Can. Ent.* **103**, 1537–1544.

Wylie W. D. (1966) Plum curculio—nonfruit hosts and survival (Coleoptera: Curculionidae). *J. Kans. ent. Soc.* **39**, 218–222.

Wyniger R. (1955) Beobachtungen über die Eiablage von *Libellula depressa* L. (Odonata, Libellulidae). *Mitt. ent. Ges. Basel* (NF) **5**, 62–63.

Xambeau V. (1902) Moeurs et métamorphoses d'insectes. *Échange* **18** (publ. 1898–1902 *Échange* **14–18**, 1–220 special pagination).

Xambeau V. (1903) Instinct de la maternité chez le *Chelidura dilatata* Lafrenoye, orthoptère du groupe des forficuliens. *Le Naturaliste* **25**, 143–144.

Yadava R. L. (1971) Note on the immature stages of *Saccharicoccus sacchari* Cockerell (Pseudococcidae, Hom.). *Indian J. Agric. Sci.* **41**, 1020–1022.

Yakimova N. L. (1968) Some factors affecting the dynamics of numbers of the currant clearwing moth *Synanthedon tipuliformis* Cl. (Lepidoptera, Aegeriidae). *Ent. Obozr.* **47**, 19–30 (English translation: *Ent. Rev.* **47**, 10–16).

Yakubovich V. Ya. (1976) Evaluation of the role of day length and temperature in reactivation of eggs of monocyclic species of the genus *Aedes* in the Moscow region. *Medskaya Parazit.* **45**, 701–704.

YAMAMOTO R. T. (1968) Mass rearing of the tobacco hornworm. I. Egg production. *J. econ. Ent.* **61**, 170–174.

YAMAMOTO R. T., JENKINS R. Y., and MCCLUSKEY R. K. (1969) Factors determining the selection of plants for oviposition by the tobacco hornworm *Manduca sexta. Entomologia exp. appl.* **12**, 504–508.

YAMAOKA K. and HIRAO T. (1971) Role of nerves from the last abdominal ganglion in oviposition behaviour of *Bombyx mori. J. Insect Physiol.* **17**, 2327–2336.

YAMAOKA K. and HIRAO T. (1973) Releasing signals of oviposition behaviour in *Bombyx mori. J. Insect Physiol.* **19**, 2215–2223.

YAMAOKA K. and HIRAO T. (1977) Stimulation of virginal oviposition by male factor and its effect on spontaneous nervous activity in *Bombyx mori. J. Insect Physiol.* **23**, 57–63.

YAMAOKA K., HOSHINO M., and HIRAO T. (1971) Role of sensory hairs on the anal papillae in oviposition behaviour of *Bombyx mori. J. Insect Physiol.* **17**, 897–911.

YANO K. (1968) Notes on Sciomyzidae collected in paddy field (Diptera). I. *Mushi* **41** (15) 189–200.

YASUMATSU K. (1942) Eggs of walking sticks (Phasmidae). *Bull. Takarazuka Insectarium* **18**, 1–20 (in Japanese).

YASUMATSU K. (1953) Some considerations on the reproductive capacity of a Kyushu race of *Anicetus ceroplastis* Ishii, an effective parasite of *Ceroplastes rubens* Maskell in Japan. *Sci. Bull. Fac. Agric. Kyushu* **14**, 7–15 (in Japanese; English summary).

YASUMATSU K. and YAMAMOTO S. (1953) Comparison on the reproductive capacity among *Anicetus ceroplastis* Ishii reared from *Ceroplastes rubens* Maskell feeding on different host plants. *Sci. Bull. Fac. Agric. Kyushu* **14**, 27–33 (in Japanese; English summary).

YATES M. G. (1974) An artificial oviposition site for tree-hole breeding mosquitoes. *Entomologist's Gaz.* **25**, 151–154.

YAZGAN S. and HOUSE H. L. (1970) An hymenopterous insect, the parasitoid *Itoplectis conquisitor*, reared axenically on a chemically defined synthetic diet. *Can. Ent.* **102**, 1304–1306.

YEARGAN K. V. and LATHEEF M. A. (1977) Ovipositional rate, fecundity and longevity of *Bathyplectes anurus*, a parasite of the alfalfa weevil. *Envir. Ent.* **6**, 31–34.

YEUNG K. C. (1934) The life history of the tortoise beetle *Metriona circumtata* Hbst. (Coleoptera, Cassididae). *Lingnan Sci. J.* **13**, 43–162.

YINON U. (1969) The natural enemies of the armored scale lady-beetle *Chilocorus bipustulatus* (Col. Coccinellidae). *Entomophaga* **14**, 321–328.

YOKOYAMA K. (1925) Studies on the Japanese Dermestidae. I. *Rep. seric. Exp. Stn Japan* **7** (2) 65–118 (in Japanese).

YOKOYAMA K. (1929) Studies on the Japanese Dermestidae. 2. Morphology and biology of *Anthrenus (Nathrenus) verbasci* L. *Rep. seric. Exp. Stn Japan* **17**, 705–706.

YOKOYAMA T. and KEISTER M. L. (1953) Effects of small environmental changes on developing silk worm eggs. *Ann. ent. Soc. Am.* **46**, 218–220.

YONCE C. E., PAYNE J. A., and PATE R. R. (1972) Feeding and oviposition preferences of female plum curculios. *J. econ. Ent.* **65**, 1206–1207.

YORK G. T. and PRESCOTT H. W. (1952) Nemestrinid parasites of grasshoppers. *J. econ. Ent.* **45**, 5–10.

YOSHIDA T. (1959) Local distribution of the eggs of the pea weevil *Bruchus pisorum* L. *Mem. Fac. liberal Arts Miyazaki Univ. nat. Sci.* **6**, 11–21.

YOSHIDA T. (1974) Rate of oviposition and effect of crowding on egg cannibalism and pre-adult mortality in *Martianus dermestoides* Chevrolat (Coleoptera, Tenebrionidae). *Sci. Rep. Fac. Agric. Okayama Univ.* **44**, 9–14.

YOSHIMURA S., NAKAMURA Y., KONDO S., NAKASHIMA K., and TADOKORO M. (1971) Position

of eggs laid by the second emergence period moth of rice stem borer *Chilo suppressalis* Walker on the rice plant. *Proc. Ass. Pl. Prot. Kyushu* **17**, 117–120 (in Japanese).
YOUNG A. M. (1965) Some observations on territoriality and oviposition in *Anax junius* (Odonata: Aeshnidae). *Ann. ent. Soc. Am. Baltimore Md* **58**, 767–768.
YOUNG A. M. (1967) Oviposition behavior in two species of dragonflies. *Ohio J. Sci.* **67**, 313–316.
YOUNG A. M. (1973a) The life cycle of *Dircenna relata* (Ithomiidae) in Costa Rica (Lepidoptera). *J. Lepid. Soc.* **27**, 258–267.
YOUNG A. M. (1973b) Notes on the biology of *Phyciodes (Eresia) eutropia* (Lepidoptera: Nymphalidae) in a Costa Rican mountain forest. *Jl NY ent. Soc.* **81**, 87–100.
YOUNG C. J. (1922) Notes on the bionomics of *Stegomyia calopus* Meigen in Brazil. *Ann. trop. Med. Parasit.* **16**, 389–406.
YOUNG E. C. (1964) The eggs of *Diaprepocoris* (Hem., Corixidae). *Entomologist's mon. Mag.* **94**, 87 (1963).
YOUNG J. C., BROWNLEE R. G., RODIN J. O., HILDERBRAND D. N., SILVERSTEIN R. M., WOOD D. L., BIRCH M. C., and BROWNE L. E. (1973) Identification of linalool produced by two species of bark beetles of the genus *Ips. J. Insect Physiol.* **19**, 1615–1622.
YOUNG J. C., SILVERSTEIN R. M., and BIRCH M. C. (1973) Aggregation pheromones of *Ips confusus* isolation and identification. *J. Insect Physiol.* **19**, 2273–2277.
YOUNG W. C. and PLOUGH H. H. (1926) On the sterilization of *Drosophila* by high temperature. *Biol. Bull. Woods Hole* **51**, 189–197.
YU C. C., WEBB D. R., KUHR R. J., and ECKENRODE C. J. (1975) Attraction and oviposition stimulation of seedcorn maggot adults to germinating seeds. *Envir. Ent.* **4**, 545–548.
YUSHIMA T., TAMAKI Y., KAMANO S., and OYAMA M. (1975) Suppression of mating of the armyworm moth *Spodoptera litura* (F.) by a component of its sex pheromone. *Appl. Ent. Zool.* **10**, 237–239.

ZACHARIAE G. (1959) Das Verhalten des Speisebohnenkäfers *Acanthoscelides obtectus* Say (Coleoptera: Bruchidae) im Freien in Norddeutschland. *Z. angew. Ent.* **43**, 345–365.
ZACHARY D., BREHELIN M., and HOFFMANN J. A. (1975) Role of the "thrombocytoids" in capsule formation in the dipteran *Calliphora erythrocephala. Cell Tissue Res.* **162**, 343–348.
ZACHER F. (1939) Der gefleckte Pelzkäfer, *Attagenus pellio* L., ein wichtiger Webwarenschädling. *Mitt. Ges. Vorratsschutz* **15** (3) 29–31.
ZAEVA I. P. (1974) Use of the radioactive marking method for studying trophic relationships of non-specialized entomophagous insects, as exemplified by predacious beetles (Coleoptera, Carabidae). *Ent. Obozr.* **53**, 73–80 (in Russian; English summary).
ZAGULYAEV A. K. and DIN-SI L. (1959) *Lacciferophaga yunnanea* Zaguljaev, gen. et sp. n., Lepidoptera, Momphidae. *Acta ent. sin.* **9**, 306–315 (in Chinese and Russian).
ZAHER M. A. (1960) Effect of adult population density on the silver Y-moth (*Plusia gamma* L.) (Lepidoptera: Noctuidae). *Bull. Soc. ent. Égypte* **44**, 235–240.
ZAHER M. A. and LONG D. B. (1959) Some effects of larval population density on the biology of *Pieris brassicae* L. and *Plusia gamma* L. *Proc. R. ent. Soc. Lond.* (A) **34**, 97–109.
ZAHER M. A. and MOUSSA M. A. (1961) Effects of population density on *Prodenia litura* (Lepidoptera: Noctuidae). *Ann. ent. Soc. Am.* **54**, 145–149.
ZAHER M. A. and MOUSSA M. A. (1963) Effect of larval crowding on the cutworm *Agrotis ypsilon* Rott. (Lepidoptera: Agrotidae). *Bull. Soc. ent. Égypte* **46**, 365–372 (1962).
ZAHER M. A. and SOLIMAN Z. R. (1972) Life history of the predatory mite *Cheletogenes ornatus* (Canestrini and Fanzano) (Acarina: Cheyletidae). *Bull. Soc. ent. Égypte* **55**, 85–89.
ZAĬTSEV V. F. (1976) Parasitic bombyliid flies. *Zashch. Rast.* **11**, 34–35.

ZAKHVATKIN A. A. (1931) Parasites and hyperparasites of the egg-pods of injurious locusts (Acridoidea) of Turkestan. *Bull. ent. Res.* **22**, 385–391.
ZAKHVATKIN A. A. (1934) Dipterous parasites of Acrididae. In Lepeshkin S. N. *et al., Acrididae of Central Asia,* pp. 150–207, Tashkent (in Russian).
ZAKHVATKIN A. A. (1954) Parasites of Acrididae of the Angara region. *Trudȳ vses. ent. Obshch.* **44**, 240–300 (in Russian).
ZANGHERI S. (1956) Un dittero minatore del riso nel Basso Ferrarese (*Hydrellia griseola* Fallen, Dept. Ephydridae). *Boll. Soc. ent. ital.* **86**, 12–16.
ZANGHERI S. and MASUTTI L. (1963) Appunti sulla biologia della *Zeiraphera rumifitrana* H.S. e de altri lepidotteri tortricidi dannosi all'abete bianco in Alto Adige. *Monti Boschi* **14**, 147–157.
ZÁVADSKY K. (1931) *Rhynchites sericeus* und *aeneovirens*—Wasmanns biologische Fremdlinge. *Zool. Anz.* **93**, 102–108.
ZÁVADSKY K. (1936) Zum Problem der vom Trichterwickler (*Deporaus betulae*) erzegten Blatschnittkurven. *Vĕst. česl. zool. Spol.* **4**, 80–85 (in Czech; German summary).
ZEČEVIĆ D. (1955) Winter feeding and development of *Lymantria dispar* under laboratory conditions in 1952–53. *Zashch. Rast.* **28**, 3–20 (in Yugoslav; English summary).
ZEČEVIĆ D. and JANKOVIĆ M. (1959) A contribution to the knowledge of variability of *Lymantria dispar* in Yugoslavia: biometrical analysis of the egg-stage with geographically distant populations. *Zašt. Bilja* **52–53**, 7–14 (in Serbo-Croat; English summary).
ZECH E. (1962) Untersuchungen über die Eiablage des Apfelwicklers (*Carpocapsa pomonella* L.). *NachrBl. dt. PflSchDienst* (NF) **16**, 7–14.
ZEHRING C. S., ALEXANDER A., and MONTGOMERY B. E. (1962) Studies of the eggs of Odonata. *Proc. Indiana Acad. Sci.* **72**, 150–153.
ZELEDON R. (1956) Anotaciones sobre una curiosa oviposición de la mosca del tórsalo en condiciones experimentales (Diptera: Cuterebridae). *Rev. biol. trop. San Josè* **4**, 179–181.
ZELEDON R., GUARDIA V. M., ZUÑIGA A., and SWARTZWELDER J. C. (1970) Biology and ethology of *Triatoma dimidiata* (Latreille, 1811). II. Life span of adults and fecundity and fertility of females. *J. med. Ent.* **7**, 462–469.
ZELENÝ J. (1969) A biological and toxicological study of *Cycloneda limbifer* Casey (Coleoptera, Coccinellidae). *Acta ent. bohemoslov.* **66**, 333–344.
ZERILLO R. T. (1975) A photographic technique for estimating egg density of the white pine weevil, *Pissodes strobi* (Peck). *Res. Paper, Northeast. For. Exp. Stn For. Serv. USDA NE* **318**, 4.
ZICKAN J. F. (1944) *Considerações sôbre a metamorphóse dos insétos,* 44 pp., Rio de Janeiro.
ZIMIN L. S. (1938) Les pontes des Acridiens. Morphologie, classification et écologie. *Tabl. anal. Faune URSS* **23**, 1–84.
ZINNA G. (1959) Ricerche sugli insetti entomofagi. I. Specializzazione entomoparassitica negli Encyrtidae: Studio morfologico, etologico e fisiologico del *Leptomastix dactylopii* Howard. *Boll. Lab. Ent. agr. Portici* **18**, 1–150.
ZINNA G. (1961) Ricerche sugli insetti entomofagi. II. Specializzazione entomoparassitica negli Aphelinidae: studio morfologico, etologico e fisiologico del *Coccophagus bivittatus* Compere, nuovo parassita del *Coccus hesperidum* L. per l'Italia. *Boll. Lab. Ent. agr. Portici* **19**, 301–357.
ZIPRKOWSKI L. and ROLANT F. (1972) Study of the toxin from poison hairs of *Thaumetopoea wilkinsoni* caterpillars. *J. Invest. Dermat.* **58**, 274–277.
ZIVOJINOVIĆ S. and VASIĆ K. (1963) A first contribution to knowledge of the Japanese oak Saturniid *A. yamamai*, a new forest pest in Yugoslavia. *Zashch. Rast.* **14**, 491–508 (in Yugoslav; English summary).
ZOCCHI R. (1957) Insetti del cipressi. I. Il gen. *Phloeosinus* Chap. (Coleoptera, Scolytidae) in

Italia. *Redia* **41**, 129–225.

ZOHREN E. (1968a) Das Eiablageverhalten der Kohlfliege *Chortophila brassicae* Bouché. *Z. angew. Ent.* **61**, 445–446.

ZOHREN E. (1968b) Laboruntersuchungen zu Massenanzucht, Lebensweise, Eiablage und Eiablageverhalten der Kohlfliege, *Chortophila brassicae* Bouché (Diptera, Anthomyiidae). *Z. angew. Ent.* **62**, 139–188.

ZWÖLFER H. (1969) Rüsselkafer mit ungewöhnlicher Lebensweise Koprophagie, Brutparasitismus und Entomophagie in der Familie der Curculionidae. *Mitt. schweiz. ent. Ges.* **42**, 185–196.

ZWÖLFER W. (1931) Studien zur Ökologie und Epidemiologie der Insekten. I. Die Kiefereule, *Panolis flammea* Schiff. *Z. angew. Ent.* **17**, 475–562.

ZWÖLFER W. (1933) Studien zur Ökologie, insbesondere zur Bevölkerungslehre der Nonne, *Lymantria monacha* L. *Z. angew. Ent.* **20**, 1–50.

Species Index

Volume I: Pages 1–474. Figs. 1–135. Plates 1–155
Volume II: Pages 475–778. Figs. 136–296

Abacetus afer 216
— *dainelli popovi* 216
Abantiades magnificus 712
Abax 272
— *ater* 650, 771; Fig. 73
— *parallelus* 649
Abedus 58, 635
— *macronyx* 635
Ablattaria laevigata 273, 298
Abraxas grossulariata 717
Acalypta parvula 611
Acanalonia 565, 769; Pl. 37
— *bonducellae* 769; Fig. 48
— *viequensis* 130
Acanaloniidae 130, 769
Acanophora compressa 270
— *femoralis* 270
— *marginata* 270, 292
— near *pallescens* 270
— *nigricornis* 270
Acanthiophilus helianthi 33, 743
Acanthocinus aediles 686
Acantholyda hypotropica 215
Acanthops falcataria 502; Fig. 158
Acanthoscelides 179, 684
— *collusus* 684
— *compressicornis* 684
— *fraterculus* 684
— *obtectus* 20, 41, 47, 61, 62, 683, 684
— *submuticus* 684
Acanthosoma 591
— *griseum* 271, 591
Acanthosomatidae 271, 272, 570, 591, 770; Fig. 212
Acanthosomidae 590
Acari 213, 226
Acarophenax tribolii 227
Acaropsis docta 227

Acaulona brasiliana 756
Acentropus niveus 714
Achaea janata 36
Achaetella (=Cordilura) 756
Achaetoneura samiae 753
Acheta 205
— *assimilis luctuosus* 228
— *domesticus* 179, 206, 209, 767; Fig. 68
Achilidae 565
Achilixiidae 565
Achlya flavicornis 774; Pl. 73
Achroia grisella 56
Acidia heraclei 79
Acilius 164
— *sulcatus* 650
Acinopterus angulatus 15
Aclerda campinensis 223
Aclerdidae 565
Acleris variana 35
Acompocoris pygmaeus 610
Acontia dacia 218
Acraea 48, 251
Acrida 519
— *bicolor* 184
— *confusa* 527
— *turrita* 527
Acrididae 4, 14, 46, 82, 98, 163, 164, 166, 169, 171, 174, 175, 184, 219, 513, 515, 518–532, 767–768
Acridinae 767
Acridium 519
Acridoidea 522; Fig. 174
Acridomorpha 513
Acroceridae 27, 54, 738–739
Acrocomia 490
Acrolepia 715
— *assectella* 33
Acromis spinifex 273, 772; Fig. 128

Acronyctinae 775
Acrosternum hilare 596
Acrotylus deustus 223
— *patruelis* 528, 767; Fig. 173
Actebia fennica 25
Aculonina 755
Acutipula 726
Acyphas 257
Acyrthosiphon pisum 569
Adactylidium 213
Adalia 675
— *bipunctata* 19, 50, 173, 219, 224, 675, 772; Fig. 257
Adapsila flaviseta 57
Adelges cooleyi 216
— *piceae* 178, 218, 219, 223
— *prelli* 219
Adelgidae 565
Adelidae 713
Adelphocoris seticornis 613
Adephaga 649–654, 771
Adleria kollari 515
Adoxophyes Fig. 21
— *orana* 91
Aechmea 490
Aedes 45, 46, 52, 75, 179, 206, 212, 216, 217, 220, 221, 314, 315, 729, 732, 733
— *aegypti* 27, 43, 45, 47, 49, 59, 75, 80, 81, 92, 175, 206, 315, 732, 733, 776; Fig. 20; Pl. 106
— *albopictus* 47
— *annandalei* 776; Pl. 111
— *annulipes* 776; Pl. 107
— *atropalpus* 776; Pl. 110
— *aurantius* 776; Pl. 111
— *cantans* 776; Pl. 107
— *caspius* 776; Pl. 109
— *cinereus* 776; Pl. 113
— *communis* 776; Pl. 108
— *detritus* 776; Pl. 108
— *dorsalis* 732, 776; Pl. 109
— *flavescens* 776; Pl. 108
— *geniculatus* 776; Pl. 113
— *hexodontus* 732
— *infirmatus* 776; Pl. 110
— *leucomelas* 776; Pl. 109
— *mediovittatus* 776; Pl. 110
— *nigripes* 34
— *nigromaculis* 776; Pl. 112
— *pembaensis* 57

— *polynesiensis* 776; Pl. 111
— *pseudoscutellaris* 776; Pl. 108
— *punctor* 776; Pl. 107
— *rusticus* 776; Pl. 107
— *serratus* 57
— *sierrensis* 732
— *sollicitans* 75
— *sticticus* 776; Pl. 109
— *taeniorhynchus* 49, 75, 776; Pl. 110
— *togoi* 776; Pl. 106
— *triseriatus* 75, 776; Pl. 112
— *vexans* 776; Pl. 113
— *vittatus* 776; Pl. 110
— *woodi* 776; Pl. 112
Aedimorphus (=*Aedes*)
Aegeria myopaeformis 51
— *pictipes* 24, 48, 51
— *tipuliformis* 51
Aegeriidae 24, 35, 51
Aelia acuminata 601, 602
— *klugi* 600
— *rostrata* 602
— *virgata* 601
Aeneolamia varia 178
Aeolothripidae 214
Aeolothrips fasciatus 214
— *mekleucus* 228
Aepophilus 571, 572, 574
— *bonnairei* 271, 571
Aeropedallus clavatus 524
Aeropus sibericus 14
Aeshna 734
— *cyanea* 766; Fig. 146
— *mixta* 766; Fig. 146
Aeshnidae 766
Aesiocopa patulana 258, 774; Fig. 102
Aetalion reticulatum 270, 769; Pl. 48
Aetalionidae 270
Afrohippus leai 768; Fig. 170
Afronurus 735
Agabus 164, 651
— *bipustulatus* 17, 651, 653, 771; Figs. 241, 242
— *chalconatus* 649, 651
— *nebulosus* 651
— *sturmii* 651
Agalliopis 315
Agapophyta 594
— *viridula* 594
Agelaius xanthomus 237

Agelastica alni 681
Ageneotettix 221
— *deorum* 524
Ageniaspis 238
Aglais 101, 715
— *urticae* 101, 107, 715, 774; Fig. 24; Pl. 64
Aglia tau 774; Fig. 87
Agonini 649
Agonoderus comma 216
Agonum dorsale 650
— *pusillum* 216
Agraptocorixa gestroi 770; Fig. 224
— *hyalinipennis* 617
Agria affinis 74
Agrilus 72
— *arcuatus* 17
— *s. populneus* 88
Agrion splendens 486
— *virgo* 486
Agriopocorinae Fig. 209
Agriopocoris froggati 769; Fig. 209
Agriotes 177
— *mancus* 17
— *ponticus* 17
Agriotypidae 53
Agriotypus 110
— *armatus* 148
Agriphila ruricollela 774; Pl. 61
Agromyza demeijeri 92
Agromyzidae 86, 762
Agropyron repens 76
Agrotidae 35, 104, 250, 714, 775
Agrotinae 775
Agrotis orthogonia 35
— *puta* 775; Pl. 77
— *repleta* 25
— *segetum* 25, 231
— *ypsilon* (=*ipsilon*) 40
Agrotoidea 256
Agrypnia picta 192
Aiolocara miribilis 231
Aiolopus 170
— *thalassinus* 175, 528
— *turnbulli plagosus* 525, 767; Fig. 171
Akiceridae 767
Alcerda 235
Alcidodes affaber 20
— *bubo* 20
Aleochara 231, 659
— *bilineata* 218

— *bipustulata* 218
— *languinosa* 231
Aleocharinae 54, 659
Alesia discolor 231
Aleurodidae (=Aleyrodidae) 231
Aleyrodidae 175, 565, 708, 769
Aleyrodoidea 170, 565, 569
Allochrysa virginica 229
Alloeotomus 105, 106
Allophorina 755
Allophyes oxyacanthae 775; Pl. 74
Allothrombium fuliginosum 213, 227
Alluaudoymia needhami 736
Alniphagus 279
Alophora hemiptera 292, 777; Fig. 292
Alsophila aescularia 258, 713
Alydidae 570, 588–589; Fig. 210
Alydinae 588
Alydus calcaratus 588
Amara equistris 216
— *impuncticollis* 216
— *obesa* 216
— *pastica* 216
Amarini 230
Amauris 48
Amblycera 549
Amblycerus 684
— *vitis* 684, 773; Fig. 264
Amblymerus bruchophagi 22
Amblyseius aleyrodis 227
— *hibisci* 227
ambrosia beetle 70
Ambrostoma quadriimpressum 19
Ambrysus 636, 637
Ameles 233
— *abjecta* 502
— *decolor* 502
Ameletus ludens 476
Amiota 136
Amorbus alternatus 769; Fig. 209
Amorphoscelidae 766
Amorphoscelinae 766
Amorphoscelis pulchra 766; Fig. 155
Amphibicorisae 164, 166, 570, 574–582, 769
Amphigerontia 546
— *bifasciata* 546
— *contaminata* 546
Amphimallon caucasicum 17
— *majalis* 17, 217
— *solstitialis* 739

Amphipyra livida 250
Amphizoidae 649
Amrasca splendens 566
Amylostereum 74
Amyrsidea 554
— *latifasciata* 554
— *minuta* 768; Fig. 186
Anablepia granulata 528, 768; Fig. 173
Anabremia inquilina 27
Anacaena 658
— *infuscata* 771
Anactinothrips 563
Anadaptus nivalis 216
Anaphe 257
— *venata* 257
Anaphes conotracheli 708
— *flavipes* 90, 238
Anaplectinae 497
Anastatus 237
— *amelophagus* 233
Anastoechus nitidulus 222
— *sibericus* 222
Anastrepha ludens 80, 743
Anatis quinquedecimpunctata 231
Anax papuensis 766; Fig. 147
Ancyrosoma leucogrammes 599
Andrena 739
Anechura bipunctata 14, 270
Aneurus laevis 769; Pl. 31
Angitia fenestralis 21, 91
Anicetus ceroplastis 37
Anisandrus 305
— *dispar* 773; Fig. 129
— *obesus* 273
Anisoctenion alacer 53
Anisodactylus californicus 216
Anisolabis eteronoma 224
— *littorea* 269
— *maritima* 269, 289
Anisomorpha 536
— *burpestoides* 533
Anisoplia 172
— *austriaca* 171
Anisops 637
Anisoptera 486, 487, 766; Figs. 146, 148
Anisota senatoria 238
Anisotomidae 4
Anobiidae 18, 37, 83
Anobium punctatum 18, 178

Anomala horticola 63
Anomalococcus 236
Anomma nigricans 758
Anommatocoris 611
— *minutissimus* 611
Anopheles 45, 75, 115, 116, 146, 266, 268, 315, 644, 728, 729, 730, 741
— *albitarsis* 731
— *atroparvus* 75
— *cinereus* 731
— *claviger* 730
— *culicifacies* 49, 731
— *donaldi* 731
— *farauti* 775; Pl. 103
— *gambiae* 730, 731, 775; Pls. 101, 103
— *hyrcanus* 731
— *labranchiae atroparvus* 27, 775; Pls. 102, 103
— *lesteri* 730
— *maculipennis* 729, 730, 731
— *melas* 75, 775; Pl. 101
— *merus* 776; Pls. 101, 102
— *multicolor* 731
— *pharoensis* 776; Pl. 101
— *philippinensis* 49
— *plumbeus* 731
— *pseudopunctipennis* 730, 775; Pl. 104
— *punctulatus* 44
— *quadrimaculatus* 732, 775; Pls. 101, 102, 103
— *sacharovi* 730
— *sinensis* 730
— *sineroides* 730
— *stephensi* 731, 776; Pl. 102
— *subpictus* 731
— *sundaicus* 731, 776; Pls. 101, 103
— *turkhudi* 731
— *walkeri* 730
— *yatsushiroensis* 730
Anophelinae 51, 775–776
Anoplocnemis 586, 594
— *phasiana* 585
Anoplura 9, 549, 551, 554, 555, 562, 764, 768
Antarctophthirius callorhini 768; Pl. 20
— *ogmorhini* 561, 768; Pl. 20
Antennata 10, 47
Antestia lineaticollis 29
Anthaxia 739
Antheraea 81, 116, 159, 177, 183

Antheraea mylitta 206, 207
— *pernyi* 159, 206, 207, 774; Pl. 92
— *yamamai* 26, 206, 207, 774; Pl. 93
Anthicidae 220
Anthicus haldemani 220
Anthocorid 770
Anthocoridae 15, 212, 215, 228, 543, 570, 610, 770
Anthocoris 610
— *musculus* 215
— *nemoralis* 229
— *nemorum* 215, 228
Anthomyia illocuta 224
Anthomyiidae 86, 124, 757, 762, 777; Fig. 294
Anthonomus 83, 313
— *grandis* 48, 49, 71, 80, 88, 315
— *pomorum* 20
Anthrax jazykovi 222
— *monarchus* 222
— *oophagus* 222
Anthrenus flavipes 82, 178
— *museorum* 17
— *verbasci* 18, 36, 49, 178
Anthribidae 84, 220
Antianthe expansa 270, 769; Fig. 2
— *foliacea* 270
Antichaeta 743, 744
Antiteuchus macraspis 272
— *mixtus* 272
— *piceus* 272
— *tripterus* 272, 294, 770; Fig. 123
— *variolosus* 272
ants 30, 56, 221, 226, 260, 269, 502
Anurogryllus muticus 269, 286, 767; Fig. 117
Anysis 220
— *agilis* 213
Aonidiella aurantii 42, 74, 234
Apalus 676
Apanteles 706
— *angaleti* 61
— *dignus* 54
— *glomeratus* 21, 92, 706, 773; Fig. 280
— *melanoscelus* 21, 74
— *thompsoni* 706, 773; Fig. 280
Apatele aceris 775; Fig. 87
— *alni* 775; Pl. 77
— *euphorbiae* 250
Apateticus 596
Aphalara nebulosa 568

Aphaniptera 10
Aphelinidae 4, 22, 52, 53, 84, 233, 237, 707, 709
Aphelinus asychis 22, 91
— *mali* 22
— *semiflavus* 22, 91
Aphelocheirus 109, 149, 636
— *aestivalis* 159
Apheta affinis 270
aphid eggs 232, 234
Aphidencyrtus 52
Aphididae 4, 15, 175, 565
Aphidius 706
Aphidoidea 170, 565, 569
aphids 63, 82, 89, 94, 216, 227, 229; Fig. 12
Aphis fabae 237, 569
— *gossypii* 236
— *pomi* 249
Aphobetoideus 220
Aphodius 277
Aphrophora 169
Aphrophoridae 169, 170
Aphycus stanleyi 91
Aphylidae 570, 607
Aphylum bergrothi 607
Aphytis 42, 233, 234
— *africanus* 42
— *chrysomphali* 42, 73, 233
— *coheni* 42, 74
— *holoxonthus* 42
— *lingnanensis* 42
— *melinus* 42, 74
— *proclia* 220
Apidae 84, 704, 709
Apiocera maritima 740
— *painteri* 740
Apioceridae 86, 740
Apiomerus 617
Apiomorphidae 565
Apion 59, 84
— *assimile* 59
— *dichroum* 59
— *seniculus* 59
— *virens* 59
Aplopus mayeri 244
Apocephalus 776; Fig. 103
Apocrita 704, 706–709
Apoderus 278
— *coryli* 773; Fig. 111
Apontanodes globosus 219

Aporia crataegi 216, 229
Apteropanorpidae 722
Apterygota 9
Arachnocoris dispar 608, 609
Aradidae 271, 294, 570, 582, 769
Aradoidea 102
Araecerus fasciculatus 225
Araschnia levana 242, 774; Fig. 95
Archips rosanus 24
Archipsocus fernandi 546
Archirileya inopinata 220
Arctia caja 775; Pl. 78
— *villica* 775; Pl. 80
Arctias selene gnoma 206, 207
Arctiidae 104, 236, 256, 716, 721, 775
Arenivaga 766; Fig. 153
Arenocorinae 588; Fig. 208
Arenocoris 58
Argia emma 765; Fig. 142
Argidae 274
Argiolaus maesa 241
Argopistes sexvittatus 677
Argrayles 710
Argynnis adippe (= *Argynnis cydippe*) 713
— *aglaja* 250
— *cydippe* 713
— *paphia* 774; Pl. 65
— *selene* 774; Pl. 65
Argyrotaenia velutinana 225
Arma custos 229, 597, 770; Fig. 215
Armigeres 732
— *flavus* 52
— *subalbatus* 776; Pl. 112
Aromia moschata 685
Arphia pseudonietana 524, 767; Fig. 171
— *xanthoptera* 524, 767; Fig. 171
Arundo 738
Ascalaphidae 643, 645, 647, 648, 771; Fig. 240
Ascaloptynx furciger 647, 771; Fig. 240
Ascogaster 238
Aseminae 688
Asilidae 86, 222, 739, 740
Asilinae 740
Asilus 740
Asolcus mitsukurii 94
Asolus basalis 709
Aspergillus 92
— *niger* 91
Aspidomorpha 188, 190, 191

— *adherens* 188
— *confinis* 772; Fig. 80
— *puncticosta* 772; Figs. 80, 81
— *tecta* 191, 772; Fig. 83
Aspidoproctus 714
— *xyliae* 221
Aspidotus 231
Aspongopus 593
Asterolecaniidae 565
Atella 251
Atemeles 226
Athalia rosae 21
Atherix 732
— *ibis* 776; Fig. 288
— *variegata* 737
Athripsodes cinereus 23
— *senilis* 23
Atractotomus mali 214, 612
Atrichomelina 744
— *pubera* 743
Atta 226, 773; Fig. 103
Attagenus alfierii 33
— *megatoma* 48, 82
— *pellio* 18, 82
— *piceus* 18, 60, 178
Attelabus 278
— *nitens* 773; Fig. 277
Atya africana 735
Auchenorrhyncha 130, 169, 565, 566, 768, 769
Aulacaspis 221
— *tegalensis* 15
Aulacigaster 745
Aulacigastridae 745
Aulacosternum nigrorubrum 769; Fig. 209
Aulicus 213
— *terrestris* 219
Austeniella cylindrica 526, 767; Fig. 173
Austracris guttulosa 224
Austroicetes cruciata 175, 222
— *pusilla* 222
Austroplatypus incompertus 310
Automeris 241, 249, 258, 774; Fig. 94
— *io* 241
Azelia 777; Fig. 296
Azteca 87
Azygophleps scalaris 24
Azyra orbigera 232

Bacillinae 536
Bacillus 534, 536, 539
— *libanicus* 536, 537, 768; Figs. 178, 179
Bactrodema 536
Badistica ornata 526, 767; Fig. 172
Baetidae 12, 477, 479, 765
Baetis 476, 735
— *dorieri* 479
— *pumilus* 12, 479
— *rhodani* 475, 479
— *subatrebatinus* 477, 479
— *vernus* 479
Bagrada stolata 602, 770; Fig. 216
Balaninus 84
banana borer weevil 230
— stalk borer 87
Barbitistes 515
bark beetles 73, 76
Basipta 188
— *stolida* 772; Fig. 82
Bathyplectes curculionis 91
Batrachedra arenosella 235
— *silvatica* 235
Beauveria 92
bed bugs 55, 71, 178
bees 3, 30, 51, 54, 183, 226, 232, 269, 285, 310, 739
beetle eggs 313
beetles 3, 48, 57, 66, 70, 71, 73, 183, 226, 282, 291, 312, 315, 739
Belaphopsocus bandonneli 768; Fig. 184
Bellura gortynoides 258
Belostoma 58, 635, 771; Pl. 46
Belostomatidae 116, 166, 170, 175, 570, 635–636, 771
Belostominae 58
Bembidion 230
— *frontale* 216
— *lampros* 216
— *litorale* 216
— *musicola* 216
— *nitidum* 216
— *oppositum* 216
— *quadrimaculatum* 216
— *trechiforme* 217
— *ustulatum* 217
Bemisia tabaci 569
Bena fagana 775; Pl. 78
Beraeidae 23, 710, 711
Bereodes minuta 23

Bergiana cyanocephala 274
Berosinae 186, 656, 771
Berosus 658
— *luridus* 186, 187, 658
— *perigrinus* 771; Fig. 245
Berothidae 643
Berytidae 215, 229
Berytinidae (= Berytidae) 570, 583, 769
Berytinus 583
— *signoreti* 583, 769; Pl. 31
Bessa 755
— *selecta* 755
Bethylidae 23, 274, 709
Bibio 733
Bibionidae 733–734
Bibioninae 733, 734
Bilimekia 292, 293
— *broomfieldi* 293
— *styliformis* 270, 291, 292, 769; Figs. 119, 121
Bimba toombi 742
Binoculus pennigerus 12, 478
Biston 717
— *atrataria* 712, 717
— *betularia* 712, 717, 774; Pl. 72
Bittacidae 10, 722, 723
Bittacus 722, 723
— *apicalis* 733
— *pilicornis* 723, 775; Pl. 99
Blaberidae 13, 285, 495
Blaberinae 201, 205
Blaberoidea 495
Blaberus 203, 205
— *craniifer* 50, 284
— *discoidalis* 201
— *giganteus* 284
black widow spider 237
Blastobasidae 221, 235
Blastobasis coccivorella 235
— *lecaniella* 235
— *thelymorpha* 235
Blastophagus (= *Myelophilus*) 67, 70
Blastothrix 704
Blatta 201, 203
— *orientalis* 184, 201, 203, 237, 284
Blattaria 9, 175, 269, 285, 495–499, 766
Blattella 203, 497, 499
— *germanica* 13, 184, 203, 284, 497, 499, 766; Figs. 152, 153
— *humbertiana* 284

Blattella vaga 203, 284
Blattellidae 13, 285, 495, 497, 766; Fig. 153
Blattellinae 497; Fig. 116
Blattidae 13, 495
Blattinae 201, 766
Blattisocius keegani 213, 227
Blattodea 9, 495
Blattoidea 495
Blattopteroidea 9
Bledius 298, 301, 772; Fig. 125
— *arenarius* 301
— *arenoides* 301, 772; Fig. 125
— *bicornis* 301
— *diota* 301, 772; Fig. 125
— *furcatus* 301
— *fuscipes* 301, 772; Fig. 125
— *longulus* 301
— *opacus* 301
— *pygmaeus* 301
— *spectabilis* 273, 301, 772; Fig. 125
— *tricornis* 301, 772; Fig. 125
Blepharidopterus angulatus 214, 612
Blepharoceridae 86
Blepharopsis mendica 503
Blissus leucopterus insularis 584
blow fly 3, 65, 255
Boarmia punctinalis 717
Bolbites 275
— *onitoides* 773; Fig. 108
Bolbonata aspidistrae 271
boll weevil 48, 75
Bombycidae 85, 206, 716, 721, 774
Bombyliidae 51, 54, 86, 213, 222, 223, 236, 739, 761
Bombylius fugax 51
— *major* 739
— *variabilis* 51
Bombyx 77, 199, 201
— *mandarina* 207
— *mori* 77, 200, 205, 206, 207, 209, 218, 219, 715, 716, 718, 774; Pl. 97
Boopedon nubilum 221
Borboridae (= Sphaeroceridae) 746–748, 777; Figs. 289, 290
Borborophilus 130
— *primitiva* 131, 622, 623, 624, 632, 634
Borborus ater 746, 747, 777; Fig. 289
Boreidae 10, 722
Boreus 722, 723
Bostrychidae 18, 83

bot fly 57
Bothynus gibbosus 230
Bourletiella hortensis 214
Brachinini 649
Brachinus cyanipennis 650
Brachycantha 231
Brachycarenus tigrinus 589
Brachycaudus klugisti 569
Brachycentrus 711
— *subnubilus* 23, 194
Brachycerus harrisella 479
Brachycrotaphus steindachneri 528
Brachylaena discolor Fig. 82
Brachymeria 52
— *femorata* 23
— *intermedia* 74
Brachypelta aterrima 271, 589
Brachystola magna 524
Brachytarsus nebulosus 220
Brachytrupes achatinus 269
Brachytrupinae 286
Bracon 314
— *hebetor* 31, 42
— *lineatellae* 21
— *mellitor* 21
— *serinopae* 42
Braconidae 21, 52, 64, 84, 94, 175, 238, 706, 709, 773; Fig. 280
Bradynotes obesa 525, 767; Fig. 171
Bradysia impatiens 27
Braula 744
— *cocca* 744
Braulidae 744–745
Brenthidae 77
Brephos perthenias 717
— *puella* 717
Brochymena 596
Brontispa longissima 677, 772; Fig. 259
Bruchidae 20, 83, 178, 179, 683–685, 689, 773
Bruchidius alfierii 49, 684
— *baudoni* 684
— *schoutendi* 684
Bruchus 684
— *analis* 683
— *brachialis* 225
— *pisorum* 20, 41, 683
— *quadrimaculatus* 20, 72
— *rufimanus* 684
Brueelia 768; Fig. 186

Brumus suturalis 219
Bryocorinae 105, 106, 166, 612, 613
Bryocoris 613
— *pteridis* 612, 613
Bucerophagus 551, 552
— *africanus* 768; Fig. 186
— *productus* 768; Pl. 21
Buenoa 637
bugs 149, 179, 312, 313
Bupalus piniarius 33, 35, 40, 41, 237, 717
buprestid 5
Buprestidae 17, 83
Bursina 169
butterflies 3, 48, 179, 226, 241, 251, 286
Byctiscus 278
— *populi* 773; Fig. 275
Byrsotria 203
— *fumigata* 284
Byturidae 19
Byturus tomentosus 19

cabbage moth 90
cabbage-root fly 230, 231
Cacoecia crataegana 76
— *oporana* 91
— *patulana* 91
Cadra cautella 60
Caecilidae 768
Caenidae 13, 479, 765
Caenis 479, 765; Fig. 139
— *horaria* 13, 479, 765; Fig. 139
— *macrura* 479
— *moesta* 13, 479, 765; Fig. 139
— *robusta* 475, 479, 765; Fig. 139
Caenocoris nerii 251
Cafius xantholoma 661, 772; Fig. 250; Pl. 53
Calerucella lineola 219
Calidea 594
Callimomidae 52, 238
Callineda testudinaria 19
Calliophrys (= Limnophora) 152
Calliphora 90, 151, 179, 213, 226, 230, 254, 255
— *erythrocephala* 28, 255, 751, 752, 777; Figs. 32, 33, 34, 35; Pl. 132
— *vicina (= erythrocephala)* 92
— *vomitoria* 159, 751, 752

Calliphoridae 28, 54, 86, 113, 151, 179, 213, 223, 751–752, 762, 777
Calliphorinae 777
Calliptaminae 175, 522, 767
Calliptamus italicus 216, 222
— *palaestinensis* 175
— *turanicus* 222, 767; Fig. 174
Callosamia 177
Callosobruchus 61, 179, 684
— *analis* 61, 62
— *chinensis* 41, 60, 61, 62, 77, 89, 683
— *maculatus* 41, 61, 62
— *rhodesianus* 41
Callostoma desertorum 222, 739
— *fascipennis* 222
— *soror* 236
Calocoris biclavatus 214
Calodexia 56, 753
Caloglyphus 287
Calosoma 753
— *calida* 217
— *externa* 217
— *scrutator* 217
— *sycophanta* 16
— *willcoxi* 217
Calvia quattuordecimguttata 219
Camnula karschi 528, 767; Fig. 173
— *pellucida* 216, 217, 222, 223
Camponotus 221, 291
— *abdominalis* 291
— *herculeanus* 221
Campoplex haywardi 21
Camposomatinae (=Clytrinae) 678
Camptobrachis lutescens 612
Camptopus lateralis 588
Campylenchia nutans 271
Campylomma verbasci 612
Campyloneura 105
— *virgula* 612, 770; Fig. 27
Cantacader infuscata 611
Cantao ocellatus 272
Cantharidae 218
Cantharsius 301
Canthon 273, 301
— *pilularius* 773; Fig. 265
Canthonina 301
Canthophorus sexmaculatus 770; Fig. 212
Capniidae 508
Capnodis tenebrionis 17
Capsidae 611–613

Capsodes 105, 106
Capus meriopterus 612
carabid beetles 34
Carabidae 16, 83, 216, 230, 234, 272, 649–650, 688, 771
Carabus 650
— *auratus* 50
— *cancellatus* 771; Fig. 115
Caraphractus cinctus 49, 52, 651, 709
Carausius 534, 539, 542, 543, 545
— *morosus* 178, 539, 768; Figs. 175, 177, 180–182; Pl. 15
Carcelia 756
— *cheloniae* 777; Fig. 292
— *gnava* 777; Fig. 293
— *obesa* 82
Carcinophoridae 269
Cardiaspina albitextura 64
— *densitexta* 77
Cardiochiles 90, 93
— *nigriceps* 56, 92, 93
Cardioperla nigrifrons 766; Fig. 162
Carphoborus 282
Carpocoris 770; Fig. 212
— *lunulatus* 601
— *pudicus* 594, 602
Carpophilus hemipterus 19, 34, 315
— *humeralis* 19
— *obsoletus* 19
Carulaspis 214
Carya ovata 87
Caryanda agomena 527
Caryedon 684
— *albonotatum* 683, 684
— *fasciatum* 683
— *fuscus* 34
— *gonagra* 33, 40, 42, 684
Caryobruchus 684
Casca 707
Cassida 188, 191
— *algerica* 189
Cassidinae 183, 188, 191, 303, 676, 772; Fig. 80
Castiniomera humboldti 87
Catantopinae 175, 522, 767
Catantops melanostictus 527, 767; Fig. 173
— *spissus* 527
Catantopsilus taeniolatus 527, 767; Fig. 173
Catharsius 273
Catoblemma sumbavensis 236

Catocala 63
Catorhintha selector 585, 586
Caudothrips buffai 563
Cecidomyiidae 27, 86, 221, 236, 761
Cedria paradoxa 274
Celama sorghiella 26
Cellia (= *Anopheles*)
Centeter 755
Centroptilum 475, 477
— *lituratum* 479, 765; Fig. 137
— *luteolum* 12, 475, 479, 765; Fig. 137
— *pennulatum* 479, 485
Centrotinae 270
Cephalonomia gallicola 23, 274
Ceraleptus gracilicornis 585
Cerambycidae 20, 77, 83, 685–688, 773
Cerambycinae 687, 773
Ceramica pisi 775; Pl. 76
Ceratina dupla 234
Ceratitis capitata 33, 79, 742
Ceratocombus coleoptratus 640, 771; Pl. 48
Ceratomegilla fuscilabris 219
— *maculata* 219
Ceratopogonidae 736, 761
Cercopidae 565
Cercotmetus 130, 162, 621, 622, 628, 634
— *asiaticus* 162, 622, 628, 629, 631, 632; Figs. 49, 232, 234
— *fumosus* 622, 628, 629, 632, 635, 771; Fig. 232
Cercyon 658
Cerobasis guestfalicus 546
Ceromasia auricaudata 777; Fig. 292
Ceropales 234
Ceroplastes 235, 236
— *egbarium* 221
— *grandis* 220
— *pseudoceriferus* 15
— *rusci* 220
Cerotoma 313
Cerura 243
— *modesta* 716, 775; Pl. 83
— *vinula* 48, 101, 718, 719, 775; Fig. 95; Pls. 81, 89
Cerurinae 775
Cerylonidae 689
Cethera musiva 614, 617
Ceuthorhynchidius 84
Ceuthorhynchus 313

Ceuthorhynchus assimilis 56
— *quadridens* 773; Fig. 274
Chaetocyptera bicolor 755
Chaetogaedia 53
Chaetophleps 57, 753
— *setosa* 57, 777; Fig. 14
Chaetopodella scutellaris 747
Chalcididae 23, 52, 84, 773
Chalcidoidea 704
Chalcodermus 84
Chamaemyiidae 223, 237
Chaonia ruficornis 718, 775; Pl. 81
Chapinia 768; Fig. 186
Charaeas graminis 775; Pl. 77
Charaxes 251
— *ethalion* 251
Chauliodes pectinicornis 642, 643
Chauliops bisontula 583
— *rutherfordi* 583
Cheilomenes lunata 224
Cheiloneurus noxius 91
Cheletogenes ornatus 227
Cheleutoptera 9, 83, 163, 178, 244, 533–545, 764, 768
Chelidura pyrenaica 270
Chelidurella acanthopygia 14, 270
Chelisoches morio 228, 270
Chelisochidae 270
Chelonus 238
— *annulipes* 21
Chepuvelia usingeri 579, 769; Fig. 204
Chermidae (=Psyllidae) 567–568
Chiasmia clathrata 25
Chilo 214
— *auricilia* 25
— *polychrysus* 80
— *suppressalis* 88, 209
Chilocorini 231
Chilocorus bipustulatus 221
— *nigritus* 19
— *rubidus* 219, 224
Chilomenes vicina 231
Chilosiinae 776
Chiloxanthus pilosus 572, 769; Fig. 198
Chion cinctus 686
Chirista compta 528, 767; Fig. 173
Chironitis pamphilus 773; Fig. 271
Chironomidae 27, 139, 735–736, 776; Fig. 287
Chironomus 237, 735, 776; Fig. 287

— *dorsalis* 776; Fig. 287
— *luridus* 92
— *plumosus* 27
— *riparius* 92
— *tentans* 27
— *tepperi* 27
Chirothrips 563
Chitonophora (= *Ephemerella*)
Chlaeniini 650
Chlaenius 230, 650
— *aestivus* 650
— *discopictus* 217
— *obesus* 217
— *quadrinotatus* 217
— *sexmaculatus* 217
— *stellatus* 217
— *transversalis* 217
Chlamisinae (=Clytrinae) 678
Chlamydinae (=Clytrinae) 678
Chlorochroa 596
— *persimilis* 215
Chlorocoris atrispinus 272
— *depressus* 272
Chlorophorus strobilcola 686
Chloropidae 86, 237, 762
Choeridium ganigerum 773; Fig. 273
Chorisia 497
Choristidae 10, 722
Choristoneura conflictana 35
— *fumiferana* 72, 315
Chorizagrotis auxiliaris 60
Chorosoma schillingii 769; Pl. 29
Choroterpes picteti 475, 478, 765; Fig. 137
Chorthippus 82, 521
— *apricarius* 237
— *biguttulus* 237
— *brunneus* 175
— *parallelus* 768; Fig. 169
Chortoicetes 82, 231
Chortophaga viridifasciata 524
Chrotogonus senegalensis 527, 767; Fig. 173
Chrysididae 234
Chrysis dichroa 234
Chrysocharis laricinellae 75
— *melaenis* 92
— *seiuncta* 76
Chrysocloa 676
Chrysolina menthastri 681, 772; Fig. 260
Chrysomela crotchi 20, 48, 220, 221
— *populi* 681

Chrysomela tremula 681
— *varians* 676
Chrysomelidae 19, 20, 37, 40, 83, 111, 188, 231, 273, 303, 676–683, 772
Chrysomelinae 772
Chrysomya bezziana 28
Chrysopa 116, 117, 119, 295, 313, 643, 648, 771; Pl. 50
— *adspersa* 215
— *albicornis* 215
— *californica* 216
— *carnea* 216, 229, 643, 645
— *chi* 643
— *downesi* 229
— *flava* 643, 645, 771; Fig. 239
— *formosa* 216
— *formosana* 16
— *harrisi* 229
— *lanata* 216
— *lineaticornis* 229
— *madestes* 16
— *majuscula* 91
— *nigricornis* 229, 643
— *oculata* 16, 49, 216, 229, 644, 645
— *perla* 224
— *plorabunda* 16, 216, 229
— *ramburi* 216
— *rufilabris* 16, 229, 643
— *septempunctata* 216
— *vulgaris* 216
Chrysophilus nubeculus 222
Chrysopidae 11, 16, 215, 643, 644, 645, 648, 771; Fig. 239
Chrysops 72, 236, 237, 738
Cicada 76, 233
Cicadellidae 565, 566, 769
Cicadetta dimissa 15
— *montana* 566
Cicadidae 15, 169, 170, 565, 566
Cicadoidea 565, 566, 768
Cicadomorpha 169, 608–617
Cicindela 650
— *campestris* 650
— *purpurea* 49, 650
Cicindelidae 83, 216, 282, 649, 650, 771; Fig. 114
Cicindelinae 282
Cidaria sociata 250
Cimbicidae 274
Cimex hemipterus 770; Pl. 35

— *lectularius* 15, 610, 770; Pl. 36
Cimicidae 570, 770
Cimicinae 570
Cimicomorpha 167, 171, 570, 608, 613, 770
Cirfula 258
Cirrospilus ovisugosus 220
Cixiidae 565
Clania cameri 712
Clastopteromyia inversa 136
Clemacantha regalis 768; Pl. 17
Clematis vitalba 242
Clemora smithi 17
clerid flies 70
Cleridae 19, 213, 219, 689
click beetle 48
Climaciella brunnea 16
Clinoceroides glaucescens 52, 139, 223, 756, 777; Pl. 129
Cloeon 476
— *dipterum* 12, 475, 476, 485
— *inscriptum* 485
— *simile* 12, 475, 479
Clonopsis gallica 533
Clytanthus pilosus 686
Clytra 679
— *laeviuscula* 678, 772; Figs. 260, 261
— *longipes* 772; Fig. 261
— *quadripunctata* 772; Fig. 261
Clytrinae 191, 678, 679, 681, 772; Figs. 261, 262
Clytus arietis 685
Cnephasia 235
Cobboldia 749
Cobboldinae 749
Coccidae 15, 220, 223, 233, 236, 293, 565
Coccidiphila gerasimovi 235
— *ledereriella* 235
Coccidiplosis 236
Coccinella 313
— *novemnotata* 19, 43, 219
— *quinquepunctata* 219
— *septempunctata* 19, 219, 224, 225
— *undecimpunctata* 228
Coccinellidae 19, 83, 219, 225, 231, 675, 689, 772
Coccinia 742
— *indica* 742
Coccoidea 565, 570
Coccomyza leefmansi 221
Coccophagus 52, 53, 707, 708

Coccophagus basalis 54
— *bivittatus* 94
— *capensis* 708
— *gurneyi* 91, 708
— *lycimnia* 708
— *ochraceus* · 708
— *trifasciatus* 708
Coccothera spissana 221
Coccotrypes carpophagus 21
Coccus 235, 236
— *africanus* 234
— *hesperidum* 91
— *magniferae* 15
— *viridis* 219
Cochliomyia hominivorax 44, 46, 47
cockroaches 44, 56, 70, 90, 184, 203, 204, 209, 237, 251, 269, 284, 312, 753
codling moth 63, 239, 314
Codophila varia 597, 603, 770; Fig. 215
Coelaenomenodera elaeidis 41
Coelinidea 238
Coelopa frigida 132
— *pilipes* 132
Coelopidae 132, 745
Coelostoma 658
Coenagriidae 765
Coenagrion mercuriale 765; Fig. 146
— *puellum* 486, 765; Fig. 141
— *pulchellum* 486, 765; Figs. 146, 150
Coenus delius 595
Colcindamyia auditrix 76
Colemania sphenaroides 223
Coleolaelaps 213
Coleomegilla 231
— *maculata lengi* 224
Coleophora laricella 33, 75
— *leucapennella* 712
— *serratella* 227
Coleoptera 10, 16, 20, 40, 41, 52, 54, 72, 80, 83, 88, 91, 100, 163, 164, 175, 178, 183, 205, 206, 209, 216, 224, 230, 272, 273, 274, 297, 649–703, 715, 763, 764, 771–773
Coleopterocoris 637
— *kleerekoperi* 637
Coleorryncha 565, 566
Colias 712
— *eurytheme* 50
— *p. eurytheme* 34
Collembola 10, 166, 213, 214

Collops 218
— *balteatus* 218, 231
— *marginellus* 218
— *vittatus* 218, 231
Collyria 238
Collyrinae 282
Collyris 650
— *bonelli* 650
— *emarginatus* 650
— *tuberculata* 650
Colobathristidae 570, 585
Colorado potato beetle 80, 229
Colpocephalum turbinatum 224
Colpurinae 588; Fig. 209
Columbicola columbae 768; Fig. 187
Colymbetinae 651
Compsilura concinnata 53, 753
Conchaspididae 565
Conchyloctenia 188
— *tigrina* 772; Fig. 80
Coniopterygidae 643, 645, 648, 771; Fig. 239
Coniopterygoidea 648
Conium maculatum 242
Conocephalinae 515, 518, 767
Conocephalus discolor 518
— *dorsalis* 518, 767; Fig. 166
— *fuscus* 13
— *longipennis* 214
Conoderus vespertinus 48, 178
Conomelus limbatus 233, 566, 567, 769; Fig. 193
Conophthorus 279
Conopidae 56, 776
Conotrachelus 84
— *nenuphar* 43, 48, 88
Contarinia nasturtii 27
— *tritici* 179
Conwentzia pienticola 230
Copidosoma 238
Copiphorinae 767
Coprina 301
Coprini 275, 276, 301, 302
Copris 276, 286, 301
— *arizonensis* 273
— *armatus* 273, 773; Fig. 126
— *fricator* 273
— *hispanicus* 227
— *hispanus* 273, 302
— *incertus* 273

Copris lugubris 273
— *lunaris* 273
— *minutus* 273
— *remotus* 273, 277
— *tullius* 273
Copromyza 132
— *equina* 116
— *pedestris* 748
— *sordidus* 747, 748, 777; Fig. 289
Copromyzidae 748
Coprophila 132, 133
— *acutangula* 132, 777; Fig. 290
— *lugubris* 132, 747, 777; Fig. 290
— *pusilla* 747
Coptocephala unifasciata 772; Fig. 262
Coptosoma 770; Fig. 218
— *cribrarium* 770; Pl. 30
— *scutellatum* 586, 770; Fig. 210
Coptosomidae (= Plataspidae) 607
Coptotomus 651
Coraebus rubi 17
Coranus 616
Corcyra cephalonica 25, 37, 214, 227, 229
Cordilura deceptiva 756, 777; Pl. 127
Cordiluridae (= Scatophagidae) 151, 223, 756
Cordulegaster boltonii 487, 766; Fig. 144
— *dorsalis* 487
Cordulegasteridae 486, 766
Cordulia aenae 766; Figs. 148, 150
Corduliidae 478, 766
Coreidae 58, 167, 271, 570, 585–588, 769; Figs. 208, 209
Coreinae 58, 588; Fig. 209
Coreoidea 102
Coreus marginatus 585, 769; Pl. 29
Coridius cuprifer 593
Corimelaeninae 589
Corioxenos antestiae 29
Corixa 617, 618, 770; Fig. 224
— *panzeri* 618, 770; Fig. 226
— *punctata* 618, 770; Figs. 72, 227
Corixidae 51, 106, 166, 167, 175, 238, 570, 617–621, 770–771
Corizinae 587
corn borer 231
Corthylus 304, 306
— *punctatissimus* 305
— *schaufussi* 305
Corydalidae 16, 642, 771

Corydalus cornutus 16, 220, 642, 643, 771; Pl. 49
Coryna 232
Corynoneura scutellata 27
Coryphosima centralis 528
Cosmolestes 585
— *bimaculata* 596
Cosmopterygidae 235
Cossidae 24, 37, 84, 717, 720, 774
Cossus cossus 24
Costelytra 315
Cozola 257
crabs 57, 735
Crambinae 714, 774
Crambus harpipterus 37
— *laqueatellus* 25
— *mutabilis* 774; Pl. 60
— *pascuellus* 774; Pl. 60
— *topiarius* 80
Craspedonotus tibialis 649
Cremastogaster 87
Cremastus 64
— *interruptor* 64
Cremifania nigrocellulata 223
Creophilus 116, 171
— *maxillosus* 662, 772; Figs. 36, 252; Pl. 55
cricket eggs 180
crickets 90, 251
Cricotopus trifasciatus 27
Crinopterygidae 713
Criocerinae 677, 681
Crioceris asparagi 677
Croce 645
Croesia semipurpurana 77
Crossotarsus saundersi 310
Crunobia (= *Pedicia*)
Cryphalus 305
— *fulvus* 231
Cryphocricos barozzi 637
Crypsinus angustatus 598
Cryptoblabes gnidiella 25
Cryptocephalinae (= Clytrinae) 678
Cryptocephalus 679
— *aureolus* 772; Fig. 261
— *bipunctata* 772; Fig. 261
— *quinquepunctata* 772; Fig. 262
Cryptocercidae 495, 766; Fig. 153
Cryptocercus punctulatus 284, 766; Fig. 153
Cryptognatha nodiceps 231
Cryptogonus orbicularis 231

Cryptolaemus montrouzieri 19, 234
Cryptolestes ferrugineus 224
— *pusillus* 34
— *turcicus* 213
— *ugandae* 34
Cryptomorpha flavoscutellaris 223
Cryptorrhynchus 84
Ctenomorphodes tessulatus 768; Pl. 17
Ctenophora guttata 27
Ctenostoma ichneumoneum 771; Fig. 114
Ctesias serra 218
Cucujidae 83
Cuculliinae 775
Culex 43, 45, 46, 116, 146, 251, 252, 253, 254, 256, 265, 266, 314, 315, 316, 729, 732
— *autogenicus* 730
— *fatigans* 75
— *pipiens* 47, 63, 220, 252, 253, 316, 715, 730, 731, 776; Pl. 105
— *salinarius* 75
— *tarsalis* 63
Culicella 732
Culicidae 3, 27, 85, 111, 117, 727–733, 751, 761, 775–776
Culicinae 732, 776
Culicoides 736
Culiseta 251, 252, 256, 732
Curculio elephas 88
Curculionidae 20, 77, 83, 278, 304, 688, 689, 773; Figs. 275–277
Cuterebria 751
Cuterebridae 751, 777
Cybister fimbriolatus 651
Cybocephalidae 19
Cybocephalus semiflavus 19
Cycloneda s. limbifer 19
— *sanguinea* 232
Cyclopelta 594
Cyclorrhapha 3, 44, 97, 260, 286, 563, 645
Cyclotornidae 54
Cydia funebrana 72, 214, 215, 216
— *inopinata* 238
— *molesta* 48
— *nigricana* 238
— *pactolana* 80
— *pomonella* 33, 80, 212, 214, 215, 228, 229
— *prunifoliae* 58
Cydnidae 271, 570, 589–590, 770; Figs. 211, 212

Cylapinae 131
Cylapofulvius 770; Fig. 50
Cylindrococcidae 565
Cylindromyia brassicaria 777; Fig. 292
Cylindromyiina 755
Cylindrotoma 724
Cylindrotominae 724, 725
Cymatia 618
— *coleoptrata* 617, 618, 771; Fig. 225
Cymatophoridae 250, 774
Cymbiodyta 658
— *marginella* 186, 187
Cynarmostis vectigalis 235
cynipid wasps 313
Cynipidae 84, 707, 773; Fig. 280
Cynipoidea 52, 704
Cynomyia 230
Cyrtacanthacridinae 175, 522, 767
Cyrtidae (=Acroceridae) 738–739
Cyrtonotum cuthbertsoni 223
Cyrtopogon 740
Cyrtorhinus lividipennis 228
— *mundulus* 213, 214
Cytherea fenestralula 223
— *infuscata* 223
— *obscura* 223
— *setosa* 739
— *transcaspica* 223
Cyzenis albicans 28

Dacnonypha 10, 713, 715
Dactylopiidae 565
Dactylotum pictum 525, 767; Fig. 171
Dacus 73, 742
— *ciliatus* 742
— *cucurbitae* 73, 92, 742
— *dorsalis* 742
— *oleae* 41, 48, 743
— *tryoni* 31, 80, 742
Dahoen goerita 244
Dalbulus 315
Damalinia bovis 549, 551, 768; Pl. 26
— *equi* 262, 549, 550, 558, 768; Fig. 189
— *ovis* 557, 558, 561, 768; Figs. 188, 191
Danaus 225, 251
— *gilippus* 50
— *plexippus* 251
Darninae 270
Dasineura citri 80

Dasylophia anguina 718, 775; Pl. 82
Dasyneura 56
— *brassicae* 56
Dasyphora 117
— *cyanella* 760
Dasypogoninae 740
Dasypoliinae 775
Datana ranaceps 718, 775; Pl. 82
Debridae 565
Decticinae 515, 518, 767
Decticus verrucivorus 518, 767; Fig. 166
Deilinia variolaria 774; Pl. 70
Delia antiqua 28
Delphacidae 565, 566, 769
Delphacodes fairmairei 769; Fig. 193
— *pellucida* 567
Deltochilum 276
Dendroctonus 66, 67, 69, 70, 227, 279
— *brevicomis* 66, 69, 70, 773; Fig. 18
— *frontalis* 66, 67, 69, 215, 773; Fig. 18
— *jeffreyi* 67
— *monticolae* 60
— *piceaperda* 773; Fig. 112
— *ponderosae* 66, 67, 68, 179, 236, 773; Fig. 19
— *pseudotsugae* 67, 68, 69, 71, 773; Fig. 112
— *terebrans* 66
— *valens* 773; Fig. 113
Dendrolimus pini 774; Fig. 87
— *punctatus* 35
— *spectabilis* 179
Dendrotettix 83
Dendrotrypum 305
Dentifibula 236
Deporaus 278
— *betulae* 773; Fig. 276
Deprana binaria 774; Pl. 73
Depranidae 774
Depressaria marcella 221
Deraeocorinae 105, 131, 612, 613
Deraeocoris 105, 106, 171, 215
— *fasciolus* 215
— *olivaceus* 613
— *ruber* 171, 612, 613
Dericorys 223
— *albidula* 767; Fig. 174
Dericorythinae 767
Dermaptera 9, 14, 214, 224, 228, 269, 287, 763, 764, 768
Dermapteroidea 9

Dermatobia 58
— *cyaniventris* 751
— *hominis* 57, 751, 777; Fig. 15
Dermestes ater 218
— *coarctatus* 218
— *erichsoni* 218
— *lardarius* 18, 218, 225
— *maculatus* 18, 49, 178
— *mustelinus* 219
Dermestidae 4, 17, 36, 83, 218, 219, 231, 689
Derobrachus brevicollis 687, 773; Pl. 57
Deroceras laeve 237
Derodontidae 17, 218
Derula flavoguttata 600
Deuteronomos alniaria 774; Pl. 69
— *fuscantaria* 774; Pl. 69
Dexia ventralis 753
Dexiinae 755
Diabrotica 57, 313, 315, 677, 681
— *balteata* 223
— *cristata* 772; Fig. 263
— *u. howardi* 34
— *virgifera* 80, 772; Fig. 263
Diadromus collaris 30
Diapheromera 534
— *femorata* 178, 245
Diaphnidia pellucida 215
Diaprepocoris zealandiae 618
Diaprepsis abbreviatus 227
Diaspididae 565
Diataraxia oleracea 181, 200, 241
Diatraea lineolata 25
— *saccharalis* 25, 214, 221, 238
Diceratothrips 563
Dichotomius carolinus 773; Fig. 270
Dichroplus arrogans 222
— *conspersus* 222
Dicranocephalus (agilus?) 769; Pl. 29
Dicranolaius villosus 231
Dicromorpha viridis 524, 768; Fig. 171
Dictya 743
Dictyophara 169
Dictyopharidae 169, 170, 565
Dictyoptera 9, 13, 56, 82, 183, 184, 201, 269, 495–507, 764, 766
Dicymolonia julianalis 211
Dicyphinae 106, 613
Dicyphus 105, 106
— *pallicornis* 613
Didymuria 179, 542, 545

Didymuria violescens 533, 539, 545, 768; Fig. 183; Pl. 16
Dielocerus 274
Digamasellus quadrisetus 227
Digelasinus 274
Dineutes americanus 49, 654
Dinidoridae 570, 593
Dinocras 509
— *cephalotes* 13, 508, 509, 766; Fig. 161; Pl. 4
Dinoderus minutus 18
Dioctria 740
Diogma 724
Diogmites 740
Dioryctria abietivorella 315
Diplacodes 490
Diplatys 269
Diplazon 238
— *fissorius* 92
Diplazoninae 706
Diplochus (= *Sphaerodema*) 635
Diplodoma 258
Diplonychus 58
Diploptera punctata 13, 184, 201, 203, 285, 286
Diplopteridae 201
Diplura 166
Diprionidae 705
Dipsocoridae (= Cryptostemmatidae) 167, 570, 640, 771
Dipsocoroidea 570, 640–641
Diptera 3, 10, 27, 28, 41, 44, 46, 52, 54, 56, 72, 80, 85, 88, 92, 93, 98, 100, 129, 144, 149, 175, 179, 183, 205, 209, 221, 227, 236, 260, 286, 645, 715, 724–762, 763, 764, 775–778
Dirhicnus alboannulatus 22
Dirphia 258
Discocoris vianai 614
Discosmoecus atripes 192
Dissosteira carolina 524, 767; Fig. 171
— *longipennis* 524, 767; Fig. 171
Ditrysia 10, 35, 44, 48, 102, 257, 713, 716, 717, 774–775; Fig. 281
Dixa 223, 756
Dixidae 52, 139
Dociostaurus 82
— *brevicollis* 768; Fig. 174
— *kraussi* 222
— *maroccanus* 175, 218, 219, 220, 222, 223, 236, 768; Fig. 168
Dolichocera 82
Dolichonabis limbatus 609
Dolichopeza (= *Oropeza*) 727, 775; Fig. 286
dolichopodid flies 70
Dolichopodidae 86, 222, 236
Doliopygus conradti 773; Fig. 134
Dolycoris baccarum 602, 770; Fig. 217
Donacia 714; Fig. 150
— *crassipes* 677
Donaciinae 677
Donusa prolixa 536
Dorcadium 686
Dorisiana 169
Dorylinae 56
Dorylus nigricans 251
Dorysthenes (= *Opisognathus*) *forficulatus* 686
Draeculacephala mollipes 566
Drepana binaria 774; Pl. 73
Drilidae 17
Drilus flavescens 17
Drino munda 28
Drosophila 33, 35, 39, 43, 45, 46, 64, 86, 110, 125, 126, 128, 129, 151, 201, 315; Fig. 57
— *algonquin* 92
— *bandeirantorum* 135
— *busckii* 92
— *euronotus* 135
— *funebris* 137
— *gibberosa* 162, 776; Fig. 55; Pl. 126
— *gibbinsi* 237
— *hypocausta* 135
— *innubila* 136
— *lebanonensis* 134
— *melanica* 135
— *melanogaster* 28, 31, 41, 46, 47, 49, 92, 137, 210, 776; Figs. 7, 8, 9, 56; Pl. 125
— *nannoptera* 135
— *paramelanica* 135
— *phalerata* 136
— *pseudoobscura* 33, 776; Pl. 125
— *quinaria* 136
— *rellima* 136
— *simulans* Figs. 7, 8
— *virilis* 776; Pl. 126
Drosophilidae 28, 129, 134, 162, 223, 237, 748, 762, 776
Drusus annulatus 23, 194
Dryinidae 53, 709
Dryomyia lichtensteinii 92

Dryomyza flaveola 132, 744, 776; Figs. 51, 52
Dryomyzidae 132, 744, 776
Dryopidae 52
Dyroderes umbraculatus 601, 770; Fig. 216
Dysaphis plantaginea 227
Dysdercus 585, 769; Fig. 207
— *cingulatus* 49
— *fasciatus* 36, 43, 585, 769; Pl. 31
dytiscid beetles 174
Dytiscidae 17, 51, 83, 649, 650–653, 771
Dytiscus 164, 651, 653, 708
— *marginalis* 17, 50, 173, 651, 771; Fig. 241
— *semisulcatus* 653

Earias vittella 88
earwigs 90, 214, 287
Eccoptopterus spinosus 305
Ecdyonurus 476
— *fluminum* 12, 488, 765; Fig. 140
— *forcipula* 478, 765; Fig. 140
— *helviticus* 12, 477, 478, 485
— *insignis* 478
Echinomyodes 53, 753
Echinophthiriidae 551, 768
Echinophthirius horridus 554, 768; Pl. 20
Echtropsis porteri 234
Eciton 753
Ecphoropsis perdistinctus 64
Ectobiinae 201, 497
Ectobius panzeri 284
Ectomyelois ceratoniae 25, 87
Ectropis bistortata 712
— *crepuscularia* 712
Egle aestiva 757, 777; Fig. 294
— *cinerella* 757, 777; Fig. 294; Pl. 144
Elachertus 53
Elaphothrips brevicornis 270
Elaphrus 749
Elasmidae 94
Elasmostethus interstinctus 272, 591
Elasmucha betulae 770; Fig. 212
Elassogaster sepsoides 223
Elassoneuria 735
Elateridae 17, 83, 175
Elgiva 744
Elipsocus hyalinus 546
— *mclachlani* 546
— *westwoodi* 546

Ellida caniplaga 718, 775; Pl. 86
elm leaf beetle 233
Elmidae 51
Elvisura 591
Embia major 13
Embidopsocus enderleini 546, 548
Embiidae 13
Embiophila 613
— *myersi* 610
Embioptera 9, 13, 269, 512, 766
Emesinae 614
Emmesomyia variabilis 757, 777; Pl. 151
Empicoris culiciformis 614, 617
— *vagabundus* 770; Fig. 221
Empididae 740–741, 776
Empis aerobatica 740
— *livida* 740
Empoasca devastans 88
— *facialis* 88
Empusa 499
Empusidae 766
Empusinae 766
Enallagma aspersum 490
— *cyathigerum* 486
Enargopelte 233
Enarmonia formosana 24
— *rufimitrana* 76
Encarsia 238
— *formosa* 94
Enchenopa bifenestrata 271
— *binotata* 271
— *caruata* 271
— near *sericea* 271
Encoptolophus costalis 524
— *sordidus* 221, 524
Encyrtidae 22, 52, 53, 84, 94, 113, 131, 144, 220, 233, 238, 707, 773
encyrtids 30, 91
Encyrtus 53
— *fuliginosus* 30
Endiastus caviceps 270
Endopterygota 9
endopterygotes 1, 2, 3
Enicocephalidae 570, 641
Enicocephaloidea 570, 641
Enithares 637
Ennomos 717
— *autumnaria* 713
— *subsignarius* 35, 713
Enochrus 658

Enochrus affinis 186, 187
— *testaceus* 186, 187
Enoclerus 71
— *barri* 19
— *lecontei* 19
Enocyla pusilla 192
Entylia bactriana 270
— *sinuata* 271
Epanaphe 257
Epaphroditic 499
Epeorus alpicola 478, 765; Fig. 140
— *assimilis* 478, 765; Fig. 140
Epepeotus luscus 20
Ephemera 479
— *danica* 475, 476, 765; Pl. 1
Ephemerella 477
— (*Eurylophella*) 765; Pl. 1
— (*Torleya*) *belgica* 479, 765; Fig. 138
— *ignita* 12, 475, 476, 479, 765; Fig. 138
— (*Chitonophora*) *krieghoffi* 479, 765; Fig. 138
Ephemerellidae 12, 479, 765
Ephemeridae 479, 765
Ephemeroptera 4, 9, 12, 47, 163, 475–485, 765
Ephestia 90, 94, 312
— *cautella* 37, 40, 77, 79, 227, 236
— *elutella* 35, 37; Fig. 5
— *kuehniella* 41, 56, 90, 179, 182, 212, 213, 229, 316; Fig. 6
Ephippiger cruciger 228, 766; Pl. 12
Ephippigeridae 184, 766
Ephippigerinae 515
Ephoron virgo 12, 477, 485, 765; Fig. 136
Ephydra 748
— *bruesi* 749
— *cinerea* 748
— *hians* 748
— *macellaria* 749
— *millbrae* 748
— *packardi* 748
Ephydridae 52, 86, 748–749, 777
Epicauta 29, 232, 251
Epicnaptera americana 774; Fig. 100
Epicoma isabella 258
— *moloneyi* 258
Epidexia 755
Epilachna 229, 675
— *chrysomelina* 19
— *28-punctata* 675

— *sparsa 26-punctata* 675
Epilachninae 231
Epilamprinae 201
Epimegastrignus 238
Epimetopus 298
— *angulatus* 298
— *hintoni* 272, 298
Epipyropidae 54
Episyrphus 741
— *balteatus* 776; Pl. 118
Erastria venustula 236
Erastriinae 236
Erax 740
— *bastardi* 222
— *femoratus* 222
Erebia 712
Erechtia abbreviata 271
— *gibbosa* 271
— *pruinosa* 271
Eretmapodites quinquevittatus 776; Pl. 114
Eretmocerus 52
Erga longitudinalis 272
Ergates faber 178
Erichsonius cinerascens 659, 772; Figs. 246, 252
Eriocera 724
— *spinosa* 724
Eriococcus 235
Eriogaster lanestris 258, 713, 717
Erioischia 313, 757
— *brassicae* 28, 35, 72, 216, 217, 218, 757, 777; Figs. 44, 294; Pls. 154, 155
Erioptera lutea 27
Eristalinae 741
Eristalis 79
— *intricarius* 741
— *tenax* 741
Ernestia ampelus 753
Erythromma najas 486, 765; Figs. 142, 146
Esakiella hutchinsoni 638
Esselenia vanduzeei 219
Esthiopterum 768; Fig. 186
Etiella zinckenella 25
Eubiomyia calasomae 753
Eublemma 236, 714
— *amabilis* 236
— *coccophaga* 221
— *communimacula* 236
— *costimacula* 236
— *dubia* 236

Eublemma gayneri 236
— *ochrochroa* 236
— *pulvinariae* 236
— *radda* 236
— *rubra* 236
— *rufiplaga* 236
— *scitula* 221, 236
— *vinotincta* 236
Euborellia annulipes 269, 289
— *cincticollis* 269, 289
— *stali* 269, 289
Euceros 54
Eucharidae 22, 53
Eucharitidae 54
Euchlaena obtusaria 774; Pl. 71
Euchloe 712
Euchloris vernaria 242
Euclemensia bassettella 221
Eucomys infeli 144
Eucoptacra basidenis 526, 767; Fig. 173
Eucosmidae 24, 85, 720
Eugaster 515
Eulecanium 221, 233, 235, 570
— *caraganae* 219
— *corni* 220
— *mali* 220, 223
— *persicae* 221
— *prunastri* 235
Euleucophaeus 258
Eulia loxonephes 216
— *sphaleropa* 238
Eulophidae 22, 52, 53, 84, 220, 233, 238, 274, 707, 708, 709
Eulophonotus myrmeleon 24
Eumastacidae 184
Eumecoptera 722
Eumenidae 709
Eumenotes obscura 594
Eumenotinae 594
Eumicromus paganus 771; Fig. 239
Eumusca (= *Musca*) 121, 142, 760
Eunotoperla 508
Eupelmidae 84, 94, 220, 233, 237
Eupelmus 233
— *cicadae* 233
Eupithecia 717
Euplectrus 53
Eupoecilia ambiguella 41
Euponera 87
Euproctis 229, 256, 257

— *chrysorrhoea* 26, 257
— *melanoma* 256
— *pauperata* 256
— *phaeorrhoea* 256
Euprosopus quadrinotatus 771; Fig. 114
Eupteromalus nidulans 22
European corn borer 231
Eurosta solidaginis 743
Eurostus hilleri 42, 178
Eurybrachidae 565
Eurycorypha 515
Eurycotis 201
— *floridana* 13, 284
Eurydema 594, 603
— *dominulus* 770; Pl. 34
— *festiva* 770; Fig. 216
— *oleracea* 770; Fig. 216
— *pulchrum* 770; Fig. 214
— *ventrale* 224
Eurygaster 593, 594
— *austriacus* 593, 770; Fig. 212
— *integriceps* 178, 230, 593, 770; Pl. 33
— *maurus* 593, 770; Fig. 212
— *testudinarius* 593
Eurylophella (= *Ephemerella*)
Eurynebria complanata 649
Eurysternina 301
Eurysternus 301
— *magnus* 273, 301, 773; Fig. 126
Eurytoma amygdali 22
— *kolobovae* 79
— *oophaga* 220
— *pini* 56
Eurytomidae 22, 84, 220
Euschistus 595
— *euschistoides* 595
— *tristigmus* 595
Eustaintonia phragmatella 235
Eustalomyia festiva 757, 777; Pl. 145
Eusthenes robustus 594
Eusthenia spectabilis 508
Eustheniopsis venosa 508
Eutanyderus 110
— *wilsoni* 110, 152
Euthypoda acutipennis 13
Euthystira 83
— *brachyptera* 82
Eutriatoma maculata 615, 770; Pl. 42
Euxanthellus 53
Euxoa ochrogaster 230

Euzophera cocciphaga 221, 713
— *osseatella* 34
Evaniidae 233
Exenterus 53
Exeristes comstockii 706
— *roborator* 21
Exochomus flavipes 19, 224
— *quadripustulatus* 219
Exopterygota 9
exopterygote 1, 2, 3
Exorista flaviceps 28
— *larvarum* 753
— *mella* 37
Exoristinae 755
Extatosoma tiaratum 267, 768; Fig. 98; Pl. 17
Eyprepocnemidinae 175, 767
Eyprepocnemis plorans 175
— — *ibandana* 527
Eysarcoris 595
— *punctatus* 597

Fannia 123, 124, 761
— *armata* 123, 761, 777; Pl. 133
— *atripes* 761, 777; Pls. 134, 135
— *canicularis* 213, 227, 761, 777; Pls. 135, 136
— *coracina* 761, 777; Pl. 137
— *manicata* 761, 777; Pl. 138
— *monilis* 761, 777; Pl. 138
— *nidica* 761, 777; Pl. 135
— *polychaeta* 761, 777; Pl. 134
— *scalaris* 761, 777; Pl. 135
Fanniinae 777
Fenusa pusilla 224
Ferribia 551
Ficalbia 729
Filippia oleae 223
Finlaya (= *Aedes*)
Flatidae 565
flea beetles 316
flies 48, 51, 149, 151, 312, 315
flour moth 73, 721
Forcipomyia 252, 254, 736
Forficula auricularia 14, 214, 228, 270, 287, 289, 290, 768; Fig. 118
— *pubescens* 270
— *tomis* 270
Forficulidae 14, 214, 768, 773
Forficulina 287

Formica 221, 647
— *fusca* 221
Formicidae 221, 234, 709
frog hoppers 213
Fucellia maritima 758
Fucus 748, 758
Fulgoraecia cerolestes 177
Fulgoridae 565
Fulgoroidea 130, 565, 566–567, 769
Fulgoromorpha 169
Fumaria casta 715
Fuscuropoda vegetans 213, 227

Gabrius appendiculatus 660, 772; Figs. 246, 248; Pl. 53
Galerita bicolor 650
Galeruca tanaceti 677, 681
Galerucella 677
— *cavicollis* 20, 677
— *lineola* 215, 219, 220
— *tenella* 772; Fig. 258
Galerucinae 677, 681, 772
Galium 78
gall midge 76, 236
Galleria 715, 717
— *mellonella* 32, 46, 91, 179, 227
Gampsocoris 583
— *punctipes* 583
Gargaphia iridescens 272
— *solani* 272, 296
— *tiliae* 272
Gasterophilidae 749–750, 777
Gasterophilinae 777
Gasterophilus 749, 750; Fig. 291
— *haemorrhoidalis* 749, 777; Fig. 291
— *intestinalis* 749, 777; Fig. 291; Pl. 130
— *nasalis* 749, 777; Fig. 291
— *pecorum* 749
Gastrimargus africanus 527
— *musicus* 218, 222, 224
— *nigericus* 527, 767; Fig. 173
Gastropacha quercifolia 774; Fig. 100
Gastrophysa raphani 681
— *viridula* 676, 677
Gazalina 257
Gelastocoridae 166, 570, 639
Gelastocoris 639
— *nebulosus* 639
— *oculatus* 639

Gelechiidae 24, 84, 713, 716, 720
Gelis transfuga 80
Geocoris pallens 215, 229
— *punctipes* 15, 215, 229
— *uliginosus* 229
Geometridae 25, 35, 37, 85, 102, 250, 714, 715, 716, 717, 721, 774
Geotrupinae 275
Gerridae 166, 175, 570, 580–582, 769
Gerridius armatus 270
Gerris 167, 581
— *thoracicus* 581, 769; Fig. 69
Gesonula punctiformis 71
Ghilianella 272
Gibbium psylloides 42
Gigantothrips elegans 270
Gitona americana 136
Globicornis nigripes 219
Glossina 46, 286
— *morsitans* 46
Glossista infuscata 739
Gloveria dolores 241
Gluphisia septentrionalis 718, 775; Pl. 82
Glyphotaelius pellucidus 195, 286
Gnathocerus cornutus 224
Gnathotrichus 304, 305
— *materiarius* 273, 305, 308
— *retusus* 305
— *sulcatus* 70, 305
Gnophos dilucidaria 250
Gnorimoscheme operculella 315
Golanudiplosis japonicus 221
Gomphas 275
Gomphidae 13, 486, 766
Gomphinae 51
Gomphocerinae 523, 768
Gomphocerus rufus 44
Gomphomacromia 487
Gomphus externus 13
Gonepteryx rhamni 774; Pl. 66
Gongylus gongylodes 766; Fig. 154
Gonia capitata 755, 777; Fig. 292
Goniodes colchici 60, 61, 556
Goniops chrysocoma 738
Gorpis diliensis 608
Gracilia minuta 686
Gracillariidae 25, 716
Graphisurus fasciatus 687
Graphoderes bilineatus 651
Grapholitha molesta 215, 216

Graphomyia maculata 777; Pl. 143
Graphopsocus cruciatus 546
Graphosoma lineatum 601
— *semipunctatum* 60, 601
grasshoppers 3, 211, 213, 216, 217, 223, 224, 232, 237
greenhouse whitefly 94
Gripopterygidae 509, 766; Fig. 162
Gromphadorhina laevigata 284
Grotea anguina 234
Gryllidae 14, 175, 184, 269, 513–514, 767
Gryllinae 513
Grylloblattidae 9
Gryllotalpa gryllotalpa 14
— *hexadactyla* 269
Gryllotalpidae 14
Gryllotalpinae 286
Gryllulus domesticus 14
Gryllus campestris 14
— *mitratus* 207
Gymnobothrus temporalis 767; Fig. 173
Gymnogryllus humeralis 14
Gymnopleurus geoffroyi 773; Fig. 265
— *moposus* 773; Fig. 108
Gymnosoma 5, 753, 754, 755
— *dolycoridis* 777; Fig. 292
— *fuliginosa* 755
Gynaikothrips ficorum 213, 215
gypsy moth 36, 44, 74, 717
Gyrinidae 51, 650, 654, 771
Gyrinus 771; Pl. 51
— *(marinus?)* 771; Pl. 51
— *natator* 49
Gyrohypnus hamatus 218
Gyrostigma 749
Gyrostigmatinae 749

Habrobracon juglandis 178
Habrocytus cerealellae 22, 773; Fig. 13
Habrophlebia auberti 478
— *lauta* 478
— *modesta* 12, 476, 478
Habrosyne pyritoides 774; Pl. 73
Hadeninae 775
Hadrotettix trifasciatus 525, 767; Fig. 171
Hadrothemis infesta 766; Fig. 143
Haemagogus 732
Haematobia 119, 121, 177, 227

SPECIES INDEX 1023

Haematobia stimulans 119, 179, 758, 761, 777; Figs. 42, 296; Pl. 143
Haematopinidae 551, 768
Haematopinus 551, 552, 555, 750
— *asini* 550, 768; Pl. 26
— *eurysternus* 551, 768; Pl. 25
— *quadripertusus* 550
— *suis* 262, 263, 768; Fig. 104; Pls. 22–24
— *tuberculatus* 768; Pl. 25
Haliplidae 649, 653–654
Haliplus 653, 654
Halobates 580
— *germanus* 580
— *hawaiiensis* 580
Halovelia 579, 580
— *hilli* 579
— *marianarum* 579
Halticinae 677
Halys 594
Haplodiplosis marginata 76
Haploembia solieri 766; Fig. 163
Haplothrips faurei 214
— *subtilissimus* 228
— *verbasci* 563
Harmolita tritici 27
Harpaglosus laevigatus 217
Harpalini 230
Harpalus 230
— *caligninosus* 217
— *hirtipes* 217
— *pennsylvanicus* 217
Harpobittacus australis 722, 723, 775; Pl. 99
— *nigriceps* 722
Hebecnema 151
— *affinis* 142, 778; Pl. 145
— *umbratica* 142, 758, 778; Figs. 62, 295; Pl. 144
Hebridae 166, 170, 570, 574–575, 769
Hebrovelia 579
Hebrus 575
— *elimatus* 769; Fig. 201
— *ruficeps* 769; Fig. 201
Hedria 744
Heizmannia 732
Heliconiinae 719
Heliconius 241, 719
— *erato* 719
— *melpomene* 719
— *sara thamar* 719
— *wallacei* 719

Helicopsychidae 711
Heliodinidae 221, 235
Heliophila 53
Heliothis 67, 89, 90, 93, 228, 229, 313
— *armigera* 25, 33, 216
— *phloxiphaga* 92
— *virescens* 56, 67, 79, 87, 92, 93, 229, 231
— *zea* 25, 50, 77, 79, 80, 87, 88, 92, 93, 215, 216, 217, 218, 219, 221, 231
Heliothrips haemorrhoidalis 563
Heliozelidae 713
Helochares 188, 656, 658
— *griseus* 186, 187, 771; Fig. 244
— *lividus* 272, 298, 771; Fig. 244
— *maculicollis* 272
— *obscurus* 272
Helomyia lateralis 755
Helomyzidae 746, 777
Helopeltis 105
— *antonii* 612
— *bergrothi* 612
— *corbisieri* 612
— *orphila* 612
— *schoutedeni* 612, 613, 770; Figs. 28, 29
Helophilus pendulus 741
Helophorinae 186, 656, 771
Helophorus 657, 658, 771; Fig. 245
Heloridae 23
Helorus paradoxus 23
Helotrephes 638
Helotrephidae 166, 570, 638
Hemerobiidae 11, 16, 216, 648, 771; Fig. 239
Hemerobioidea 648
Hemerobius 771; Fig. 239
— *humuli* 16
— *stigma* 216
Hemerocampa leucostigma 177, 179
— *pseudotsugata* 315
Hemerodromia 740
Hemiacridinae 767
Hemiempusa capensis 766; Fig. 155
Hemileuca 199, 258
— *oliviae* 713, 753, 774; Fig. 243
Hemileucidae 256, 258
Hemimerus 9
Hemiptera 5, 9, 15, 39, 41, 46, 58, 83, 88, 91, 100, 102, 129, 144, 149, 163, 166, 178, 209, 214, 224, 228, 291, 499, 565–641, 715, 768–771

Hemisarcoptes coccisugus 213
— *dzhashii* 213
— *malus* 213
Hemistola immaculata (= *Euchloris vernaria*) 242, 717, 774; Fig. 95
Henicopus pilosus 219
hepialid moths 51
Hepialidae 23, 24, 37, 84, 197, 715, 720, 774
Hepialus 197
— *humuli* 23, 774; Pl. 59
Heptagenia 477, 478
— *lateralis* 12, 765; Fig. 140
— *sulphurea* 477, 765; Figs. 136, 140
Heptageniidae 12, 478, 765
Herpystis cuscutae 24
Hesperia comma 713
Hesperiidae 25, 85, 716
Hesperocimex sonorensis 770; Pl. 36
Hesperotettix viridis praetensis 525, 767; Fig. 171
Heteracris (= *Thisoicetrus*) 175
— *littoralis* 175
Heterocampa astarte 718, 775; Pl. 85
— *cubana* 718, 775; Pl. 85
— *manteo* 718, 775; Pl. 85
— *umbrata* 718, 775; Pl. 86
Heteroceridae 273, 303, 773
Heterocerus flexuosus 273, 773; Fig. 127
Heterocordylus tibialis 770; Fig. 220
Heterocorixa 617
Heterogaster urticae 584
Heteronotinae 270
Heteronotus vespiformis 270
Heteropelma 52
Heteroptera 9, 163, 164, 166, 167, 168, 170, 175, 271, 499, 565, 570–641, 763, 764, 769–771
Heteropternis thoracica 767; Fig. 173
Heterostylum robustum 51
Heterotylenchus autumnalis 92
Hetrodinae 515
Hexatoma 724
Hexatomini 724
Hierodula 237
— *crassa* 506, 766; Fig. 160
— *notata* 269
— *patalifera* 201
Hieroglyphus 82
Hilara 740

Hille herbicola 271
Himacerus apterus 608, 609
— *lativentris* 609
— *major* 609
Hippiscus rugosus 524, 767; Fig. 171
Hippoboscidae 286
Hippodamia convergens 19, 219, 296
— *tredecimpunctata* 219
Hippodamiini 231
Hirmoneura articulata 739
— *obscura* 739
Hirtodrosophila 136
— *sexvittata* 136
Hispinae 677, 772
Hister abbreviatus 230
— *striola* 230
Histeridae 218, 230, 688
Holcocera icaryaella 221
— *phenacocci* 235
— *pulverea* 221, 714
Holcostethus vernalis 601, 770; Fig. 217
Holopercna gerstaeckeri 528, 767; Fig. 173
Holoquiscalus brachypterus 237
Homoptera 9, 54, 58, 130, 169, 170, 175, 234, 236, 270, 565–570, 763, 764, 768–769
Homorocoryphus nitidulus 767; Pl. 5
— — *vicinus* 515
honey bees 11, 30, 47, 744
Hoplocampa crataegi 21
Hoplocerambyx spinicornis 20
Hoplocorypha nigerica 766; Fig. 155
Hoplodictya 744
Hoplophorioninae 271
Hoplopleura cryt 768; Fig. 185
— *oenomydis* 768; Fig. 185
Horiola picta 271
Hornia 676
— *boharti* 224
— *minutipennis* 232
— — *occidentalis* 233
horse flies 79
housefly 46
Humbe tenuicornis 175, 528
Hyaliodes 105, 106
— *harti* 215
Hyalophora 176, 199, 206, 753
— *cecropia* 206, 209, 774; Figs. 90, 92; Pl. 91
Hydara tenuicornis 586
Hydaticus 651

SPECIES INDEX

Hydraena 657
— *pennsylvanica* 771; Fig. 245
Hydraenidae (=Limnebiidae) 52, 186, 656, 657, 771
Hydrellia griseola 749
Hydrobasileus croceus 489
Hydrobiinae 186, 656, 771
Hydrobius 658
— *fuscipes* 187, 771; Fig. 77
— *globosus* 771; Fig. 245
Hydrochara caraboides 186, 187
Hydrochidae 186, 656, 771
Hydrochus 186, 658
— *squamifer* 771; Fig. 245
Hydrocorisae 166, 570, 617–641, 770–771
Hydrocyrius 58
Hydrometra 581
— *albolineata* 166
— *gracilenta* 577, 769; Fig. 203
— *martini* 577
— *stagnorum* 577, 769; Fig. 203; Pl. 28
Hydrometridae 166, 570, 577–579, 769
Hydromya 744
Hydrophilidae 51, 147, 148, 183, 186, 187, 188, 217, 272, 298, 656, 657, 771–772
Hydrophilinae 186, 656, 772
Hydrophiloidea 186, 656–658, 771–773; Fig. 245
Hydrophilus 188, 658
— *fuscipes* 772
— *piceus* 159, 186, 187, 651, 772; Figs. 64, 244
Hydrophorus depressus 651
— *latescens* 651
— *planus* 651
Hydropsyche angustipennis 33
— *colonica* 710
Hydropsychidae 710, 711
Hydroptilidae 710, 711
Hydroscapha natans 654, 771; Fig. 243
Hydroscaphidae 51, 654, 771; Fig. 243
Hydrotaea albipuncta 777; Fig. 296
— *tuberculata* 759, 760, 777; Fig. 296
Hydrotaeini 777
Hydrous 188, 658
— *piceus* 148
Hygrobia hermanii 654
Hygrobiidae (=Pelobiidae) 654
Hylastes nigrinus 71
Hylecoetus dermestoides 219

Hylemya 46
— *brassicae* 73
— *cilicrura* 224
— *coarctata* 79
— *liebermanni* 224
— *seneciella* 77
Hylesia 258
Hylesinus 281
— *fraxini* 70
Hylobius 84
— *abietis* 33, 47
Hyloicus pinastri 250, 774; Fig. 87
Hylotrupes bajulus 60, 178, 179, 685
Hymenarcys aequalis 596
Hymenolepis diminuta 90
Hymenopodidae 499, 766
Hymenopodinae 766
Hymenoptera 10, 21, 22, 30, 52, 54, 56, 80, 84, 88, 91, 94, 163, 175, 205, 209, 213, 220, 224, 233, 238, 245, 274, 286, 310, 311, 704–709, 752, 753, 763, 764, 773; Fig. 280
Hymenopteroidea 10
Hypena proboscidalis 775; Pl. 76
Hypeninae 775
Hypera 84
— *brunneipennis* 91
— *nigrirostris* 20, 59
— *postica* 35, 61, 71, 91, 233
— *punctata* 91
— *variabilis* 20, 48, 88, 91
Hyperalonia morio 51
Hyperaspini 231
Hyperaspis 231
— *lateralis* 216, 219, 675
— *signata* 220, 237
Hyphantria cunea 34, 258
Hypocala andremona 88
Hypochlora alba 525, 767; Fig. 171
Hypoderma 749, 750, 777; Fig. 291
— *bovis* 750, 751; Fig. 291
— *lineatum* 750, 751
Hypolimnas 251
Hyponomeuta 72
— *mallinellus* 214, 215
Hyposoter disparis 21
— *ebeninus* 92
— *exiguae* 92
Hypselosoma 641, 771; Fig. 238

Ibalia 74
— *drewseni* 74
Ibaliidae 84, 238
Ibidoecus bisignatus 554
Icerya 223, 232, 234, 236
— *purchasi* 219, 220, 232
ichneumonid wasp 30
Ichneumonidae 21, 52, 53, 54, 64, 84, 94, 234, 237, 706, 709, 773
Ichthyura 116
— *albosigma* 718, 774; Pl. 84
Icterus portoricensis 237
Ictinogomphus 492
— *ferox* 492
Idiolomorpha lateralis 766; Fig. 155
Idionotus siskiyou 767; Pl. 10
Idiostatus aequalis 767; Pl. 8
— *californicus* 767; Pl. 9
— *fuscopunctatus* 767; Pl. 9
— *gurneyi* 767; Pl. 8
— *hermanni* 767; Pl. 10
— *inyo* 767; Pl. 8
— *rehni* 767; Pl. 10
— *sarcobatoides* 761; Pl. 9
— *wymorei* 767; Pl. 8
Idiostolidae 570, 582
Illiesoperla 508
Ilybius 651
— *ater* 651
— *fuliginosus* 651, 771; Fig. 241
Ilyocoris 637
— *cimicoides* 167, 637
Incurvariidae 713
Inocelliidae 642
Insecta 213
Ipidae (= Scolytidae) 304
Ipinae 274
Ipomoea 189
Iponemus 227
Ips 65, 66, 70, 227, 281
— *acuminatus* 773; Fig. 278
— *amitinus* 773; Fig. 279
— *calligraphus* 65
— *confusus* 227
— *grandicollis* 65
— *pini* 21, 213
— *typographus* 282
Iridomyrmex 70
— *humilis* 234
Iris oratoria 766; Fig. 154

Irochrotus lanatus 592
Ischnocera 549
Ischnopsyllus octactenus 775; Pl. 100
Ischnoptera panamae 766, 775; Fig. 153
Isophya 515
— *pyrenaea* 13, 767; Fig. 164
Isoptera 9, 286, 507
Issidae 169, 170, 565
Issus 169
Isturgia truncataria 774; Pl. 70
Isyndrus heros 614
Ithonidae 643, 721
Itoplectis conquisitor 46, 73, 315

Jalla dumosa 596, 770; Fig. 215
Jalysus spinosus 215, 229, 260, 583
Jamaicana flava 766; Pl. 6
Jassidae (= Cicadellidae) 15, 565, 566
Jewettoperla munoani 508
Julistus floralis 218

Kalotermes flavicollis 11, 507
Kapala terminalis 22
Kelerimenopon 555, 768
Kermes 221, 235
— *galliformis* 221
Kermidae 565
Kermococcus 233
Kinnaridae 565
Kleidocerys resedae 584, 769
Knutsonia 744
— *lineata* 743
Krizousacoris azteca 238

Labidostomis taxicornis 772; Fig. 261
Labidura riparia 228, 270, 287, 289, 768; Fig. 118
Labiduridae 270, 768
Labiidae 270
Laccifer 236
— *aurantica* 236
— *lacca* 221, 235
Lacciferidae 565
Lacciferophaga yunnanea 235
Laccobius 658
Laccoptera 191, 622
Laccotrephes 130, 623, 634, 771; Fig. 235

Laccotrephes fabricii 131, 622, 624, 632, 634
— *maculatus* 622, 634
— *robustus* 622, 632, 634
Lachesilla forcepata 547, 768; Pl. 18
Lachnocnema bibulus 225
Laelia coenosa 241
Laemobothriidae 551, 768
Laemobothrion tinnunculi 768; Pl. 21
Laetilia coccidivora 221
Lagenaria 743
Lagynotomus 595
Lamachus xiphias 768; Fig. 175
Lamia textor 687
Lamiinae 686, 687, 688, 773
Laminaria 745, 758
Lampromyia 737
— *fortunata* 80
Lamprophorus tenebrosus 17
Lamprosominae (= Clytrinae) 678
Lampyridae 17
Langsdorfia franckii 243, 774; Fig. 96
Laphria 740
Laphriinae 740
Laphygma frugiperda 25, 219
lappet moth 240; Fig. 100
Largidae 570, 585
Largus fulvipes 585
Laricobius erichsonii 17, 218
Lasiocampa quercus 193, 713, 716, 774; Pl. 95
— *trifolii* 241
Lasiocampidae 26, 35, 85, 256, 717, 774
Lasiochalcidia igiliensis 773; Fig. 11
Lasioderma 232, 315
— *serricorne* 18, 37, 214
Lasius 234, 254
— *flavus* 221, 253
— *niger* 253, 254; Fig. 101
— *sitkaensis* 221
Laspeyresia pomonella 209
Latheticus oryzae 224
Lathridiidae 4
Lauxaniidae 745, 762, 776
Lebia analis 217
— *viridis* 217
lecaniine 220
Lecanium 235, 236
— *nigrofasciatum* 221, 271, 293
Legnotus limbosus 271, 294
Leioscyta spiralis 271
Leis conformis 19

Lema melanoplus 238
— *praeusta* 20
— *trilineata* 72
— — *trivittata* 90
Lepidophthirus macrorhini 554, 562, 768; Pl. 21
Lepidoptera 3, 10, 23, 24, 25, 26, 35, 40, 41, 44, 48, 52, 72, 80, 84, 88, 91, 93, 101, 102, 179, 183, 200, 206, 209, 221, 225, 235, 242, 256, 258, 312, 706, 712–721, 739, 763, 764, 774–775; Figs. 281–282
Lepidosaphes beckii 213
— *ulmi* 213, 214
Lepisma 3
— *saccharina* 12, 765; Pl. 1
Lepismatidae 12, 765
Leprosoma inconspicuum 599
Leptacinus parumpunctatus 663, 772; Fig. 256
Leptidae (= Rhagionidae) 222, 737–738
Leptideala brevipennis 686
Leptinotarsa 225
— *decemlineata* 20, 41, 72, 87, 213, 215, 220
Leptocentrus altifrons 270
Leptocera 132, 133, 747, 777; Fig. 290
— *caenosa* 748
— *clunipes* 747
— *moesta* 133, 746, 747, 777; Fig. 290
Leptoceridae 23, 710, 711
Leptocerinae 777
Leptocoris augur 589
Leptocorisa 588
Leptogaster 740
Leptogastrinae 740
Leptoglossus 585
Leptohylemyia 125, 313
— *coarctata* 119, 124, 181, 757, 777; Fig. 38; Pls. 152, 153
Leptomastix 144
— *dactylopii* 91, 773; Fig. 63
Leptopalpus rostratus 49
Leptoperla bifida 766; Fig. 162
Leptophlebia marginata 478
— *vespertina* 12, 478
Leptophlebiidae 12, 478, 765
Leptophyes punctatissima 518, 766; Fig. 166; Pl. 12
Leptopodidae 166, 570, 573, 769
Leptopodoidea 570–574, 769
Leptopterna dolabrata 770; Pl. 31
Leptopus 573

Leptopus hispanus 573
— *marmoratus* 573, 769; Fig. 199
Leptothorax canadensis 221
Leptothrips mali 214
Lepturinae 686
Leptynia 533
Leschenaultia adusta 28
— *exul* 755
Lestes congenera 486
— *dryas* 486
— *sponsa* 486, 765; Figs. 141, 146
— *viridis* 486, 494
Lestidae 765
Lestinogomphus 492, 766
— *africanus* 491; Fig. 149
Lestodiplosis 221
— *aonidiellae* 221
Lethocerinae 52, 636
Lethocerus indicus 125, 636, 771; Pl. 47
Lethrus apterus 773; Fig. 268
Leucania litoralis 713
— *separata* 179
Leucaspis 231
Leucoma salicis 258, 775, 777; Fig. 88
Leucophaea 203
— *maderae* 44, 184, 284
Leucophaga 136
Leucopis 223
— *bella* 223
Leucoptera meyricki 37
Leucopterygidae 37
Leucostomatina 755
Leucozona 741
— *lucorum* 77, 776; Pl. 117
Leuctra hippopus 508
Libellula depressa 766; Fig. 148
Libellulidae 486, 487, 489, 490, 766
Liberonautes latidactylus 735
Limnebiidae (=Hydraenidae) 186
Limnebius 657
Limnephilidae 23, 52, 192, 710, 711, 774
Limnephilus bipunctatus 192
— *flavicornis* 23
— *griseus* 192
— *indivisus* 192, 193, 195, 710
— *lunatus* 23, 774; Fig. 84
— *stigma* 192
— *vittatus* 774; Fig. 85
Limnerium alkae 92
— *crassifemur* 91

Limnogeton 58
Limnophora 142
— *riparia* 142, 759, 777; Pl. 146
Limnophorinae 777
Limnoxenus niger 186, 187
Limonia flavipes 724
— *macrostigma* 726, 775; Fig. 285
— *nubeculosa* 725, 775; Figs. 283, 285
— *quadrinotata* 724
— *stigma* 725, 775; Fig. 284
— *tripunctata* 775; Fig. 283
— *trivittata* 775; Fig. 285
Limoniinae 725, 775
Limoniini 724
Limosina (=*Leptocera*) 133
— *moesta* 133
Linognathus breviceps 768; Fig. 185
— *ovillus* 559
— *pedalis* 559, 768; Fig. 190; Pl. 21
Lioadalia flavomaculata 224
Liocoris tripustulatus 613
Liorhyssus hyalinus 769; Fig. 210
Liothrips oleae 15
Liposcelidae 14, 214, 768
Liposcelis (=*Troctes*) 228, 584
— *diviniatoria* 214
— *granicola* 14
Lipsothrix 724
Lisarda vandenplasi 614, 616
Lissorhoptrus 84
Listronotus oregonensis 60
Litaneutra minor 502
Lithina chlorosata 717
Lithocolletis blancardella 25
Lithophane laticinerea 25
Litus krygeri 709
Loboptera decipiens 233
locust eggs 739
Locusta 82, 184, 282, 283, 520, 521
— *migratoria* 34, 44, 75, 216, 217, 219, 220, 222, 223, 224, 261, 282, 520, 767; Fig. 169
— — *manilensis* 175
— — *migratorioides* 14, 175, 520
Locustana 83
— *pardalina* 14, 175, 177, 222, 223, 519
locusts 51, 222, 232, 282
Lomechusa 226
Lonchaea 745, 777; Pl. 121
— *corticis* 745

Lonchaeidae 745, 777
Longistigma caryae 242, 249
Lophoblatta 497
Lophodonta angulosa 718, 775; Pl. 82
Lophopidae 565
Lopodiplosis 236
Lorichius umbonatus 608
Loricula elegantula 610, 770; Fig. 211
— *pselaphiformis* 610
Loxodon 749
Lozogramma subaequaria 774; Pl. 73
Lucilia 71, 151, 179, 230
— *cuprina* 28
— *sericata* 28, 751
Luffia lapidella 715
Lycaena phlaeas 774; Pls. 62, 63
Lycaenidae 226, 721, 774
Lyciella rorida 745, 776; Pl. 120
Lycosidae 230
Lyctidae 18, 83
Lyctocoris elongatus 215
Lyctus 5, 72, 77
— *africanus* 77
— *brunneus* 18
— *linearis* 18, 60
Lydella nigripes 753
Lydus viridissimus 50
Lygaeidae 15, 36, 167, 215, 229, 570, 583–584, 769; Figs. 206, 207
Lygaeoidea 102, 590
Lygris populata 717
Lygus campestris 612
— *cervinus* 613
— *hesperus* 48, 60, 61
— *lineolaris* 87
— *rugulipennis* 215
Lymantria 199, 229, 258
— *ampla* 26
— *dispar* 26, 35, 40, 209, 212, 218, 219, 258, 314, 315, 775; Fig. 90; Pl. 96
— *monacha* 35, 258, 775; Fig. 87
Lymantriidae 26, 35, 85, 219, 256, 258, 721, 775
Lymnaea ollula 744
Lyneborgia ammodyta 739
Lyonetidae 721
Lyperosia 119, 177
— *irritans* 119, 179, 761, 777; Fig. 296
Lypha dubia 74
Lysandra bellargus 774; Pl. 62

Lytta 232
— *sphaericollis* 232
— *viridana* 676
Lyttinae 232

Maccevethus lineola 769; Fig. 208
Machadonannus ocellatus 641
Machaeridia bilineata 527
Macrocentrus homonae 73
Macrocheles glaber 227
— *merdarius* 227
— *muscaedomesticae* 213, 227
— *peniculatus* 227
— *preglaber* 227
— *subbadius* 227
Macroglossum stellatarum 63, 78
Macrogrotea gayi 234
Macrolophus 105
— *nubilus* 613, 770; Fig. 27
Macromantis ovalifolia 766; Fig. 154
Macroplea 111
Macrorileya oecanthi 220
Macrosiagon tricuspidata 17
Macrosteles 315
Macrotheca unipunctata 236
Macrotoma securiformis 222
Macrovelia horni 579
Malachiidae 218, 231
Malachius bipustulatus 218
Malacocoris chlorizans 612
Malacosoma 229, 713, 714
— *americanum* 197
— *californicum* 714
— *castrensis* 713, 717
— *constrictum* 713
— *disstria* 26, 200, 713, 714
— *incurvum* 714
— *neustria* 241, 713
— *tigris* 713, 714
Malcidae 570, 583
Malcus 583
Mallodon downesi 687, 773; Pl. 56
Mallophaga 9, 549, 551, 554, 555, 562, 750, 764, 768; Fig. 186
Mamestra 81
— *brassicae* 33, 40, 775; Pl. 76
Manduca sexta 73, 78, 215, 229, 260
Mansonia 732
— *uniformis* 63, 776; Pl. 106

Mantidae 13, 213, 499, 766
mantids 201, 209
Mantinae 506, 766
Mantis 219, 499, 502
— *religiosa* 13, 202, 503
Mantispa 116, 117
— *interrupta* 16, 771; Pl. 50
Mantispidae 11, 16, 54, 229, 643, 648, 771
Mantodea 9, 184, 269, 499–507, 766
Marava arachidis 270, 289
Margarodidae 565
Margaronia unionalis 25
Martianus dermestoides 225
Mastotermes darwiniensis 507
Mayetiola destructor 88
mayfly 51
Meadorus lateralis 272
mealworm 93
mealybugs 231, 235
Mecas saturina 686
Mecistogaster 490
— *modestus* 490
Meconema thalassina 515, 518, 766
Meconematinae 515, 518, 766
Mecoptera 3, 10, 116, 209, 651, 722–723, 763, 764, 775
Mecopteroidea 10
Medama 257
Medetera 70
— *aldrichii* 236
Meenoplidae 565
Megacephala euphratica 216
Megachilidae 709
Megacoelum 105, 106
— *infusum* 613, 770; Fig. 25
Megacrania wagneri 533, 768; Fig. 176
Megaloblatta blaberoides 766; Fig. 153
Megaloptera 9, 16, 215, 642–643, 771
Megarhyssa 74
Megaselia aequalis 237
— *aspersa* 223
— *leucozona* 223
— *parvula* 223
Megastigmus pistaciae 88
Megathopa 275
— *bicolor* 773; Fig. 108
Megathrips lativentris 563
Megatoma pici 219
— *pubescens* 219
— *undata* 219

— *variegata* 219
Megistocera longipennis 724
Megochterus nasutus 639
Megymenum purpurascens 593
Meigenia 755
Meinertzhageniella 549
Melaleucopsis ortheziavora 237
— *simmondsi* 237
Melanargia galathea 713
Melanchrinae 775
Melangyna triangulifera 77
Melanoplus 45, 83, 173, 513, 520, 521
— *alpinus* 221
— *augustipennis* 525
— *bilituratus* 221, 222
— *bivittatus* 175, 221, 525
— *bowditchi* 525
— *confusus* 525
— *differentialis* 169, 175, 209, 525
— *femoratus* 202
— *femur-rubrum* 526
— *foedus* 526
— *gladstoni* 525
— *lakinus* 526
— *mexicanus* 526
— *occidentalis* 526
— *packardii* 525, 767; Fig. 171
— *sanguinipes* 39, 47
— *spretus* 216, 217, 222, 223, 224
Melanostoma 741
Melanotus fuscipes 17
— *longulus* 17
Melaphoninae 774
Melasina 258
Melasoma aenea 214, 215, 219
— *crotchi* 237
— *populi* 178
Melichares dentriticus 213
— *tarsalis* 213
Melittobia acasta 22, 274
Meloe 251
— *autumnalis* 29, 676
— *cavensis* 29
— *majalis* 29
— *violaceus* 29
Meloidae 29, 54, 83, 213, 232, 676, 689
Melolontha melolontha 17, 61, 178
Melyridae 219, 231
Membracidae 270, 291, 565, 566, 769
Membracinae 271

Membracis c-album 271
— *tectigera* 271
Menacanthus cornutus 554
— *stramineus* 555, 556
Menaccarus arenicola 597, 770; Fig. 216
Menecles insertus 596
Menoponidae 768
Mermiria maculipennis macclungi 525
— *neomexicana* 525, 768; Fig. 171
Merobruchus julianus 684
Meromyza americana 88
Meropidae 10, 722
Mescina peruella 238
Mesembrina 117
— *meridiana* 122, 760, 777; Figs. 43, 296; Pl. 143
Mesochorus 52
Mesoleius aulicus 91
— *tenthredinis* 91
Mesomorphus villiger 19
Mesopsocus immunis 547
— *unipunctatus* 547
Mesovelia furcata 575, 769; Figs. 70, 202
Mesoveliidae 166, 168, 570, 571, 574, 575–577, 769
Mestocharis bimacularis 708
Metaphycus luteolus 30
Metapterus banksii 617
Metarhizium 92
Metasyrphus corollae 237
Metator nevadensis 222
Metratropis rufescens 583
Metrioptera brachyptera 518
— *roeselii* 518
Metura elongata 712
Mezira rugosa 582
Mezium affine 178
Microcentrum rhombifolium 515
Microcorthylus castaneus 305
Microgaster 238
Microgasteridae 220
Microlarinus lareynii 71, 215
— *lypriformis* 71, 215
Micromus posticus 16, 216, 224
Micronecta 620
— *quadristrigata* 620
— *scutellaris* 617, 771; Fig. 223
Microphysidae 570, 610, 770; Fig. 211
Microplitis croceipes 92
— *plutellae* 21

Microterys 53, 144
— *chalcostomus* 233
— *flavus* 705, 773; Pl. 58
— *sylvius* 220
Microvelia pulchella 579
— *reticulata* 579
— *signata* 579, 769; Fig. 204
Mictis metallica 585
— *profana* 769; Fig. 209
midge 34, 76
Milesiinae 741
Miltogramma 752
Mimemodes japonus 231
Mimomyia 116
— *aurea* 776; Pl. 114
Mimosestes 683
— *sallaei* 685
Miomantis savignyi 502, 505
Mirabilis jalapa 244
mirid bugs 79, 174
Miridae (=Capsidae) 15, 105, 106, 131, 166, 167, 171, 175, 212, 214, 228, 570, 611–613, 770; Fig. 220
Mirinae 105, 613
Mirperus jaculus 588
mites 213, 236
Molanna angustata 194
Molannidae 710, 711
Mollusca 237
Molops piceus 649
Monalocoris filicis 612
Moneilema appressa 687, 773; Pl. 56
Moniliformis dubius 90
Monochamus 687
Monoctonus paludum 91, 94
Monopis 286, 712, 715
— *chrysogramma* 712
— *meliorella* 286, 712
— *rusticella* 179
— *trimacullella* 712
Monotrysia 10, 712, 715, 773
Monotylota ramburi 13, 269
Montandoniella moraguesi 215
Morellia 117
— *hortorum* 760, 777; Pl. 143
Mormidea lugens 596
Mormolyce phyllodes 272, 297
Mormoniella 30
Morphacris fasciata sulcata 528
Morseiella 221

mosquito eggs 313
mosquitoes 48, 51, 52, 75, 79
moth eggs 248, 313
moths 48, 51, 54, 84, 147, 179, 228, 229, 312, 313, 315, 316
Mucidus (= *Aedes*)
Murgantia histrionica 596
Musca 39, 117, 121, 142, 151, 227, 312
— *autumnalis* 121, 142, 230, 231, 758, 760, 777; Figs. 60, 61; Pl. 141
— *domestica* 28, 35, 46, 47, 92, 119, 149, 214, 218, 227, 230, 231, 315, 758, 760, 777; Fig. 40
— *sorbens* 117, 121, 777; Pl. 139
— *vetustissima* 50, 117, 121, 777; Pl. 140
— *vincina* 92
— *xanthginelas* 121, 142
muscid fly Fig. 39
Muscidae 4, 28, 35, 86, 113, 115, 120, 128, 132, 142, 151, 224, 739, 751, 758–761, 762, 777–778; Figs. 295, 296
Muscinae 113, 119, 120, 123, 124, 142, 760, 777
Mycodiplosis 221
Mydaea scutellaris 778; Fig. 295
— *urbana* 142, 778; Fig. 295; Pl. 147
Mydaeinae 777–778
Myelophilus 70, 71
— *piniperda* 67, 71, 273, 308
Myeolis grossipunctella 236
Mylabris 29, 232
— *circumflexa* 29
— *wagneri* 29
Myllocerus maculosus 20
Mymaridae 52, 238, 708–709
Myospila 142
— *meditabunda* 142, 758, 778; Fig. 295; Pl. 148
Myrmedobia coleoptrata 610
Myrmeleontidae 643, 645, 647
Myrmeleontoidea 647, 648
Myrmeleotettix 521
Myrmica l. fracticornis 221
— *rubra* 11, 226
Myrmus miriformis 589, 769; Figs. 206, 207
Mystacides longicornis 23
Myxophaga 654–655, 771; Fig. 243
Myzus ascalonicus 91
— *circumflexus* 91

Nabidae 215, 228, 570, 608–609, 610, 770; Fig. 219
Nabididae 15
Nabis 166, 215
— *alternatus* 15
— *americoferus* 215
— *ericetorum* 609
— *feroides* 228
— *ferus* 215, 608, 609
— *flavomarginatus* 608, 609, 770; Fig. 219; Pl. 48
— *rugosus* 609, 770; Fig. 219; Pl. 48
Nadata gibbosa 718, 720, 775; Pl. 84
Nagusta punctaticollis 614
Nannochorista 10
Nannochoristidae 10, 722
Nannocoris 641
Nannomecoptera 10, 722
Nascioides enysi 79
Nasonia vitripennis 65, 773; Fig. 280
Naucoridae 51, 166, 167, 570, 636–637
Naucoris 637
Nauphoeta 203
— *cinerea* 13, 284
Necrobia rufipes 19
Necrophorus 49, 226, 227, 286, 298, 301
— *germanicus* 273, 298
— *humator* 226, 273
— *interruptus* 273
— *investigator* 226, 273
— *orbicollis* 273
— *tomentosus* 273
— *vespillo* 273, 298, 772; Fig. 124
— *vespilloides* 273, 298
Necroscia sparaxes 534
Necrosciinae 536
Neduba 515, 767; Pl. 11
Neides tipularis 583
Neididae (= Berytinidae) 583
Nematocera 761
Nematus ribesii 705
Nemeritis 56, 90, 93, 706, 773; Fig. 21
— *canescens* 56, 60, 90, 91
Nemestrinidae 54, 221, 222, 739
Nemognatha 251
— *chrysomelina* 50
Nemognathinae 232
Nemopoda 132
Nemoptera bipennis 647, 771; Fig. 239
Nemopteridae 643, 771; Fig. 239

SPECIES INDEX

Nemorilla 754
— *maculosa* 221
Nemoura columbiana 34, 60
Neoaplectana carpocapsae 92
Neoblattella detersa 766; Fig. 153
Neodiprion fulviceps 59
— *lecontei* 91
— *swainei* 59
Neohermes concolor 642, 643
Neomecoptera 10, 722, 723
Neomyzus circumflexus 91
Neorhynchocephalus sackenii 222
— *sulphureus* 222
— *vitripennis* 222
Neostylopyga 201
— *rhombifolia* 237
Neottiglossa 596
— *leporina* 600, 770; Fig. 215
— *pusilla* 600
Nepa 126, 130, 170, 624, 632
— *apiculata* 622
— *cinerea* 162
— *rubra* 621, 622, 623, 624, 631, 632, 633, 634, 771; Figs. 45, 46, 71, 228
Nepenthes 236
Nepidae 52, 97, 113, 128, 129, 130, 149, 162, 166, 170, 570, 621–635, 636, 771
Nepinae 131, 621, 622, 623, 630, 634
Nepticula ruficapitella (= *Stigmella ruficapitella*) 774; Pl. 59
Nertha 639
— *colaticollis* 639
— *laticollis* 639
— *martini* 639
Netelia 706
Neuroctena anilis 132, 744
Neuroctenus pseudonymus 271, 294
Neuroptera 9, 16, 54, 91, 116, 117, 209, 215, 224, 229, 240, 315, 642, 643–648, 764, 771; Fig. 239
Neuropteroidea 9
Neuroterus lenticularis 707
Neurotoma nemoralis 91
Nezara 94, 595
— *viridula* 39, 41, 56, 74, 770; Fig. 214
Niamia 607
Nicotina 89
Nidicola marginata 229
Nigronia fasciatus 643
— *serricornis* 642, 643

Niptus hololeucus 42
Nitidulidae 19, 689
noctuid moths 70
Noctuidae 25, 26, 85, 221, 225, 236, 715, 721, 774
Noctuoidea 717
Nogodinidae 565
Nola sorghiella 236
Nolidae 26
Nomadacris 83
— *septemfasciata* 14, 175, 217, 220, 223
Nomophila noctuella 25
Noteridae 649
Notidobia ciliaris 23
Notiothauma reedi 722
Notiothaumidae 10, 722
Notiphila brunnipes 749
— *riparia* 777; Pl. 122
— *sekiyai* 749
Notodonta dromedarius 718, 775; Pl. 87
Notodontidae 26, 85, 101, 104, 256, 257, 716, 718, 774–775
Notodontinae 775
Notonecta 314, 637
— *maculata* 637
— *unifasciata* 238
Notonectidae 51, 166, 238, 570, 637–638
Notonemouridae 508
Notoptera 9
Notosaurus albicornis 222
Notostira 166, 175
Notoxus calcaratus 220
Numica viridis 228
Nupserha 686
— *antennata* 687
Nycteribiidae 286
Nyctiborinae 497
Nymphalidae 85, 102, 250, 721, 774
Nymphalis io 774; Pl. 65
Nymphidae 648
Nymphomyiidae 734
Nymphula 714, 715, 717
— *maculalis* 714
Nymphulinae 715
Nysius huttoni 15, 48
— *lineatus* 583
— *vinitor* 61
Nythobia insularis 21

Oberea 686
Ocellovelia germari 579, 769; Fig. 204
Ocenteter 238
Ochlerotatus (=Aedes) 731
— *stimulans* 731
Ochrogaster contraria 257
Ochropleura plecta 775; Pl. 77
Ochteridae 166, 570, 638–639, 771
Ochterus 639
— *marginatus* 771; Fig. 236
Ochthebius 657
Ochthiphilidae 223
Ocydromia glabricula 740
Ocypus 116, 171
— *brunnipes* 772; Fig. 261
— *fuscatus* 662, 772; Fig. 36
— *globulifer* 662, 772; Fig. 261
— *olens* 152, 171, 173, 662, 772; Figs. 36, 75, 252; Pl. 55
Odezia atrata 717
Odoiporus longicollis 20
Odonata 9, 12, 82, 206, 209, 486–494, 765–766
Odontoscelis fuliginosa 592
— *hispidula* 592
Odontotarsus purpureolineatus 592
— *robustus* 592
Odozana obscura 199, 775; Fig. 89
Oecanthinae 513
Oecanthus 228, 514
— *latipennis* 514
— *nigricornis* 514
— *niveus* 220, 767; Fig. 164
— *pellucens* 220, 767; Fig. 164
Oeceticus platensis 183
Oecetis lacustris 23
Oecophoridae 221
Oedaleus decorus 767; Fig. 174
Oedematopoda cypris 235
— *venusta* 235
Oedipoda coerulescens 767; Fig. 174
Oedipodinae 175, 523, 767–768
Oedoparea buccata 132, 745
Oeme costata 687
— — *abietus* 773; Pl. 56
Oeneini 231
Oestridae 86, 750–751, 777
Oestrus ovis 751
Oiketicus kirbyi 24
Okanagana rimosa 76

Olethreutidae 221, 242
Oligembia melanura 512
Oligoneuriella rhenana 475, 479, 765; Fig. 136
Oligoneuriidae 479, 765
Oligotoma nigra 13
Oligotomidae 13
Omania marksae 574, 769; Fig. 200
— *samoensis* 574
Omaniidae 166, 570, 574, 769
Omaspides pallidipennis 273, 772; Fig. 128
Omocestus 83, 521
Oncideres 686
Oncodes pallipes 739
Oncopeltus 251
— *fasciatus* 43, 251, 769; Fig. 207; Pl. 29
Oncopera 51
— *intricata* 37
Oniticellini 275
Ontholestes 171
— *murinus* 116, 662, 772; Fig. 253
Onthophagini 275
Onthophagus nuchicornis 773; Figs. 108, 272, 273
Ontini 275
Onymocoris hackeri 613
Oocassida pudibunda 189
Ooencyrtus 144
— *johnsoni* 773; Fig. 63
— *kuwanai* 22
Operophtera brumata 37, 74
Ophelosia crawfordi 220
Ophyra 230
Opius 30, 73
— *fletcheri* 73
— *oophilus* 742
— *tryoni* 92
Opsolasia meadei 757, 777; Pl. 150
Orasema 53
Oravelia pege 579
Orbillus coeruleus 518, 767; Fig. 172
Orgilus lepidus 64
— *obscurator* 64
Orgyia antiqua 716, 775; Pl. 96
— *postica* 26
Orina 676
Orius albidipennis 229
— *insidiosus* 215, 229
— *laevigatus* 229
— *minutus* 229

Orius rufilabris 215
— *tristicolor* 215, 229
Ornithacris turbida 527
Oropeza (= *Dolichopeza*) Fig. 286
Orphulella pelidna 524
Orsillinae 584
Orsillus depressus 584, 769; Fig. 206
Orthellia 117, 151
— *caesarion* 107, 217, 758, 760, 777; Figs. 30, 31; Pl. 142
— *cornicina* 777; Pl. 142
Orthezia 237
— *insignis* 237
Ortheziidae 565
Orthochtha brachycnemis 527, 767; Fig. 173
Orthocladiinae 735
Orthodera ministralis 766; Fig. 154
Orthoderinae 766
Ortholomus 584, 769; Fig. 206
Orthophrys pygmaeum 571, 769; Fig. 196
Orthopodomyia albipes 776; Pl. 115
— *kummi* 776; Pl. 115
— *signifera* 75, 776; Pl. 115
— *wilsoni* 776; Pl. 114
Orthoptera 9, 13, 39, 46, 56, 71, 82, 163, 166, 175, 183, 184, 206, 209, 214, 228, 232, 513–532, 763, 764, 766–768
Orthopteroidea 9
Orthosia gothica 775; Pl. 75
— *gracilis* 775; Pl. 75
— *hibisci* 25
— *stabilis* 775; Pl. 75
Orthotylinae 105, 613
Orthotylus adenocarpi 770; Fig. 220
— *concolor* 770; Fig. 220
— *marginalis* 215, 612
Orus punctatus 217
Oryctes boas 17
— *elegans* 17
— *monoceros* 34
— *owariensis* 213
— *rhinoceros* 17, 205
Orygama 132
— *luctuosa* 745
Oryza 738
Oryzaephilus surinamensis 34, 41, 213, 232
Oscinella 72
— *frit* 58, 213, 216, 217, 218, 230
Osmia 29
— *leucomeleana* 234

— *rufohirta* 234
Osmylidae 648
Osmylus fulvicephalus 643
Ostrinia nubilalis 73, 76, 92, 315
Othreis materna 25
Otiorrhynchus 84
— *lavandus* 60
— *ligustici* 72
— *sulcatus* 59
— *turca* 33
Oulema melanopus 49, 219, 228
Oxya hyla 767; Fig. 173
— *japonica* 218
— *velox* 218
Oxycanus cervinata 24
Oxyinae 767
Oxylipeurus 768; Fig. 186
Oxyothespinae 499
Oxytelinae 659
Oxytelus sculpturatus 299
Ozaenini 649
Ozopemon brownei 273, 773; Fig. 133

Pachybrachys hieroglyphicus 772; Fig. 262
Pachycoleus 640
— *rufescens* 640, 771; Fig. 237
Pachycolpura manca 769; Fig. 209
Pachycoris fabricii 272
— *torridus* 272
Pachydiplax longipennis 490
Pachylister chinensis 230
Pachylota 274
Pachymerus 684
— *cardo* 684, 685
— *chinensis* 48
Pachynomidae 570, 610
Paederinae 659
Paederus 231
— *alfierii* 218
— *fuscipes* 218
Palaeodipteroa walkeri 734
Palingenia longicauda 475, 476, 765; Pl. 1
Palingeniidae 475, 765
Palloptera saltuum 745, 777: Pl. 20
Pallopteridae 745, 777
Palomena prasina 601, 770; Fig. 217; Pl. 32
Palophinae 536
Palorus subdepressus 34
Pammene isocampa 235

Pammene juliana 24
Pamphagidae 175, 184, 522, 767
Pamphilidae 705
Pamphilus 705
— *betulae* 705
— *gyllenhali* 705
— *histrio* 705
— *latifrons* 705
— *pillipes* 705
Panaxia dominula 775; Pl. 80
Panchlora 203
— *nivea* 50
Pandanus 533
Pandemis heparana 80, 91
— *ribeana* 24
Panesthia australis 284
— *laevicollis* 284
Panolis flammea 33
— *grisseovariegata* 775; Figs. 87, 88
Panonychus ulmi 213, 228
Panorpa 315, 651, 722, 723
— *anomala* 116, 124, 723, 775; Pl. 98
— *communis* 722
— *germanica* 722
— *nuptialis* 722, 723
Panorpidae 722, 723
Panorpodidae 722
Panorpoidea 10, 93, 722
Pantala flavescens 490
Papilio 48, 251
— *demoleus* 25
— *machaon* 50, 774; Fig. 87
— *zolocaon* 216
Papilionidae 25, 85, 197, 715, 717, 774
Papilionoidea 715, 717
Parabryocoropsis 105
— *typicus* 612
Paracinema tricolor 527, 767; Fig. 173
Paraclunio 735
Paracoenia 748
— *turbida* 749
Paracrypta microptera 222
Paracymus 658
Parakanchia 257
Paraleptophlebia submarginata 478
Paramecops farinosa 21
Paramyelois transitella 31
Parandra brunnea 686, 687
Paranleptes 687
— *reticulata* 687

Parasemidalis annae 771; Fig. 239
Parasitus copridis 227
Parasosibia parva 533
Parasymmictus clausus 739
Paratamia conicola 686
Paratenodera sinensis 201, 219
Paratentranychus pilosus 229
Paratrechina longicornis 221, 291
Parlatoria blanchardii 216
— *oleae* 227
Parnassius 48
— *apollo* 48, 712, 713
Paroplapoderus 278
Paropsis 677
— *atomaria* 37, 677
— *dilatata* 677
Parthenolecanium corni 220
Parus caeruleus 237
— *major* 237
Pasammodes carbonarius 220
Pasimachus californicus 217
— *elongatus* 217
Paspalum 738
Passalidae 273, 303
Passalus 273, 303
— *interstitialis* 303
Passiflora auriculata 241
— *cyanea* 241
Passilidae 273
Paussidae 649
Paxillus recticarinatus 303
Pectinophora gossypiella 33, 34, 88, 179, 215, 218, 220, 315
— *malvella* 72
Pediasia teterrella 179
Pedicia (Crunobia) littoralis 775; Fig. 283
— — *straminea* 775; Fig. 285
Pedicinus piageti 555
Pediculidae 15, 551, 768
Pediculoides 213
Pediculus 549, 555
— *humanus* 15, 549, 554
— — *corporis* 768; Pl. 27
— *mjobergi* 768; Fig. 185
Pediobius foveolatus 22
Pegohylemyia fugax 757, 777; Pl. 150
— *seneciella* 72
Pegomyia hyoscyami 72
Pelobiidae (=Hygrobiidae) 654
Pelocoris 636

SPECIES INDEX 1037

Pelodytes 653, 654
Peloridiidae 565
Pelurga comitata 250
Pentacora leucographa 571
— *signoreti* 571, 769; Fig. 197
Pentatoma rufipes 597, 770; Fig. 217; Pl. 32
Pentatomidae 4, 167, 215, 229, 272, 570, 590, 594–606, 770; Figs. 212, 214–217
Pentatomoidea 102, 590
Pentatomomorpha 167, 570, 582–617, 769
Percosia obesa 217
Peregrinus maidis 214
Perga 705
— *affinis* 311
— *lewisi* 274, 311
Pergamasus longicornis 213
Pergidae 274, 705
Peribalus limbolarius 596
Peridesmia 233
— *discus* 233
Perilampidae 53, 54
Perilampus hyalinus 91
Perilloides bioculatus 215
Perillus bioculatus 229, 596
Periphetes duivenbodei 768; Fig. 175
Periplaneta 201, 203, 205
— *americana* 67, 184, 201, 202, 203, 284, 285, 766; Fig. 76
— *australasiae* 13
Perisceles integer 768; Fig. 175
Perisphaerus glomeriformis 284
Peritrechus sylvestris 769; Fig. 207
Perkinsiella 228
Perla 509
— *carlukiana* 766; Fig. 161
Perlidae 13, 509, 510, 766
Perlodes 509
— *microcephala* 766; Pl. 3
— *mortoni* 13, 509
Perlodidae 13, 509, 510, 766
Pero honestarius 774; Pl. 71
Petaluridae 486
Pexicopia malvella 31, 48
Phaedon cochleariae 214, 219
— *tumidulus* 681
Phaenacantha 585
Phaeogenes nigridens 21
Phaeoura kirkwoodi 774; Pl. 68
Phalacrocera replicata 724
Phalacrodira nioritarsis 237

Phalacrotophora (nedae?) 223
Phalces 534
Phalera bucephala 26, 241, 716, 718, 720, 775; Pl. 81
Phaleriinae 775
Phaloniidae 85
Phanaeus 275, 276, 302
— *daphnis* 773; Fig. 107
— *milon* 773; Fig. 108
— *palliatus* 773; Fig. 269
Phaneroptera falcata 516, 518
— *nana* 516
— *quadripunctata* 515
Phaneropteridae 767
Phaneropterinae 515, 516, 766
Phanerotoma 238, 316
Phaniina 755
Phanuropsis 295, 296
— *semiflaviventris* 294
Phanurus beneficiens 23
Phaonia 745
— *variegata* 777; Pl. 149
Phaoniinae 142, 777
Phasiinae 755
Phasmatidae 536, 768
Phasmida 9
Phasmidae 97, 116
Phaulacridium vittatum 222
Pheidole flavens 87
— *punctulata* 234
Phenacoccus 235
Phenacoleachiidae 565
Pheosia 116
— *gnoma* 718, 775; Pl. 88
— *rimosa* 718, 720, 775; Pl. 83
— *tremula* 718, 775; Pl. 88
Pherbellia 743, 744
Pheropsophus hispanicus 217
Philia 733
Philobostoma quadrimaculatum 524
Philonicus albiceps 740
Philonthus 218, 231, 659
— *cognatus* 661, 772; Figs. 250, 251
— *cruentatus* 661, 772; Figs. 23, 249
— *decorus* 661, 772; Fig. 250
— *fimetarius* 660, 772; Fig. 246; Pl. 52
— *fumarius* 660, 772; Fig. 249
— *fuscipennis* 772; Pl. 52
— *immundus* 660, 772; Fig. 249
— *intermedius* 661, 772; Fig. 250

Philonthus laminatus   772; Pl. 52
— *marginatus*   660, 772; Figs. 246, 247
— *nigrita*   661, 772; Fig. 249
— *ochropus*   660, 772; Figs. 246, 247
— *puella*   660, 772; Fig. 246
— *quisquiliarius*   661, 772; Fig. 249
— *sanquinolentus*   661, 772; Fig. 249
— *sordidus*   660, 772; Fig. 249
— *splendens*   661, 772; Figs. 250, 251
— *tenuicornis*   661, 772; Fig. 249
— *theveneti*   231
— *varians*   661, 772; Fig. 250
— *varius*   660, 772; Fig. 246
Philopotamidae   710
Philopteridae   551, 768
Philosamia   199
— *pryeri*   206, 207
— *ricini*   206, 207, 774; Pl. 90
Philotarsus picicornis   546
Philudoria potatoria   241
Phimodera nodicollis   591
Phlaeophana corticata   272
— *longirostris*   272, 296
Phlaeothripidae (= Phloeothripidae)   15
Phlebonotus pallens   284
Phloeidae   272, 296
Phloeophthorus   282, 773; Fig. 279
Phloeosinus   279, 281
— *thujae*   773; Fig. 279
Phloeothripidae   212, 214, 564
Phlogophora meticulosa   775; Pl. 78
Phloridosa   136
— *flavicola*   136
Phoetaliotes nebrascensis   526
Pholadoris   136
Pholidoptera griseoaptera   518, 767; Fig. 166; Pl. 12
Phoridae   86, 223, 776
Phormia   39, 179
— *regina*   46, 315
Phortica   136
Phosphuga atrata   17
Photininae   766
Phragmatobia   159
— *fuliginosa*   775; Pl. 79
Phrenapates bennetti   273
Phryganea grandis   23, 194, 710, 711, 774; Fig. 84
— *striata*   192
Phryganeidae   23, 710, 711, 774

Phthiraptera   5, 9, 15, 224, 549–562, 768, 775
Phthiridae   15
Phthirus   555
— *pubis*   15, 554, 768; Pl. 27
Phthorimaea operculella   24, 34, 212
Phylliidae   536
Phyllium   534
— *crurifolium*   244
— *pulchrifolium*   244, 768; Fig. 176
— *scythe*   244
Phyllocrania paradoxa   766; Fig. 155
Phyllodecta laticollis   772; Fig. 258
Phyllodectinae   772
Phyllognathus silenus   36
Phyllomorpha   58
— *laciniata*   58
Phyllovates chlorophaea   502
Phylloxera   89
Phylloxeridae   565
Phymata crassipes   769; Fig. 208
Phymatapoderus   278
Phymateus viridipes   527, 767; Fig. 173
Physcus   707
— *intermedia*   708
Physomerus grossipes   271
Phytocoris dimidiatus   215
— *tiliae*   612
Phytodecta   286, 303
— *olivacea*   214
— *rufipes*   303
— *viminalis*   303, 676
Phytoecia cylindrica   687
Phytonomus   84
— *meles*   21
Picromerus bidens   596, 770; Fig. 215; Pl. 32
Pieridae   25, 85, 104, 774
Pieris   179, 706
— *brassicae*   40, 78, 79, 92, 96, 181, 200, 209, 216, 719
— *melete*   92
— *napi*   774; Pl. 67
— *rapae*   25, 73, 92, 313, 774; Pl. 67
— — *crucivora*   92
Piesma quadrata   34, 582, 769; Figs. 205, 206
Piesmatidae   570, 582–583, 584, 769; Fig. 206
Piezodorus lituratus   602, 770; Fig. 217; Pl. 34
Piezosternum calidum   770; Fig. 213
Pilophorus perplexus   215, 612

Pimelia rugosa 220
— *subglobosa* 220
Pimpla turionellae 91, 94
pink bollworm 229
Pinus radiata 717
— *resinosa* 73
— *sylvestris* 73
Piophila 46
Pipizella varipes 741
Piratinae 614
Pisilus tipuliformis 272, 297
Pissodes 313
— *approximatus* 70
— *strobi* 34, 745
Pitedia pinicola 603, 770; Fig. 215
— *sayi* 755
Pitrufquenia coypus 554
Pityogenes 282
Pityokteines 227, 279
Pityophthorus 281, 282
— *bisulcatus* 227
— *micrographus* 773; Fig. 279
Plaesius javanus 230
Plagiolepis 226
— *pygmaea* 226
Plagiostira gilletti 767; Pl. 11
Plastophora rufa 223
Plataspidae 167, 570, 607, 770; Figs. 207, 210
Plataspis 607
Plathypena scabra 87
Platybrachys 769; Pl. 48
Platycentrus acuticornis 271
Platycheirus 741
— *peltatus* 741, 776; Pl. 117
Platycleis denticulata 518, 767; Fig. 166
— *sabulosa* 767; Pl. 12
Platycnemididae 52
Platycnemis pennipes 486
Platycotis vittata 271, 291
Platygaster dryomyiae 92
— *equestris* 88
Platygasteridae 52, 238
Platynota 238
Platynus cupreus 217
Platypalpus pallidiventris 740, 776; Pl. 116
Platypodidae 274, 286, 304, 308, 773
Platypternodes savannae 528
Platyptilia carduidactyla 213
Platypus 309

— *severini* 773; Fig. 135
Platysma punctulatum 217
Platystethus 286, 299, 659
— *arenarius* 273, 299, 301
Platystomatidae 223
Plea atomaria 638
— *leachi* 638
— *striola* 638
Pleciinae 733
Plecoptera 9, 13, 52, 116, 117, 163, 507, 508–511, 764, 766; Figs. 161, 162
Plectopterinae 497
Pleidae 166, 570, 638
Pleolophus basizonus 65
Plesiocoris rugicollis 215
Pleurococcus 546
Plodia interpunctella 25, 43, 79, 214, 718
Plokiophilidae 570, 609, 610
Plusia gamma 40, 41
Plutella 715
— *maculipennis* 33, 40, 77
Plutellidae 85, 715; Fig. 281
Podacanthus viridiroseus 768; Pl. 17
Podagrion 237, 502
Podisus maculiventris 595
Podops inuncta 598
Podotachina 755
Poduridae 214
Poecilips 273, 308
Poecilocapsus lineatus 220
Poecilochirus necrophori 213, 226
Poecilogonales henicospili 29
— *thwaitesii* 29
Poekilocerus 251
— *bufonius* 251
Pogonocherus 685
Pogonognathellus flavescens 214
Pogonosoma 740
Poissonia longifemorata 636
Poliaspis 235
Polietes 121, 151
— *lardaria* 755, 761, 777; Figs. 41, 296; Pl. 144
Pollenia rudis 751
Polochrum repandum 234
Polycentropidae 710
Polydrusus mollis 278
Polyglypta dorsalis 271, 291, 769; Fig. 120
— *lineata* 271
Polygonia c-album 242, 774; Pl. 66

Polygonum 738
Polygraphus 279
Polymitarcidae 12, 765
Polyommatus 48
— *icarus* 774; Pl. 62
Polyphaga 656–688, 771–773
— *aegyptica* 766; Fig. 153
Polyphagidae 495, 499, 766; Fig. 153
Polyplax affinis 555
— *serrata* 551
Polyploca flavicornis 250
Polyporus 297
Polystoechotidae 648
Pompilidae 234
Pompilus 234
— *analis* 234
Pontania proxima 706
Porthemia 257
Potamanthidae 479, 768
Potamanthus luteus 479, 765; Fig. 136
Potamogeton illinoiensis 654
Potamonautes 735
Potamophylax stellatus 23
Premnobius cavipennis 305
Prepusa punctum 771; Fig. 114
Primicimex cavernis 770; Pl. 36
Prionidae 773
Prioninae 687
Prionoxystus robiniae 33, 37, 249
Prionus 686
— *californicus* 20
— *coriarius* 688
Pristiphora erichsonii 91, 215
Proboscidia 749
Proceras sacchariphagus 35
Proctacanthus 740
proctotrupids 502
Proctotrupoidea 704
Prodenia 231
— *litura* 40, 72, 218
Prodoxidae 713, 719, 774
Promachus 740
— *yesonicus* 740
Promecotheca reichi 20, 677
Propylea quatuordecimpunctata 224
Prosena siberita 753
Prosierola bicarinata 274
Prosimulium 734
— *exigens* 734
— *ursinum* 34

Prosopistomatidae 12, 478
Prospaltella 707, 708
— *perniciosi* 22
Prossierola bicarinata 311
Prostemma guttula 609
Protoboarmia porcelaria indicataria 774; Pl. 68
Protocylas coimbatorensis 21
Protomecoptera 10, 722
Protoparce maculata 26
— *sexta* 26
Protophormia terrae-novae 159, 751, 752, 777; Pl. 132
Protostegana 136
Psacasta exanthematica 593
— *neglecta* 593, 770; Fig. 212
Psallus ambiguus 15, 612
Psephenidae 111
Pseudacraea 251
Pseudeucoila bochei 92
— *trichopsila* 92
Pseudochaeta syngamiae 59
Pseudochelidura sinuata 270
Pseudococcidae 565
Pseudococcus 6, 220, 232, 235, 570
— *citri* 216
— *comstocki* 221
— *gahani* 91
— *maritimus* 91
— *sequoiae* 219
— *vastator* 15
Pseudodiamensa arctica 34
Pseudodoniella 105
— *laensis* 612
Pseudogaurax achora 237
— *signata* 237
Pseudoharpax virescens 766; Fig. 155
Pseudohylesinus 279
Pseudoleria 746
Pseudomorellia albolineata 777; Fig. 296
Pseudomorphinae 201
Pseudophasmatinae 536
Pseudophloeinae 58
Pseudophoraspis nebulosa 284
Pseudophyllinae 515, 766
Pseudorhyssa 74
— *sternata* 74
Pseudostenophylax edwardsi 192
Pseudoterpna pruinata 717
Psila rosae 73, 743, 776; Pl. 123
Psilidae 86, 743, 776

Psilocephala lateralis 222
Psilopa leucostoma 749
— *petrolei* 748
Psocidae 4
psocids 213
Psocoptera 5, 9, 14, 102, 214, 228, 546–548, 768
Psocopteroidea 9
Psocus gibbosus 547
— *nebulosus* 547
Psorophora 75, 729, 732, 733, 776; Fig. 58
— *confinis* 75
— *ferox* 57
Psyche 183
— *unicolor* 197
psychid moths 313
Psychidae 24, 197, 198, 713, 720
Psychodidae 761
Psychomyiidae 710
Psychopsidae 643, 648
Psylla alni 567
— *buxi* 567
— *melanoneura* 567
— *pyricola* 48, 567, 568, 769; Fig. 194
— *pyrisuga* 568, 769; Fig. 194
Psyllidae 15, 175, 565, 567–568, 769
psyllids 228
Psyllipsocidae 768
Psyllipsocus 547
Psylloborini 231
Psylloidea 169, 170, 565, 567–568
Psyllopsis fraxinicola 567, 568, 769; Fig. 194
Pteleobius 279
Pterocyclon 305
— *bicallosum* 305
— *brasiliense* 305
— *nudum* 305
Pterodonta flavipes 27, 738
Pteromalidae 22, 84, 94, 220, 233, 707, 709, 773; Fig. 280
Pteromiera 744
Pteronarcidae 510, 766
Pteronarcys 117, 509, 510
— *dorsata* 117, 508, 766; Pl. 2
— *proteus* 509, 510, 766; Fig. 161
Pterophoridae 716
Pterostichus 650
— *kugelanni* 217
— *lucublandus* 217
— *metallicus* 649

— *multipunctatus* 272, 649
— *occidentalis* 217
— *purpurascens* 217
Pterygota 9
Ptilomacra senex 713
Ptilomera 581
Ptinidae 18, 83
Ptinus fur 18
— *sexpunctatus* 18
— *tectus* 42, 60, 178, 183, 225
Publilia concava 271
Pulex irritans 26
Pulicidae 26
Pulicophora pectinata 223
— *rhodesiana* 223
Pulvinaria 216, 220, 221, 236
— *polygonata* 221
— *vitis* 219
Pycanum ruber 594
Pycnoscelus 203, 284
— *surinamensis* 13, 44, 284
Pyemotes parviscolyti 227
— *ventricosus* 213
Pygidicranidae 269
Pygiopachymerus lineola 684
Pyralidae 25, 35, 37, 85, 102, 221, 236, 242, 715, 716, 720, 774; Fig. 281
Pyrameis 251
Pyrausta nubilalis 215, 216, 219
Pyraustidae 720
Pyrellia 117
— *cyanicolor* 760
Pyrgomorpha cognata 527, 767; Fig. 173
— *dispar* 184
— *kraussi* 527, 767; Fig. 173
Pyrgomorphidae 175, 184, 522, 767
Pyrgota undata 57
Pyrgotidae 57
Pyroderces 235
— *bicincta* 235
— *falcatella* 235
— *gymnocentra* 235
— *holoterma* 235
— *philogeorga* 235
— *rileyi* 235
Pyrrhocoridae 570, 585, 769; Fig. 207
Pyrrhocoris 585
Pyrrhocoroidea 102

Quadraspidiotus perniciosus 213, 220
Quedius boopoides 662, 772; Figs. 23, 254
— *cinctus* 663, 772; Figs. 254, 255
— *cruentus* 662, 772; Fig. 254; Pl. 55
— *fumatus* 662, 772; Fig. 254
— *laevicollis* 661, 662, 772; Figs. 254, 256
— *maurorufus* 663, 772; Fig. 254
— *mesomelinus* 662, 772; Figs. 254, 255; Pl. 55
— *picipes* 772; Pl. 55
— *ventralis* 663, 772; Fig. 254
Quinta cannae 25

Racodineura pallipes 755, 777; Fig. 292
Ramburiella turcomana 768; Fig. 174
Ranatra 130, 621, 622, 634
— *dispar* 622, 629, 632, 635
— *elongatula* 622, 628, 629, 632, 635, 771; Fig. 231
— *fusca* 622, 629, 632, 635, 771; Figs. 49, 233
— *linearis* 162, 621, 622, 625, 628, 629, 632, 633, 634, 771; Figs. 47, 230, 233; Pl. 45
— *parvipes* 621
— *sordidula* 622, 628, 629, 635
— *vicina* 621, 622, 628, 629, 635
Ranatrinae 130, 621, 622, 628, 630, 634, 771
Raphidia (= *Agulla*) 215
— *astuta* 642
— *bractea* 642
— *notata* 642
— *ophiopsis* 215
Raphidiidae 215, 642
Reduviidae 229, 272, 296, 570, 614–617, 770; Figs. 221, 222
Reduviinae 614
Reduvius 229, 616
— *personatus* 770; Fig. 221
Renocera 743
Reticulitermes flavipes 315
Reuterella helvimacula 546
Rhabdoscelus 84
Rhagadotarsus 581
Rhagionidae 27, 737–738, 776; Fig. 288
Rhagium 687
Rhagoletis 742
— *cerasi* 28, 65, 742, 776; Pl. 121
— *completa* 742
— *fausta* 65
— *pomonella* 65, 80, 742

Rhantus (*calidus?*) 651
Rhaphigaster nebulosa 598
Rhingia campestris 27, 741, 776; Pl. 117
Rhinoceros bicornis 749
— *simus* 749
— *sumatrensis* 749
Rhinocoris 296, 614, 616, 617, 770; Fig. 221
— *albopilosus* 272, 296, 297, 614
— *albopunctatus* 272, 296, 297
— *annulatus* 770; Fig. 221
— *bicolor* 614
— *carmelita* 272, 296, 614
— *iracundus* 614
— *niger* 770; Fig. 221
— *obtusus* 614
— *rapax* 614
Rhinoestrus purpureus 136, 223, 751
Rhinoleucophaga obesa 223
Rhinophora lepida 752, 777; Pl. 130
Rhinophoridae 752, 777
Rhipiceridae 17
Rhipidia maculata 27
Rhipiphoridae 54
Rhithrogena 476, 478, 765; Fig. 140
— *semitincta* 476
Rhizobius ventralis 232
Rhizopertha dominica 18, 178
Rhizophagidae 219, 231
Rhizophagus dispar 219
Rhodia fugax 774; Pl. 94
Rhodnius 29, 90, 97, 180, 181, 614, 616, 617
— *prolixus* 35, 45, 47, 178, 616, 623, 770; Pls. 38, 39
Rhodococcus bulgariensis 15
Rhopalidae 570, 584, 589, 769; Figs. 207, 210
Rhopalosiphum fitchii 249
Rhopalus parumpunctatus 769; Pl. 30
Rhyacionia 315
— *buoliana* 72, 73, 215
— *duplana* 717
— *simulata* 59
Rhyacophila nubila 710
Rhynchaenus fagi 183
Rhynchites 278, 282
— *auratus* 773; Fig. 110
— *bacchus* 48
Rhynchitinae 278
Rhynchophorus ferrugineus 228
— *palmarum* 21

Rhynchophthirina 9, *549, 555, 562*
Rhyopsocus phillipsae 547, 768; Pl. 18
Rhyparochrominae 584
Rhysodidae 649
Rhyssa 74
— *lineolata* 706
— *persuasoria* 74, 706
Rhyzopertha 232
Ricaniidae 169, 565
Ricinus communis 244
Riekoperla alpina 766; Fig. 162
— *tuberculata* 766; Fig. 162
— *williamsi* 766; Fig. 162
Riodinidae 226
Rioxa pornia 50
Rodolia cardinalis 19, 220, 232
— *chermesina* 234
Romalea 521
Rubiconia intermedia 602
Ruepellia 739

Sabethes chloropterus 49
Sabethini 51
Sagaritis 238
Saginae 515
Sahlbergella 105, 166
— *singularis* 79, 166, 612
Saissetia 221, 236
— *oleae* 220, 221, 232
Sakuntula ravana 611
Salda littoralis 571, 769; Fig. 198
Saldidae 117, 271, 570–573, 574, 575, 769
Saldula 571
— *madonica* 571, 769; Fig. 197
— *palustris* 572, 769; Fig. 198
— *scotica* 571
Salix 713, 738
Salpinogaster nigra 741
Saltella 126, 132
— *scutellaris* 125, 126, 132, 776; Fig. 54; Pl. 124
Salticella 744
Salyavatinae 616
Samia 177
Saperda carcharias 5, 20, 76
— *scalaris* 686
Saprinus aeneus 230
— *cuspidatus* 230

— *elegans* 218
— *ornatus* 218
— *semistriatus* 230
Sapromyzidae (=Lauxaniidae) 745
Sapygidae 234
Sarcophaga bullata 752, 777; Pl. 131
— *carnaria* 752
— *hunteri* 224
— *opifera* 224
Sarcophagidae 86, 752, 777
Sarcophaginae 777
Sarrothripinae 236
Saturnia pavonia 774; Pl. 92
Saturniidae 26, 206, 207, 210, 248, 258, 715, 716, 717, 721, 774
sawfly 3, 59, 74, 312, 313
scale 73, 234, 235, 236, 237, 708
Scarabaeidae 17, 36, 83, 175, 273, 275, 301, 688, 773; Figs. 108, 265–273
Scarabaeinae 275, 276, 301
Scarabaeus 276
— *semipunctatus* 265, 266, 267, 773; Fig. 109
Scatella silacea 748
— *thermarium* 748
Scatophaga (=*Scopeuma*) 139, 140, 151, 756
— *lutarium* 139
— *stercoraria* (=*Scopeuma stercorarium*) 129, 231, 756, 777; Figs. 58, 59, 65; Pl. 231
Scatophagidae 756–777
Scatophaginae 777
Scelionidae 23, 64, 84, 237, 709, 773
scelionids 502
Schedius kuvanae 705
Schidium callipygum 617
Schistocerca 45, 71, 83, 184, 519, 521, 522
— *gregaria* 14, 34, 47, 90, 175, 205, 216, 217, 218, 219, 220, 222, 223, 520, 522, 767; Pl. 14
— *lineata* 524, 525, 526
— *obscura* 524, 525, 526
— *paranensis* 218, 223
Schistosoma 743
Schizaphis graminum 224
Schizolachnus piniradiatae 15
Schizophora 760
Schizopteridae 570, 640–641, 771
Schizura ipomoeae 775; Pl. 83
Schoenobius incertellus 233
— *sordidellus* 240
Sciaridae 27

Sciocoris 594
— *cursitans* 599, 770; Fig. 216
— *deltocephalus* 599, 770; Fig. 216
— *distinctus* 601, 770; Fig. 216
— *homalonotus* 600, 770; Fig. 216
— *macrocephalus* 598, 770; Fig. 216
— *sulcatus* 599, 770; Fig. 216
Sciomyza 744
Sciomyzidae 28, 237, 743–744, 762
Sciopus euzonus 222
Scirophaga 256
— *nivella* 80
Scirpus 738
Scleroderma immigrans 274
— *macrogaster* 274
Scolia manilae 23
Scoliidae 23
Scoloposcelis flavicornis 215
Scolytidae 21, 70, 77, 84, 227, 273, 274, 279, 282, 286, 304, 688, 689, 773; Figs. 130, 278, 279
Scolytoplatypodinae 274
Scolytoplatypus acuminatus 305
— *daimio* 305
— *kivuensis* 273, 773; Fig. 130
— *nikada* 305
— *shogun* 305
— *tycon* 305
Scolytus 70, 72, 279, 281
— *multistriatus* 87
— *rugulosus* 21
Scopeuma (= *Scatophaga*)
Scotomedes alienus 609
— *minor* 609
Scotusa cilens 518
Scudderia 515
Scutelleridae 271, 570, 590, 591–593, 594, 770; Fig. 212
Scutellista cyanea 220, 233
Scylaticus 222
Scymnus 231
— *impexus* 220
— *sieverini* 232
Sehirinae 589
Sehirus 589
— *bicolor* 271, 589, 770; Figs. 122, 211
— *luctuosus* 271
— *sexmaculatus* 271
Seilus 213
Selenia bilunaris 774; Pl. 72

Selepa leucogonia 236
Semidalis aleyrodiformis 771; Fig. 239
Semiothisa signaria dispuncta 774; Pl. 70
Sennius 684
— *fallax* 684
— *morosus* 684
— *simulans* 684
Sepedon 237, 743, 744
— *macropus* 744
— *neili* 28
— *plumbella* 743, 744
— *sphegea* 28
Sepsidae 125, 126, 129, 132, 151, 162, 776
Sepsidimorpha 132
Sepsis 132, 776
— *violacea* 132, 162, 776; Fig. 53
Sericostomatidae 23, 192, 710, 711
Sessidae 713, 720
Sevtshenkia 726
Sextus virescens 270
Sherifuria hanningtoni 223
Siagona dejeani 217
Sialidae 16, 642, 771
Sialis 642
— *lutaria* 16, 642, 771; Pl. 49
Sialoidea 642
Sibylla limbata 766; Fig. 155
Sibyllinae 766
Sigara 620
Silphidae 17, 273, 298, 772
Silvanidae 83
Similiinae 270, 271
simuliid eggs 313
Simuliidae 27, 85, 734–735
Simulium 27, 237, 315, 734, 735
— *arcticum* 734
— *berneri* 735
— *copleyi* 735
— *damnosum* 734
— *dukei* 735
— *erthrocephalum* 734
— *fuscum* 734
— *mixtum* 734
— *neavei* 735
— *ornatum* 64, 159
— *ovazzae* 735
— *simile* 734
Sinophthalmus pictus 136
Siphlonuridae 12, 479, 765
Siphlonurus aestivalis 479

Siphlonurus lacustris 12, 479, 765; Fig. 140
Siphnus 594
— *alcides* 594
Siphonaptera 26
Siphonella palposa 237
Sipyloidea 536
— *sipylus* 768; Pl. 17
Sirex 74
— *noctilio* 88
Siricidae 84, 304, 705
Sisyphus schaefferi 773; Fig. 265
Sisyra (fuscata (F.), *terminalis* Curtis) 643
Sisyridae 648
Sitarobrachys buigasi 49
Sitona 60, 61, 177
— *crinitus* 61
— *hispidulus* 60, 61
— *lineatus* 61, 80
— *lividipes* 61
Sitophilus 84
— *granarius* 21, 41, 42, 60, 212
— *oryzae* 21, 48, 77, 88, 179, 315
— *zeamais* 211
Sitotroga 231
— *cerealella* 33, 41, 88, 211, 214, 220
Sminthuridae 214
Smittia gracilis 27
Solanum luteum 87
Solenopsis 227
Solubea pugnax 596
Sophophora 134
Spalangidae 64
Spaniocerca 508
— *tasmanica* 508
Sparganothis pilleriana 80
Spathosternum pygmaeum 528, 767; Fig. 173
Spercheidae 186, 272, 297, 656, 658
Spercheus 186, 658
— *emarginatus* 272
Sphaeridiinae 186, 656, 658
Sphaeridium 658
— *scarabaeoides* 217
Sphaeriidae 654
Sphaerius 654
Sphaerocera 748
— *curvipes* 777; Fig. 289
— *pusilla* 777; Fig. 289
Sphaeroceridae 113, 116, 132, 133, 746–748, 777; Figs. 289, 290
Sphaerocerinae 777

Sphaerodema 635
— *rusticum* 636
Spharagemon collare 524
— *equale* 524
Sphecia bembeciformis 713
Sphecidae 709
Sphedanovarus camerunensis 617
Sphidia obliqua 258
Sphingidae 26, 85, 250, 715, 717, 721, 774
Sphodromantis 209, 499, 505
— *centralis* 202
— *lineola* 202, 209
— *viridis* 499, 503, 504, 766; Figs. 157, 159
spider eggs 230
Spilonota ocellana 24, 213, 214, 215
Spilopsyllus cuniculi 775; Pl. 100
Spilosoma 314
Spintherus 233
Spodoptera 67
— *littoralis* 40, 228, 229
Stachys silvaticus 597
Stagmomantis carolina 503
— *dimidiata* 766; Fig. 154
— *limbata* 502, 503
Stagonomus amoenus 600
— *bipunctatus* 600
— *pusillus* 600
Stalia boops 609
Staphylinidae 4, 54, 98, 100, 102, 171, 175, 218, 230, 231, 234, 273, 298, 659–674, 688, 772
Staphylininae 659–663, 772
Staphylinus 116, 171
— *caesareus* 662, 772; Fig. 253
— *erythropterus* 662, 772; Fig. 253
— *parumtomentosus* 662, 772; Figs. 37, 253
— *pubescens* 772; Pl. 54
Staria lunata 602
Stathmopoda 235
— *basiplectra* 235
— *callichrysa* 235
— *crocophanes* 235
— *conioma* 235
— *melanochra* 235
— *ovigera* 235
— *theoris* 235
Stator 684
Steganinae 136
Stegobium 232, 315
— *paniceum* 37; Fig. 16

Stegomyia (= *Aedes*)
Steiroxys (borealis?)   767; Pl. 13
Stelidae   234
Stelis   234
— *minuta*   234
Stenidea   686
Stenobothrus   521
— *eurasius*   768; Fig. 174
— *lineatus*   82
— *nigromaculatus*   237
— *theryi*   237
Stenocephalidae   570, 584–585, 769
Stenocera   5
Stenoelmis   617
Stenopogon   222, 740
Stenopsocus immaculatus   546
Stenoria analis   232
Stephanitis rhododendri   611
Stephanoderes hampei   21
Sternodontus obtusus   598, 770; Fig. 216
Sternorrhyncha   163, 169, 565, 567–570
Sterrha   717
— *aversata*   717
— *emarginata*   717
— *humiliata*   717
— *serpentata*   713
— *strigilaria*   717
— *virgularia*   713
Stethomezium squamosum   42
Stethorus   231
Sticholotis   231
stick insects   90
Stictococcidae   565
Stictococcus   235
Stictopleurus abutilon   589
Stigmella aurella   774; Pl. 59
— *ruficapitella* (= *Nepticula ruficapitella*)   774;   Pl. 59
Stigmellidae   84, 774
Stilbula cynipiformis   22
Stilpnotia salicis   26
Stolas thalassina   273
Stollia aenea   602, 770; Fig. 217
— *inconspicua*   603
— *venustissima*   597
Stomorhina lunata   224
Stomoxyinae   760, 761, 777
Stomoxys   57, 119, 121, 227, 231
— *calcitrans*   28, 119, 761, 777; Fig. 42
— *ochrosoma*   758

— *sitiens*   28
stone flies   509, 510
Stranglia melanura   686
Stratioboroborus fimetarius   748
Stratiomyidae   761
Streblidae   286
Strepsiptera   10, 29, 54
streptococci   92
Stringomyia   724
Strongylocoris   105, 106
— *luridis*   613, 770; Fig. 103
Strophosomus   84
Strymon   717
Sturmia   29
— *bohemica*   74
— *cilipes*   29
— *harrisinae*   28, 29
— *scutellata*   28, 29
Stylogaster   56, 57
— *westwoodi*   776; Pl. 119
Stylops aterrima   29
Sulpicia distincta   586
Supella supellectilium   284
Suriana   244
— *maritima*   244
Symmerista albifrons   718
Symmictus costratus   222
Sympetrum   208, 489
— *frequens*   207, 208
— *infuscatum*   207, 208
— *sanguineum*   766; Fig. 148
— *striolatum*   490, 766; Figs. 144, 145
Sympha   238
Sympherobiidae   11, 16, 216
Sympherobius amicus   16
— *angustatus*   216
Symphyta   311, 704, 705–706
Synanthedon tipuliformis   35
Synapsis   301, 302
— *tmolus*   273, 302
Syneura cocciophila   223
Synharmonia conglobata   220
Synlestes weyersi   765; Fig. 147
Synonycha grandis   224
Synthemia primigenia   487
Synthemidae   486
Syrbula admirabilis   524
Syrphidae   27, 77, 86, 116, 119, 237, 644, 706, 741, 762, 776
Syrphidis   741

Syrphidis ribesii 776; Pl. 118
Syrphinae 741, 776
Syrphus luniger 63
— *ribesii* 77, 92
Systoechus acridophagus 223
— *aurifacies* 223
— *autumnalis* 223
— *gradatus* 223
— *marshalli* 223
— *oreas* 223
— *pallidulus* 223
— *socius* 223
— *somali* 223
— *sulphureus* 223
— *vulgaris* 223
— *waltoni* 223
— *xerophilus* 223

Tabanidae 27, 52, 86, 738, 749, 761, 776; Fig. 288
Tabanus 738
— *macer* 27
— *quaturnotatus* 738, 776; Fig. 288
— *rubidus* 27
— *striatus* 738
— *tenens* 27
— *townsvilli* 738
tachinid fly 55, 57
Tachinidae 4, 37, 53, 54, 56, 86, 211, 739, 752–756, 777; Fig. 292
Tachopteryx thoreyi 486
Tachyporus 218
— *hypnorum* 218
Tachys ruscarius 217
Tachythorax maculicollis 533
Taeniogonalos venatoria 706
Taeniorhynchus fuscopennatus 80
Taenoglyptus 70
Taphrophila 724
Tarachodes afzellii 269
— *maurus* 269
Tarisa pallescens 598
Tectocoris diophthalmus 272
Tegeticula 717
— *yuccasella* 774; Fig. 282
Telenomus sphingis 73
Teleogryllus commodus 179
Telmatotrephes 130, 625, 630
— *breddini* 622, 632, 634, 771; Fig. 229

— *grandicollis* 771; Pl. 44
Temnochila 70, 71
Tenanocera 744
— *elata* 743
Tenanura 744
Tenebrio molitor 19, 60, 90, 207, 212
— *obscurus* ·19
Tenebrionidae 4, 19, 83, 219, 232, 273, 689
Tenebroides mauritanicus 19
Tenodera 505
— *angustipennis* 503
— *sinensis* 204, 503
Tenthredinidae 21, 705–706
Tenthredinoidea 84
Tentyria nomas 220
Tephritidae 86, 741–743
Tephrochlamys rufiventris 746, 777; Pl. 120
Terebrantia 563
Termatophylidae 125
— *opaca* 131, 612, 770; Pl. 37
termites 223, 269, 285
Termitoxenia 223
Tessaratoma 594
— *javanica* 594
Tessaratomidae 272, 570, 593–594, 770
Tetanocera elata 743
Tethea ocularis 774; Pl. 73
Tetragoneuria 766; Fig. 150
Tetrastichus 233, 238
— *asparagi* 233
— *chrysopae* 91
— *ovivorax* 237
— *pachymerus* 233
— *pallipes* 708
— *schoenobii* 233
— *xanthomelaenae* 233
Tetrathemis 486
Tetrigidae 57, 164, 175, 184, 513, 514–515, 659, 768
Tetrigoidea 768
Tetrix 57, 175, 514
— *subulata* 514
— *vittata* 514, 768; Fig. 74
Tetropium cinnamopterum 688
Tettigidea parvipennis 768; Fig. 167
Tettigometra atrata 769; Fig. 48
— *sulphurea* 769; Fig. 48
Tettigometridae 130, 565, 769
Tettigonia 159
— *viridissima* 14, 516, 767; Fig. 165; Pl. 5

Tettigoniidae 4, 13, 83, 116, 175, 177, 184, 214, 513, 515–518, 766–767; Fig. 166
Tettigoniinae 515, 516, 767
Thalessa 706
Thanasimus 71
— *dubius* 70
— *formicarius* 215
— *undulatus* 70
Thaumaglossa 213, 502
— *oothecobia* 219
— *ovivorus* 219
— *rufocapillata* 231
Thaumaleidae 85
Thaumastella aradoides 582
Thaumastellidae 570, 582
Thaumastocoridae 570, 613–614
Thaumastocoris australicus 614
Thaumatoperla albina 508
— *robusta* 508
Thaumatopsis edonis 774; Pl. 61
Thaumetopoea 51, 257
— *pityocampa* 218, 228, 257
— *processionea* 26
Thaumetopoeidae 26, 256, 257
Themos olfersii 274, 311
Therevidae 222, 739
Therioaphis trifolii 569
Therion 52
Thermobia domestica 12, 34, 91, 178, 765; Pl. 1
Thermonectes 651
Thespinae 766
Thisoicetrus (=*Heteracris*) 175
— *pulchripes guineensis* 526
— — *jeanneli* 526, 767; Fig. 173
Thoracochaeta zosterae 748
Thraulus bellus 475, 478
Thyanta custator 596
Thyreocoris scarabaeoides 589
Thyridopteryx ephemeraeformis 211
Thysanoptera 9, 15, 214, 228, 270, 563–564, 764; Fig. 192
Thysanopteroidea 9
Thysanura 9, 12, 47, 91, 165, 166, 765
Tiarothrips subramanii 563
Tinea 715
— *pellionella* 33, 37, 42, 179
— *viviparis* (=*Monopis meliorella*) 712
Tineidae 24, 37, 84, 715, 716; Fig. 281
Tineola 79

— *bisselliella* 24, 178, 197
Tingidae 16, 272, 296, 570, 611
Tingis ampliata 178
Tiphia popilliavora 23
Tiphiidae 23, 84, 94, 709
Tipnus unicolor 60
Tipula 724, 726
— *aino* 33
— *czizeki* 726
— *flavolineata* 27
— *luna* 27
— *oleracea* 179
— *paludosa* 724, 726
Tipulidae 27, 85, 116, 175, 724–727, 775; Figs. 283–286
Tipulinae 725, 726, 775
Tiracola plagiata 25
Titanochaeta 136
Titanus giganteus 5, 687
Tmenthis tartarus 767; Fig. 174
Tomocera californica 220
Tongamya 740
— *miranda* 740
Torleya (=*Ephemerella*) 479, 765; Fig. 138
Torridincolidae 654, 771; Fig. 243
Tortricidae 24, 35, 77, 84, 720, 774
Tortricodes tortricella 74
Tortrix callopista 235
— *viridana* 40, 74, 179
Torymidae 84, 502, 707
torymids 502
Toxorhynchites 729
Toxorhynchitinae 51
Trabutina 235
Trachodinae 499
Trachyrhachis kiowa 222, 525, 768; Fig. 171
— *plattei* 524, 768; Fig. 171
Tragidion armatum 686
Tragocephalus 687
— *guerni* 687
Tragopinae 271
Tramea onusta 490
Trauxalinae 768
Trechus obtusus 217
— *quadripustulatus* 217
Tremax 706
Triaenodes bicolor 192, 194
Trialeurodes vaporariorum 569, 769; Fig. 195
Triatoma 614, 617
— *infestans* 178, 614, 770; Fig. 222; Pl. 40

SPECIES INDEX

Triatoma phyllosoma   770; Pl. 41
— *proteinus*   615, 770; Pl. 41
Triatominae   614
Tribolium   224, 225, 232, 315
— *castaneum*   19, 60, 61, 76, 212, 213, 220, 224, 225, 227, 232
— *confusum*   90, 178, 213, 224, 225
Trichardis   222
Trichiaspis equina   747, 748, 777; Fig. 289
Trichiocampus viminalis   21
Trichodectidae   551, 768
Trichodes   213
— *affinis*   219
— *amnios*   219
— *apiarus*   219
— *crabroniformis*   219
— *laminatus*   219
— — *cyprius*   219
— *umbellatarum*   219
— *x-littera*   219
Trichogramma   53, 73, 238, 239, 313, 316, 708, 773; Fig. 17
— *cacoeciae*   22, 41, 55, 73
— *embryophagum*   22, 41, 238
— *evanescens*   22, 41, 73, 239
— *minutum*   22, 73, 238, 239
— *semblidis*   22
Trichogrammatidae   22, 64, 84, 237, 708, 709, 773
Trichopepla semivittata   595
Trichophaga   715
— *tapetzella*   179
Trichoplusia ni   26, 33, 47, 77, 315
Trichopoda   755
— *pennipes*   56, 74
Trichopsidea clausa   222
— *ostracea*   222
Trichopsocus dalii   546
Trichotanannus dundo   640
Tricondyla cyanea   650
Tricoptera   10, 23, 29, 52, 84, 140, 163, 183, 192, 193, 194, 209, 710–711, 735, 764, 774
Tricrania sanguinipennis   29, 233
Tridactylus apicalis   269
Trigonalidae   29, 53, 54, 706
Trigonocorypha   515
Trigonogenius globulus   42
Trigonosoma philalyssum   599
— *trigonum*   599

Trilophidia conturbata   528
Trimerotropis citrina   524
Trinodes rufescens   18, 50
Triogma   724
Triommata   236
Triophtydeus triophthalmus   227
Trioza erytreae   15
— *urticae*   567
Triphaena pronuba   63
Triplectides australis   192
— *vivipara*   192
Trissolcus   295
— *basalis*   48, 773; Fig. 17
— *bodkini*   294
Tristaria grouvellei   77
Tristria conops   528, 767; Fig. 173
— *discoidalis*   528, 767; Fig. 173
Trochilum apiforme   51
Troctes bostrychophilus   214, 228
Trogidae   218
Trogiidae   214
Trogoderma angustum   36
— *anthrenoides*   61
— *granarium*   18, 48, 61, 82, 178, 227
— *parabile*   34, 213
— *varium*   18, 219
— *versicolor*   18, 82, 219
Trogositidae   19
Troilus luridus   597, 770; Pl. 33
Tropaea   176
Tropidaspis carinatus   270
— *cornuta*   270
— *minor*   270
Tropiduchidae   130, 565
Trox   213
— *gemmatus*   218
— *procerus*   218
— *squalidus*   218
— *suberosus*   218
Trybliographa rapae   707, 773; Fig. 280
Trypetidae (=Tephritidae)   28, 741–743, 762, 776
Tryphon semirufus   773; Fig. 10
Tryphoninae   706
Trypodendron   67, 305, 306
— *betulae*   273
— *lineatum*   29, 70, 273, 773; Fig. 129
tsetse fly   79
Tubulifera   563
Tylopsis thymifolia   515

SPECIES INDEX

Tylotropidius gracilipes 526
— *speciosus* 526
Typha 738
Typhlocyba tenerrima 59
Typhlodromus baccettii 214
— *sudanicus* 227
— *tineivorus* 214
Tyria jacobaeae 225, 315
Tyroglyphus longior 214
Tyrophagus putrescentiae 227
Tytthus mundulus 228
— *parviceps* 228

Ula macroptera 724
Ululodes 647
— *macleayana* 647
— *mexicana* 647, 771; Fig. 240
Umbonia ataliba 271, 293
— *crassicornis* 271, 291, 293
Uracis ovipostrix 487
— *siemensi* 487
Uranotaenia 116
— *anhydor* 776; Pl. 114
Urentius echinus 16
Urocrus gigas 706
Uropetala 486
Urostylidae 570, 591, 607
Urostylis flavomaculata 591
— *striicornis* 591
— *westwoodi* 591
Urothemis 489

Valleriola moesta 573
Vanessa atlanta 774; Pl. 64
— *cardui* 774; Pl. 66
Varioperla 508
Velia 579
Veliidae 166, 570, 579–580, 769
Velleius 659, 661
— *dilatatus* 663
Velocipedidae 570, 609
Ventidius 581
Vermilio 738
— *comstocki* 738
— *degeeri* 738
— *vermilio* 27, 737
Vertebrates 237
Vesperus xatarti 686

Vespidae 704
Vestula lineaticeps 614
Vianaididae 570, 611
Vilhenannus angolensis 641
Villa decipula 223
— *quinquefasciata* 51
Vilpianus galii 598
Vitula edmandsii serratilineella 25
— *saissetiae* 236
Volucella 741
Voria ruralis 28
Vyburgia lounsburyi 15

wasps 183, 269, 285, 291, 310, 313
weevils 43, 55, 59, 111, 177
Wesmaelius concinnus 771; Fig. 239
Westermanninae 775
wheat-bulb fly 230, 231
Winthemia 753, 755
— *fumiteranae* 28
wireworms 177
Wohlfahrtia balassogloi 224
— *erythrocera* 224
— *euvittata* 224
wood-boring beetles 563
woodwasps 73, 74.

Xanthampulex luzonensis 234
Xantholinus fracticornis 663, 772; Fig. 22
— *glabratus* 218, 663, 772; Fig. 22
— *hamatus* 218
— *linearis* 663, 772; Fig. 22; Pl. 55
Xanthorhoe montanata 774; Pl. 72
— *munitata* 774; Pl. 72
Xenodusa 226
Xenopsylla 176
Xylastodoris luteolus 614
Xyleborus 305
— *agnaticeps* 274, 773; Fig. 130
— *alluaudi* 274, 773; Fig. 130
— *ambasiusculus* 305
— *asperatus* 274
— *compactus* 274, 305
— *comparabilis* 274, 773; Fig. 130
— *cruciformis* 274, 773; Fig. 130
— *discolor* 305
— *dispar* 305
— *ferrugineus* 305

SPECIES INDEX

Xyleborus fornicatus 274, 305, 306, 773; Figs. 131, 132
— *germanus* 21, 274, 305
— *gracilis* 305
— *integer* 274, 773; Fig. 130
— *mascarensis* 305
— *monographus* 274, 305
— *mutilatus* 773; Fig. 135
— *saxeseni* 305
— *schreineri* 305
— *semigranosus* 773; Fig. 130
— *semiopacus* 305
— *sentosus* 305
— *septentrionalis* 773; Fig. 135
— *subscribosus* 274
— *subtuberculatus* 305
— *xanthopus* 274
— *xylographus* 274
Xylechinus pilosus 773; Fig. 278
Xyletinus peltatus 178, 228
Xyleutes durvillei 712
Xylocopa violacea 234
Xylocoris flavipes 228, 610, 770; Pl. 31
Xyloterinus 305
— *politus* 305
Xyloterus domesticus 304
— *lineatus* 304
— *scabricollis* 304
— *signatus* 304
Xylotrechus 688
Xylotrupes gideon 210

Yanga guttulata 15
Yersinella raymondi 124, 767; Pl. 7
Yponomeutidae 85

Ytu zeus 654, 771; Fig. 243

Zabrini 230
Zabrotes 179
— *subfasciatus* 20, 61, 62, 213
Zacycloptera atripennis 767; Pl. 13
Zeiraphera diniana 81
Zelius 617
— *exsanguis* 614
Zelus renardii 229
Zenillia 5, 755
— *caudata* 59
— *pullata* 755
Zenodochium coccivorella 221
Zenophassus 51
Zetzellia mali 227
Zeugloptera 10
Zeuxevania splendidula 233
Zeuzera coffeae 24
— *pyrina* 24, 712
Zicrona caerulea 597, 770; Fig. 214
Zonabris (= *Mylabris*) 29
Zonitis 676
Zonocerus 520
— *variegatus* 527, 767; Fig. 173
Zonosemata vittigera 743
Zootermopsis 507
Zoraptera 9, 507
Zygaena 199, 251
— *filipendulae* 251
— *lonicerae* 251
— *trifolii* 251
Zygaenidae 198, 199, 717, 721
Zygonyx 489
Zygoptera 52, 486, 765; Fig. 146

Author Index

Volume I: Pages 1–474.   Volume II: Pages 475–778

Abasa R. O.   86
Abbas H. M.   312
Abbott C. E.   659
Abdel-Fattah M. I.   see El-Saadany G. 85
Abdel-Malek A.   84, 732
Abd El-Rahman H. A.   see Taher El-Sayed M. 34
Abdelrahman I.   763
Abderhalden   183
Abdinbekova A. A.   85
Ables J. R.   84
Abouzeid N. A.   see Azab A. K. 83, 689
Abraham C. C.   see Das N. M. 235
Abraham V. A.   228
Abrahamson L. P.   273
Abul-Nasr S.   85: see Tawfik M. F. S. 270, 287
Acheson R. M.   see Pau R. N. 201
Achtelig M.   642
Adair E. W.   502, 505
Adam D. S.   37
Adams C. H.   21
Adams T. S.   46, 47: see Ercelik T. M. 764; Nelson D. R. 46, 47
Adarsh H. S.   61
Adashkevich B.   237
Addy N. D.   see Jennings D. T. 316
Adiyodi K. G.   204
Adkisson P. L.   33, 179: see Williams C. M. 82
Adlakha V.   46
Aeschlimann J. P.   313: see Delucchi V. 720
Afify A. M.   228
Afzal M.   88
Agapova E. G.   72
Agarwal P. N.   see Khalsa H. G. 77
Agarwala S. B. D.   518
Agee H. R.   50
Agrawal N. S.   see Singh K. 84

Ahluwalia P. J. S.   see Venkatesh M. V. 83
Ahmad R.   19, 231
Ahmad S. H.   see Quraishi M. S. 733
Ahmad T.   179
Ahmed M. K.   218, 659
Ainslie C. N.   258, 753
Ainslie G. G.   25
Aitken T. H. G.   728
Akey D. H.   49
Akhmedov R. M.   see Abdinbekova A. A. 85
Alayo D.   636
Al-Azawi A.   see Barr A. R. 732, 733
Albrecht F. O.   14
Alden C. H.   220
Aldrich J. M.   748
Alex J. F.   see Nair K.-S. S. 86
Alexander A.   see Zehring C. S. 491
Alexander A. J.   269
Alexander C. P.   724, 725
Alexander R. D.   719: see West M. J. 269
Alfieri A.   497
Alford D. V.   84
Alger N. E.   315
Allemand R.   86
Allen D. C.   721: see Hardy Y. J. 84
Allen V. A. L.   see Fenemore P. G. 84
Allison V. F.   see Ubelaker J. E. 90, 562
Allison W. E.   see Adkisson P. L. 33
Alonso de Medina F. J.   see Robredo F. 717
Alpatov W. W.   731
Alpin R.   see Kay D. 763
Altahtawy M. M.   709
Altson A. M.   5
Altwegg P.   85, 312
Amante E.   88
Ambler K. D.   see Wearing C. H. 85
Amman G. D.   84, 178
Ammar E. D.   228

Amos T. G.   83: *see* Waterhouse F. L. 83
Ananthakrishnan T. N.   *see* Ananthasubramanian K. S. 497, 502; Viswanathan T. R. 563
Ananthasubramanian K. S.   497, 502
Ancona L. H.   238
Andarelli L.   *see* Sevenet G. 729, 730
Andersen K. T.   60
Andersen N. M.   581, 582
Anderson D.   *see* Nevill E. M. 213
Anderson D. S.   184, 751, 762
Anderson D. T.   1, 4, 5, 166
Anderson H. V.   *see* Mangum C. L. 315
Anderson J.   257
Anderson J. F.   84, 731: *see* Horsfall W. R. 731
Anderson J. M.   177: *see* Ludwig D. 179
Anderson J. R.   213
Anderson L. D.   *see* Cameron J. W. 88; Morrow J. A. 735
Anderson R. C.   238
Anderson R. F.   *see* Hertel G. 66
Andreadis T. G.   92: *see* Hall D. W. 92
Andres A.   502
Andres L. A.   71: *see* Frick K. E. 72; Shorey H. H. 33
Andrewartha H. G.   222: *see* Birch L. C. 175
Angalet G. W.   *see* Andres L. A. 71
Angus R. B.   656, 657
Ankersmit G. W.   56, 720
Annila E.   279, 282
Anon.   25, 215, 216, 534
Ansari A. K.   *see* Kumar R. 612
Ansari M. H.   86
Antonenko O. P.   *see* Grivanov K. P. 230
Antoniou A.   520
Antony J.   *see* Nirula K. K. 17
Anwar M.   85
Anwar M. S.   *see* Abbas H. M. 312
Aplin R.   *see* Rothschild M. 256, 257, 258
Appert J.   34
Apple J. W.   *see* Litsinger J. A. 84; Patel K. K. 83, 677
Applebaum S. W.   89: *see* Avidov Z. 77
Arbogast R. T.   228: *see* Press J. W. 228
Areekul S.   *see* Bhuangprakone S. 743, 744
Arias Giralda A.   84
Arif M. D.   *see* Anwar M. 85
Arkhipov G. E.   238
Armanius N. E.   *see* Hammad S. M. 764

Armitage H. M.   *see* Smith H. S. 234
Armstrong E. A.   490
Armstrong J. S.   491
Arnason A. P.   *see* Fredeen F. J. H. 86, 734
Arnoud Br.   *see* Cobben R. H. 610
Aronson M. D.   *see* Tesh R. B. 213
Arora G. L.   83, 502, 683
Arrand J. C.   613, 764
Arrow G. J.   219
Arru G. M.   88
Arruda H. V.   *see* Rossetto C. J. 689
Arthur A. P.   47, 56, 64, 73, 84, 94: *see* Hegdekar B. M. 84
Ascher K. R. S.   85, 86: *see* Meisner J. 84
Ashby D. G.   743
Ashby J. W.   212
Ashby K. R.   224
Ashenhurst J. B.   *see* Riddiford L. M. 764
Asher W. C.   85
Ashraf M.   *see* Anwar M. 85; Berryman A. A. 87
Askew R. R.   84, 92, 239
Aspöck H.   642
Aspöck U.   *see* Aspöck H. 642
Asquith D.   *see* Horsburgh R. L. 83
Assem J. van den   85, 709
Astiaso F.   *see* Dafauce C. 84
Ata A. M.   *see* Tawfik M. F. S. 229, 764
Ata M. A.   *see* Salama H. S. 85
Atkin E. A.   732
Atkins E. D. T.   191
Atkins M. D.   *see* McMullen L. H. 279
Atkinson P. W.   209
Atwal A. S.   40, 83, 84: *see* Grewal S. S. 33
Atyeo W. T.   681
Aubert J. F.   94
Audemard H.   312: *see* Féron M. 56, 86; Guennelon G. 76
Austara O.   85, 721
Autuori M.   *see* Fonseca J. P. C. da 223
Avidov Z.   15, 25, 77, 94
Awadallah A. M.   *see* Hafez M. 312
Awadallah K. T.   568
Axtell R. C.   227: *see* Karandinos M. G. 86; O'Donnell A. E. 227
Ayatollahi M.   236
Ayertey J. N.   720
Ayyappa P. K.   33
Ayyar T. V. R.   221, 235, 236, 272, 714
Azab A. K.   83, 569, 689

AUTHOR INDEX

Azaryan A. G.   81, 86
Azaryan G. Kh.   31
Azim A.   84
Azuma K.   92

Babayan A. S.   *see* Azaryan G. Kh. 31
Baccetti B.   184, 214: *see* Guerritore A. 184
Bachiller P.   *see* Dafauce C. 84
Bachvalova T. T.   *see* Alpatov W. W. 731
Bacon O. G.   *see* Dunbar D. M. 229
Bacot A. W.   732, 733: *see* Atkin E. A. 732
Badawi A.   *see* Habib A.   566
Badawy A.   48
Badcock R. M.   23, 84, 711
Badgley M. E.   659: *see* Flanders S. E. 213
Badonnel A.   547
Baer W.   752
Bagdavadze A. I.   25
Baggiolini M.   51
Bagnall R. S.   270
Bailey J. C.   83
Bailey P. T.   *see* Monro J. 31
Baisas F. E.   729
Baker B. H.   70
Baker C. R. R.   314
Baker G. T.   *see* Chen P. S. 46
Baker J. E.   *see* Gilbert B. L. 87
Baker J. L.   709
Baker J. R.   642
Baker W. A.   21, 219
Bakke A.   66: *see* Christiansen E. 33; Vité J. P. 65
Bakkendorf P.   233, 267
Bakker K.   313
Balachowsky A.   221
Balachowsky A. S.   279
Balay G.   85
Balduf W. V.   16, 17, 19, 83, 211, 216, 232, 235, 237, 648, 651, 675, 711: *see* Houser J. S. 57
Balfour-Browne F.   187, 494, 654, 656
Ball H. J.   80, 83
Ball J. C.   *see* Bartlett B. R. 144
Ballard E.   219, 272
Balobeshko V. S.   80
Baloch G. M.   24
Balogun R. A.   46, 279
Baltensweiler W.   *see* Bos J. van den 721
Balter R. S.   549, 554, 555, 562

Bandal S. K.   *see* Gordon H. T. 43
Bandyopadhyay M. K.   *see* Choudhury A. K. S. 227
Banegas A. D.   *see* Mourier H. 86, 751
Banerjee A. C.   85
Banham E. J.   212
Banks C. G.   175, 224, 225, 620
Banks C. J.   *see* Way M. J. 237
Baranowski R. M.   614
Barata J. M. S.   *see* Forattini O. P. 617
Barber G. W.   720
Barber M. A.   728, 732
Barbier R.   179, 715: *see* Chauvin G. 178, 179, 614, 715
Barbour J. F.   *see* Castek K. L. 70; Gray B. 70
Bardner R.   86, 119: *see* Kempton R. A. 86
Bare C. O.   213, 638, 709
Barfield C. S.   764: *see* Vinson S. B. 709
Barker P. S.   213
Barlow A. R.   87
Barlow C. A.   86
Barnard K. H.   510
Barnes H. F.   221, 236, 676, 724
Barnes J. H.   82
Barnes J. K.   762
Barnor J. L.   *see* Kumar R. 209
Baronio P.   227
Barr A. R.   732, 733
Barras D. J.   *see* Vinson S. B. 92
Barraud P. J.   *see* Christophers S. R. 729, 731
Barrera A.   83
Barrett C. F.   *see* Westdal P. H. 86
Barrosa P.   315
Bartell J.   *see* Browne L. B. 71
Barter G. W.   73, 83
Barth R.   199
Bartholomai C. W.   229
Bartlett B. R.   73, 84, 144: *see* Orphanides G. M. 215, 218, 220
Bartlett F. J.   *see* Gross H. R. 312
Bartoloni P.   24
Barton L. C.   315
Bar-Zeev M.   85
Basio R. G.   *see* Reisen W. K. 314
Bass J. A.   227
Basu B. C.   738
Bates M.   58, 85, 729, 730
Bath J. L.   *see* Morrow J. A. 735
Batiste W. C.   79, 314

Batra L. R.   304
Batra R. C.   *see* Verma J. P. 231
Batsch W. W.   *see* Arthur A. P. 84
Baudot E.   *see* Seguy E. 51
Bauer J.   689
Baumann H.   46
Baumhover A. H.   *see* Smith C. N. 312
Baurant R.   *see* Pierrard G. 85
Bay D. E.   86
Bayer M.   *see* Heymons R. 298
Bazire-Bénazet M.   84, 226
Beament J. W. L.   96, 180, 181, 200, 623, 763: *see* Wigglesworth V. B. 97, 98, 114, 129, 499, 534, 539, 542, 568, 675, 751
Beamer R. H.   271, 291
Beams H. W.   491
Beards G. W.   *see* Grigarick A. A. 84
Beatty A. F.   491: *see* Beatty G. H. 82
Beatty G. H.   82: *see* Beatty A. F. 491
Beaucournu J.-C.   562
Beaucournu-Saguez F.   *see* Doby J.-M. 313
Beaver R. A.   279
Beavers J. B.   227
Beck S. D.   89, 315: *see* Chippendale G. M. 315
Beckel W. E.   85, 728, 732
Becker P.   736
Beckwith R. C.   35
Bedard W. D.   66: *see* Lindquist E. E. 227
Bedford G. O.   533, 764: *see* Readshaw J. L. 533
Bee M. J.   *see* Jones S. L. 764
Beebe W.   258
Beeden P.   85
Beegle C. C.   709
Beeson C. F. C.   20, 83, 235, 274
Begg M.   315
Behrenz W.   44, 199
Behura B. K.   *see* Sengupta G. C. 20
Beier M.   502: *see* St. Quentin D. 491
Beilmann A. P.   271, 293
Beingolea G. O.   237
Beique R.   21
Beirne B. P.   *see* McKenzie L. M. 566
Bell K. O.   212, 215, 217, 218, 219
Bell R.   *see* Whitcomb W. H. 217
Bell R. A.   *see* Villacorta A. 312
Bell V. A.   *see* Hill D. L. 315
Bell W. J.   30, 31, 71: *see* Burk T. 71; Liechti P. M. 284

Bellamy R. E.   729: *see* Corbet P. S. 314
Bellas T. E.   *see* Silverstein R. M. 70
Belton P.   85
Belur N. V.   215, 216, 219
Benassy C.   220
Benedek P.   224, 594, 609
Benedetto L. A.   508
Bengtsson S.   477
Benjamin D. M.   *see* Schenk J. A. 21
Bennett C. B.   269
Bennett E. W.   *see* Prokopy R. J. 80
Bennett F. D.   59: *see* Clements A. N. 740
Bennett L. W.   *see* Harden F. W. 79
Benois A.   235
Ben Saad A. A.   71
Benson J. F.   230
Benson R. B.   274, 311
Benson W. W.   83
Bentley M. D.   85, 764: *see* McDaniel I. N. 85
Bentur J. S.   46
Benz G.   84
Bequaert J.   272
Berberet R. C.   *see* Gibson W. P. 91
Bérenguier P.   515
Berg C. O.   743, 744: *see* Bratt A. D. 744; Eckblad J. W. 743; Foote B. A. 744; Knutson L. V. 743, 744; Neff S. E. 28, 743, 744
Berg K.   85, 711, 714
Berg M. A. van den   84
Bergerard J.   534
Beri S. K.   762
Berisford C. W.   70
Berkowitz S.   256, 257
Berlepsch A. von   11
Berlese A.   3
Berlinger M. J.   *see* Avidov Z. 77
Berlowitz A.   *see* Batiste W. C. 314
Bernard F.   502, 503
Bernard J.   *see* Chauvin G. 178, 614
Bernays E. A.   175, 522
Bernhard C.   485
Beroza M.   *see* Jones R. L. 84, 85; Leonard D. E. 74; Sekul A. A. 67
Berry I. L.   *see* Kunz S. E. 764
Berry R. A.   *see* Mallack J. 733
Berry S. J.   199
Berryman A. A.   19, 84, 87
Bertrand H.   683

AUTHOR INDEX

Bess H. A.   26, 91, 93, 755
Betten C.   711: see Needham J. G. 642
Betz N. L.   see Nettles W. C. 315
Bevan D.   273, 308
Bevier G. A.   see Evans B. R. 314; Jakob W. L. 314
Bevis A. L.   221
Beyer A. H.   82
Bezrukov J. G.   237
Bhanot J. P.   227: see Kapil R. P. 227
Bharadwaj R. K.   269, 289
Bhatia B. M.   see Beeson C. F. C. 20
Bhatia K. R.   83
Bhatia M. L.   741
Bhatia S. K.   742: see Srivastava B. K. 83
Bhuangprakone S.   743, 744
Bianchi U.   see Laudani U. 85
Bianchi Bullini A. P.   see Bullini L. 46
Bibolini C.   189
Bick G. H.   491
Bick J. C.   see Bick G. H. 491
Bickley W. E.   86: see Mallack J. 733; Shriver D. 733
Bier K.   709
Bierl B. A.   see Jones R. L. 84, 85; Leonard D. E. 74; Sekul A. A. 67
Biermann G.   764
Bigger J. H.   60
Bigger M.   37
Biliotti E.   51, 59, 257
Billings R. F.   70: see Gray B. 70
Billingsley C. H.   see Burger T. L. 313
Bills G. T.   227
Bingham J.   87
Birch L. C.   18, 21, 175, 178, 179, 224
Birch M. C.   66: see Lanier G. N. 70; Young J. C. 66, 67
Birk Y.   see Applebaum S. W. 89
Bírová H.   see Jasič J. 34
Bishara S. I.   84
Bishop G. W.   see Ben Saad A. A. 71
Bishop J. L.   see Woodside A. M. 84
Bjegović P.   86, 228
Blackith R. E.   see Albrecht F. O. 14
Blagoveshchenskiĭ D. I.   551, 554, 562
Blake G. M.   36
Blakeley P. E.   see Jacobson L. A. 35, 85
Blanchard E. E.   223
Blanco-Marco E.   see Robles-Chillida E. M. 742

Blewett M.   315: see Fraenkel G. 182
Bliss C. I.   60
Bliss M.   83, 570
Blount V. N.   see Greene G. L. 85
Blowers J. R.   15
Blume R. R.   79: see Kunz S. E. 86
Blunck H.   17, 80, 165, 651
Bobb M. L.   51, 638
Bodenheimer F. S.   11, 175
Bodenstein D.   204: see Shaaya E. 204
Bodine J. H.   see Thompson V. 175
Bodkin G. E.   235
Bodnaryk R. P.   see Smith B. C. 83
Boethel D. J.   see Criswell J. T. 689
Bogach A. V.   see Gensitskiĭ I. P. 314
Bogawat J. K.   see Srivastava R. P. 25
Bogoescu C.   476
Bogoescu M.   549
Bogush P. P.   215
Bohart G. E.   51
Bohlen E.   73
Bohm M. K.   see Bell W. J. 31
Boldori L.   272, 649
Boldt P. E.   84: see Marston N. 312
Boldȳrev M. I.   63, 83, 224
Boles H. P.   84, 314, 689
Boling J. C.   85
Bollaerts D.   see Bruel W. E. van den 212
Bollen W. B.   see Gjullin C. M. 733
Boller E. F.   312, 742: see Hurter J. 65; Katsoyannes B. I. 65; Prokopy R. J. 86, 312
Bombosch S.   86
Bond E. J.   312
Bondar G.   685
Bongers W.   83: see DeWilde J. 72, 87
Bonnanfant-Jaïs M. L.   286
Bonnemaison L.   35, 82, 83
Booth D. C.   70
Bordas L.   184
Borden J. H.   67, 70, 273, 279: see Byrne K. J. 70
Bordon J.   86
Boreham P. F. L.   213: see Braverman Y. 213
Borg A. F.   733
Borkhausen M. B.   257
Bornemissza G. F.   230, 722
Bos J. van den   721

Bosch R. van den 91: *see* Puttler B. 92; Salt G. 91, 93; Schlinger E. I. 270
Boscher J. 85
Bose K. C. 594
Boselli F. B. 271
Boswell A. L. *see* Smith F. F. 314
Bottiger G. T. *see* Smith C. N. 312
Bottrell D. G. *see* Barfield C. S. 764
Boulard M. 83
Bouligand Y. 209
Boulton R. *see* Klemperer H. G. 764
Bourgogne J. 197, 198
Bournier A. 563
Bouvier G. 562
Bovey P. 738
Böving A. 654
Bøving A. G. 272, 656, 657: *see* Parker J. B. 233
Bowditch F. C. 656
Bowers W. S. *see* Waldbauer G. P. 26
Boyce A. M. 742
Boyce J. M. 224
Boyé R. 258
Boyer J. 257
Brack-Egg A. 278
Bracke J. W. *see* Brand J. M. 66
Bracken G. K. 706
Brader L. 273
Bradley G. H. 728: *see* King W. V. 732
Bradley J. R. *see* Rabb R. L. 64, 73
Bradley W. G. *see* Baker W. A. 219
Brammanis L. 83
Branch H. E. 735
Brand J. M. 66
Brandt A. 582, 620
Branson T. F. 88
Bratt A. D. 744
Brauer F. 643, 739
Braun A. F. 235
Braverman Y. 213
Bray A. D. *see* Judson C. L. 732
Brazzel J. R. 84, 88
Bready J. K. 315
Bredo H. J. 224
Breeland S. G. 85
Breese M. H. 83, 84
Brehelin M. *see* Zachary D. 92
Breland O. P. 499, 502, 729
Bremer H. 72
Brenière J. 315

Brennan B. M. 90
Brereton J. le Gay 224
Bresslau E. 729
Brewer F. D. 92
Brian M. V. 11, 226
Briand L. J. 215, 216
Bridges E. J. 566
Bridwell J. C. 274, 683
Brien P. 272, 296
Briggs E. M. 679
Bright D. E. 304
Brinck P. 508
Brindley M. D. H. 572
Brindley T. A. *see* Klun J. H. 89; Larson A. O. 20, 83, 683
Brinkhurst R. O. 582
Bristowe W. S. 238
Britton E. B. 654
Britton W. E. 737
Broadhead E. 546, 547, 548
Broadhead E. C. *see* Waloff N. 35, 37
Broadhead F. 14
Brocher F. 109, 485, 579, 656
Brock M. L. 236, 748, 749
Brock T. D. *see* Brock M. L. 236, 748, 749
Bro Larsen E. 271
Bromley S. W. 490, 740
Brongniart C. 244, 505
Brönnimann H. 228
Broodryk S. W. 84
Brooke M. M. 212
Brooks F. E. 17
Brossut R. 71
Brothers D. J. *see* Moran V. C. 709
Brovdy V. M. 676
Brower L. P. 225
Brown A. W. A. *see* Singh K. P. R. 315
Brown C. E. *see* Shepherd R. F. 313
Brown F. S. 83
Brown H. D. 224
Brown H. P. 648
Brown J. B. *see* Hoffman J. D. 312
Brown J. G. 175
Brown K. W. 689
Brown N. R. *see* Clark R. C. 218, 220
Brown R. W. 497
Brown S. N. 556
Brown W. V. *see* Atkinson P. W. 209
Browne F. G. 273, 274, 304, 308, 310
Browne L. B. 51, 71, 80, 86

Browne L. E.   *see* Brand J. M. 66; Silverstein R. M. 70; Wood D. L. 66; Young J. C. 66
Browning F. R.   83
Browning T. O.   175, 179, 180
Brownlee R. G.   *see* Bedard W. D. 66; Silverstein R. M. 70; Young J. C. 66, 67
Brubaker R. W.   28
Bruce-Chwatt L. J.   729
Brückner W. R.   648
Bruel W. E. van den   212
Bruijn E. Feuth-de   *see* Assem J. van den 709
Brun P.   *see* Daumal J. 312
Bruner L.   218
Brunet P. C. J.   184, 201, 203: *see* Kent P. W. 209; Mercer E. H. 202; Pau R. 202, 203; Whitehead D. L. 201; Willis J. H. 204
Brunson M. H.   94
Brust R. A.   179, 727: *see* Costello R. A. 175; Horsfall W. R. 731; Kalpage K. S. 85, 729; Lefkovitch L. P. 728
Bryan D. E.   *see* Jackson C. G. 28
Bryant D. G.   313: *see* Otvos I. S. 313
Bryant E. H.   86
Bryantseva I. B.   184: *see* Bulyginskaya M. A. 72, 84
Bryk F.   48
Buchan P. R.   *see* Moran V. C. 83, 764
Buchanan W. D.   77
Buchner G. D.   537
Buck H.   278
Buck J.   *see* Keister M. 152
Buckell E. R.   213: *see* Treherne R. C. 217, 222, 223, 224
Buéi K.   35
Bueno J. K. de la Torre   636, 638
Buffam P. E.   218
Bugnion E.   279, 502, 763
Bühler R.   *see* Chen P. S. 45, 47
Buleza V. V.   84
Bull D. L.   *see* Adkisson P. L. 33
Bull J. O.   *see* Howe R. W. 83
Bull R. M.   228
Bullière D.   *see* Sengel P. 44
Bullini L.   46
Bullock H. R.   315
Bullock R.   727
Bulyginskaya M. A.   48, 72, 84
Bunnett E. J.   729
Buranday R. P.   85
Burden G. S.   *see* Smith C. N. 312

Burdick D. J.   84
Burger T. L.   313: *see* Maltby H. L. 90
Burges H. D.   61, 720: *see* Howe R. W. 18, 42, 83, 178
Burgess A. F.   16
Burk T.   71: *see* Bell W. J. 71
Burke H. B.   16
Burke H. E.   83
Burkholder W. E.   *see* Lanier G. N. 67; Vick K. 83
Burnet B.   46
Burnett J.   213, 763
Burns A. N.   508
Burns J. M.   50
Burov V. N.   41
Burrage R. H.   *see* Gyrisco G. G. 17, 83, 217
Burton, C. J.   257
Burton G.   *see* Kain W. M. 761
Burton G. J.   727
Burton R. L.   *see* Harrell E. A. 312; Jones R. L. 85; Widstrom N. W. 313
Burts E. C.   48
Busching M. K.   721
Busck A.   235
Bush G. L.   *see* Prokopy R. J. 80, 86
Bushing R. W.   87
Bushland R. C.   522
Bussart J. E.   57
Butani D. E.   *see* Sikka S. M. 88
Butcher J. W.   *see* Haynes D. L. 72; Ross R. H. 315
Butler E. A.   609, 610, 620
Butler G. D.   229: *see* Jackson C. G. 28
Butler L.   502
Butovitsch V. von   83, 685, 687
Butt B. A.   *see* Westigard P. H. 85
Buxton J. H.   270, 287
Buxton P. A.   169, 176, 549, 732, 733: *see* Williams C. B. 499, 504, 506
Buzicky A. W.   *see* Cook F. 313
Byas   *see* Singh H. N. 85
Byers G. W.   722, 727
Bynum E. K.   *see* Ingram J. W. 214
Byrne H. D.   84
Byrne K. J.   70

Cabal Concha A.   225
Cadahia A.   84
Caffrey D. J.   258: *see* Wildermuth V. L. 243

Cain G. D.   *see* Ingram M. J. 285
Calaby J. H.   547
Calcote V. R.   689
Calderon M.   689
Caldwell R. L.   81: *see* Rankin M. A. 81, 83
Calkins C. O.   *see* Kirk V. M. 83
Callahan P. S.   78, 85, 763
Callan E. McC.   272
Callenbach J. A.   *see* Villacorta A. 312
Calori L.   485
Caltagirone L. E.   *see* Carrillo J. L. 709; Laing D. R. 21
Calvert P. P.   490
Cameron A. E.   724, 734, 745, 751
Cameron A. L.   *see* Smittle B. J. 316
Cameron E. A.   *see* Ellenberger J. S. 85, 764
Cameron J. W.   88
Cameron M. D.   *see* Eidt D. C. 313
Cammell M. E.   *see* Burges H. D. 61
Camors F. B.   70
Campan M.   86, 741
Campbell B.   *see* Marston N. 312
Campbell D.   *see* Klug W.-S. 762
Campbell F. L.   169
Campbell I. M.   *see* Wallace D. R. 705
Campbell J. B.   231
Campbell J. W.   547
Campbell K. G.   78
Campbell R. W.   313
Campbell W. V.   34, 88
Campodonico M. J.   522
Canard M.   224
Canerday T. D.   *see* Chalfant R. B. 84
Cañizo J. del   219, 221
Cappe de Baillon P.   515, 533, 534, 541
Caprotti M.   *see* Grigolo A. 46
Carayon J.   30, 607, 609, 610, 641
Carbonell E.   83: *see* Orozco F. 83
Cardé R. T.   67
Cardona C.   54
Cardoso J. G. A.   220
Carfagna M.   86
Carle P.   70, 689
Carlson O. V.   83, 223, 566: *see* Hibbs E. T. 83
Carlyle S. L.   312: *see* Leppla N. C. 313
Carlyle T. C.   *see* Leppla N. C. 315
Carne P. B.   37, 311, 677
Carnegie A. J. M.   228: *see* Symmons P. 83
Carrel J. E.   251

Carrillo J. L.   709
Carthon M.   *see* Arbogast R. T. 228
Carton Y.   84
Carvalho R. P. L.   20
Case J. J.   *see* Moran V. C. 709
Castek K. L.   70
Castillo R. L.   729
Catley A.   25
Catling H. D.   568
Caudell A. N.   507
Causey O. R.   *see* Deane L. M. 729; Deane M. P. 729, 730, 731, 732
Caussanel C.   270, 287, 288
Cauwer P.   *see* Lhoste J. 720
Cavicchi S.   312
Cawthon D. A.   *see* Mertz D. B. 225
Cazier M. A.   740
Center T. D.   684
Cervone L.   729, 730
Chabora P. C.   84
Chacko M. J.   64
Chadwick L. E.   *see* Hill D. L. 315
Chai Chi-hui   *see* Chin Chu-teh 175
Chalam B. S.   732
Chalfant R. B.   84, 312
Chamberlain W. F.   764: *see* Hopkins D. E. 549
Chamberlin T. R.   27, 84
Chambers D. L.   *see* Keiser I. 312
Chambers V. H.   705
Chambon J. P.   84
Champion G. C.   676
Champion H. G.   686
Champlain R. A.   15
Chan Tak-ming   *see* Chen Chu-teh 175
Chandler A. E. F.   63, 77, 86, 741
Chandler L.   *see* Levine E. 258
Chandra H.   *see* Venkatesh M. V. 83
Chandrashekaran M. K.   *see* Loher W. 81, 82
Chang G. S.   *see* Meng H. C. 80
Chang V. C. S.   84: *see* Hovanitz W. 85
Chaniotis B. N.   *see* Tesh R. B. 213
Channa Basavanna G. P.   *see* Puttarudriah M. 224; Thontadarya T. S. 83, 567, 611
Chapman H. C.   *see* Petersen J. J. 75
Chapman J. A.   30: *see* Dyer E. D. A. 70
Chapman R. F.   518, 522, 523
Chapman R. N.   224

Chapman T. A.   270, 289, 712, 739: *see* Champion G. C. 676
Chararas C.   66, 279: *see* Balachowsky A. S. 279
Charpentier L. J.   *see* Ingram J. W. 214
Chatterjee N. C.   *see* Imms A. D. 235
Chatterjee S. N.   316: *see* Beeson C. F. C. 274
Chatterji S. M.   *see* Rajasekhara K. 610; Sharma V. K. 85
Chaudhry H. S.   569
Chaudhuri R. P.   *see* Narayanan E. S. 94
Chauthani A. R.   28
Chauvin G.   178, 179, 614, 715: *see* Barbier R. 179, 715
Chauvin R.   84
Cheatham J. S.   *see* Sparks M. R. 85
Cheema P. S.   37
Chemsak J. A.   83, 686
Chen An-kuo   179
Chen C. C.   83
Chen Chih-hui   *see* Chen An-kuo 179
Chen Chu-ying   *see* Chin Chu-teh 175
Chen P. S.   45, 46, 47
Chen S. H.   58, 648
Cheng C.-H.   566
Cheng H. H.   224
Cheng L.   74, 582
Cheng T. C.   *see* Brennan B. M. 90
Cheong W. H.   729
Cherevatova A. S.   764
Chesnut T. L.   211
Cheu S. P.   76, 224
Chiang H. C.   28, 39, 313, 762: *see* Hsiao T. H. 86; Palmer D. F. 683; Windels M. E. 85
Chiba T.   *see* Hasegawa T. 86
Chin Chun-teh   175
Chippendale G. M.   315
Chittenden F. H.   83, 683, 684
Chock Q. C.   744
Chodjai M.   279
Choi S. Y.   721
Choi Y. H.   721
Cholodkovsky N. A.   198, 199
Chong M.   *see* Chock Q. C. 744
Chopard L.   269, 284, 499, 515, 533, 764
Choudhuri J. C. B.   82, 83
Choudhury A. K. S.   227: *see* Mukherjee A. B. 229
Chow Sze-chun   83
Chrestian P.   17, 83
Christensen C. M.   688
Christenson D. M.   688: *see* Lewis L. F. 85
Christiansen E.   33, 84
Christophers S. R.   251, 729, 731
Chu H.   *see* Norris D. M. 305
Chu H.-M.   *see* Abrahamson L. F. 273
Chun M. W.   85
Chung Hsiang-chen   *see* Chen An-kuo 179
Church N. S.   83, 676: *see* Gerber G. H. 689; Gerrity R. G. 676; Rempel J. G. 676; Sweeney P. R. 676
Chutter F. M.   85
Cierniewska B.   569
Ciesielska Z.   41
Ciesla W. M.   78
Ciglioli M. E. C.   75
Cirio U.   86
Claassen P. W.   258, 650
Clancy D. W.   91, 755
Claridge M. F.   566
Clark A. F.   677
Clark A. M.   42
Clark C. A.   *see* Baker W. A. 219
Clark D. P.   82
Clark G. N.   *see* Wingo C. W. 230, 231
Clark J. T.   764
Clark L. R.   64, 83, 175, 568
Clark N.   178
Clark R. C.   218, 220: *see* Bryant D. G. 313
Clarke K. U.   36
Clausen C. P.   21, 22, 23, 29, 52, 57, 84, 94, 144, 221, 223, 224, 239, 274, 649, 709, 740
Clavel M. F.   *see* David J. 43
Clemens W. A.   476
Clements A. N.   256, 257, 733, 740
Cleveland L. R.   284
Cleveland M. L.   *see* Wong T. T. Y. 48
Clifford J. R. S.   257
Clift A. E.   *see* Howell J. F. 312
Coad B. R.   85
Coaker T. H.   216, 217, 230, 313: *see* Finch S. 35
Cobben R. H.   106, 117, 130, 131, 166, 167, 168, 169, 170, 171, 568, 570, 572, 573, 574, 575, 579, 582, 585, 586, 591, 594, 608, 609, 610, 613, 614, 616, 617, 618, 620, 621, 622, 636, 637, 638, 639, 640, 641, 764

Cobelli R.   718
Cockerell T. D. A.   257
Colas-Belcour J.   *see* Roubaud E. 732
Colbo M. H.   313
Cole F. R.   738
Cole S.   *see* Hinton H. E. 124, 757
Coleman E.   712
Coleman L. C.   518
Colhoun E. H.   218, 659
Collett G. C.   *see* Winget R. N. 748
Collier J.   *see* Cleveland L. R. 284
Collins C. W.   *see* Burgess A. F. 16
Collins J. M.   *see* LeCato G. L. 763
Collyer E.   230, 612, 648
Colthrup C. W.   257
Colthurst I.   764
Coluzzi M.   85, 729: *see* Bullini L. 46
Colyer C. N.   749
Combs G. F.   *see* Smith F. F. 314
Common I. F. B.   218, 712
Compere H.   232: *see* Smith H. S. 220
Comstock J. H.   106, 221
Conci C.   562
Conde M.   *see* Robredo F. 717
Condrashoff S. F.   313
Conil P. A.   218
Connell W. A.   313
Connin R. V.   83: *see* Wellso S. G. 83
Connolly K.   *see* Burnet B. 46
Connor P. J.   *see* Wearing C. H. 85
Conrad M. S.   231
Consoli R. A. G.   85
Cook E. F.   *see* Stehr F. W. 713
Cook F.   313
Cook J. M.   *see* McWilliams J. M. 313
Cook R.   *see* Burnet B. 46
Cooling L. E.   732
Coombs C. W.   42, 83, 225
Cooper B. P.   *see* Thurston R. 89
Cooper K. W.   709, 723, 763
Cooper N. B.   *see* Ubelaker J. E. 90
Cooper R.   *see* Gothilf S. 721
Copeland E. L.   764
Copland M. J. W.   *see* King P. E. 706, 707, 708, 763
Copony J. A.   70
Coppel H. C.   28, 74: *see* Klein M. G. 83; Valek D. A. 83
Corbet P. S.   34, 63, 81, 85, 314, 487, 490, 710, 734, 735: *see* Gillett J. D. 81;

Haddow A. J. 81
Corbet S. A.   56
Corby H. D. L.   584
Corey R. A.   *see* Rimando L. C. 312
Cornet M.   314
Cornic J. F.   83
Corporaal J. B.   21
Corrêa C.   *see* Pessôa S. B. 284
Corrêa R. R.   729
Correia R.   *see* Galvão A. A. 729
Coscarón S.   216
Costa M.   227
Costello R. A.   175: *see* Brust R. A. 179
Coster J. E.   66
Cott H. B.   247, 248
Cotton R. T.   19
Coulson R. N.   70
Coutin R.   84, 86, 88
Coutts M. P.   88
Couturier A.   83
Cova-Garcia P.   729
Craig D. A.   761
Craig G. B.   45, 47, 727, 729: *see* Fuchs M. S. 44, 45, 47; Gwadz R. W. 45; Horsfall W. R. 44, 45, 47; Leahy M. G. 45, 47
Craighead F. C.   686
Cramer E.   27, 725
Crampton G. C.   722
Crauford-Benson H. J.   556
Crawford C. S.   37, 80, 81, 85: *see* Morrison W. P. 179
Crawshay L. R.   17
Crewe W.   86
Criddle N.   82, 217
Crisp D. J.   149, 160, 162: *see* Thorpe W. H. 109, 149, 159, 160
Criswell J. T.   689
Crocker R. L.   229
Crombie A. C.   83
Crook L. J.   *see* Banham E. J. 212
Cros A.   29, 49, 50, 219
Crosasso C.   *see* Chararas C. 66
Crosby C. R.   220
Cross E. A.   *see* Moser J. C. 227
Cross W. H.   *see* Adams C. H. 21; McGovern W. L. 314; Mitchell H. C. 49, 80
Crosskey R. W.   735: *see* Lewis D. J. 734, 735
Crossman S. S.   21

Crowe T. J. 83
Cruickshank W. J. 721
Cruz C. 85
Crystal M. M. 44
Culver J. J. 753
Cummings M. R. 715, 762: see Klug W.-S. 762
Currie G. A. 86
Currie J. E. 34: see Howe R. W. 60, 61, 62, 83, 179; Lefkovitch L. P. 37
Cussac E. 272, 656
Cuthbert R. A. see Peacock J. W. 70; Pearce G. T. 70
Cuthbertson A. 223, 725
Cutright C. R. 16, 648
Cutten F. E. A. 734
Cymorek S. 77

Daanje A. 278
D'Abrera V. St. E. 729
Dafauce C. 20, 84
Dahl R. 744
Dahlsten D. L. see Luck R. F. 313; Stark R. W. 59; Stephen F. M. 70
Dallas E. D. 258
Damasceno R. G. see Galvão A. L. A. 729
Dampf A. 522
Danilevskiĭ A. S. 235
Danks H. V. 753: see Corbet P. S. 85
Danthanarayana W. 313
Darrow E. M. 732
Das N. M. 235, 566
Das S. R. see Paul C. F. 178
Dasgupta B. see Roy P. 86
Daterman G. E. see Furniss M. M. 70; Rudinsky J. A. 70
Daugherty D. M. see Wilkinson J. D. 27
Daumal J. 312
Davatchi A. 88
Davey K. G. 45, 47, 617, 684
Davey P. M. 88
Davich T. B. see Smith C. N. 315; Vanderzant E. S. 84, 313
David J. 43, 86, 762
David K. 648
David M. H. 83
David W. A. L. 78
Davidson W. M. 741

Davies B. R. 764
Davies D. M. 85, 734: see Golini V. I. 86
Davies J. B. 85, 734: see Lewis D. J. 734
Davis C. 764
Davis C. C. 620, 638, 711
Davis C. J. 235: see Chock Q. C. 744
Davis D. G. see Wong T. T. Y. 48
Davis E. E. 85
Davis E. G. 520
Davis J. J. 57, 740
Davis K. C. 642
Davis R. see Press J. W. 228
Davis R. B. 86, 314
Dawson P. S. 225: see Ho F. K. 225
Dean R. L. 764
Dean R. W. 763
Deane L. M. 729: see Causey O. R. 729
Deane M. P. 729, 730, 731, 732: see Causey O. R. 729; Deane L. M. 729
DeBach P. 234, 238, 239: see Gordh G. 709
Debey M. 278
De Buck A. 75, 729, 730, 731
Décamps H. 192
Decker G. C. see Banerjee A. C. 85
Deco M. D. see Coluzzi M. 85
De Coursey J. D. 728
Déduit Y. 85
Deegener P. 3
Deeming J. C. 748
DeFoliart G. R. see Loor K. A. 314
De Geer C. 192, 270, 271
Degrange C. 12, 13, 475, 476, 477, 483, 485, 491, 510
Degrugillier M. E. see Leopold R. A. 46, 47; Terranova A. C. 46, 47
Delanoue P. see Féron M. 312
Delattre P. 227
Delcourt A. 638
Delgado A. see Pesce H. 256
Delmas H. G. 83
Delmas J. C. see Demolin G. 228
Delpech M. see Merle P. du 86
Delsman H. C. 582
Delucchi V. 220, 223, 720
De Meillon B. 729, 732: see Gillies M. T. 731; Smith C. N. 312
Demolin G. 228: see Biliotti E. 51
Dempster J. P. 212, 214, 220
Denisova T. V. 89

Denmark H. A.   see Beavers J. B. 227
Denny A.   see Miall L. C. 184
Depner K. R.   see McLintock J. 86
Deseö K. V.   72
Despommier D. D.   see Fuchs M. S. 45, 47
De Stefani P. T.   219
DeTar J. E.   see Batiste W. C. 314
Dethier V. G.   56, 721
DeVaney J. A.   see Eddy G. W. 86
Devecioğlu H.   see Tokmakoğlu C. 25
DeWitt P. R.   see Maltby H. L. 90
Dharmaraju E. A.   742
Diakonoff A.   712
Dias B. F. de S.   274, 311
Dick J.   60
Dick R. D.   24
Dickason E. A.   225
Dickens J. C.   70
Dicker G. H. L.   650
Dickerson W. A.   see Hoffman J. D. 312
Didlake M.   502
Diekman J. D.   see Rudinsky J. A. 70
Diem C.   see Chen P. S. 45
Dieng P. Y.   see Cornet M. 314
Dietrick E. J.   see DeBach P. 234; Schlinger E. I. 270
Dimetry N. Z.   83, 224
Dimitrov A.   see Kharizanov A. 648
Dingle H.   see Caldwell R. L. 81; Rankin M. A. 81, 83
Din-si L.   see Zagulyaev A. K. 235
Dinther J. B. M. van   212, 214, 216, 217, 218, 230
Dinulescu G.   749
Dirimanov M.   24
Dirsch V. M.   522
Disney R. H. L.   237, 735: see Lewis D. J. 735
Dispons P.   617
Distant W. L.   636
Ditman L. P.   see Bickley W. E. 86
Dixit R. S.   see Kochhar R. D. 85
Dixon T. J.   741
Djamin A.   88
Doane C. C.   see Cardé R. T. 67; Leonard D. E. 315
Doane J. F.   83, 175, 313: see Wishart G. 216, 218.
Dobosh I. G.   82

Dobrovol'skiĭ B. V.   278
Dobson J. W.   see Breland O. P. 499, 502
Dobson R. C.   see Christensen C. M. 688
Doby J.-M.   313
Dockerty A.   see Alford D. 84
Docters van Leeuwen W.   650
Dodd C. W. H.   see Saunders D. S. 46
Dodd F. P.   272
Dodd G. D.   see Coaker T. H. 313
Dodge H. R.   273, 279
Dodson M.   199
Dogiel V. A.   751
Doĭnikov A.   84
Dolidze G. V.   33
Donahaye E.   see Calderon M. 689
Donaldson J. M. I.   520
Donia A.   see El-Minshawy A. M. 570
Donisthorpe H. St. J. K.   679
Donley D. E.   83
Doom D.   273, 304, 308
d'Orchymont A.   272, 656
Doreste E.   25
Doria R. C.   83
Doring E.   243, 250, 715, 717
Douglas J. W.   763
Douglas W. A.   see Chesnut T. L. 211
Doull K. M.   see Standifer L. N. 84
Doutt R. L.   239, 274: see Rahalker G. W. 316
Douwes P.   85
Dowden P. B.   753
Downe A. E. R.   212: see Hall R. R. 212
Downes J. A.   34
Drake C. J.   614
Draudt M.   715
Drea J. J.   48
Dresner E. A.   231
Drooz A. T.   35, 85, 91, 312, 713
Druger M.   33
Dubinin V. B.   551
Du Bois A. M.   16: see Geigy R. 642
Dubois J.   15
Dubois P.   see Brossut R. 71
Dudley C. O.   279, 313
Dudley J. W.   see Campbell W. V. 88
Duerden J. C.   594
Duffey S. S.   251
Duffy E. A. J.   83, 686, 687, 688
Dufour L.   739
Dulizibarić T.   85

Dumbleton L. J.   738
Du Merle P.   312: see Biliotti E. 51
Dunbar D. M.   229
Dunn P. H.   see Smith O. J. 28
Duperrex H.   see Baggiolini M. 51
Du Plessis C.   224
Dupree J. W.   733
Dupuis C.   86, 752, 755
Durchon M.   225
Durr H. J. R.   60
Dusaussoy G.   see Coutin R. 84, 88
Dustan G. G.   49
Dyar H. G.   236
Dyer E. D. A.   see Lawko C. M. 70
Dysart R. J.   see Gruber F. 233
Dzhibladze K. N.   213

Earle N. W.   see Everett T. R. 84
Eastham J. W.   220
Eastham L. E. S.   84
Eaton A. E.   476
Eberhard W. G.   272, 294
Eberhardt G.   224
Ebora P.   see Reyes P. V. 636
Eckblad J. W.   743
Eckenrode C. J.   86: see Yu C. C. 86
Eddy G. W.   86
Edel'man N. M.   35, 36
Edman J. D.   see Lea A. O. 43
Edmunds G. F.   see Koss R. W. 764
Edney E. B.   176, 181
Edwards C. A.   23
Edwards C. A. T.   see Ibbotson A. 84
Edwards D. K.   85, 314
Edwards F. W.   729, 731
Edwards J. S.   272, 296, 617
Edwards P.   see Rolston L. H. 191
Edwards R. L.   83, 84, 707
Edwards W. D.   235
Ege R.   106
Egger A.   229
Egglishaw H. J.   745, 748, 758
Egorov N. N.   76
Eguagie W. E.   178, 764
Eichhorn O.   220
Eichler W.   83, 239, 515, 549, 551, 562
Eidmann H.   196, 197
Eidt D. C.   313
Eifsackers H. J. P.   see Bakker K. 313

Eikenbary R. D.   see Criswell J. T. 689; Jackson H. B. 22; Rogers C. E. 224
Ekblom T.   572, 609
Ekkens D.   271, 293
Elbadry E. A.   212, 213
El-Borollosy F. M.   709
El-Deeb A.   see El-Kady E. 34; Hammad S. M. 764
Eldefrawi M. E.   24
El-Gayer F. H.   see El-Helaly M. S. 569
El-Halfawy M.   764
El-Halfawy M. A.   see Bishara S. I. 84
El-Hefny A. M.   see Borollosy F. M. 709
El-Helaly M. S.   569
El-Husseini M. M.   see Tawfik M. F. S. 270, 287, 610
Eliason D. A.   see Thaggard C. W. 314
El-Kady E.   34
El Khidir E.   19
Elkind A.   534
Ellenberger J. S.   85
Ellington G. W.   see Simmons P. 19, 689
Elliott E. W.   70
Elliott J. M.   711
Elliott K. R.   237
Ellis R. O.   see Shaw J. T. 764
Ellis T. L.   see Wallner W. E. 314
El-Minshawy A. M.   570
El-Mirsawi D. H.   see Azab A. K. 569
El-Moursy A. A.   see Hafez M. 27, 738
El-Saadany G.   85
El-Sawaf S. K.   41: see Altahtawy M. M. 709; El-Minshawy A. M. 570
Elsey K. D.   28, 215, 228, 229, 261
El-Shazli A. Y.   see El-Helaly M. S. 569
El-Shazli N. Z.   91
El-Sherif A. R. A.   24
El-Sherif S. I.   see Abul-Nasr S. 85
El-Titi A.   86
Eltringham H.   63, 256, 257
Eluwa M. C.   13
Elzghari M.   312
El Zoheiry M. S.   see Ballard E. 219
Embleton A. L.   144
Embree D. G.   28
Emden F. van   218, 274, 483
Emery D. A.   see Campbell W. V. 34
Emschermann F.   76
Ene J. C.   269, 499
Engel H.   191

Engelmann F.   44, 184
Englert D. C.   225: *see* Fogle T. A. 689
English K. M. I.   740
Ennis W. R.   *see* Harlan D. P. 523
Enock F.   638
Entwistle P. F.   24
Epp H. T.   *see* Edwards R. L. 83
Eppley R. K.   *see* Bohart G. E. 51
Ercelik T. M.   764
Ernst R. L.   *see* Boles H. P. 689
Ertle L. R.   *see* Hoffman J. D. 312
Esah S.   *see* Miller T. A. 761
Esaki T.   641
Esau K.   238
Escandor N. B.   *see* Reyes P. V. 636
Eschle J. L.   *see* Blume R. R. 79
Eskafi F. M.   312
Espejo M.   *see* Orozco F. 83
Espínola H. N.   *see* Consoli R. A. G. 85
Esselbaugh C. O.   595
Essig E. O.   216, 219, 221
Etcheverry M.   25
Étienne J.   312
Euw J. von   251: *see* Reichstein T. 251; Rothschild M. 256, 257, 258
Evans A. C.   179: *see* Duerden J. C. 594
Evans A. M.   729
Evans B. R.   314
Evans D. G.   *see* Lea A. O. 43; O'Meara G. F. 764
Evans H. E.   310: *see* Gordh G. 709
Evans H. F.   83
Evenden J. C.   279
Everett T. R.   75, 84
Ewen A. B.   *see* Pickford R. 45, 47
Ewer D. W.   60, 83
Ewer R. F.   84: *see* Ewer D. W. 60, 83
Eyles A. C.   48
Ezzat M. A.   *see* Hafez M. 312
Ezzat Y. M.   83

Fabre J. H.   502, 679
Falkenström G.   651
Falleroni D.   729, 730
Fang Y. C.   764
Farahat A. Z.   *see* Okasha A. Y. H. 35
Fares F. M.   314: *see* Hafez M. 312
Farghaly H. T.   *see* Afify A. M. 228
Farlinger D. F.   *see* Alden C. H. 220

Farner D. S.   247
Farquharson C. O.   236
Farrag S. M.   *see* Ammar E. D. 228
Farrell J. A. K.   *see* Wightman J. A. 312
Farrier M. H.   *see* Hunt T. N. 84; Schneider I. 274
Farrow R. A.   231
Fatzinger C. W.   315
Faure J. C.   14, 269
Favrelle M.   96, 515, 534
Fay R. W.   *see* Jakob W. L. 314; Perry A. S. 85
Fedorov S. M.   518
Feeny P. P.   *see* Read D. P. 84
Fellows A. G.   502
Felt E. P.   764
Fenard G.   203
Fenemore P. G.   84
Feng Wei-hsiung   *see* Chen An-kuo 179
Fentiman A. F.   *see* Kinzer G. W. 67; Pitman G. B. 67; Rudinsky J. A. 67; Vité J. P. 67
Fernald C. H.   257
Fernando C. H.   620
Fernando W.   546
Féron M.   79, 86, 312, 742, 743
Ferrer F. R.   *see* Hower A. A. 312
Ferris C. D.   721
Ferris G. F.   562
Fertone C.   234
Fewkes D. W.   178: *see* Southwood T. R. E. 609
Fiebrig K.   273, 679
Fiedler O. G. H.   236
Fielding J. W.   732, 733
Figueroa M.   *see* Walker D. W. 25
Filipek P.   20
Filipponi A.   227
Finch S.   35, 86
Fink D. E.   272, 296: *see* Chittenden F. H. 83
Finlayson L. H.   547
Finlayson L. R.   214
Finnegan R. J.   304
Finney G. L.   315
Fiori G.   689
Fischer M.   274
Fischer O. von   184
Fischer W. R.   *see* Burts E. C. 48
Fishelson L.   *see* Euw J. von 251
Fisher C. K.   *see* Larson A. O. 20
Fisher R. C.   64, 84, 709

Fisher T. W.   19, 744: *see* Bartlett B. R. 73, 84; Flanders S. E. 54
Fitch E. A.   515
Fitzgerald T. D.   71
Flaherty B. R.   *see* Lum P. T. M. 43, 79, 85; Press J. W. 228
Flake H. W.   *see* Lyon R. L. 315
Flanagan T. R.   *see* Hall D. W. 92
Flanders S. E.   30, 54, 84, 94, 144, 213, 234, 707, 708, 709
Fleschner C. A.   *see* Badgley M. E. 659; DeBach P. 234
Flessel J. K.   *see* Niemczyk H. D. 84
Fletcher B. S.   86
Fletcher K. E.   *see* Bardner R. 86; Kempton R. A. 86
Fletcher L. W.   83, 86
Fletcher T. B.   223, 235
Flint W. P.   *see* Metcalf C. L. 239
Flitters N. E.   80, 743
Florence L.   562
Flower N. E.   *see* Atkins E. D. T. 191; Kenchington W. 202
Fockler C. E.   *see* Borden J. H. 273
Foerster K. W.   *see* Kunz S. E. 764
Fogle T. A.   689
Foltz J. L.   *see* Coulson R. N. 70
Foltz R. L.   *see* Kinzer G. W. 67; Rudinsky J. A. 67
Fonseca J. P. C. da   33, 40, 83, 223
Foote B. A.   742, 743, 744: *see* Bratt A. D. 744; Garnett W. B. 746; Miller R. M. 762; Robinson W. H. 237; Trelka D. G. 743
Foott W. H.   743
Forattini O. P.   617
Forbes W. T. M.   236
Force D. C.   72
Ford H. R.   *see* Lowe R. E. 316
Fordy M. R.   *see* King P. E. 709
Forgash A. J.   *see* Riley R. C. 762
Forister G. W.   684
Forrest W. W.   *see* Browning T. O. 175, 179
Förster G.   *see* Klausnitzer B. 681
Foster C. H.   86
Foster W. A.   49
Fouillet P.   *see* David J. 43, 86
Fourie G. J. J.   48
Fowler H. W.   *see* Horsfall W. R. 732
Fowler M.   *see* Wightman J. A. 315

Fox A. S.   45
Fox C. J. S.   83, 85
Fox H.   503
Fox L.   237
Fox R. C.   763
Fraenkel G.   72, 182: *see* Blewett M. 315; Hsiao T. H. 83
Francis E.   732
Francke-Grosmann H.   278, 304, 305
Francois J.   91
Frank J. H.   212, 230
Frank M. B.   *see* Birch L. C. 224
Frankenberg G.   272, 656
Frankenstein P. W.   *see* Kugler O. E. 184
Frankie G. W.   85
Franklin R. T.   *see* Berisford C. W. 70
Franssen C. H. J.   683
Franz J.   17, 215, 218
Fraser F. C.   82, 487, 489, 491
Fredeen F. J. H.   86, 314, 315, 734, 748
Freeborn S. B.   *see* Herms W. B. 732
Freeman B. E.   84
Freeman G. H.   *see* Finch S. 86
Freeman P.   735
French J.   *see* Retnakaran A. 315
Fresneau M.   *see* Hurpin B. 17
Frick K. E.   72, 77
Friedel T.   46
Friedman S.   *see* Breddy J. K. 315
Frill F.   64
Frison T. H.   508
Fritze A.   502
Froeschner R. C.   589
Froggatt W. W.   236, 274
Fröhlich G.   86
Fronk W. D.   215
Frost F. M.   *see* Herms W. B. 729
Frost S. W.   83, 272, 296
Fuchs M. S.   44, 45, 47, 761: *see* Hiss E. A. 45, 46, 47; Schlaeger D. A. 206
Fuentes M. del C.   61: *see* Carbonell E. 83
Führer E.   709
Fujisaki K.   610
Fuldner D.   659
Fuller M. E.   222
Fulton B. B.   22, 270, 289, 513
Funkhouser W. D.   270, 271
Furneaux P. J. S.   165, 175, 179, 205, 209, 210, 513: *see* McFarlane J. E. 175, 179; Neville A. C. 206

Furniss M. M.   66, 70: *see* Lanier G. N. 67, 70; Ringold G. B. 70; Rudinsky J. A. 67, 70
Fuseini B. A.   184

Gabaldon A.   727
Gadd C. H.   74
Gadeau de Kerville H.   270
Gagnepain C.   86
Gahan J. B.   *see* Smith C. N. 312
Galford J. R.   313
Galichet P. F.   *see* Grison P. 257
Galindo P.   49, 85
Gallard L.   648
Galliker P.   720
Galun R.   *see* Braverman Y. 213
Galvão A. L. A.   729
Gamal-Eddin F. M.   *see* Hafez M. 28, 86
Gambles R. M.   492
Gameel O. I.   83, 227
Gander R.   732: *see* Geigy R. 733
Ganesalingam V. K.   56
Gangrade G. A.   534
Gara R. I.   *see* Billings R. F. 70; Gray B. 70; Pitman G. B. 70
Garcia M. F.   764
Garcia R.   79, 314
Garcia-Bellido A.   45, 47, 49
Gardiner B. O. C.   50: *see* David W. A. L. 78
Gardiner L. M.   687, 688, 706
Gardner A. E.   82, 491: *see* Gambles R. M. 492
Gardner J. C. M.   218
Gardner T. R.   *see* Clausen C. P. 57
Gardoš J.   761
Garham P.   490
Garman H.   216, 656
Garms R.   735
Garnett W. B.   746
Garrett-Jones C.   86
Gaskin R. C.   *see* Harden F. W. 79
Gasperi G.   *see* Grigola A. 46
Gassner G.   723
Gast R. T.   84, 312, 313, 315: *see* Smith C. N. 315
Gaufin A. R.   *see* Knight A. W. 510; Stark B. P. 764
Gaumont R.   568

Gebert S.   732
Geer B. W.   43
Gehring R. D.   85
Geigy R.   642, 733: *see* DuBois A. M. 16
Geissler K.   86
Geisthardt G.   178
Gelperin A.   *see* Stay B. 82
Genchev N.   25
Genieys P.   233
Gensitskii I. P.   314
Gentner L.   *see* Westigard P. H. 85
Geoff A. M.   *see* Nault L. R. 270, 271
George B. W.   *see* Smith C. N. 313
George C. J.   763
Gerber G. H.   689
Gerberg E. J.   727
Gerling D.   84, 94
Germain M.   735
Gerrity R. G.   676: *see* Sweeney P. R. 676
Gerwen A. C. M. van   *see* Browne L. B. 51
Gestetner B.   *see* Applebaum S. W. 89
Getzin L. W.   315
Geyer J. W. C.   19, 83, 224
Ghani M. A.   547: *see* Afzal M. 88; Ahmad R. 231; Baloch G. M. 24
Ghent A. W.   59, 84
Ghose S. K.   570
Ghosh C. C.   269, 643, 648
Ghouri A. S. K.   *see* McFarlane J. E. 175, 179
Gianotti J. F.   *see* Coscarón S. 216
Giardina A.   502, 504, 505, 518
Giardino J.   *see* Brust R. A. 727
Gibbins E. G.   729
Gibson A.   217, 218
Gibson R. W.   89, 91
Giese R. L.   *see* Kabir A. K. F. M. 77
Gifford J. R.   316
Gilbert B. L.   72, 87
Gilbert L. E.   241
Gilby A. R.   *see* Atkinson P. W. 209
Giles E. T.   269, 762
Gill R. W.   *see* Irwin M. E. 220, 229
Gillett J. D.   43, 81, 733, 764: *see* Haddow A. J. 81
Gillett S. D.   71
Gillette C. P.   738
Gillies M. T.   79, 729, 731: *see* Omer S. M. 79
Gillon Y.   499, 503

Gillot C.   *see* Friedel T. 46; Pickford R. 45, 47
Gilmer P. M.   256, 257, 258
Gingrich A. R.   *see* Chamberlain W. F. 764
Giorgi D.   200
Girault A. A.   221, 503
Girish G. K.   *see* Singh K. 84; Tyagi A. K. 689
Gironi A.   *see* Coluzzi M. 85
Gisin H.   *see* Murphy D. H. 727
Gjullin C. M.   313, 728, 732, 733
Glasgow J. P.   710
Glen D.   228, 229
Glen D. M.   763
Glendenning R.   26
Glover P. M.   235, 236, 714
Glumac S.   *see* Vukasović P. 20
Gninenko Yu. I.   85
Goantsa I. K.   220, 223
Godden D. H.   81
Godwin P. A.   689
Goeden R. D.   215, 743
Goeldi E. A.   245, 246, 252, 729
Goff A. M.   *see* Nault L. R. 270, 271
Goidanich A.   191
Goiny H.   57
Golberg L.   *see* De Meillon B. 732
Goldberg M.   *see* Hackman R. H. 201
Goldsmith M. R.   *see* Paul M. 210
Golebiowska Z.   85
Golini V. I.   86
Golub V. B.   *see* Khitsova L. N. 753
Golubenko N. N.   214, 215, 216
Goma L. K. H.   733
Goma P. C.   *see* Reyes P. V. 636
Gomaa A. A.   721
Gómez Ruano R.   *see* Orozco F. 76
Gonzalez B. J. E.   238
Gonzalez D.   *see* Irwin M. E. 220, 229; Orphides G. M. 214, 218, 220
Goodwin J. A.   85
Goonewardene H. F.   83, 314
Goot P. van der   583
Gopinath K.   273: *see* Subba Rao B. R. 61
Gordh G.   709
Gordon C. C.   *see* Shands W. A. 312, 313
Gordon G.   *see* Murray M. D. 556
Gordon H. T.   43
Gore W. E.   70: *see* Peacock J. W. 70; Pearce G. T. 70

Görg J.   506
Goryunova Z. S.   22
Goseco F. P.   223
Gospodinov G.   *see* Grigorov S. 595
Gothilf S.   87, 721: *see* Avidov Z. 25; Peleg B. A. 316
Gotz P.   *see* Vey A. 92
Gotz R.   80
Gowdey C. L.   257
Gower A. M.   193, 711
Graber V.   553
Graf E.   *see* Delucchi V. 720; Needham J. G. 486
Graham A. J.   *see* Smith C. N. 312
Graham H. M.   315: *see* Hendricks D. E. 315
Graham M. W. R. de V.   708
Grandi M.   475, 485
Grassé P. P.   219, 515, 518
Grassia A.   313
Gravelle P. J.   *see* Ringold G. B. 70
Graves T. M.   *see* Rudinsky J. A. 67
Gray B.   70
Gray K.   *see* Edwards W. D. 235
Gray T. G.   313: *see* Shepherd R. F. 313
Grbić V.   60
Greany P. D.   64, 84: *see* Oatman E. R. 84
Greathead D. J.   211, 217, 218, 220, 221, 222, 223, 224, 232: *see* Stower W. J. 83, 217, 222, 224
Green E. E.   274, 306
Green W.   *see* Swain R. B. 84
Greenberg B.   315
Greenberg L.   *see* Streams F. A. 92
Greene G. L.   85, 313
Greenfield M. D.   720
Grellet P.   175, 180
Grenacher H.   477
Grenier P.   735: *see* Germain M. 735
Gressitt J. L.   17
Grewal S. S.   33
Griffin J. G.   313
Griffith K. H.   *see* Fox R. C. 763
Griffith M. E.   620
Griffiths D. C.   84, 91, 94
Griffiths K. J.   84
Grigarick A. A.   84, 749: *see* Knabke J. J. 269
Grigolo A.   46
Grigorov S.   595
Grimble D. G.   83: *see* Nord J. C. 77, 83

Grimpe G.   245
Grimstone A. V.   90
Griot M.   220
Grison P.   41, 80, 257
Griswold G. H.   18, 24, 60, 178
Grivanov K. P.   230
Grodzinski W.   see Park T. 225
Gromova A. A.   764
Gromovaya E. F.   51
Grosch D. S.   see Valcovic L. R. 31
Gross F.   490
Gross H. R.   312, 314: see Nordlund D. A. 313, 709
Gross J.   554
Grove A. J.   see Barnes J. H. 82
Grover P.   221, 763
Gruber F.   233
Gruner L.   36
Grunin K. Y.   740
Gruwez G.   81
Gruys P.   35, 40, 41
Guardia V. M.   see Zeledon R. 617
Gubb D. C.   209
Gubler D. J.   85
Guennelon G.   76, 85, 265
Guerra A. A.   see Bullock H. R. 315
Guerritore A.   184
Guidicelli J.   86
Guilding L.   647
Guillot F. S.   64
Gunstream S. E.   see Hagstrum D. W. 75
Gunther K.   287, 534
Guppy P. L.   741
Guppy R.   269
Gupta A. P.   681
Gupta B. P.   20: see Srivastava A. S. 24
Gupta P. C.   see Chaudhry H. S. 569
Gupta P. D.   78, 85, 184
Gupta S. B. L.   see Lal R. 741
Gupta S. N.   see Misra M. P. 714
Gurney A. B.   237, 502, 507: see Parfin S. I. 648
Gurr L.   15
Gur'yanova T. M.   84
Gwadz R. W.   45, 47
Gyrisco G. G.   17, 83, 217

Haber V. R.   see Frost S. W. 272
Habib A.   566

Hackett L. W.   729, 730: see Bates M. 729, 730
Hackman R. H.   201
Hackman W.   748
Hadaway A. B.   18
Haddow A. J.   80, 81: see Gillett J. D. 81
Hadley C. H.   see Herrick G. W. 520
Hadlington P.   533
Hadwen S.   750
Haegermark U.   86
Hafeez M. A.   83
Hafez M.   27, 28, 49, 83, 86, 312, 659, 684, 738, 764
Hagemann-Meurer U.   36
Hagen H. R.   184
Hagen K. S.   232: see Rajendram G. F. 312; Shimizu J. T. 312
Hagley E. A. C.   21, 720
Hagstrum D. W.   75, 81, 83, 85
Hahn J.   534, 542
Hain F. P.   see Coulson R. N. 70; Hertel G. D. 66
Haisch A.   see Boller E. F. 742
Hale H. M.   620
Hale R. L.   see Shorey H. H. 33
Haley W. E.   see Ingram J. W. 214
Halffter G.   273, 275, 276, 277, 302, 688
Halffter V.   see Halffter G. 276
Hall A. E.   see Bryant E. H. 86
Hall D. W.   92: see Andreadis T. G. 92
Hall J. C.   see Schlinger E. I. 84
Hall P. M.   see Dyer E. A. D. 70
Hall R. R.   212
Hall S. R.   see Cleveland L. R. 284
Hallez P.   203
Halstead D. G. H.   34, 689
Hamid A.   712
Hamilton A.   642
Hamilton A. G.   34
Hamilton M. A.   634
Hamm A. H.   736
Hamm J. J.   see Chauthani A. R. 28
Hammad S. M.   764: see Eldefrawi M. E. 24; El-Kady E. 34; El-Minshawy A. M. 570
Hammer O.   27, 107, 133, 217, 747, 757, 758
Hammerschlag G.   see Tangel F. 537
Hammond A. M.   see Hensley S. D. 312
Hammond C. O.   see Colyer C. N. 749
Hamon C.   566
Hampson G. F.   236

Hanbal I.   see Eldefrawi M. E. 24
Hancock J. L.   82, 175, 514
Handke A. D.   see Eddy G. W. 86
Handlirsch A.   739
Hanitsch R.   284
Hanna H. M.   23, 711
Hansell M. H.   79
Hansen G.   748
Hansen H. L.   see Shade R. E. 219
Harada F.   85
Harakly F. A.   34
Haramoto F. H.   see Newell I. M. 742
Harcharan Singh   48
Harcourt D. G.   see Matthewman W. G. 86
Hardeland R.   see Rensing L. 81
Harden F. W.   79
Harding J. A.   763
Hardwick D. F.   85
Hardwood R. F.   see Logen D. 85
Hardy D. E.   742
Hardy G. A.   20
Hardy J.   547
Hardy R. J.   see Grassia A. 313
Hardy Y. J.   84
Hare W. W.   see Harrell E. A. 312
Hargreaves H.   218
Harjai S. C.   see Bhatia K. R. 83
Harlan D. P.   522
Harman D. M.   745
Harman G. E.   see Eckenrode C. J. 86
Harman M. T.   184
Harman R. R. M.   see Rothschild M. 256, 257, 258
Harrell E. A.   312: see Gross H. R. 312; Nordlund D. A. 313
Harris F. A.   see Hines B. M. 212; Moore S. T. 316
Harris K. M.   236
Harris M. K.   689
Harris R. H. G.   see Carnegie A. J. M. 228
Harris W. V.   507
Harrison F. P.   215, 216: see Bickley W. E. 86
Hart W. G.   see Reed D. K. 91
Hartley J. C.   57, 98, 164, 171, 175, 177, 514, 515, 516, 520, 521, 522, 613, 659, 679, 741: see Deane R. L. 764
Hartman R. D.   see Burke H. E. 83
Hartung W. J.   see Séverin H. H. P. 53
Hartzell F. Z.   20, 83, 677

Harvey G. W.   635
Harwood R. F.   see Logen D. 85
Harwood W. G.   71
Harz K.   13, 14, 270
Hase A.   562
Hasegawa T.   86
Haskell P. T.   82
Hassan A. I.   567
Hassan S. A.   230
Hassanein A. M. M.   see Okasha A. Y. K. 35
Hassanein M. H.   33
Hassell M. P.   211
Hathaway D. O.   212, 312
Hausermann W.   49
Haviland M. D.   270, 271
Hawkes R. B.   315
Haydock I.   see Judson C. L. 733
Haynes D. L.   72
Hays D. B.   56
Hays K. L.   see Foster C. H. 86; Johnson A. W. 236, 237; Knox P. C. 79; Tidwell M. A. 72
Hays S. B.   see Bass J. A. 227; Johnson A. W. 43
Hayslip N. C.   269
Hayward K. J.   218
Hazard E. I.   85
Hearle E.   751
Heberday R. F.   534
Hedden R.   66
Hedden R. L.   see Pitman G. 70
Hegarty C. P.   see Gjullin C. M. 733
Hegdekar B. M.   84: see Arthur A. F. 47, 84
Heidemann O.   582
Heie O. E.   569
Heihata K.   see Hirose Y. 709
Heinrichs E. A.   see Matheny E. L. 714
Heisch R. B.   see Goiny H. 57
Helms T. J.   314
Helson G. A. H.   510
Hemmingsen A. M.   80, 85, 271, 724, 725, 726, 727, 738, 763
Henderson L. M.   see Horsfall W. R. 733
Hendricks D. E.   67, 315
Henking H.   719
Henneberg B.   676
Henneberry T. J.   26, 31, 85
Henneguy L. F.   244, 534
Henriksen K. L.   see Bøving A. G. 272, 656, 657

Henry C. S.   645, 647
Hensley S. D.   312
Henson R. D.   see Vinson S. B. 709
Henstra S.   see Cobben R. H. 616
Hepner L. W.   see Hormchan P. 763
Herakly F.   see Habib A. 566
Herewege J. van   86: see David J. 43, 86
Herfs A.   178
Hermanussen J. F.   see Campbell J. B. 231
Herms W. B.   85, 729, 732
Hernandez N. S.   see Hendricks D. E. 315
Hernández Roque F.   569
Heron R. J.   315
Herrebout W. M.   82
Herrick G. W.   258, 520
Herring J. L.   582
Hertel G. D.   66
Hertel R.   583
Herter K.   269, 270, 289: see Günther K. 287
Hesse A. J.   223
Hetrick L. A.   83, 533
Hewer H. R.   198, 199
Heymer A.   82
Heymons R.   270, 273, 298, 303, 483
Heywood H. B.   see Needham J. G. 13, 486, 489
Hibble J.   see Brian M. V. 11
Hibbs E. T.   see Carlson O. V. 83, 566; Miller R. L. 83
Hickey W. A.   see Gwadze R. W. 45, 47
Hiehata K.   see Hirose Y. 709
Hightower B. G.   see Davis R. B. 314
Hilderbrand D. N.   see Young J. C. 66, 67
Hill A. R.   610
Hill D. L.   315
Hill M. A.   736
Hill R. B.   see Rivera J. 729, 730
Hill W. R.   256, 258
Hillhouse T. L.   85
Hilliard J. R.   522
Hillyer R. S.   85
Hilsenhoff W. L.   735
Hinckley A. D.   85
Hincks W. D.   270
Hines B. M.   212
Hinko E.   728
Hinks C. F.   91
Hinman E. H.   732
Hinman F. G.   see Larson A. O. 20, 83, 683
Hinton H. E.   3, 4, 5, 9, 10, 31, 46, 52, 82, 95, 96, 98, 106, 107, 108, 110, 111, 112, 113, 114, 115, 119, 120, 124, 125, 129, 132, 137, 139, 142, 144, 146, 147, 149, 150, 151, 152, 159, 164, 170, 171, 173, 177, 181, 182, 187, 211, 221, 223, 225, 226, 229, 238, 241, 252, 253, 261, 266, 270, 271, 273, 274, 275, 286, 291, 292, 293, 298, 475, 552, 554, 556, 570, 612, 613, 614, 622, 623, 624, 629, 632, 634, 636, 647, 648, 649, 659, 688, 716, 719, 720, 722, 728, 729, 732, 734, 737, 741, 744, 751, 756, 757, 758, 759, 760, 761
Hinton T.   315
Hirakoso S.   85
Hirana C.   see Ishii S. 89
Hirao T.   see Yamaoka K. 78, 85, 764
Hirata S.   40, 85: see Suzuki K. 197
Hirose Y.   709
Hiss E. A.   45, 46, 47: see Fuchs M. S. 44, 45, 47
Hitchcock S. W.   238
Ho Chung   see Tsai Pang-hua 35
Ho F. K.   225
Hobby B. M.   740: see Broadhead E. 14, 548
Hobgood J. M.   see Knott C. M. 312
Hochmut R.   76
Hocking H.   74
Hodek I.   232, 675
Hodges J. D.   66
Hodjat S. H.   40
Hodson A. C.   183, 200: see Chiang H. C. 28, 39, 762; Witter J. A. 26
Hoffman B. L.   85
Hoffman C. H.   16
Hoffman J. A.   see Zachary D. 92
Hoffman J. D.   312
Hoffman W. E.   582
Hoffmeyer S.   246
Hogan B. F.   see Kunz S. E. 86
Hogan T. W.   175
Hohorst W.   551, 562
Hokama Y.   727: see Judson C. L. 732, 733
Hokyo N.   94: see Kiritani K. 595
Holdaway F. G.   689; see Ballard E. 272; Belur N. V. 215, 216, 219; Hsiao T. H. 86; Loan C. 64; Raros R. S. 312; Schurr K. 73, 85
Holler G.   82
Holling C. S.   229
Hollinger A. H.   221

Holmes M. C.   see Burger T. L. 313; Maltby H. L. 90
Holmes R. L.   see Shands W. A. 312
Holst Christensen P. J.   721
Holt G. G.   47: see Ercelik T. M. 764
Honda K.   85
Hood J. D.   563
Hooper R. L.   86
Hopkins A. D.   279, 282
Hopkins A. R.   see Taft H. M. 48
Hopkins C. R.   30, 707, 709: see King P. E. 30
Hopkins D. E.   549, 551
Hopkins G. R. E.   see Buxton P. A. 732, 733
Horber E.   see Nwanze K. F. 83
Hormchan P.   763
Horn D. J.   84
Horsburgh R. L.   83
Horsfall W. R.   232, 729, 731, 732, 733: see Anderson J. F. 731; Borg A. F. 733; Craig G. B. 729; Harwood W. G. 732; Trpiš M. 733; Wilson G. R. 733
Horton J. R.   216
Hoshino M.   see Yamaoka K. 78
Hosler G. W.   see Burger T. L. 313
Hosny M. M.   72
Hosoi M.   656
Hosoi T.   731
Hoste C.   see Gruwez G. 81
Houillier M.   79
House H. L.   see Coppel H. C. 74; Yazgan S. 315
Houser J. S.   57
Houseweart M. W.   764: see Coulson R. N. 70
Houyez P.   312
Hovanitz W.   85
How T'au-ch'ien   see Tsai Pan-hua 35
Howard B.   see MacFarlane W. V. 557
Howard L. O.   220
Howe R. W.   18, 19, 42, 60, 61, 62, 83, 178, 179, 764
Howe W. L.   312: see Smith C. N. 313
Howell J. F.   312
Hower A. A.   312
Howland A. F.   see Stone M. W. 17
Hoxie R. P.   see Wells S. G. 83
Hoyt C. P.   213
Hrdý I.   221
Hrutfiord B. F.   see Billings R. F. 70

Hsiao T. H.   83, 86: see Saunders R. C. 22
Hsu Y. C.   see Needham J. G. 477
Hu S. M. K.   see Baisas F. E. 729
Hubbard H. C.   304
Huber F.   see Loher W. 44
Hudon M.   312
Hudson A.   85
Hudson G. V.   235
Huffaker C. B.   see White E. G. 213
Hughes P. R.   66, 70: see Renwick J. A. A. 67, 69, 70; Strong F. E. 48; Vité J. P. 70
Hughes R. D.   216, 217: see Wright D. W. 216, 217, 218
Huie L. H.   83, 547, 650
Huignard J.   47
Hungerford H. B.   16, 58, 572, 575, 576, 577, 582, 617, 620, 636, 638, 639, 648
Hunsley J. R.   see Paul M. 210
Hunt T. N.   84
Hunter P. E.   227
Hunter-Jones P.   175, 520: see Antoniou A. 520
Hurlbut H. S.   729, 730: see Matheson R. 730
Hurpin B.   17, 34
Hurter J.   65
Husain M. A.   522
Husbands R. C.   728
Husman C. N.   see Fletcher L. W. 86; Smith C. N. 312
Hussein E. M. K.   see Strong F. E. 48
Husseiny M. M.   31
Hussey N. W.   33, 86
Hussey R. F.   272
Hutchings R. F. N.   see Wearing C. H. 85
Hyde G. E.   84
Hyer A. H.   see Tingley W. M. 83
Hynes H. B. N.   13, 223, 508, 509, 510, 735

Ibbotson A.   84, 86
Ibrahim M. M.   520
Idris B. E. M.   251, 729
Ignoffo C. M.   315: see Hoffman J. D. 312
Ihering R. von   272
Iizima N.   see Tamura I. 749
Ijima H.   710
Ikeda J.   see Rozeboom L. E. 85
Ikeshoji T.   85
Ilan A. R. B.   see Levinson H. Z. 71
Iltis W. G.   252, 253, 729

Imms A. D.   2, 3, 13, 183, 235, 272, 512, 643, 745
Ingle S. J.   *see* Reed D. K. 91
Ingram J. W.   214
Ingram M. J.   285
Injac M.   *see* Vasiljević L. 212
Iperti G.   83
Irwin M. E.   86, 220, 229, 739, 740, 764
Irzykiewicz H.   *see* Powning R. F. 209
Isaac P. V.   27, 738
Isaacson D. L.   225
Isaev V. A.   761
Isart J.   235
Ishii G.   18, 219
Ishii S.   77, 89
Islas S. F.   *see* Vité J. P. 70
Isle D.   79
Israel P.   *see* Misra B. C. 83
Ito H.   184
Itô Y.   212, 227, 313
Ives W. G. H.   215, 313
Iwata Y.   *see* Ishii S. 89
Iyengar M. O. T.   729

Jacklin S. W.   48
Jackson C. G.   28
Jackson D. J.   17, 49, 83, 94, 165, 175, 651, 653, 708, 709, 763
Jackson H. B.   22: *see* Rogers C. E. 224
Jacob S. A.   232
Jacob W.   83
Jacobson E.   236
Jacobson L. A.   35, 85
Jaillet I.   *see* Le Berre J. R. 75
Jakob W. L.   314
Jalil M.   227: *see* Wicht M. C. 227
James C. R.   *see* Furneaux P. J. S. 165, 209, 513
James H. G.   212, 216, 217, 221, 502
James J. D.   *see* Mayer M. S. 79
James P. E.   *see* Magnum C. L. 315
James S. P.   729, 761
James W.   *see* Taft H. M. 48
Jancke O.   15, 26, 562
Janisch E.   179
Janković M.   *see* Zečević D. 718
Jannone G.   222, 522
Jansen D. H.   683, 684
Janson O. E.   502
Jantz O. K.   71: *see* Connin R. V. 83

Janvier H.   84, 234
Jara P. B.   86
Jarrett P.   *see* Burges H. D. 720
Jasič J.   34: *see* Hrdý I. 221
Jasmin J. J.   *see* Perron J. P. 86
Jaynes H. A.   *see* Clausen C. P. 57
Jeannel R.   641
Jenkins J. N.   *see* Bailey J. C. 83; Maxwell F. G. 71, 88; Parrott W. L. 88
Jenkins L.   *see* Oatman E. R. 720
Jenkins R. Y.   *see* Yamamoto R. T. 78, 85
Jennings D. T.   85, 316
Jensen B.   *see* Hemmingsen A. M. 725, 726, 727
Jensen-Haarup A. C.   271
Jentsch S.   546
Jepson W. F.   234
Jermy T.   763: *see* Muschinek G. 689
Jewett H. H.   *see* Garman H. 216
Jobling B.   85
Johannsen O. A.   735, 736
John O.   563
Johnsey R. L.   *see* Gray B. 70
Johnson A. W.   43, 236, 237
Johnson B.   89, 764
Johnson C.   82, 491
Johnson C. D.   684: *see* Center T. D. 684; Forister G. W. 684; Pfaffenberger G. S. 684, 685
Johnson C. G.   166, 175, 178, 611
Johnson J. A.   *see* Miller B. S. 88
Johnson J. R.   *see* Terranova A. C. 46, 47
Johnson K. M.   *see* Tesh R. B. 213
Johnson M. W.   85
Jöhnssen A.   50
Johnston A. F.   233
Jolivet P.   677
Jolly M. S.   85
Joly N.   485
Jones B. J.   748
Jones B. M.   261, 764
Jones C. R.   741
Jones D. A.   251
Jones D. W.   *see* Smith W. W. 85
Jones E. L.   27
Jones E. T.   *see* Miller B. S. 88
Jones F. G. W.   *see* Thorpe W. H. 56
Jones F. M.   502
Jones J. C.   49: *see* Akey D. H. 49; Wheeler R. E. 49

Jones L. G.   *see* Baker W. A. 21
Jones M. G.   59, 86, 213, 216, 217, 218, 230, 757: *see* Kempton R. A. 86
Jones R. L.   84, 85: *see* Lewis W. J. 84; Norlund D. A. 709
Jones S. L.   764: *see* Lingren P. D. 229, 231
Joos J. L.   *see* Batiste W. C. 314
Jordan K. H. C.   572, 575, 579
Jörg M. E.   258
Joshi B. G.   19
Joubert P. C.   763
Jourdheuil P.   *see* Bonnemaison L. 83
Journet P.   51
Joyce R. J. V.   520
Juarez E.   316
Judenko E.   274
Judge P. J.   181
Judson C. L.   85, 732, 733: *see* Hokama Y. 727
Judulien F.   273
Juillet J. A.   215
Junnikkala E.   214, 215
Jura C.   166, 764

Kabanov V. A.   17
Kabir A. K. M. F.   77
Kabos W. J.   741
Kaczor W. J.   *see* Hall D. W. 92
Kadyi H.   203
Kafatos F. C.   *see* Paul M. 210
Kahn N. H.   86
Kaĭdanov L. Z.   *see* Sapunov V. B. 762
Kain W. M.   761
Kajita H.   84
Kalandadze L.   49
Kalpage K. S. P.   85, 729
Kalshoven L. G. E.   24, 274
Kaltenbach A.   512
Kamal M.   23, 595
Kamano S.   *see* Yushima T. 67
Kamiyama Y.   *see* Sumimoto M. 70
Kaneko T.   273, 274
Kanervo V.   214, 215, 219, 220
Kangas E.   60, 70, 71: *see* Oksanen H. 71; Perttunen V. 71; Selander J. 67
Kapil R. P.   227: *see* Bhanot J. P. 227
Kaplan W. D.   312
Karadzhov S.   229
Karandinos M. G.   86: *see* Greenfield M. D. 720

Karashina J.   *see* Tomita M. 537
Karelin V. D.   *see* Adashkevich B. 237
Karpenko C. P.   85
Kartman L.   732
Kaschef A. H.   84
Kastner A.   642
Katanyukul W.   229
Katayama E.   84
Katiyar K. N.   82, 523
Katiyar O. P.   *see* Lal L. 227
Kato K.   184, 202
Katô M.   732
Kato S.   749
Katsoyannos B.   65: *see* Hurter J. 65
Katsuno S.   *see* Takizawa Y. 199
Kaufmann T.   713
Kaup J. J.   534
Kaur R. B.   *see* Narayanan E. S. 84
Kawahara S.   33, 42, 179
Kawasaki H.   205, 206, 207, 208, 210, 721
Kay D.   763
Kaya H. K.   *see* Anderson J. F. 84
Kearby W. H.   *see* Bliss M. 83, 570
Kearney M.   *see* Burnet B. 46
Kearns C. W.   23
Kehat M.   61
Keilin D.   142, 562
Keiser I.   312
Keister M.   152
Keister M. L.   *see* Yokoyama T. 721
Kéler S. von   551, 562
Kellen W. R.   579
Keller B. L.   *see* Ubelaker J. E. 751
Kellog F. E.   79
Kelner-Pillault S.   192
Kelsey J. M.   83
Kemner N. A.   659
Kempster R. H.   *see* Corbet P. S. 314; Osgood C. E. 63
Kempton R. A.   86
Kenchington W.   202, 764: *see* Atkins E. D. T. 191; Rudall K. M. 183
Kennard C. P.   *see* McFarlane J. E. 175, 179
Kennedy C. H.   486
Kennedy G. G.   43, 566
Kennedy J. S.   82
Kent P. W.   209: *see* Brunet P. C. 201; Whitehead D. L. 201
Kenton J.   *see* Bardner R. 119
Kephart C. F.   257

Keremidchiev M.   see Stefanov D. 218
Kerenski J.   171, 175
Kerr W. E.   see Sakagami S. F. 84
Kershaw J. C.   502
Kershaw W. E.   see Hynes H. B. N. 735
Kessel E.   551, 562
Kessel R. G.   see Beams H. W. 491
Kettle D. S.   730
Keuchenius P. E.   236
Kevan D. K. McE.   25, 734
Keys J. H.   271
Khalifa A.   82, 83, 170, 175
Khalsa H. G.   77
Khan A. A.   79
Kharizanov A.   648
Khitsova L. N.   753
Kidwell A. S.   see Pratt H. D. 314
Kikuchi T.   66
Kilgore W. W.   see Painter R. R. 762
Kilincer N.   91
Killick-Kendrick R.   764
Killingsworth B. F.   see Hoffman B. L. 85
Killington F. J.   642, 648
Kim B. K.   see Kim C. W. 212
Kim C. W.   212
Kimoto H.   see Hirose Y. 709
Kimura K.   see Kiritani K. 39, 41, 48
Kindler S. H.   see Weissman-Strum A. 733
King A.   see Stay B. 203, 204
King A. B. S.   313
King H. E.   see Parnell F. R. 88
King J. L.   27, 650, 738, 739: see Clausen C. P. 649
King P. E.   30, 65, 649, 706, 707, 709, 763: see Copland M. J. W. 708; Hopkins C. R. 30, 707, 709; Richards J. G. 709
King R. C.   709
King W.   732
Kingsolver J. G.   764
Kingsolver J. M.   see Johnson C. D. 684
Kinn D. N.   227
Kinzer G. W.   67: see Pitman G. B. 67; Rudinsky J. A. 67; Vité J. P. 67
Kinzer H. G.   279
Kirichenko A.   216
Kiritani K.   39, 41, 48, 83, 219, 595: see Hokyo N. 94; Ito Y. 219
Kirk V. M.   83
Kirkaldy G. W.   638
Kirkland A. H.   see Fernald C. H. 257

Kirkpatrick T. W.   29, 177, 221, 272
Kishaba A. N.   see Henneberry T. J. 26, 85
Kishi Y.   94, 231
Kishimoto T.   see Otsura M. 761
Kishino K.   see Tamura I. 749
Kishore P.   see Kundu G. G. 86
Kitamura B.   85
Kitano H.   89, 92, 706: see Azuma K. 92; Takada M. 92
Kitching R. L.   270
Klatt B.   44
Klausnitzer B.   675, 681
Klein M. G.   83
Kleine R.   191
Klein-Krautheim F.   741
Klemperer H.-G.   764
Kliefoth R. A.   see Vité J. P. 70
Kliewer J. W.   see Judson C. L. 733
Kline L. N.   see Furniss M. M. 70; Rudinsky J. A. 67, 70
Klingler J.   33, 59
Klingstedt H.   711
Kloft W.   see Shepard M. 270, 289
Klomp H.   40
Klug W.-S.   762
Klun J. A.   89
Knab F.   729
Knabke J. J.   269
Knight A. W.   510
Knight F. B.   279: see Grimble D. G. 83; Nord J. C. 77, 83
Knight G. H.   569
Knoll F.   63, 78
Knopf J. A. E.   70
Knott C. M.   312
Knox P. C.   79
Knutson L. V.   743, 744: see Bratt A. D. 744; Deeming J. C. 748; Foote B. A. 743; Rozkošný R. 744
Kobayashi S.   85, 313
Kobayashi T.   see Oku T. 85
Kochetova N. I.   94
Kochhar R. D.   85
Koehler C. S.   see Frankie G. W. 85
Koeppe J. K.   see Lake C. R. 209
Kogan M.   see Waldbauer G. P. 313
Kohno M.   510
Koide T.   225
Koizumi K.   34
Kojima T.   18, 60, 179

Kok L. T.   see Ward R. H. 84
Kokhmanyuk F. S.   85
Kolmakova V. D.   58, 238
Komárek J.   649
Komp W. H. W.   729
Kondo S.   82: see Yoshimura S. 85
Kondo T.   see Sumimoto M. 70
Konig A.   738
Kôno H.   278
Kormondy E. J.   491
Korsakoff M. N.   503
Korschelt E.   633, 638, 718
Koss R. W.   477, 764
Kot J.   see Sandner H. 41; Wiackowski S. K. 214
Kotby F. A.   764: see Hosny M. M. 72
Koura A.   see Bishara S. I. 84
Kovacs J.   see Hill W. R. 256, 258
Koval' Yu. V.   213, 215, 221
Kovaleva M. F.   84
Kovtun I. V.   20
Kowalska T.   312
Koyama K.   314: see Mitsuhashi H. 312
Koyama T.   80
Kozhanchikov I. V.   258
Kozulina O. V.   551
Kraemer G. D.   279
Kramer K. J.   204
Kramer P.   553
Kramer S.   659
Krausse A.   see Wolff M. 236
Kreyenberg J.   18, 49, 225
Krieg A.   238
Kriegl M.   see Eichhorn O. 220
Krishna K.   see Sankaran T. 71
Krishnaswami S.   see Jolly M. S. 85
Kristensen N. P.   see Achtelig M. 642
Krnjaić S.   83
Krzymánska J.   see Wegorek W. 89
Kubomura K.   see Kato K. 184, 202
Kudler J.   see Tichý V. 237
Kugler O. E.   184
Kühlow F. (= Kühlhorn)   729, 730, 731
Kuhr R. J.   see Yu C. C. 86
Kulkarny H. L.   see Patel R. C. 16
Kullenberg B.   612, 613
Kulman H. M.   84: see Houseweart M. W. 764; Witter J. A. 26, 313
Kulshrestha S. K.   83, 86
Kumar R.   209, 499, 505, 612: see Fuseini B. 184, 202; Oppong-Mensah D. 763
Kumm H. W.   729
Kummer H.   47
Kunckel d'Herculias J.   518, 522
Kundu G. G.   86
Kung K. S.   20
Kunike G.   17, 49
Kunitskaya N. T.   764
Kunou I.   see Matsuo K. 761
Kunz S. E.   86, 764
Kupershtein M. L.   212
Kurbanova D. D.   764
Kurian C.   see Abraham V. A. 228
Kurir A.   84
Kurodo H.   see Okada J. 23
Kuroko H.   648
Kurosawa T.   18, 50
Kurstak E.   60
Kusnezov N. J.   712
Kusui K.   183
Kuwayama S.   568, 648, 711, 749
Kuznetzova Y. I.   314

Laabs A.   187
Laan P. A. van der   see Turnhout H. M. T. van 612
Labeyrie V.   33, 79, 83, 86
Labine P. A.   85
Lachmajer J.   213
Lacroix J. L.   648
Ladduwahetty A. M.   83
Lafever H. N.   see Maxwell F. G. 88
Lafrance J.   see Perron J. P. 86
Laidlaw W. B. R.   216
Laigo F. M.   see Sanchez F. F. 26
Laing D. R.   21
Laing J.   see Garcia R. 79
Lake G. R.   209
Laker A.   656
Lal K. B.   567
Lal L.   227
Lal R.   732, 741: see Beament J. W. L. 96, 181, 200
Lall B. S.   22
Lamb R. J.   764
Lambert J. G.   see Hunter-Jones P. 175
Lamborn W. A.   235, 236, 270
Lancieri M.   see Carfagna M. 86
Landau   see Abderhalden 183
Landin M.   see Gast R. T. 315

Landis B. J.   749: *see* Fox L. 237
Lane C.   721
Lane J.   *see* Galvão A. L. A. 729
Lang C. A.   *see* Wallis R. C. 85
Lange W. H.   213
Lanier G. N.   66, 67, 70: *see* Booth D. C. 70; Elliott E. W. 70; Peacock J. W. 70; Pearce G. T. 70; Piston J. J. 66
Lapie G. E.   258
Lara E. F.   87
Larsen E. B.   119, 179, 273, 301, 303
Larsen J. R.   *see* Parks J. J. 763
Larsén O.   637
Larson A. O.   20, 72, 83, 683
Latheef M. A.   *see* Yeargan K. V. 84
Lattin J. D.   589
Laubmann M.   83
Lauck D.   58
Laudani U.   85
Laudon   256
Laughlin R.   175, 179, 724
Launois M.   *see* Le Berre J. R. 82
Laurence B. R.   64, 85, 217, 231
Lauverjat S.   184
Lavabre E. M.   273
Lavie D.   *see* Gothilf S. 721; Meisner J. 84
Lavigne R. J.   740
Laviolette R.   *see* Sharma M. L. 15
Lavoipierre M. M. J.   728: *see* De Meillon B. 732
Law J. H.   *see* Kramer K. J. 204
Lawko C. M.   70
Lawlor W. K.   730
Lawson D. E.   313: *see* Atyeo W. T. 681
Lawson F. A.   497, 683
Lawson F. R.   215: *see* Hoffman J. D. 312; Knott C. M. 312
Lazarević B.   48, 278
Lea A. O.   43: *see* Foster W. A. 49
Leahy M. G.   45, 47: *see* Spielman A. 45
Leal M. P.   *see* Hendricks D. E. 315
Leaney A. J.   *see* Killick-Kendrick R. 764
Le Berre J. R.   75, 82, 215
Lebrun D.   575, 582, 620, 637
Lécaillon A.   678, 679, 681
LeCato G. L.   35, 61, 84, 232, 763
Leclercq J.   84, 178
Leclercq-Smekens M.   721
Ledesma L.   228
Ledoux A.   84, 269

Lee D. J.   729
Lee H.   *see* Kaplan W. D. 312
Lee H. P.   *see* Bentley M. D. 85, 764; McDaniel I. M. 85
Lee H. R.   *see* Choi S. Y. 721
Lee Hong Suk   *see* Stephens S. G. 88
Lee J.-O.   *see* Choi S. Y. 721
Lee R. D.   83
Lees A. D.   82
Lees A. H.   568
Leeson H. S.   556: *see* Evans A. M. 729
Le Faucheux M.   738
Lefkovitch L. P.   34, 37, 728
Lefroy H. M.   272
Legay J. M.   721
Leger M.   258
Legner E. F.   86: *see* White E. B. 659
Lehmensick R.   715
Leigh T. F.   *see* Tingey W. M. 83
Lekander B.   279, 689
Lekić M.   15, 17
Lelup N.   *see* Vincke I. 729
Lengerken H. von   273, 278, 301, 650: *see* Heymons R. 273, 298
Lenteren J. C. van   *see* Bakker K. 313
Lenz F.   725
Leon R. R. de   *see* Hibbs E. T. 83
Leonard D. E.   74, 313, 315
Leong C. Y.   638: *see* Fernando C. H. 620
Leong J. K. L.   21
Léonide J. C.   222
Leopold R. A.   46, 47: *see* Terranova A. C. 46, 47
Lépiney J. de   219, 739
Leppla N. C.   315: *see* Carlyle S. L. 312
Lergenmüller E.   34
Lerner I. M.   *see* Sokoloff A. 225
Leroi B.   86
Le Roux E. J.   *see* Cheng H. H. 224; Paradis R. O. 313
Leroy Y.   513
Leska W.   19, 84
Leslie P. H.   *see* Park T. 225
Lespéron L.   186
Lesse H. de   85
Lestage J. A.   711
Leston D.   589, 593, 595, 610, 611: *see* Southwood T. R. E. 609, 620
Leuckhart R.   477, 553, 582, 620, 718
Leuzinger H.   534, 542

Levine E. 258
Levinson H. Z. 71
Levy E. C. *see* Gothilf S. 721
Lewallen L. 762
Lewis D. J. 734, 735
Lewis F. B. 84
Lewis L. F. 85
Lewis T. 563
Lewis W. J. 64, 84, 92, 93: *see* Brewer F. D. 92; Copeland E. L. 764; Jones R. L. 84, 85; Nordlund D. A. 313, 709; Todd J. W. 86; Vinson S. B. 92
Leydig F. 98, 718
Lhoste J. 270, 720
Li Chin-yin *see* Tsai Pang-hua 35
Li S. S. *see* Cheu S. P. 224
Li Ya-tse *see* Tsai Pang-hua 19
Libbey L. M. 70: *see* Rudinsky J. A. 67, 70
Liebermann J. 224, 518
Liebers R. *see* Lehmensick R. 715
Liecht P. M. 284
Lieftinck M. A. *see* Meer Mohr J. C. van der 689
Lien J. C. *see* Matsuo K. 761
Likventov A. V. 33
Lin-Chow S. H. 85
Lincoln D. C. R. 98, 152, 171, 175, 253, 650, 651, 675, 729
Lindberg H. 709
Lindgren D. L. 178
Lindig C. H. *see* Griffin J. G. 313
Lindner E. 724, 725
Lindquist A. W. 273, 277
Lindquist E. E. 227
Lingren P. D. 229, 231: *see* Jones S. L. 764
Linsley E. G. 83, 219, 224, 232, 233, 685, 686, 763
Linstow von 197, 257
Lints C. V. *see* Gruwez G. 81
Lints F. A. *see* Gruwez G. 81
Lipa J. J. 229, 314
Lipkow E. 688
Listov M. V. 83
Lite S. W. *see* Wallis R. C. 315
Litsinger J. 84
Liu Yu-ch'iao *see* Tsai Pan-hua 35
Livingstone D. 611
Lloyd D. C. 30, 64, 84
Lloyd E. P. 84
Loan C. 64

Lobanov A. M. 86, 762
Lobatón Márquez M. 238
Locke M. 181
Lofty J. R. *see* Bardner R. 86
Logan D. M. *see* Proverbs M. D. 314
Logen D. 85
Logothetis C. *see* Gyrisco G. G. 17, 83, 217
Loher W. 44, 81, 82: *see* Gordon H. T. 43
Loi G. 725
Long D. B. 40, 81: *see* Zaher M. A. 40
Long J. S. *see* Fletcher L. W. 83
Long S. H. *see* Hoffman J. D. 312
Longfield C. *see* Corbet P. S. 487, 490
Loor K. A. 314
Lopes H. De Souza 56, 57
Lopez A. W. 223
Lopez E. *see* Davis R. D. 86
Lopez I. G. *see* Halffter G. 276
Lopez Y. G. *see* Halffter G. 688
Lord F. T. 213
Loschiavo S. R. 34, 83
Löser S. 272, 649
Loughton B. C. 212
Lovitt A. E. 214
Lowe M. L. *see* Leahy M. G. 45, 47
Lowe R. E. 316: *see* Smittle B. J. 316
Lower H. F. 175
Lozinskiĭ V. A. 35
Lu L. C. *see* McDonald J. L. 85
Lubbock J. 3
Luca Y. de 83
Lucas F. 645
Lucas W. J. 270
Luck R. F. 313
Luckmann W. H. *see* Shaw J. T. 764; Singh Z. 595
Ludvik G. F. 312
Ludwig D. 175, 177, 179
Luff M. L. 688
Luhmann M. 303
Lukasiak J. 729, 731
Luke B. M. *see* Neville A. C. 202, 209
Lukefahr M. J. 87, 89
Lum P. T. M. 43, 79, 85: *see* Horsfall W. R. 733
Lumaret J. P. 83
Lumsden W. H. R. 729
Lundbeck W. 580
Lundie A. E. 689
Luong H. *see* LeBerre J. R. 82

Lupton F. G. H.   *see* Bingham J. 87
Lutze G.   *see* Wetzel T. 313
Lydin L. V.   *see* Hathaway D. O. 312
Lyle G. T.   642
Lyle P. T. W.   *see* Service M. W. 212
Lyman F. E.   483
Lynn D. C.   764
Lyon R. L.   315
Lyonet P.   656
Lyons G. R. L.   *see* Lewis D. J. 734, 735

McCabe J. M.   *see* Rose R. I. 315
McClendon J. F.   647
McClure H. E.   271, 294
McCluskey E. S.   82
McCluskey R. K.   *see* Yamamoto R. T. 78, 85
McCulley S. B.   *see* Eastham L. E. S. 84
McDaniel I. N.   85: *see* Bentley M. D. 85
McDonald J. L.   85
McDuffie W. C.   *see* King W. V. 732
McEwen F. L.   *see* Nair K. S. 86, 762
McFarlane J. E.   165, 175, 179, 513: *see* Furneaux P. J. S. 175, 179
MacFarlane W. V.   557
McFeely J.   85
McGaughey W. H.   75
McGovern T. P.   *see* Jones R. L. 85
McGovern W. L.   314
MacGregor E. A.   642: *see* Woglum R. S. 642
McGregor E. A.   *see* Simmons P. 19
McGregor M. D.   *see* Furniss M. M. 70
MacGregor M. E.   732
McGuffin W. C.   714
McKenzie H. L.   216, 219, 675
McKenzie L. M.   566
McKittrick F. A.   497
McKnight M. E.   84
McLachlan R.   216, 711
MacLellan C. R.   214, 215: *see* Hall R. R. 212
McLeod J. M.   84: *see* Tostowaryk W. 313
McLeod W. S.   *see* Moreland C. R. 312
McLintock J.   86, 251: *see* Hudson A. 85
MacLoughlin M.   *see* Carrel J. E.   251
McMahon E.   179
McMahon H.   *see* Arrand J. C. 613, 764
McMeans J. L.   *see* Lloyd E. P. 84
McMillian W. W.   313: *see* Wiseman B. R. 313
McMullen L. H.   279, 764: *see* Walters J. 279

McMullen R. D.   43
McMurtry J. A.   83
McNeil J. N.   721
McNew G. L.   69
MacPhee A. W.   214, 215
McPherson J. E.   595: *see* Walt J. F. 764
McRae T. M.   *see* Rice M. J. 763
MacSwain J. W.   *see* Linsley E. G. 224, 232, 233, 763
McWilliams J. M.   313
Macdonald W. W.   52, 732
Macdougall R. S.   751
Machado A. B. M.   491
Mackay A. L.   *see* Furneaux P. J. S. 205, 209, 210
Mackerras I. M.   86, 738
Mackerras M. J.   28: *see* Mackerras I. M. 86
Madden J. L.   73, 74, 84
Madge D. S.   *see* Buxton J. H. 270, 287
Madge P. E.   720
Madhavan M. M.   175
Madhukar B. V. R.   *see* Pillai M. K. K. 75
Madrid F.   71
Madsen H. F.   *see* Gehring R. D. 85; Goodwin J. A. 85; Husseiny M. M. 31
Maelzer D. A.   83
Maercks H.   178
Magalhães Bastos J. A.   83
Magalhães P. S. de   272
Mahany P. G. A.   *see* Boles H. P. 314
Maharaj S.   312
Mahdihassan S.   236
Mahto Y.   *see* Bhatia S. K. 742
Maibach H. I.   *see* Khan A. A. 79
Maillard Y.-P.   186, 187, 656
Main H.   269, 289
Mainardi D.   85
Mainardi M.   64: *see* Mainardi D. 85
Makarenko G. N.   *see* Shapiro V. A. 314
Maki T.   86: *see* Okada J. 23
Makings P.   85
Makky A. M. M.   *see* Hafez M. 83
Malhotra C. P.   589
Malicky H.   192, 193
Mallack J.   733
Mallea A. R.   25
Malone R. S.   *see* McMurtry J. A. 83
Maltby H. L.   90
Mamet R.   *see* Moutia L. A. 17, 214
Mammen M. L.   *see* Wattal B. L. 86

Mandaron P.   85
Mangum C. L.   *see* Bullock H. R. 315; Graham H. M. 315
Mani E.   80
Mani M. S.   270
Mank H. G.   659
Mannheims B. J.   86
Manning A.   46
Mansingh A.   764
Mansour M. H.   86, 761: *see* Dimetry N. Z. 83
Maple J. D.   22, 30, 144
Maramorosch K.   *see* Mitsuhashi J. 315
Marchand W.   738
Mariau D.   *see* Hurpin B. 34; Morin J. P. 41
Markin G. P.   234
Markkula M.   21, 59, 60, 84
Markovetz A. J.   *see* Brand J. M. 66
Marliér G.   711, 735
Marlin J. C.   *see* Webb D. W. 722
Marques L. A. de A.   274
Marr J. D. M.   734: *see* Lewis D. J. 734
Marro J. P.   *see* Benois A. 235
Marshall J. F.   729, 730
Marshall W. S.   199
Marston N.   312
Martelli G.   223: *see* Silvestri F. 220
Martelli G. M.   80
Martin D. F.   *see* Brazzel J. R. 84, 88; Burke H. B. 16; Lukefahr M. J. 87, 89; Richmond C. A. 312, 315
Martin J. H.   *see* Buchner G. D. 534
Martin J. O.   577
Martin J. T.   617
Martínez-Beríngola M. L.   33
Martini E.   256, 258: *see* Hackett L. W. 729, 730
Martinko E. A.   *see* Bell W. J. 71
Martinovich V.   33
Martyn E. J.   37, 235, 764
Marucci P. E.   224
Masaki J.   33
Mashhood Alam S.   84
Masi L.   *see* Silvestri F. 220
Masirg R. A.   86
Maskell F. E.   *see* Kempton R. A. 86
Maskell W. M.   738
Masutti L.   *see* Zangheri S. 76
Mathad S. B.   *see* Bentur J. S. 46
Matheny E. L.   714

Matheson R.   220, 654, 730
Mathew K. P.   *see* Das N. M. 235
Mathis M.   27
Mathos R.   *see* Ingram J. W. 214
Mathur C. B.   175
Mathur R. N.   502
Mathur Y. K.   *see* Verma J. P. 231
Matsuda R.   3: *see* Usinger R. L. 582
Matsumoto Y.   84, 86
Matsuo K.   761
Matter J. J.   *see* Blume R. R. 79; Kunz S. E. 86
Matteson J. W.   313: *see* Smith C. N. 312
Matthée J. J.   169, 175, 177, 519
Matthewman W. G.   86
Matthews E. G.   273: *see* Halffter G. 273, 275, 277, 302
Matthysse J. G.   551
Mattingly P. F.   728, 761
Mau R. F. L.   *see* Mitchell W. C. 56, 74
Maudlin J. K.   *see* Williams L. H. 83
Maulik S.   191, 677
Maw M. G.   85, 91: *see* Coppel H. C. 74
Max M. G.   91, 763
Maxwell F. G.   71, 88: *see* Bailey J. C. 83; Parrott W. L. 88
May C. J.   *see* Butler G. D. 229
Maybee G. E.   *see* Wishart G. 216, 218
Mayer K.   736
Mayer M. S.   79: *see* Hazard E. L. 85
Mayes R.   *see* Rolston L. H. 191
Maynard-Smith J.   763
Mayo Z. B.   569
Mayyasi A. M.   *see* Coulson R. N. 70
Mazzini M.   643, 645, 648, 716, 718, 763, 764
Mead C. G.   *see* Fox A. S. 45
Mead-Briggs A. R.   43
Meats A.   175, 179
Mecznikov E.   582, 620
Medem F.   547
Meek C. L.   85
Meelis E.   *see* Bakker K. 313
Meer Mohr J. C. van der   689
Megahed M. M.   312: *see* Azab A. K. 569
Megušar F.   656
Mehra B. P.   16
Mehta R. C.   88
Meijere J. C. H. de   56
Meinert F.   656
Meisner J.   84

Meissner G. 718
Melamed-Madjar V. 20, 61
Meldola R. 712
Melis A. 51
Mellini E. 84, 221, 755
Melnikov N. 553
Mel'nikova N. I. 279
Melville A. R. 234
Mende F. see Wetzel T. 86, 313
Mendes Ferreira A. 213
Meng H. L. 80
Menke A. S. see Lauck D. 58
Menon K. P. V. see Nirula K. K. 17
Menon M. A. U. 729
Menon P. B. see Basu B. C. 738
Mensink F. T. see Dinther J. B. M. van 212, 214, 217, 218
Menzer R. E. see Walker W. F. 206
Mer G. 730
Mercer E. H. 202
Merino M. G. 49
Merkl M. E. see Lloyd E. P. 84
Merle J. 45, 47
Merle P. du 86, 235
Merrill J. C. 11
Merton L. F. H. 219, 220
Mertz D. B. 225: see Park T. 225
Merwe C. P. van der 175
Metcalf C. L. 239, 741
Meyer D. 81
Meyer J. R. see Lukefahr M. J. 87
Meyer N. F. 21
Meyer O. E. 83
Meyners H. H. see Crystal M. M. 44
Miall L. C. 4, 184, 656, 677, 724, 735
Michael R. R. 69: see Rudinsky J. A. 67, 69
Michener C. D. 226, 310, 729
Mickoleit G. 763
Middleton M. I. 764
Miger F. 656
Migunda J. see Austara O. 721
Mijušković M. 763
Mika G. 44
Miles M. 28
Miles P. W. 89
Miles R. C. see Horsfall W. R. 729
Miller A. 509, 510
Miller B. S. 88
Miller D. see Ringold G. B. 70
Miller G. W. 764

Miller J. A. see Blume R. R. 79
Miller M. C. 233
Miller N. C. E. 585, 594, 609, 614, 764
Miller R. L. 83
Miller R. M. 762
Miller R. S. 39
Miller T. A. 761
Millot J. 738, 739
Mills R. B. see Qureshi Z. A. 212
Mills R. D. see David M. H. 83
Mills R. O. 257
Mills R. R. 312: see Lake C. R. 209; Taylor N. 209
Milne A. 63, 83, 763
Milne L. J. 273
Milne M. see Milne L. J. 273
Milum V. G. see Wilson H. F. 11
Milyanovskiĭ E. S. 51
Mimeur J. M. see Lépiney J. de 739
Minder I. F. 24
Mineo G. 85, 312
Minis D. H. 81
Minks A. K. see Noordink J. P. W. 212
Miskimen G. W. 81, 238
Misra B. C. 83
Misra C. S. 236
Misra M. P. 714
Missiroli A. see Hackett L. W. 729, 730
Mistikawi A. M. see Ballard E. 219
Mitchell E. R. see Carlyle S. L. 312
Mitchell H. C. 49, 80: see Adams C. H. 21
Mitchell J. A. see Gross H. R. 314
Mitchell R. 83
Mitchell S. 312: see Smith C. N. 312
Mitchell W. C. 56, 74
Mitić-Mužina N. 28
Mitlin N. see Hines B. M. 212
Mitrofanov P. I. see Milyanovskiĭ E. S. 51
Mitrokhin V. U. 86
Mitsuhashi H. 312
Mitsuhashi J. 312, 315
Miura T. 85, 313
Miyagawa H. see Ishii S. 89
Miyagawa M. see Otsuru M. 729
Miyahara Y. 721
Miyamoto S. 579: see Esaki T. 641
Mizuta K. 40
Mjöberg E. 551
Mkrtumyan K. L. see Azaryan G. Kh. 31
Mochida O. 83

Modéer A.   271
Modi B. N.   see Rawat R. R. 189
Moeck H. A.   70
Moen D. J.   see Riemann J. G. 46, 47
Mohammad Ali S.   15
Mohan M. S.   see Jacob S. A. 232
Mohyuddin A. I.   84: see Baloch G. M. 24
Moiseeva T. S.   92
Moloo S. K.   175
Monadjemi N.   84
Monro H. A. U.   see Bond E. J. 312
Monro J.   31
Monroe R. E.   see Ross R. H. 315
Montana A. C.   see Reyes P. V. 636
Monteiro L.   see Lopes H. De Souza 57
Monteith E.   see Wishart G. 707
Monteith L. G.   74, 86
Montenegro M. J.   see Sakagami S. F. 84
Montgomery B. E.   82: see Zehring C. S. 491
Mookherjee P. B.   see Tuli S. 60
Moore D. V.   see Ubelaker J. E. 562
Moore H.   82
Moore H. B.   178
Moore I.   24
Moore K. M.   568
Moore N. W.   82, 490: see Corbet P. S. 487, 490
Moore S. T.   316
Moorhouse D. E.   see Colbo M. H. 313
Morais H. C. de   see Benson W. W. 83
Morales A. E.   502
Moran V. C.   15, 83, 709, 764
Moreland C. R.   312
Moretti G. P.   711
Morgan A. H.   476, 477
Morgan C. E.   see Wingo C. W. 230, 231
Morgan F. D.   79
Morgan H. A.   see Dupree J. W. 733
Morgan M. E.   see Libbey L. M. 70; Rudinsky J. A. 67, 70, 689
Morgan N. C.   711
Morgan N. O.   see Pickens L. G. 313
Mori H.   166, 167, 175
Mori K.   see Hedden R. 66
Moriarty F.   170, 173, 175
Morin J. P.   41
Moriya K.   see Harada F. 85
Moriyama T.   715
Morlan H. B.   see Smith C. N. 312

Morris C. L.   see Copony J. A. 70
Morris H. M.   733
Morris R. F.   721, 764
Morris R. J. H.   see MacFarlane W. V. 557
Morrison F. O.   see Richards P. G. 751
Morrison R. D.   see Criswell J. T. 689
Morrison R. K.   764
Morrison W. P.   179: see Crawford C. S. 85
Morrow J. A.   735
Mortenson E. W.   315, 728
Morton K. J.   711
Moscona A.   534, 536, 537, 539, 542
Mosebach-Pukowski E.   298
Moser J. C.   227, 763, 764: see Hunter P. E. 227; Woodring J. P. 227
Mosna B.   see Filipponi A. 227
Mote D. C.   751: see Edwards W. D. 235
Mouchet J.   see Germain M. 735; Grenier P. 735
Mourier H.   86, 751
Moussa M. A.   see Zaher M. A. 40
Moutia L. A.   17, 214
Mouzels P.   see Leger M. 258
Mrkva R.   37
Mueller J. F.   550
Muesebeck C. F. W.   21
Muir F.   189, 190
Muirhead Thomson R. C.   729
Mukerji M. K.   72
Mukharji S. P.   see Lal L. 227
Mukherjee A. B.   229: see Choudhury A. K. S. 227; Somchoudhury A. K. 227
Muldrew J. A.   91
Mulkern G. B.   see Onsager J. A. 522, 523
Mulla M. S.   86: see Ikeshoji T. 85
Mullen G. R.   763
Müller A.   491
Müller F.   711
Müller H. J.   169
Müller J.   86
Müller K.   718
Müller O.   24
Muma M. H.   227
Munchberg P.   491
Mundy A. T.   735
Mungomery R. W.   224
Muñiz M.   see Robles-Chillida E. M. 742
Munsterhjelm G.   735, 736
Munyon I. L.   see Fox A. S. 45
Murai S.   218

Murbach R. 313
Murphy D. H. 727
Murphy H. E. 476, 711
Murphy P. W. 313
Murray A. 244
Murray M. D. 261, 547, 549, 551, 556, 557, 558, 559, 561, 562: see Norris K. R. 28
Murtfeldt M. E. 271
Muschinek G. 689
Muspratt J. 729
Muth H. see Barth R. 617
Muzafarov S. S. see Saakyan-Baranova A. A. 233, 764
Myers C. M. 729
Myers J. G. 641
Myers K. 743
Myllymäki S. see Markkula M. 21, 59, 84

Nabrotzky F. V. 735
Nagashima Y. see Otsura M. 761
Nagatomi A. 738, 744
Nagel W. P. see Fitzgerald T. D. 71
Nagui A. see Tawfik M. F. S. 215
Naguib M. A. see Abul-Nasr S. 85
Nair K.-S. S. 86, 762
Nair M. R. G. K. see Das N. M. 235, 566; Oommen C. N. 21
Naito A. 85
Nakamura H. A. 41, 83
Nakamura Y. see Otsura M. 761; Yoshimura S. 85
Nakanishi A. see Shiga M. 94
Nakasawa M. see Ishii S. 89
Nakashima K. see Yoshimura S. 85
Nakashima M. 85
Nakashima T. 83
Nakashima Y. 764
Nakasuji F. see Hokyo N. 94; Ito Y. 212
Nakata G. 75
Nam S.-H. see Choi Y. H. 721
Napompeth B. see Nishida T. 85
Nappi A. J. 92, 762
Narayanan E. S. 84, 94
Narayandas M. G. 729
Nasonova L. I. 24
Nasr E. S. A. 85
Nassif F. see Nasr E. S. A. 85
Nathanson M. see Park T. 225
Natskova V. 26

Nault L. R. 270, 271
Navarro S. see Calderon M. 689
Navon A. see Moore I. 24
Nawrot J. 83
Neander A. 686
Nebeker A. V. see Knight A. W. 510
Neboiss A. 192: see Burns A. N. 508
Nechols J. R. 84, 764
Needham J. G. 13, 476, 477, 486, 489, 490, 491, 636, 642
Neel W. W. see Solomon J. D. 84
Neff S. E. 28, 743, 744, 756: see Foote B. A. 744; Knutson L. V. 744
Negi P. S. see Glover P. M. 236
Neilson W. T. A. see Wood G. W. 25
Neiswander R. B. 713
Nelson B. C. 224
Nelson D. R. 46, 47: see Adams T. S. 46, 47; Ercelik T. M. 764
Nelson E. L. see O'Donnell A. E. 213, 227
Nemec S. J. 79
Nerya A. Ben see Bodenheimer F. S. 11
Nettles W. C. 315
Neubecker F. 312
Neuffer G. 313
Neumann F. G. 764
Neumann K. W. 226
Neunzig H. H. see Baker J. R. 642
Nevill E. M. 213
Neville A. C. 201, 202, 206, 209, 490: see Smith D. S. 205, 206, 209
New T. R. 546, 547, 647
Newell I. M. 742
Newkirk M. R. 722, 729
Newstead H. 732
Nicholls D. G. see Murray M. D. 562
Nicolas H. U. 686
Nielsen A. 23, 192, 711
Nielsen B. R. see Hemmingsen A. M. 80, 738
Nielson J. C. 752
Nielson M. W. 15
Niemczyk E. 15, 228
Niemczyk H. D. 84
Nieto Calderon J. see Arias Giralda A. 84
Nietzke G. 191
Nigam B. S. see Khalsa H. G. 77
Nigmann M. 714
Nijholt W. W. 67
Nijhout H. F. see Hausermann W. 49

Nijveldt W.   236: see Barnes H. F. 221
Nikol'skii V. V.   224, 763
Ninomiya E.   741
Nirula K. K.   17
Nishida T.   73, 84, 85
Nishigaki J.   688
Nishijima Y.   85
Nishikawa Y.   502
Nitsche H.   257
Nixon G. E. J.   234
Nogami T.   see Nakashima M. 85
Noll J.   33
Nonveiller G.   218, 219
Noordink J. P. W.   212
Nord J. C.   75, 83: see Grimble D. G. 83
Nordlund D. A.   313, 709
Norrevang A.   see Hemmingsen A. M. 85, 724
Norris D. M.   305: see Abrahamson L. P. 273; Gilbert B. L. 72, 87; Kingsolver J. G. 764
Norris K. R.   28
Norris M. J.   14, 18, 71, 82, 83: see Waloff N. 35, 37
North D. T.   see Holt G. G. 47; Karpenko C. P. 85
Novák K.   192: see Hrdý J. 221
Novak V.   see Rudinsky J. A. 66
Nozato K.   59
Nunome J.   718
Nuorteva P.   230
Nuttall G. H. F.   see Keilin D. 562
Nutting W. L.   50, 284
Nuzzaci G.   220
Nwanze K. F.   83
Nyiira Z. M.   272, 297

Oakley J. N.   762
Oatman E. R.   84, 720: see Beegle C. C. 709; Cardona C. 54; Greany P. D. 64, 84; Leong J. K. L. 21; Odebiyi J. A. 764
O'Connor B. A.   51
Odebiyi J. A.   764
Odell T. M.   315: see Godwin P. A. 689
Odhiambo T. R.   43, 297
O'Donnell A. E.   213, 227
Oester P. T.   66
O'Farrell A. F.   490
O'Gower A. K.   85

O'Grady J. J.   see Davis R. B. 314
Ogura K.   see Kikuchi T. 66
O'Halloran T. J.   see Cummings M. R. 762
Ohaus F.   273, 303
Ohi S.   see Okabe T. 558
Ohmori Y.   see Otsura M. 729, 730
Ohno M.   191
Ojima N.   see Kawasaki H. 721
Ojima Y.   see Kaplan W. D. 312
Okabe T.   558
Okada J.   23
Okasha A. Y. K.   35
Okelo O.   83
Oksanen H.   71: see Kangas E. 67, 70, 71; Perttunen V. 71; Selander J. 67
Oksenov V.   278
Oku T.   85
Okumura G. T.   25
Okuni T.   19
Okuno T.   224, 762
Oldberg G.   310
Oldiges H.   32
Oldroyd H.   223, 737, 740
Oliveria F. L.   see Coulson R. N. 70
O'Loughlin G. T.   742
Oloumi-Sadeghi H.   see Showers W. B. 313
Olson J. K.   see Meek C. L. 85
Olson W. H.   see Batiste W. C. 314
O'Meara G. F.   764
Omer S. M.   79
Ong J.   see Kramer K. J. 204
Ongaro D.   181
Oniki Y.   see Sakagami S. F. 84
Ono T.   85
Onsager J. A.   522, 523
Onyearu A. K.   see Waterhouse F. L. 83
Oommen C. N.   21
Oppong-Mensah D.   763
Örösi-Pál Z.   744
Orozco F.   76, 83: see Barrera A. 83; Fuentes M. del C. 61
Orphanides G. M.   215, 218, 220
Orphanidis P. S.   41
Orth R. E.   see Fisher T. W. 744
Osburn M.   see Tedders W. L. 763
Osgood C. E.   63, 312: see Starratt A. N. 63
Osman F. H.   see Hafez M. 49, 83, 684
Osmani M. H.   see Quraishi M. S. 733
O'Sullivan P. J.   see Roberts F. H. S. 44
Ota A. K.   see Chang V. C. S. 84

Otsura M.   729, 730, 761
Ottaviani L.   *see* Mainardi D. 85
Otvos I. S.   313, 764
Ouchi M.   595
Ouchi Y.   *see* Tanaka A. 85
Oudemans A. C.   562
Overmeer W. P. J.   568
Owusu-Manu E.   764
Oyama M.   *see* Yushima T. 67

Packard A. S.   7
Padgett G. R.   *see* Gross H. R. 314
Pag H.   72
Paganelli C. V.   *see* Rahn R. 107
Pagayon A. U.   *see* Baisas F. E. 729
Pagden H.   233
Page T. F.   *see* Kinzer G. W. 67
Pagenstecher A.   506
Paillot A.   91, 92, 216
Painta F.   278
Painter R. H.   25, 87: *see* Winburn T. F. 216
Painter R. R.   762
Pajni H. R.   *see* Arora G. L. 83, 683
Palmen J. A.   477
Palmer D. F.   683
Palmer M.   709
Palomino H.   *see* Solar E. del 41
Pande Y. D.   568
Pandey R. C.   *see* Teotia T. P. S. 78
Pandey Y. D.   *see* Srivastava R. P. 36
Panis A.   15, 236
Panizzi A. R.   764
Pankanin M.   *see* Sandner H. 83
Pantel J.   536, 752, 753, 754, 763
Pantelouris E. M.   33
Pantyukhov G. A.   219, 224
Paoli G.   222, 223, 232, 581
Papillon M.   520
Papp C. S.   675
Paradis R. O.   88, 313
Paramanov S.   223
Parfin S. I.   648
Parihar D. R.   175
Pariser K.   16
Park J. S.   *see* Choi S. Y. 721
Park T.   225: *see* Birch L. C. 224; Prus T. 225
Parker A. H.   272, 297, 736
Parker D. L.   *see* Muesebeck C. F. W. 21
Parker F. D.   92

Parker G. A.   86
Parker H. L.   22, 234
Parker J. B.   233
Parker J. R.   216
Parker K. D.   202, 645
Parkin E. A.   18
Parks H. B.   *see* Hollinger A. H. 221
Parks J. J.   763
Parnell F. R.   88
Parrott W. L.   88: *see* Maxwell F. G. 71
Parsons C.   *see* Bell W. J. 71
Parsons J. A.   *see* Euw J. von 251; Jones D. A. 251; Reichstein T. 251
Paschke J. D.   315: *see* Anderson R. C. 238
Pasquali A.   *see* Mainardi D. 85
Pass B. C.   71, 84, 313: *see* Bridges E. T. 566; Morrison W. P. 179
Passera L.   226
Patana R.   *see* Jackson C. G. 28
Pate R. R.   *see* Yonce C. E. 84
Patel J. C.   *see* Patel R. C. 312
Patel J. K.   *see* Patel R. C. 312
Patel K. K.   83, 677
Patel R. C.   16, 312
Patel R. M.   80
Paterson N. F.   681, 682
Pathak M. D.   *see* Cheng C. H. 566; Djamin A. 88
Patsakes P. G.   *see* Orphanidis P. S. 41
Patten W.   711
Patterson H. E.   729, 731
Patterson N. A.   *see* Pickett A. D. 763
Patterson R. S.   *see* Smittle B. J. 316
Patton R. L.   *see* Wood T. K. 271
Pau R. N.   201, 202, 203: *see* Neville A. C. 201
Paul C. F.   178
Paul M.   210
Paul P. K.   *see* Ghose S. K. 570
Paulian R.   659
Pavan M.   83, 687, 688
Pawan J. L.   729
Pawar A. D.   228
Pawlowsky E. N.   256, 257
Payne E.   *see* Ubelaker J. E. 562
Payne E. O.   *see* Radcliffe J. E. 83
Payne J. A.   *see* Yonce C. E. 84
Payne N. M.   177, 179
Payne T. L.   70: *see* Camors F. B. 70; Coulson R. N. 70; Dickens J. C. 70

Peacock J. W.   70: see Pearce G. T. 70
Peake F. G. G.   684
Pearce G. T.   70: see Gore W. E. 70; Peacock J. W. 70
Pearl R.   41
Pearman J. V.   547, 548
Pedigo L. P.   87
Peet W. B.   229
Peleg B. A.   312, 316
Pelerents C.   212: see Vereecke A. 212
Pemberton C. E.   92
Peña G. L. E.   722
Pendergrast J. G.   591
Pener M. P.   175: see Perez Y. 82; Shulov A. 175
Pengelly D. H.   85
Pennington K. M.   764
Penny N. D.   see Webb D. W. 722
Percival E.   13, 476, 508, 509
Perez M. Q.   21
Perez Y.   82
Perkins P. V.   15
Perkins W. D.   see Gross H. R. 312; Harrell E. A. 312
Perrier E.   505
Perris E.   219, 686, 689
Perron J. P.   86: see Hudon M. 312
Perry A. S.   85
Persson B.   85
Perti S. L.   see Ayyappa P. K. 33; Paul C. F. 178
Perttunen V.   71: see Kangas E. 67, 70, 71; Oksanen H. 71; Selander J. 67
Pesce H.   256
Peschken D.   86
Peschken D. P.   see Riordan D. F. 316
Pessôa S. B.   284
Pessozkaja F. S.   270
Peter A. M.   see Buchner G. D. 534
Peterlík Z.   72
Peters D. C.   see Rowley W. A. 681
Peters L. L.   312
Peters T. M.   see Barrosa P. 315
Petersen B.   85
Petersen J. J.   75
Petersen W.   48
Peterson A.   240, 242, 249, 315, 714
Peterson B. V.   86, 734: see Davies D. M. 85, 734
Petrelli G.   see Filipponi A. 227

Petrusewicz K.   see Park T. 225
Pétsikou N. A.   see Orphanidis P. S. 41
Pettinger L. F.   see Furniss M. M. 70
Peus F.   725, 729
Peyerimhoff P. de   548
Peyron J.   715
Pfaffenberger G. S.   684, 685
Philippe R.   645
Philips F. M.   see Burger T. L. 313
Phillips J. R.   80
Phillips R. H.   see Lum P. T. M. 85
Phillips V.   see Gillett J. D. 761, 764
Phillips W. J.   see Webster F. M. 175
Phillips W. M.   83
Phipps J.   184, 502, 520
Picado C.   490
Pick F.   617, 727
Pickard L. S.   see Hodges J. D. 66; Moser J. C. 227, 764
Pickens L. G.   313
Pickens M. O.   see Blume R. R. 79
Pickett A. D.   763
Pickford R.   45, 47, 764
Pienkowski R. L.   224, 225: see LeCato G. L. 35, 61, 84; Woodside A. M. 84
Pierce F. N.   712
Pierrard G.   85
Pierre Abbé   486, 494
Pierre C.   725
Pijnacker L. P.   764
Pillai J. S.   728: see O'Connor B. A. 51
Pillai M. K. K.   75: see Adlakha V. 46
Ping C.   749
Piston J. J.   66
Pitman G. B.   65, 66, 67, 70, 71: see Kinzer G. W. 67; Knopf J. A. E. 70; Renwick J. A. A. 66, 70; Vité J. P. 66, 70
Pitre H. N.   see Boling J. C. 85; Hillhouse T. L. 85
Pitts C. W.   see Bay D. E. 86; Hooper R. L. 86; Nwanze K. F. 83
Plank H. K.   18
Platner G. R.   see Oatman E. R. 84
Plaut H. N.   22, 720
Pleskot G.   476
Pletsch J. D.   see Breeland S. G. 85
Plichet F.   220
Plough H. H.   see Young W. C. 35
Plugaru S. G.   26, 77
Plyater-Plokhotskaya V. N.   184

Po-Chedley D. S.   212
Podoler H.   *see* Avidov Z. 94
Podoplelov I. I.   *see* Soboleva-Dokuchaeva I. 212
Poisson R.   572, 576, 579, 582, 620, 636, 638
Pol B. C. van der   *see* Ankersmit G. W. 720
Polak R. A.   269
Polhemus J. T.   *see* Andersen N. M. 581
Polivanova E. N.   178
Pollard D. G.   83
Polnik A.   224, 225
Pomeroy A. W. J.   257
Pomonis J. G.   *see* Nelson D. R. 46, 47
Pond D. D.   *see* Sherwood R. C. 312
Ponnaiya B. W. X.   *see* Miller B. S. 88
Pontin A. J.   709
Poole A. F.   86
Poole H. K.   *see* Standifer L. N. 84
Poolson B. J.   *see* Harden F. W. 79
Pope P.   497
Popham E. J.   106
Popov G. B.   14, 83, 216, 217, 520: *see* Stower W. J. 83, 217, 222, 224
Popov P. A.   17, 20
Popova V.   25
Popovich A. P.   761
Portier G.   *see* Le Berre J. R. 215
Portier P.   656
Portman R.   *see* Swain R. B.   84
Post F. J.   *see* Kirk V. M. 83
Potgieter J. T.   223, 224
Potter C.   *see* Salkeld E. H. 181, 200
Potter E.   763
Potter S. A.   *see* Furneaux P. J. S. 165, 209, 513
Poujade G. A.   656
Poulton E. B.   83, 257
Poutiers R.   234
Pouzat J.   689
Powell J. A.   709: *see* Chemsak J. A. 83, 686
Powning R. F.   209
Poyarkoff E.   475
Prasad S. K.   73
Prasad S. N.   *see* Grover P. 221, 763
Pratt H. D.   314
Pratt R. W.   *see* Davis R. B. 86
Predtechenskii S. A.   217
Preece W. H. A.   *see* Hardy G. A. 20
Prell H.   278, 764

Prentice J. H.   *see* Vincent J. F. V. 283
Prescott H. W.   221, 222: *see* York G. T. 222
Press J. W.   228
Prevett P. F.   21, 677, 683, 684
Price P. W.   65
Prigge M.   689
Prince G. J.   709
Prine J. E.   *see* Mortenson E. W. 315
Pringle G.   730
Priore R.   83, 569
Pritchard G.   50, 86, 742
Prokop'ev V. N.   *see* Kunitskaya N. T. 764
Prokopy R. J.   65, 80, 86, 312, 742, 762: *see* Boller E. F. 742
Proper A. B.   22
Proske H. O.   *see* Brooke M. M. 212
Protensko A. I.   217, 219, 220
Proverbs M. D.   314
Provine R. R.   285: *see* Gross H. R. 312; Harrell E. A. 312
Pruess K. P.   60, 689
Prus T.   225: *see* Park T. 225
Pruszyński S.   21
Pruthi H. S.   284
Pryor M. G. M.   201
Przibram H.   503
Pschorn-Walcher H.   *see* Delucchi V. 223
Puchov V. G.   582
Puchova L. V.   582, 584, 585, 587, 590, 591, 596: *see* Puchov V. G. 582
Pukowski E.   49, 273, 298
Putman P.   *see* Shannon R. C. 733
Putman W. L.   213, 228, 229
Putnam L. G.   21
Putnam T. B.   *see* Libbey L. M. 70; Rudinsky J. A. 67, 689
Puttarudriah M.   224: *see* Veeresh G. K. 26
Puttler B.   91, 92: *see* Thewke S. E. 313
Pyenson L.   179

Quarterman K. D.   *see* Breeland S. G. 85
Quayle H. J.   742
Quednau F. W.   33, 42, 75, 84
Quednau W.   22, 55, 73, 316
Quintana V.   *see* Walker D. W. 721
Quintana-Muñiz V.   85
Quo F.   *see* Chin Chun-teh 175
Quraishi M. S.   733
Qureshi A. H.   83

Qureshi Z. A.   212

Raatikainen M.   83
Rabb R. L.   64, 73: *see* Elsey K. D. 28; Johnson M. W. 85; McNeil J. N. 721; Turnipseed S. G. 48, 178
Radcliffe J. E.   83
Radjabi G.   79
Rafai J.   *see* King P. E. 65
Rafferty R. A.   *see* Kugler O. E. 184
Rafiq Ahmad   233
Ragonot E. L.   236, 648
Ragusa S.   84
Rahalkar G. W.   316
Rahman M.   28, 85
Rahn R.   *see* Chauvin C. 107, 715
Raine J.   59
Raisbeck B.   67
Rait L.   508
Rajasekhara K.   610
Rajendram G. F.   312
Rajulu G. S.   209
Ram R. D.   *see* Chatterjee S. N. 316
Rammner W.   191
Ramos A. S.   *see* Corrêa R. R. 729
Ramponi G.   *see* Guerritore A. 184
Ramsay G. W.   763
Rankin K.   636
Rankin M. A.   81, 83
Rao A. M. S.   *see* Rao B. A. 729, 731
Rao B. A.   729, 731: *see* Sweet W. C. 729, 731
Rao S. N.   *see* Mani M. S. 270
Rao T. R.   *see* Russell P. F. 49
Rao V. P.   221
Rao Y. R.   218
Raros R. S.   312: *see* Buranday R. P. 85; Doria R. C. 83
Raske A. G.   227
Rasmussen D.   *see* Chiang H. C. 313
Rasmussen L. A.   70
Ratanov K. N.   523
Ratcliffe N. A.   *see* King P. E. 706, 707
Rattan L.   689
Rau N.   *see* Rau P. 502
Rau P.   82, 272, 502
Rauch F.   *see* Lhoste J. 720
Raulston J. R.   85
Rault B.   *see* Doby J.-M. 313

Raun E. S.   315: *see* Helms T. J. 314; Smith C. N. 312
Rawat B. L.   637
Rawat R. R.   189
Rawlins S. C.   *see* Mansingh A. 764
Rawlins W. A.   17, 86
Ray C.   31
Ray J. O.   *see* Everett T. R. 75
Ray R. M.   *see* Crocker R. L. 229
Raybould J. N.   *see* Wenk P. 734
Read D. C.   218
Read D. P.   84
Readio P. A.   617
Readshaw J. L.   27, 533
Ready P. A.   *see* Ward R. D. 761
Ready P. D.   *see* Killick-Kendrick R. 764
Réaumur R. A. F. de   129, 164
Redlinger L. M.   *see* Lewis W. J. 84
Redmond R. D.   *see* Landis B. J. 749
Reed D. K.   91
Reed G. L.   *see* Showers W. B. 313
Reed W.   33
Reed W. D.   *see* Simmons P. 19
Reeks W. A.   709
Rees D. M.   *see* Nabrotzky F. V. 735; Petersen J. J. 75; Winget R. N. 748
Regier J. C.   *see* Paul M. 210
Régnier P. R.   217
Reichardt H.   654
Reichart G.   17
Reichstein T.   251: *see* Euw J. von 251; Rothschild M. 256, 257, 258
Reid E. T.   *see* Lewis D. J. 734
Reid J. A.   729, 731
Reid J. C.   *see* Greene G. L. 85
Reid R. W.   60, 179, 213, 279
Reikhardt A. N.   216, 217
Reinert J. F.   729
Reinhard H. J.   26
Reisen W. K.   314
Remamony K. S.   *see* Das N. M. 235, 566
Remaudière G.   222, 224, 595
Remmert H.   735
Rempel J. G.   676: *see* Church N. S. 676; Fredeen F. M. 86, 734; Gerrity R. G. 676; Sweeney P. R. 676
Ren S. Z.   *see* Meng H. L. 80
Renaud G. D.   *see* Foster C. H. 86
Renganathan K.   *see* Rajulu G. S. 209
Renner M.   82

Rensing L.   81
Renwick J. A. A.   65, 66, 67, 69, 70, 763: see Madrid F. 71; Vité J. P. 65, 66, 69, 70
Repass R. P.   see Bellamy R. E. 729; Kartman L. 732
Retamal T.   see Etcheverry M. 25
Rethfeldt C.   676
Retnakaran A.   315
Rettenmeyer C. W.   56
Reuben R.   see Soman R. S. 85
Reyes A. V.   25
Reyes P. V.   636
Reynaud P.   see Lavoipierre M. M. J. 728
Reynolds F. H. K.   see Stone W. S. 732
Reynolds W. J.   see Claridge M. F. 566
Rhode R. H.   743: see Peleg B. A. 312
Rice L. A.   638
Rice M. J.   763
Rice R. E.   19, 71, 312
Rich E. R.   225
Richards J. G.   709: see King P. E. 707, 709
Richards O. W.   21, 25: see Waloff N. 681
Richards P. G.   751
Richardson C. D.   see Vanderzant E. S. 84
Richardson C. H.   763
Richardson E. G.   see Jacklin S. W. 48
Richmond C. A.   312, 315
Richmond E. A.   272, 656
Richter G.   61, 273
Richter H. C.   551
Richter P. O.   273
Ricker D. W.   see Goeden R. D. 215, 743
Riddiford L. M.   764
Riddle T. C.   see Greene G. L. 85
Ridgill B. J.   see Kinzer H. G. 279
Ridgway R. L.   see Lingren P. D. 229, 231; Morrison R. K. 764; Wilson D. D. 84
Riek E. F.   648
Riemann J. G.   46, 47, 721
Ries E.   562
Rigaud J.   see Brossut R. 71
Riley C. V.   16, 216, 217, 222, 223, 224, 515, 642, 650, 713, 717
Riley R. C.   762: see Gupta A. P. 681
Rimando L. C.   312
Rimes G. D.   571
Ringold G. B.   70
Rings R. W.   25
Riordan D. F.   316
Risbec J.   27, 217, 218

Riser G. R.   see Willis E. R. 13, 284
Ritter R.   see Grison P. 41
Rivera J.   729, 730
Rivière J. L.   86
Rizki T. M.   92
Robert P.   see Couturier A. 83
Robert P. A.   491
Roberts F. H. S.   44, 550
Roberts J. R.   see Arbogast R. T. 228
Roberts R. A.   502, 503
Roberts R. H.   79
Robertson F. W.   39, 40
Robertson I. A. D.   520: see Chapman R. F. 522, 523
Robertson J. R.   see Mertz D. B. 225
Robinson C. H.   60
Robinson D. M.   726
Robinson I.   745
Robinson R. J.   see Miller B. S. 88
Robinson S. H.   see Hendricks D. E. 315
Robinson W. H.   237
Robles-Chillida E. M.   742
Robredo F.   717
Rock G. C.   225
Rodin J. O.   see Bedard W. D. 66; Silverstein R. M. 66; Wood D. T. 66; Young J. C. 66, 67
Rodriguez J. G.   227: see Jalil M. 227; Wade F. 227; Wallwork J. H. 227; Wicht M. C. 227
Roelofs W. L.   see Cardé R. T. 67
Roepke W.   512
Roffey J.   218
Rogacheva T. V.   see Gromova A. A. 764
Rogers C. E.   224, 230
Rogers D.   84, 763
Rogers J. S.   724
Rohatgi K. K.   see Siddiqi Z. A. 25
Roivainen S.   see Markkula M. 60
Rolant F.   see Ziprkowski L. 257
Rollins L.   see Arthur A. P. 47
Rollinson W. D.   see Odell T. M. 315
Rolston L. H.   191
Roman E. A.   see Gillett J. D. 761, 764
Rommel E.   273, 303
Roonwal M. L.   175, 258, 520, 522, 764: see Husain M. A. 521
Root R. B.   see Read D. P. 84
Rosay B.   175, 727
Rose A. H.   83

Rose R. I.   315
Rosel A.   83
Rosen D.   *see* Avidov Z. 25
Rosen L.   *see* Rozeboom L. E. 85
Rosenberger J. H.   *see* Smith O. J. 28
Ross E. S.   512
Ross H. H.   497
Ross R. H.   315
Rossetto C. J.   689: *see* Carvalho R. P. L. 20
Rosskothen P.   278
Roth L. H.   541
Roth L. M.   44, 49, 50, 82, 84, 175, 205, 284, 285, 495, 496, 497, 502, 763: *see* Stay B. 201, 202, 203, 204; Willis E. R. 13, 284
Rotherham S.   706: *see* Grimstone A. V. 90
Rothschild G. H. L.   212, 214
Rothschild M.   256, 257, 258, 764: *see* Euw J. von 251; Jones D. A. 251; Kay D. 763; Reichstein T. 251
Roton L. M.   *see* Moser J. C. 227
Rotramel G. L.   *see* Mortenson E. W. 315
Roubaud E.   729, 730, 732
Rowley W. A.   681
Roy D. N.   729
Roy P.   86
Roy R.   *see* Gillon Y. 499, 503
Rozeboom L. E.   85, 729, 730, 731
Rozkošný R.   744
Ruano R. G.   *see* Barrera A. 83
Rubenstein A. D.   *see* Hill W. R. 256, 258
Rubtsov I. A.   735
Rubtsova N. N.   179
Rudall K. M.   183, 202: *see* Parker K. D. 202, 645
Rudinsky J. A.   66, 67, 69, 70, 71, 689: *see* Castek K. L. 70; Furniss M. M. 70; Harwood W. G. 71; Jantz O. 71; Kinzer G. W. 67; Libbey L. M. 70; Madrid F. 71; Michael R. R. 69, 70; Oester P. T. 66; Ryker L. C. 689; Schneider I. 70; Swaby J. A. 689
Rudnev D. F.   35
Rühm W.   64, 86, 734
Rummel R. W.   312, 313
Rupérez A.   84
Russ K.   312
Russell M. P.   88
Russell P. B.   *see* Pryor M. G. M. 201
Russell P. F.   49
Russev B.   476

Rust R. W.   709
Ruston D. F.   *see* Parnell F. R. 88
Ruud R. L.   *see* Riemann J. G. 721
Ruzaev K. S.   33
Růžička Z.   *see* Hrdý I. 221
Ryan J.   *see* Ryan M. F. 230
Ryan M. F.   230
Ryan R. B.   84
Ryckman A. E.   *see* Smith C. N. 312
Ryckman R. E.   *see* Smith C. N. 312
Rygg T.   86: *see* Sömme L. 86
Ryker L. C.   689: *see* Rudinsky J. A. 70

Saakyan-Baranova A. A.   233, 764
Saba F.   223
Sabrosky C. W.   237
Sacchi L.   *see* Grigolo A. 46
Saccuman G.   26
Safavi M.   *see* Remaudière G. 595
Safranyik L.   *see* Dyer E. D. A. 70
Sahni S. L.   *see* Sharan R. K. 585
Sahni V. M.   *see* Sikka S. M. 88
Sailer R. I.   34, 60
St. Quentin D.   486, 491
Saito K.   *see* Ikeshoji T. 85
Sakagami S. F.   84
Sakanoshita A.   85
Salaĭmanov Kh. A.   *see* Sinadskiĭ Yu. V. 566
Salama H. S.   85
Salavin R. G.   *see* Santoro de Crouzel I. 222
Salavin R. J.   86
Saliba L. J.   83
Saliternik Z.   729, 730
Salkeld E. H.   181, 200, 584, 617, 715, 721: *see* Church N. S. 676
Salpeter M. M.   *see* Wigglesworth V. B. 114, 751
Salt G.   64, 89, 90, 91, 93, 175: *see* Grimstone A. V. 90
Salzen E. A.   175
Samaniego A.   312
Samarawickrema W. A.   *see* Laurence B. R. 64
Sampaio M. M.   *see* Causey O. R. 729
Sams G. R.   *see* Bell W. J. 71
Sanchez F. F.   26
Sander K.   315
Sanders E. P.   *see* Cleveland L. R. 284
Sanders W.   86
Sandner H.   41, 83

Sands D. P. A.   *see* Spradbery J. P. 762
Sanford J. W.   763
Sanford K. H.   613: *see* MacPhee A. W. 214, 215
Sang J. H.   39: *see* Begg M. 315
Sankaran T.   15, 71
Santis L. de   713
Santoro F. H.   *see* Campodonico M. J. 522
Santoro de Crouzel I.   222
Sapunov V. B.   762
Sardesai J. B.   41: *see* Clarke K. U. 36
Sáringer G.   21
Sarlet L.   715
Sartwell C.   *see* Rudinsky J. A. 67
Sato H.   *see* Kawasaki H. 205, 206, 207, 208, 210, 721
Sauer D. B.   *see* David M. H. 83
Saunders D. S.   46
Saunders R. C.   22
Saunders S. S.   222
Saunt J. W.   725
Saupe R.   284
Saussure H. de   506
Savage K. E.   *see* Hazard E. I. 85
Savilov A. I.   582
Sawyer W. H.   *see* Steele C. W. 257
Saxena K. N.   *see* Mehta R. C. 88
Saxod R.   654
Sborshchikova M. P.   749
Scales A. L.   *see* Stadelbacher E. A. 85
Schaefer C. H.   279
Schaefer H. A.   568
Schaller F.   271
Schätz W.   85
Schedl K. E.   44, 273, 274
Scheidter F.   278
Scheltema R. S.   582
Schenk J. A.   21
Scherf H.   551, 562, 677, 681
Schieferdecker H.   41
Schilder F. A.   231
Schilder M.   *see* Schilder F. A. 231
Schiödte J. C.   656
Schlaeger D. A.   206: *see* Fuchs M. S. 761
Schlick W.   272, 656
Schlinger E. I.   84, 270
Schmid J. M.   86, 236
Schmidt C. T.   19
Schmidt G. T.   85
Schmiege D. C.   35

Schmitt J. B.   *see* Lin-Chow S. H. 85
Schmitz G.   83, 612
Schmitz H.   223
Schmitz R. F.   *see* Furniss M. M. 66, 70; Lanier G. N. 70; Rudinsky J. A. 67
Schneider E. L.   *see* Keiser I. 312
Schneider F.   30, 92
Schneider H.   546, 547
Schneider I.   70, 274
Schoenleber L. G.   *see* Hathaway D. O. 312
Schoeppner R. F.   *see* Whitsel R. H. 312
Schonherr J.   70
Schoof H. F.   *see* Jakob W. L. 314
Schoonhoven A. van   *see* Wilde G. 83
Schoonhoven L. M.   *see* Chun M. W. 85; Rothschild M. 764
Schorr H.   271, 589
Schoute E.   *see* De Buck A. 75, 729, 730, 731
Schoutenden H.   271
Schrader F.   569
Schremmer F.   684
Schroeder W. J.   85
Schubert W.   34
Schuch K.   178
Schuder D. L.   313
Schultz V. G. M.   225
Schumacher F.   271
Schurr K.   73, 85
Schuster M. F.   *see* Hormchan P. 763; Moore S. T. 316
Schütte F.   40
Schvester D.   21, 83
Schwardt H. H.   *see* Gyrisco G. G. 17, 83, 217
Schwarz E.   63
Scott A. W.   712
Scott E. I.   741
Scott J. R.   *see* Vinson S. B. 90
Scott M. T. S.   556
Scriven G. T.   *see* McMurtry J. A. 83
Scudder G. G. E.   585, 763: *see* Duffey S. S. 251; Southwood T. R. E. 764
Sechriest R. E.   20
Sechser B.   *see* Eichhorn O. 220
Secrest J. P.   *see* Tardif R. 312, 313
Seddiqi P. M.   83
Seeley D. C.   *see* Stahler N. 85
Séguy E.   51, 237, 562
Sehnal F.   *see* Novák K. 192
Sein F.   284

AUTHOR INDEX

Seitz A.   715
Sekeris C. E.   *see* Shaaya E. 204
Sekhon S. S.   *see* Slifer E. H. 169, 171, 175, 209
Sekido S.   764
Sekul A. A.   67
Selander J.   67
Selhime A. G.   219: *see* Beavers J. B. 227
Sellick G.   *see* Kettle D. S. 730
Senevet G.   729, 730
Sengalevich G.   *see* Dirimanov M. 24
Sengalewitsch G.   24
Sengel P.   44
Sen Gupta C. M.   *see* Basu B. C. 738
Sengupta G. C.   20
Sepulveda R.   272
Sergeeva T. K.   212
Sergent E.   729, 730
Serre P.   258
Service M. W.   85, 212, 313, 728: *see* Bruce-Chwatt L. J. 729; Hinton H. E. 729
Seshagiri Rao D.   25
Setty L. R.   722
Séverin H. C.   *see* Séverin H. H. P. 53, 178
Séverin H. H. P.   53, 178, 245, 534, 542
Sha Cha-yun   *see* Chin Chun-teh 175
Shaaya E.   204: *see* Bodenstein D. 204
Shade R. E.   83, 219
Shaffi M.   *see* Bhatia M. L. 741
Shagov E. M.   229
Shalaby F. F.   *see* Altahtawy M. M. 709
Shands W. A.   312, 313
Shannon R. C.   733
Shapinskiĭ D. V.   237, 522
Shapiro I. D.   58
Shapiro V. A.   28, 314, 709
Shapovalov A. A.   51
Sharan R. K.   585
Sharifi S.   25, 316
Sharma G. C.   *see* Tuli S. 60
Sharma M. L.   15
Sharma S. K.   *see* Verma J. P. 231
Sharma V. K.   85
Sharp D.   244: *see* Muir F. 189, 190
Shaver T. N.   *see* Hendricks D. E. 67
Shaw E.   284
Shaw J. G.   86
Shaw J. T.   764
Shaw J. T. B.   *see* Lucas F. 645
Shaw M. W.   86

Shaw Z. A.   *see* Gross H. R. 314
Sheldahl J. A.   *see* Strong F. E. 48, 60, 61
Shelford R.   272, 284, 297, 497, 499, 650: *see* Miall L. C. 724
Shelford V. E.   49
Shepard M.   270, 289: *see* Ables J. R. 84; Waddill V. 229
Shepherd R. F.   84, 313: *see* Gray T. G. 313
Sherman M.   84
Sherwell I. R.   *see* Browne L. B. 51
Sherwood R. C.   312
Shewell G. E.   *see* Soper R. S. 76
Shields K. S.   709
Shiga M.   94: *see* Hokyo N. 94
Shimizu J. T.   312
Shimizu K.   *see* Nakashima Y. 764
Shipp E.   533: *see* Haddlington P. 533
Shiraga M.   *see* Sumimoto M. 70
Sholdt L. L.   *see* Champlain R. A. 15
Shono T.   *see* Matsumoto Y. 84
Shorey H. H.   33, 78, 85: *see* Browne L. B. 71
Showers W. B.   313
Shri Ram   *see* Siddiqi Z. A. 25
Shriver D.   733
Shteĭnberg D. M.   91
Shukla G. N.   *see* Paul C. F. 178
Shulov A.   175: *see* Bodenheimer F. S. 175; Pener M. P. 175
Shute P. G.   27
Siddiqi Z. A.   25
Siddons L. B.   *see* Roy D. N. 729
Siebold C. T. E. von   485
Sijazov M.   273
Sikes E. K.   555
Sikka S. M.   88
Sikora H.   547
Silfvenius A. J.   192, 710
Siltala A. J.   192
Silva E. P. de C. E.   *see* Juarez E. 316
Silva G. de M.   312
Silva W. J. da   *see* Rossetto C. J. 689
Silver G. T.   313
Silverstein R. M.   66, 70: *see* Bedard W. D. 66; Byrne K. J. 70; Gore W. E. 70; Peacock J. W. 70; Pearce G. T. 70; Wood D. L. 66; Young J. C. 66, 67
Silvestre R. de S.   *see* Grison P. 257
Silvestri F.   92, 144, 220, 233, 235, 507
Simanton F. L.   221

Simeone J. B.   *see* Elliott E. W. 70; Pearce G. T. 70
Simmonds F. J.   64, 219
Simmonds H. W.   230
Simmons G. A.   *see* Leonard D. E. 313
Simmons P.   19, 689
Simpson G. W.   *see* Shands W. A. 312, 313
Sinadskiĭ Yu. V.   566
Singh B.   14
Singh H. N.   85
Singh I.   *see* Arora G. L. 502
Singh J.   *see* Lal L. 227
Singh K.   84
Singh K. R. P.   315
Singh O. P.   743
Singh P.   *see* Rodriguez J. G. 227
Singh R. P.   315
Singh S. R.   43: *see* O'Connor B. A. 51
Singh V. S.   *see* Teotia T. P. S. 78, 83
Singh Z.   595
Sinha P. B.   *see* Bose K. C. 594
Sinha R. N.   *see* Loschiavo S. R. 83
Sisojevic P.   *see* Embree D. G. 28
Sisson V.   *see* Chiang H. C. 313
Siverly R. E.   213
Skaff V.   *see* Spielman A. 45
Skinner G.   *see* Finch S. 86
Skuf'in K. V.   86
Slater C. E.   *see* Borden J. H. 70
Slater F. W.   635
Slater J. A.   584: *see* Drake C. J. 614
Slifer E. H.   164, 166, 169, 171, 173, 175, 209, 520
Slobodchikoff C. N.   84
Slooff R.   *see* Wilde J. de 72, 87
Smee C.   217
Smereka E. P.   20, 48, 221
Smirnoff W.   216
Smirnov E. S.   *see* Valdimorova M. A. 39
Smirnova I. M.   *see* Gromovaya E. F. 51
Smit B.   234
Smit C. J. B.   83
Smith B. C.   83, 689: *see* Boldȳrev M. I. 63, 83; Coppel H. C. 28
Smith C.   *see* Miller M. C. 233
Smith C. N.   312: *see* King W. V. 732
Smith D. S.   39, 83, 205, 206, 209, 228
Smith E. H.   24
Smith F. F.   314: *see* Webb R. E. 314
Smith G. J. C.   84

Smith H. S.   220, 234
Smith J. G.   *see* Panizzi A. R. 764
Smith J. W.   *see* Barfield C. S. 764
Smith K. G. V.   57, 741: *see* Hobby B. M. 740
Smith L. B.   689
Smith L. M.   220, 279
Smith L. W.   85: *see* Mallock J. 733
Smith M. S. R.   *see* Murray M. D. 558
Smith O. J.   28
Smith R. C.   49, 216, 224, 642, 643, 644, 648: *see* Tuck J. B. 522
Smith R. E.   *see* Clark A. M. 42
Smith R. F.   *see* Stern V. M. 34
Smith R. H.   5
Smith S. G.   *see* Lucas F. 645
Smith W. T.   *see* Thurston R. 89; Wicht M. C. 227
Smith W. W.   85
Smithers C. N.   547
Smittle B. J.   316: *see* Lowe R. E. 316; Vick K. 83
Snapp O. I.   20
Snieckus V.   *see* Nair K. S. S. 762
Snow J. W.   *see* Lewis W. J. 64
Snow W. F.   85: *see* Boreham P. F. L. 213
Snyder T. E.   83: *see* Burke H. E. 83
Soans A. B.   *see* Soans J. S. 225
Soans J. S.   225
Soboleva-Dekuchaeva I. I.   212
Sofner L.   547
Sogawa K.   *see* Sekido S. 764
Sohi G. S.   *see* Adarsh H. S. 61
Sokatch J. T.   *see* Horsfall W. R. 729
Sokoloff A.   39, 225
Sol R.   79, 86
Solar E. del   41, 64
Soliman M. H.   86
Soliman Z. R.   *see* Zaher M. A. 227
Solinas M.   27
Solomon J. D.   33, 37, 84, 85: *see* Drooz A. T. 85, 713
Solozhenikina T. N.   *see* Egorov N. N. 76
Soman R. S.   85
Somaya C. I.   *see* Kochhar R. K. 85
Somchoudhury A. K.   227
Someren E. C. C. van   *see* Goiny H. 57
Somerhalder B. R.   *see* Pruess K. P. 689
Sömme L.   86: *see* Rygg T. 86
Sommerman K. M.   547

Sonderstrom E. L. *see* Lovitt A. E. 214; Singh S. R. 43
Sonleitner F. J. 225
Soo Hoo C. F. *see* Browne L. B. 51
Soper R. S. 76
Soria F. *see* Féron M. 312
Soto P. E. 86
Soucek Z. *see* Murray M. D. 558
Southwood T. R. E. 582, 589, 591, 594, 607, 609, 610, 611, 612, 620, 764: *see* Leston D. 589
Soylu O. Z. *see* Tokmakoğlu C. 25
Sparks A. N. *see* Harrell E. A. 312; Jones R. L. 84, 85; Lewis W. J. 84; Sekul A. A. 67
Sparks M. R. 85, 312
Spencer G. J. 222, 739
Speyer E. R. 273, 274, 306, 307
Speyer W. 175, 568
Spielman A. 45
Spiller D. 18, 178, 312
Spittler H. 88
Spooner J. D. 83
Spradbery J. P. 74, 84, 762
Sprague I. B. 579: *see* Bodenstein D. 204
Springett B. P. 213, 225
Srinath D. *see* Sankaran T. 71
Srivastava A. S. 25
Srivastava B. K. 83
Srivastava P. D. 82
Srivastava R. P. 25, 36
Šrot M. 51, 76
Stabins V. *see* King P. E. 649
Stadelbacher E. A. 85, 312
Städler E. 85, 86
Stage H. H. *see* Gjullin C. M. 732, 733
Stäger R. 270
Stahler N. 85
Stainer J. E. R. *see* Arthur A. P. 64
Staley J. *see* Marshall J. F. 729, 730
Standifer L. N. 84
Staniland L. N. 741
Stanley J. 225, 315
Stark B. P. 764
Stark R. W. 59: *see* Bright D. E. 304
Starks K. J. *see* Mayo Z. B. 569
Starr D. F. *see* Shaw J. G. 86
Starratt A. N. 63: *see* Smith B. C. 83
Stauffer J. F. *see* Beck S. D. 315
Stavraki H. G. 709
Stavraki-Paulopoulou H. G. 30
Stay B. 44, 82, 201, 202, 203, 204: *see* Ingram M. J. 285; Roth L. M. 44
Stebbing E. P. 235
Steele C. W. 257
Steele R. W. 79, 85
Steer W. 609
Stefani R. 512
Stefanov D. 218
Stehr F. W. 713: *see* Barton L. C. 315
Stein A. K. *see* Pawlowsky E. N. 257
Stein E. 86
Stein J. D. 85
Stein W. 84
Steiner E. *see* Wyl E. von 46
Steiner L. F. *see* Mitchell S. 312; Smith C. N. 312
Steiner P. 179
Steinhaus E. A. 238
Steinhausen W. 191
Stellwag F. 278
Stenseth C. 569
Stepanov P. T. 223
Stephen F. M. 70
Stephen W. P. *see* Bohart G. E. 51
Stephens C. S. 24
Stephens G. S. 764
Stephens S. G. 84, 88
Stephenson J. W. *see* Knutson L. V. 744
Stern V. M. 34
Sternlicht M. 74
Sterringa J. T. *see* Samaniego A. 312
Stevanovic D. 14
Stevenson J. H. 86
Stewart K. W. 85
Stiehl B. *see* Bentley M. D. 85, 764
Stinner R. E. *see* Elsey K. D. 215; Johnson M. W. 85; Morrison R. K. 764
Stitz H. 198, 648
Stockard C. R. 244
Stockel J. 30, 81, 84, 88, 314
Stoffolano J. G. *see* Nappi A. J. 92, 762
Stokkink E. *see* Byrne K. J. 70
Stone M. W. 17, 25
Stone W. S. 732
Stower W. J. 83, 217, 222, 224
Straatman R. 85: *see* Stride G. O. 687
Strassen R. 659
Straus-Durckheim H. 106
Strawinski D. 609

Streams F. A.   92: *see* Nappi A. J. 762
Stretton G. B.   278
Strickland E. H.   755
Stride G. O.   83, 315, 675, 686, 687
Striebel H.   507
Strong F. E.   48, 60, 61
Strübing H.   567
Stryker R. G.   *see* Miller T. A. 761
Stuardo C.   739
Stuart A. M.   84
Stubbs A. E.   86
Stuckenberg B. R.   56, 57, 737: *see* Irwin M. E. 86, 740
Stultz H. T.   213, 214, 215
Sturm H.   191
Štušak J. M.   611
Štys Z.   *see* Peterlík Z. 72
Suárez J. H.   *see* Mallea A. R. 25
Subba Rao B. R.   61: *see* Narayanan E. S. 84; Venkatraman T. V. 84
Subba Rao S.   *see* Jolly M. S. 85
Subra R.   85
Subramaniam V. K.   220
Subramanian T. R.   20, 21
Su-Fang M.   729, 730
Sugimoto T.   735
Sugiyama T.   *see* Okabe T. 558
Sulaĭmanov Kh. A.   *see* Sinadskiĭ Yu. V. 566
Sullivan C. R.   34, 77
Sumaroka A. F.   94
Sumimoto M.   70
Sun Y.   *see* Rimando L. C. 312
Sundby R. A.   19
Surtees G.   60, 732
Suzuki K.   197
Suzuki M.   *see* Kawasaki H. 205, 206, 207, 208, 721
Suzuki T.   *see* Sumimoto M. 70
Svec H. J.   26
Svensson S. A.   722
Švihra P.   *see* Rudinsky J. A. 66
Swaby J. A.   689
Swadener S. O.   617
Swailem S. M.   *see* Awadallah K. T. 568
Swailes G. E.   46, 86, 312
Swain R. B.   84
Swaine J. M.   279, 305
Swartzwelder J. C.   *see* Zeledon R. 617
Sweeney P. R.   676: *see* Gerrity R. G. 676

Sweet M. H.   584
Sweet W. C.   729, 731: *see* Rao B. A. 729, 731
Sweetman H. L.   12, 34, 47, 178, 763: *see* Ghani M. A. 547; Pyenson L. 179
Swellengrebel N. H.   *see* De Buck A. 75, 729, 730, 731
Swigar A. A.   *see* Byrne K. J. 70
Swilley E. M.   *see* Leopold R. A. 47
Swynnerton C. F. M.   251
Sychevskaya V. I.   92
Syed R. A.   86
Syme P. D.   84: *see* Davies D. M.   734
Symmons P.   83
Syms E. E.   722
Szczytko S. W.   *see* Stark B. P. 764
Szentesi Á.   83, 312: *see* Muschinek G. 689
Szmidt A.   22: *see* Franz J. 215
Szujecki A.   659
Szumkowski W.   219

Tadokoro M.   *see* Yoshimura S. 85
Taft H. M.   48
Taher El-Sayed M.   34
Takada M.   92
Takagi K.   *see* Kaneko T. 273, 274
Takahashi F.   37, 40
Takahashi R.   272
Takahashi R. M.   *see* Miura T. 85
Takahashi Y.   715
Takaie H.   84
Takizawa Y.   199
Taksdal G.   87
Tamaki Y.   85: *see* Yushima T. 67
Tamanini L.   582
Tamashiro M.   *see* Sherman M. 84
Tampi M. R. V.   *see* Menon M. A. U. 729
Tamura I.   749
Tanaka A.   85: *see* Miyahara Y. 721; Nagatomi A. 744
Tanaka K. K.   *see* Kaplan W. D. 312
Tanaka N.   312: *see* Mitchell S. 312
Tanaka T.   *see* Kaplan W. D. 312
Tanaka Y.   82
Tandon S. K.   86
Tang C. C.   764
Tangel F.   537
Tanimoto V.   *see* Chang V. C. S. 84
Tanletin D. T.   *see* Renwick J. 69

Tanner G. D. 314
Tantawy A. O. 33
Taranukha M. D. 83
Tardif R. 312, 313
Tashiro H. 720
Tauber M. J. *see* Nechols J. R. 84, 764
Tawfik M. F. S. 92, 215, 229, 270, 287, 610, 764: *see* Azab A. K. 83, 689; Elbadry E. A. 213
Taylor B. *see* Rodriguez J. G. 227
Taylor J. S. 191, 677
Taylor M. E. *see* Fredeen F. J. H. 748
Taylor N. 209
Taylor R. L. 745
Taylor T. A. 84
Taylor T. H. C. 233, 677
Tedders W. L. 763
Teichert M. 273
Tejada A. 258
Telfer W. H. *see* Smith D. S. 205, 206, 209
Telford A. D. 732
Tempelis C. H. 213: *see* Anderson J. R. 213
Teotia T. P. S. 78, 83: *see* Singh O. P. 743
Teranishi C. *see* Clausen C. P. 649
Ter-Grigoryan M. A. 570
Terranova A. C. 46, 47: *see* Leopold R. A. 46, 47
Terry R. J. 748
Terry T. W. 269, 270
Tesh R. B. 213
Teyrovský V. 620
Thaggard C. W. 314
Thalenhorst W. 84: *see* Biermann G. 764; Hagemann-Meurer U. 36
Thalmann J. 83
Thanh-Xuan N. 568
Thatcher R. C. *see* Moser J. C. 227, 764
Theisen B. F. *see* Hemmingsen A. M. 727
Theobald F. V. 732
Theodor O. 730
Theowald B. 725
Theron P. P. A. 709
Thewke S. E. 313
Thiagarajan K. B. 269, 289
Thienemann A. 27, 711
Thomas A. 83
Thomas D. C. 271, 294, 610
Thomas G. D. *see* Wingo C. W. 230, 231
Thomas H. D. 733
Thomas H. W. *see* Newstead H. 732

Thomas R. T. S. 188
Thomas V. 49: *see* Smith C. N. 312
Thomas W. H. *see* Englert D. C. 225
Thompson C. B. 507
Thompson H. E. *see* Weber R. G. 83
Thompson R. J. *see* Hinton H. E. 159
Thompson V. 175
Thompson W. *see* Carrel J. E. 251
Thompson W. R. 752, 753, 755
Thomsen L. C. 736
Thomson M. G. 313
Thomson V. 18
Thontadarya T. S. 83, 567, 611
Thorn R. W. *see* Rust R. W. 709
Thorpe W. H. 56, 109, 144, 149, 159, 160, 748, 758: *see* Crisp D. J. 160
Thorson B. J. *see* Leopold R. A. 46, 47; Riemann J. G. 46, 47, 721
Thorsteinson A. J. *see* Gupta P. D. 78, 85; Hillyer R. J. 85; Matsumoto Y. 86
Throckmorton L. H. 136
Thurston R. 89: *see* Greene G. L. 85; Katanyukul W. 229
Thygesen T. 86
Tichomiroff A. 205, 206
Tichý V. 237
Tidwell M. A. 72
Tiegs O. W. 166
Tilden J. W. 83
Tilden P. E. *see* Bedard W. D. 66
Tillyard R. J. 235, 486, 491, 494, 643, 648
Tingey W. M. 83
Tinnilä A. *see* Markkula M. 59, 84
Tisseuil J. 258
Titschack E. 15, 79
Tiwari N. K. 80
Tobias V. I. 709
Tod M. E. 212
Todd A. R. *see* Pryor M. G. M. 201
Todd D. H. 86
Todd J. W. 86
Tokmakoğlu C. 25
Toles S. L. *see* Nielson M. W. 15
Toll S. von 712
Tomašević B. *see* Mijušković M. 763
Tomblin C. F. *see* Hagstrum D. W. 81, 85
Tomita M. 537
Tonapi G. T. 582
Tonkes P. R. 256
Topolovskii V. A. *see* Shapiro V. A. 314

Toriumi M.   *see* Katô M. 732
Torre-Bueno J. R. de la   272
Tort M. J.   *see* Guennelon G. 84, 265
Tostowaryk W.   313
Tothill J. D.   213, 214, 753
Touzeau J.   41
Townsend C. H. T.   753, 756
Townshend B. G.   *see* Goonewardene H. F. 83, 314
Toye S. A.   520
Trägårdh I.   685
Trager W.   732
Trahan G.   *see* Everett T. R. 84
Trahan G. B.   *see* Gifford J. R. 316
Traub R.   *see* Macdonald W. W. 732
Travassos filho L.   502
Traver J. R.   *see* Needham J. G. 477
Travis B. V.   *see* Bradley G. H. 728
Travis V. B.   733
Traynier R. M. M.   73, 84, 86
Treece R. E.   *see* Sechriest R. E. 20
Trehan K. N.   569
Treherne R. C.   217, 222, 223, 224: *see* Gibson A. 217, 218
Trelka D. G.   743
Trembley H. L.   727
Trensz F.   146: *see* Sergent E. 729, 730
Triggiani O.   613, 764
Trostle G. C.   *see* Baker B. H. 70; Furniss M. M. 70
Trpiš M.   179, 313, 733: *see* Horsfall W. R. 732
Truckenbrodt W.   507
Tsacas L.   740
Tsai N.   *see* Woo W. 313
Tsai Ning-hua   *see* Wu Wei-chün 85
Tsai Pan-hua   19, 35
Tsalev M.   15
Tsiropoulos G. J.   48, 743
Tsujita M.   199
Tuck J. B.   522
Tuft P. H.   97
Tuli S.   60
Turnbull A. L.   *see* Arthur A. P. 64
Turner D. A.   79
Turner E. C.   *see* Rummel R. W. 312, 313
Turner J. P.   *see* Davis R. B. 86; Fletcher L. W. 86
Turner W. J.   *see* Powell J. A. 709
Turnhout H. M. T. van   612

Turnipseed S. G.   48, 178
Turpin F. T.   *see* Busching M. K. 721
Turtaut P.   *see* Stockel J. 81, 314
Tutt W. J.   242, 713
Tuxen S. L.   748
Tyagi A. K.   689
Tyndale-Biscoe M.   50
Tynegar M. O. T.   761
Tyrell D.   *see* Soper R. S. 76
Tychen P. H.   283
Tyzzer E. E.   257
Tzanakakis M. E.   *see* Tsiropoulos G. J. 48

Ubelaker J. E.   90, 562, 751
Uematsu H.   84, 709
Ueno H.   21
Uhler L. D.   86, 743
Ullyett G. C.   37, 41
Uncles J. J.   *see* Oakley J. N. 762
Undeen A. H.   *see* Alger N. E. 315
Unzicker J. D.   193
Urbino C. M.   729
Urquhart F. A.   83, 269
Usinger R. L.   582, 636, 637, 639
Usman S.   37
Utida S.   61, 83
Uvarov Sir B.   513, 518, 521, 523

Vaidya V. G.   85
Valadarès da Costa M.   86
Valade M.   *see* Cornet M. 314
Valcovic L. R.   31
Valdimorova M. A.   39
Valek D. A.   83
Valery-Mayet M.   232
Valle K. J.   491
Vallo V.   *see* Hrdý I. 221
Vance A. M.   21, 706
Van den Brande J.   *see* Pelerents C. 212
Van Derwerker G. K.   *see* Leonard D. E. 313
Vanderzant E. S.   84, 313
Van Leeuwen E. R.   312
Van Meter C. L.   *see* Pass B. C. 313
Van Someren E. C. C.   *see* Goiny H. 57
Vargas L.   729, 731
Varma B. K.   189, 681
Vasev A.   84
Vasić K.   *see* Zivojinović S. 26

AUTHOR INDEX

Vasiljević L.   212
Vasilyan V. V.   31: *see* Azaryan G. Kh.   31
Vaughan J. A.   *see* Mead-Briggs A. R. 43
Vedy J.   550
Veer J. van der   *see* Herrebout W. M. 82
Veeresh G. K.   26
Velasquez C. C.   550
Venkatesh M. V.   83
Venkatraman T. V.   84
Vepsalainen K.   582
Verdier M.   *see* Albrecht F. O. 14; Perez Y. 82
Vereecke A.   212
Verhoeff C.   234, 270
Verma J. P.   231
Verma J. S.   214
Verson E.   718
Vey A.   91, 92
Vichet G. de   *see* Grassé P. P. 515
Vick K. W.   83
Vidaud J.   *see* Féron M. 743
Vieira R. M. S.   742
Vilkova N. A.   *see* Shapiro I. D. 58
Villacorta A.   312
Villiers A.   272, 686
Vincent J. F. V.   83, 283: *see* Tychen P. H. 283
Vincent L. E.   *see* Lindgren D. L. 178
Vincke I.   729
Vinson S. B.   90, 92, 709: *see* Guillot F. S. 64; Hays D. B. 56; Lewis W. J. 92, 93; Lynn D. C. 764; Wilson D. D. 84
Viswanathan T. R.   563
Vité J. P.   65, 66, 67, 69, 70, 87: *see* Bauer J. 689; Coster J. E. 66; Hedden R. 66; Kinzer G. W. 67; Madrid F. 71; Pitman G. B. 66, 67, 71; Renwick J. A. A. 65, 66, 67, 69, 70
Vlasblom A. G.   148
Voegele J.   60: *see* Daumal J. 312
Voelkel H.   18, 61, 82
Vogel R.   184
Vogel W.   *see* Klingler J. 33
Voigt E.   549
Volk S.   86: *see* Bombosch S. 86
Vollmar H.   *see* Sander K. 315
Vonderheyden F.   *see* Touzeau J. 41
Voorhees F. R.   *see* Horsfall W. F. 729
Vorhies C. T.   711
Voris R.   659

Voss F.   555
Vosseler J.   221, 519, 522
Voukassovitch P.   681
Voy A.   184, 533
Vukasović P.   20

Wachter S.   548
Waddill V.   229: *see* Shepard M. 270, 289
Wade C. F.   227: *see* Rodriguez J. G. 227
Wadley F. M.   *see* Davis E. G. 520
Wadsworth J. T.   659
Wafa A. K.   *see* El-Borollosy F. M. 709
Wagner W.   567
Wakeland C.   *see* Parker J. R. 216
Wakikado T.   *see* Miyahara Y. 721; Tanaka A. 85
Wal Y. C.   *see* Ayyappa P. K. 33
Walch E. W.   729, 731
Walch-Sorgdrager G. B.   *see* Walch E. W. 729, 731
Walcott C.   *see* Williams C. M. 82
Waldbauer G. P.   26, 313
Walker D. W.   25, 721: *see* Quintana-Muniz V. 85
Walker E. M.   513
Walker J. K.   84, 218
Walker W. F.   206
Wallace D. R.   705: *see* Ghent A. W. 59, 84; Neff S. E. 756
Waller J. B.   312: *see* Harman D. M. 745
Wallis R. C.   75, 85, 315
Wallis R. L.   *see* Landis B. J. 749
Wallner W. E.   314
Wallwork J. H.   227
Waloff N.   35, 37, 184, 523, 681, 709
Walsingham Lord   235
Walt J. F.   764
Walters J.   279
Walton G. A.   620, 638
Walton R. R.   *see* Stewart K. W. 85
Walton W. R.   57
Wandolleck B.   764
Wapshere A. J.   *see* Broadhead E. 546, 547
Ward G.   *see* Bay D. E. 86
Ward R. D.   761
Ward R. H.   84
Ward V. K.   *see* Hunter-Jones P. 520
Warwick E. P.   *see* Stride G. O. 83, 675, 686
Watanabe M.   86

Water J. K.   see Ankersmit G. W. 720
Waterhouse C. O.   502
Waterhouse F. L.   83
Waterston A. R.   219
Waterston J.   551
Watson C. A.   see Fletcher B. S. 86
Watson T. F.   see Adam D. S. 37; Perkins P. V. 15
Watt M. N.   715
Wattal B. L.   86
Watts J. G.   see Kinzer H. G. 279
Way M. J.   237
Wearing C. H.   85
Webb D. R.   see Eckenrode C. J. 86; Yu C. C. 86
Webb D. W.   722
Webb R. E.   314: see Smith F. F. 314
Webber L. G.   28
Weber H.   175, 551, 553, 555, 568, 620
Weber R. G.   83
Webster A. P.   see DeCoursey J. D. 727
Webster F. M.   175, 725
Weekman G. T.   see Atyeo W. T. 681; Lawson D. E. 313; Pruess K. P. 689
Wefelscheid H.   638
Wegner A. M. R.   533
Wegorek W.   89
Weidemann G.   649
Weidhaas D. E.   see Lowe R. E. 316
Weidner H.   256, 257, 270, 507, 547, 548
Weiman H. L.   681
Weinman C. J.   see Hodson A. C. 183, 200
Weir J. S.   709
Weiser J.   238
Weismann L.   see Hrdý I. 221
Weiss A.   656
Weiss H. B.   272, 502
Weissman-Strum A.   733
Welch J.   224
Welch P. S.   714
Wells C. N.   see Rodriguez J. G. 227
Wellso S. G.   83
Wendt A.   see Jordan K. H. C. 572
Wenk P.   734
Werner R. A.   66
Weseloh R. M.   84, 91
Wesenberg-Lund C.   49, 50, 476, 491, 579, 582, 651, 656, 711, 724, 735
West A. S.   see Downe A. E. R. 212; Hall R. R. 212; Loughton B. G. 212

West M. J.   269
Westdal P. H.   86
Westfall J. A.   see Hooper R. L. 86
Westfall M. J.   see Needham J. G. 491
Westigard P. H.   85
Westwood J. O.   648
Wetmore A.   237
Wetzel T.   86, 313
Weyer F.   30, 268, 729, 730
Weyrauch W. K.   270, 289
Whang W. Y.   see Chapman R. N. 224
Wharton R. H.   49
Wheeler R. E.   49: see Jones J. C. 49
Wheeler R. M.   134
Wheeler W. M.   274, 303, 499, 738
Whitcomb W. D.   60
Whitcomb W. H.   217, 221: see Bell K. O. 212, 215, 217, 218, 219; Crocker R. L. 229; Gyrisco G. G. 17, 83, 217; Phillips J. R. 80
White C. E.   see Singh Z. 595
White E. B.   659
White E. G.   213
White R.   see Miller M. C. 233
White T. C. R.   78, 83, 163, 175
White W. B.   313
Whitehead D. L.   201
Whitehead H.   see Percival E. 13, 476, 508, 509
Whitman L.   see Wallis R. C. 85
Whitsel R. H.   312
Whitten J. M.   9
Wiąckowski S. K.   214
Wichmann H. E.   215, 274
Wicht M. C.   227
Widstrom N. W.   313: see Wiseman B. R. 313
Wiegert R. G.   see Brock M. L. 236, 748, 749
Wigglesworth V. B.   9, 29, 97, 98, 114, 129, 499, 534, 539, 542, 555, 568, 569, 675, 751: see Sikes E. K. 555
Wiggins G. B.   192
Wightman J. A.   83, 312, 315, 613, 764
Wiklendt M.   see Lipa J. J. 314
Wiklund C.   85
Wilbert H.   64, 91
Wilbur D. A.   see Qureshi Z. A. 212
Wilcke J. L.   175
Wildbolz T.   85: see Wille H. 17, 83
Wilde G.   83

Wilde J. de   72, 87
Wilde W. H. A.   568: *see* Boldȳrev M. I. 63, 83, 224
Wildermuth V. L.   243
Wiley G. O.   572
Wilkes A.   *see* Salkeld E. H. 584
Wilkes T. J.   *see* Gillies M. T. 79
Wilkinson D. S.   257
Wilkinson J. D.   27
Wilkinson R. C.   84
Wilkinson R. N.   *see* Miller T. A. 761
Willard H. F.   *see* Pemberton C. E. 92
Wille H.   17, 83: *see* Klingler J. 33
Wille H. P.   175
Williams C. B.   499, 504, 506, 642, 676
Williams C. E.   502
Williams C. M.   82
Williams D. A.   216: *see* Coaker T. H. 216, 217, 230
Williams F. X.   23, 234, 651
Williams J. L.   197, 198, 681, 763
Williams J. R.   15, 35, 567
Williams L. H.   83, 228
Williams M. J.   *see* Pau R. N. 202, 203
Williams P.   *see* Crewe W. 86
Williams R. R.   *see* Smith B. C. 689
Williams R. T.   60, 61, 556
Williams T. R.   735: *see* Hynes H. B. N. 735
Williamson D. L.   71: *see* Vité J. P. 70
Williamson E. B.   491
Willis E. R.   13, 44, 284: *see* Roth L. M. 44, 50, 82, 84, 175, 284, 763
Willis J. N.   204
Willis O. R.   *see* Petersen J. J. 75
Wilson B. R.   210
Wilson C. B.   654, 656
Wilson D. D.   84
Wilson F.   48, 706
Wilson G. R.   733
Wilson H. F.   11
Wilson L. F.   72, 84, 85, 313
Wilson M. C.   *see* Shade R. E. 83, 219
Wilson M. E.   21, 83
Wilson M. R.   *see* Claridge M. F. 566
Wilson R. L.   312
Wilton D. P.   85
Winburn T. F.   216
Windeguth D. L. von   *see* Jakob W. L. 314
Windels M. B.   85: *see* Palmer D. F. 683

Winget R. N.   748
Wingfield M.   *see* Rolston L. H. 191
Wingo C. W.   230, 231
Wirth W. W.   748
Wirz P.   *see* Hurter J. 65
Wiseman B. R.   313: *see* McMillian W. W. 313
Wishart G.   216, 218, 707
Withycombe C. L.   643, 645, 646, 647
Witter J. A.   26, 313
Woglum R. S.   642
Wohlgemuth R.   36
Wojnarowska P.   *see* Pruszyński S. 21
Woke P. A.   85
Wolcott G. N.   284
Wolfe L. S.   486
Wolff M.   236
Wollaston T. V.   686
Wollberg Z.   84
Wolvekamp H. P.   *see* Vlashlom A. G. 148
Wong H. R.   *see* Elliott K. R. 237
Wong T. T. Y.   48
Woo W.   313
Wood C. S.   *see* Gray T. G. 313
Wood D. L.   66, 87: *see* Bedard W. D. 66; Birch M. C. 66; Brand J. M. 66; Bushing R. W. 87; Lanier G. W. 66; Silverstein R. M. 66, 570; Vité J. P. 87; Young J. C. 66, 67
Wood G. W.   25
Wood S. D. E.   *see* Vincent J. F. V. 82, 283
Wood T. K.   271, 291: *see* Nault L. R. 270, 271
Woodhill A. R.   *see* Lee D. J. 729
Wood-Mason J.   192
Woodring J. P.   227
Woodroffe G. E.   *see* Coombs C. W. 42, 83, 225
Woodrow D. F.   75, 83
Woods W. C.   683
Woodside A. M.   84
Wool D.   225
Worrall J.   *see* Wright D. W. 216, 217, 218
Wright D. W.   216, 217, 218: *see* Ashby D. G. 743
Wroblewski A.   620
Wu Wei-chün   85
Wulker G.   84
Wygodzinsky P.   640, 641, 764: *see* Cobben R. H. 764

Wyk L. E. van 184
Wyl E. von 46
Wylie H. G. 84, 94, 721: see Arthur A. P. 94
Wylie W. D. 84
Wyndham M. see Kehat M. 61
Wyniger R. 82

Xambeu V. 270, 686

Yabe T. see Harada F. 85
Yadava R. L. 570
Yago M. see Kawasaki H. 210
Yagunga A. S. K. see Raybould J. N. 735
Yakimova N. L. 35
Yakubovich V. Ya. 761
Yamamoto R. T. 78, 85, 312: see Waldbauer G. P. 26
Yamamoto S. see Yasumatsu K. 37
Yamanaka H. see Ito Y. 212
Yamaoka K. 78, 85, 764
Yanagita Y. see Sakanoshita A. 85
Yano A. see Ikeshoji T. 85
Yano K. 744
Yasumatsu K. 37, 64, 534
Yatagai M. see Bentley M. D. 85, 764; McDaniel I. N. 85
Yates M. G. 761
Yates W. W. see Gjullin C. M. 732, 733
Yazgan S. 315
Yeargan K. V. 84
Yen Yü-hua see Wu Wei-chün 85
Yeung K. C. 191
Yien Y. see Woo W. 313
Yinon U. 221
Yokoyama K. 18, 218
Yokoyama T. 721
Yonce C. E. 84: see Jacklin S. W. 48
Yonke T. R. see Swanender S. O. 617
York G. T. 222
Yoshida T. 83, 225
Yoshida Y. see Matsua K. 761
Yoshimura S. 85
Young A. M. 82, 491, 721
Young B. see Chen S. H. 58, 648

Young C. J. 732
Young E. C. 620
Young J. C. 66, 67
Young J. R. 764
Young W. C. 35
Yu C. C. 86
Yushima T. 67

Zachariae G. 83
Zachary D. 92
Zacher F. 18, 82
Zaeva I. P. 212
Zagulyaev A. K. 235
Zaher M. A. 40, 41, 227: see Long D. B. 40
Zaĭtsev V. F. 761
Zaitzov A. see Avidov Z. 15
Zakhvatkin A. A. 222, 223, 739, 763
Zangheri S. 76, 749
Zarea N. see Sharifi S. 25
Závadsky K. 278
Zazou H. see El-Kady E. 34
Zdarkova E. see Howe W. L. 312
Zečević D. 26, 718
Zech E. 80
Zehring C. S. 491
Zeigler D. L. see Park T. 225
Zeigler J. R. see Park T. 225
Zeledon R. 617, 751
Zelený J. 19: see Hrdý I. 221
Zerillo R. T. 313
Zethner-Møller O. see Rudinsky J. A. 71
Zickan J. F. 506
Zimin L. S. 522, 523
Zinna G. 94, 144, 707
Ziprkowski L. 257
Zivojinović S. 26
Zocchi R. 279
Zohren E. 28, 86
Zucas S. M. see Juarez E. 316
Zucchi R. see Sakagami S. F. 84
Zuñiga A. 279: see Zeledon R. 617
Zweig G. see Iltis W. G. 252, 253, 729
Zwölfer H. 278
Zwölfer W. 33, 35

Subject Index

Volume I: Pages 1–474. Figs. 1–135. Plates 1–155
Volume II: Pages 475–778. Figs. 136–296

abdominal pouch 51
absorption 29, 30
accessory gland substance 46, 47, 275
accessory glands 44, 45, 46, 47, 48, 49, 183, 196, 197, 202, 240, 251, 264, 634, 647. Figs. 83; 87
acoustic signals 66, 67, 68, 69, 71, 76, 82
adhesive disk 52, 478, 479, 739
adhesive organs 477, 478, 479
adhesive substance 479
adult 6, 7, 10, 31, 33, 34, 35, 36, 38, 41, 51, 52, 107, 111, 142, 159, 225, 228, 229, 230, 232, 475, 507, 536, 561, 567. Fig. 2
aero-micropylar cup 583
aero-micropylar projections 583, 589, 594, 607. Figs. 206A; 207B
aero-micropyles 583, 584, 585, 589, 594, 607, 614. Fig. 207D
aeropylar canals 640
aeropyles 96, 97, 98, 100, 102, 104, 105, 107, 110, 111, 114, 115, 116, 121, 123, 125, 126, 128, 134, 151, 152, 159, 171, 497, 499, 507, 539, 545, 547, 551, 552, 567, 568, 570, 571, 573, 578, 582, 586, 608, 609, 610, 611, 612, 614, 616, 617, 626, 628, 630, 631, 632, 633, 634, 635, 636, 637, 639, 640, 641, 644, 649, 651, 659, 660, 661, 662, 663, 675, 677, 681, 687, 715, 720, 723, 742, 743, 744, 745, 760, 761. Figs. 23; 25; 29B; 36; 37; 41C; 43C; 47; 52D; 219A; 222; 230; 231; 233; 234; 247A; 251; 252; 255; 256; 263B. Pls. 5A, B; 12A–H; 18C, D; 19D; 22F; 23C–E; 24F; 25E, F; 29C, D; 31A, B, F; 32A–F; 34A–D, F; 38A–C; 40C; 41C; 42C, D; 44D; 45A, B; 46A, B; 48F; 52B–D; 53E, F; 54A; 55A, B, D–F; 56B, D; 60C, F; 64C, F; 65E, F; 66E; 67D; 68D; 69D, F; 70B, C; 71B–F; 72A–D, F; 73A, B, D–F; 74B, C; 76A–C; 77A, B; 78; 79C, D; 80B–E; 81B, C, E, F; 82B–D; 83E; 84C, D; 85D, F; 86C; 87B; 88B–E; 89A; 90D; 91A; 92A, B, E; 93A, B, E, F; 95A; 121A, C, D; 123C, D; 130A, B
aeroscopic plate 144, 145. Fig. 63A, C, E
aggregation 65–67, 70, 71
alanine 183, 203, 204, 205, 206, 207
algae 231, 301
alkaloids 89
ametabolous insects 7
amino-acids 205, 206, 207, 208, 210, 645
ampulla 575
anabasine 89
anal gland 678, 679
anal tuft 256, 257
anterior filament 568
anterior horn 175. Fig. 74
anterior of egg Fig. 22C. Pls. 9A, B; 13B; 14A, B, F; 16A; 23B; 26E, F; 28A–C; 31F; 35D; 36A, E; 38A, B; 40A, B; 41B, C; 48D; 51A; 53C, D; 65A, B, G; 68B, C; 69D–F; 74A; 82E; 85E; 86D; 109A; 122; 123B, C; 127A; 136A; 138D; 146A, B; 148A; 149A; 151A
anterior pole 478, 479, 510, 539, 547, 573, 574, 575, 579, 581, 583, 584, 585, 588, 607, 613, 618, 637, 639, 640, 642, 645, 653, 675, 676, 687, 710, 718, 742. Figs. 198A; 201A, B; 204. Pls. 7A; 10F; 42A, B; 49E, F; 50B; 70A, B, E, F; 96A; 99B, C; 104A; 105B; 135B, C, D; 151F
anterior projection 661. Figs. 152; 249D
apical cup 591
apical filament 568
apical projection Fig. 227

apolysis 5, 6, 7, 8, 89, 98, 142, 170, 173, 250, 475, 522. Figs. 2; 3
appendages 3, 4
attachment disk 509, 553. Pls. 2C; 3C, F; 4A, C, D
attachment stalk 589

bacteria 202, 238, 314, 651, 652
basal stalk 567, 568, 617, 643. Fig. 225
benzoic acid 89
biological control 229, 230, 232, 234, 235, 238, 239, 752
black body 263, 264
blastokinesis 166, 167
body cells Fig. 171F–M; O; Q
brevicomin 69, 70
brood balls 275, 276, 277, 302. Figs. 107–109
brood burrows 275, 276. Fig. 107
brood chambers 288, 289, 299, 301, 303, 306, 307. Figs. 117; 126; 127; 278
brood galleries Figs. 112; 273B; 279
brood pouch 753
brooding 292, 296, 302, 705. Fig. 120
burrows Figs. 269; 270

calcium 163, 536, 537
calcium carbonate layer 537, 539. Fig. 179
calcium citrate 202
calcium oxalate 203, 204, 247, 539
calcium oxalate layer 537, 542, 545. Fig. 179
camphene 71
cannibalism 224–226, 231
cantharidin 250, 251
cap 523, 524, 525, 549, 588. Figs. 30B; 193. Pl. 24A, D–E
cap cells Fig. 171E, N, P
capitulum 536, 539, 541. Fig. 181A. Pl. 17C, D
cardiac glycosides 251
carvone 70
cement 549, 551. Fig. 185
cercal cables 506, 507
chemical control 239
chemoreceptors 55, 56, 65, 73, 245
chitin 185, 201, 205, 209
chorion 30, 53, 54, 96, 97, 117, 118, 119, 120, 124, 165, 171, 174, 175, 176, 179, 180, 200, 205, 206, 208, 209, 210, 211, 249, 252, 260, 261, 264, 477, 478, 479, 491–492, 499, 506, 508–510, 512, 513, 515–516, 520–522, 527, 528, 534–536, 539–545, 547, 551–553, 554, 563–564, 565, 568, 571–572, 574, 575, 577–578, 579, 581, 582, 588, 589, 590, 591, 592, 593, 594, 595, 596, 597, 598, 599, 602, 603, 607, 610, 612, 614, 615, 617–618, 625, 626, 630–632, 636, 637, 638, 639, 640, 641, 642, 643, 644, 648, 649, 650, 651, 654, 656, 659, 663, 675, 676, 678, 681, 687, 706, 707, 710, 714–716, 723, 725–727, 729, 736, 740, 741, 745, 754, 755, 756, 760, 761. Figs. 23B; 31; 34B; 41; 42; 43B; 61A; 68; 165; 166; 179; 180; 182; 183; 192; 236; 247; 251; 252; 255; 256; 281; 284B; 285. Pls. 30C, D; 57C–F; 59D, F; 92F; 97F, G; 98D–F; 127C, D; 131C, D; 142A, B; 153A, B
chorionic hairs 598, 599, 600, 601, 602
chorionic hydropyles 164, 166, 170–171, 264, 551, 553–554, 568, 574, 632, 636, 637. Fig. 71
chorionic layers of *Bacillus libanicus* 537–538. Fig. 179
chorionic plastron 97, 112–127
chorionic plug 542
chorionic respiratory system 512, 534, 547, 612, 681, 683, 687, 742, 743, 744, 745, 746, 748, 749, 753, 754, 755, 756, 758
chorionic spines 595, 596, 600, 602
chorionin 120, 537, 539, 541, 542, 554, 741
Circadian rhythms 81
cleaning eggs 288, 289
cocoon 147, 148, 186, 187, 188, 211, 257, 656. Fig. 244
colleterial glands 183–200, 201, 209, 240, 504, 513, 515, 518, 520, 534, 537, 644, 676, 679, 681. Figs. 76; 83; 85; 90; 260A
colour 78, 243, 518, 569, 577, 591, 592, 593, 595, 596, 597, 598, 599, 600, 601, 602, 603, 643, 705, 738
  aposematic 240, 250, 251, 675
  changes 249–250
  cryptic 240, 248, 296
  disruptive 240, 247–249. Figs. 99; 100
  protective 146
  pseudosematic 240
  semi-cryptic 250

colour (*cont.*)
    warning   251
compressible gill   106, 107
copulation   43, 49–50, 251
cremastral cables   507
crypt   298. Fig. 124
cryptobiosis   95, 177
cuticle   201
cytoplasm   Pl. 133A, D, E

defaecation chamber   287. Fig. 117
defensive behaviour   291, 296, 297, 299, 300, 302, 306, 311
defensive devices   240–268
defensive fluid   251–256. Pl. 105F
    ant response   253–256
    apical drop   252–256
density   38–42, 225. Figs. 7–9
desiccation   119, 121, 181, 194, 196, 200, 519, 536, 633, 732, 733
    resistance to   176–182, 735
detoxication   72, 251
diapause   169, 170, 173, 177, 485, 518, 533, 608, 611, 612, 713, 734
diploid eggs   94
dispersal of larvae   29
distortion of surface film   265–268. Figs. 105; 106
distribution of eggs
    by birds   246–247
    by grooming   551
dopamine   179
dorsal organ   165, 166
dyes   315

ecdysis   2, 7, 29, 31, 47, 122, 142, 233, 256, 282, 289, 310, 475, 522
ecdysone   89
eclosion line   575, 579, 581, 584, 586, 637
eclosion splits   582, 583, 586, 588, 591, 613, 618, 620, 637, 639, 640
ectoparasites   708, 739
effect of rain   561–562
egg-bursters   483, 494, 507, 513, 548, 551, 555, 563, 565, 571, 573, 574, 575, 579, 581, 582, 583, 585, 586, 588, 589, 590, 591, 592, 593, 594, 607, 613, 614, 616, 637, 641, 642, 646–647, 650, 653, 659, 681, 685, 687–688, 711, 719, 723, 725.

Figs. 140; 187A; 215B, D; 216L; 217
egg capsule   676. Fig. 172A, B
egg-cases   657, 658. Fig. 245
egg-cocoons   11, 186, 187, 651
egg-mass   257, 258, 259, 294, 295, 297, 314, 476, 526, 527, 528, 597, 598, 599, 600, 601, 602, 603, 642, 651, 658, 710, 711, 713, 734, 735, 736, 737, 738. Figs. 84; 122; 212; 287; 288
egg-pods   29, 232, 519, 520, 523, 526. Figs. 170; 172–174
egg rafts   252, 253, 254, 266–268
egg-sacs   237
egg-stalk   644–645
eistigma   553, 554
electrophoresis   65
elytra   504, 505
embryo   7, 212, 249, 250, 261, 263, 285, 536, 537, 546, 552, 585
embryogenesis   497
embryonic cuticle   171, 485, 494, 506, 507, 513, 522, 548, 551, 555, 563, 575, 585, 594, 608, 611, 613, 616, 642, 646, 653, 654, 711
embryonic membranes   537. Fig. 140S
embryonic tissue   495
emissivity   263, 264
encapsulation   54, 90–94, 211
endochorion   509, 520, 538, 732
endoparasites   706, 752, 753
endopterygotes   1, 3, 4, 5, 6, 7, 40
enemies of eggs   211–237, 763
enzymes   170, 174, 249, 250, 522, 732
epembryonic membrane   181
epembryonic ring   181
epichorion   679
epithema   477, 478, 479
equatorial band   171, 594
equatorial organ   477
equilibrium temperature   263, 264
exochorion   509, 520–521, 536, 538, 676
exopterygotes   1, 2, 3, 4, 5, 6, 7, 40
extrachorion   520, 678, 679, 681. Figs. 169; 258B; 261B
extra-chorionic membrane   509, 595
extraction of eggs   313

fecundity   11–50, 82, 89, 225, 520, 706, 709. Fig. 5

feeding   227, 229, 230, 231, 238
female reproductive system   Fig. 86
fertility   42, 43, 48
fibroins   183
filaments   492, 555, 726, 727. Figs. 149A, C; 286
floating of egg-rafts   226–228
floats   730, 731. Pls. 103; 104B–D
follicular cells   113, 123, 126, 491, 507, 520, 522, 547, 563, 631, 632, 634, 650. Fig. 234B. Pl. 133B
follicular epithelium   513, 536
follicular pits   614, 615. Pls. 38F, G; 40A, B; 41A, D
food-balls   Figs. 266; 267
food burrows   282
foregut cuticle   507
frills   730, 731. Pl. 103
frontalin   66, 67, 69, 70
fungal spores   546
fungi   73, 74, 194, 202, 231, 238, 239, 275, 287, 301, 304, 305, 314, 724, 746

galleries   279–282, 298, 301, 303, 304, 305, 306, 307, 308, 309, 310, 512, 563. Figs. 112–114; 124; 129–132; 134; 278; 279
genitalia   Fig. 144
genotypes   225
glycine   183, 203, 204, 205, 206, 207, 208
gonapophyses   486, 508
gonopods   549
gonopore   476
gossypol   71, 89

haemocoele   52, 64
haemolymph   536
haploid eggs   94
hatching   29, 53, 177, 192, 259, 285, 299, 485, 494, 502, 506–507, 513, 520, 522, 551, 558, 561, 566, 569, 574, 575, 590, 593, 638, 641, 654, 676, 681, 683, 684, 705, 711, 719, 728, 732, 733, 738, 749, 755. Figs. 158; 177; 191; 241
hatching lines   113, 114, 119, 120, 121, 497, 499, 552, 555, 749, 752, 755, 758, 759, 760, 761. Figs. 31A; 33B; 34A; 35; 39; 43A; 44; 61A. Pls. 24B, C; 35F; 127B; 129F; 131B; 132C–F; 133A, C, F; 136C;

138B, C, F; 139A, B; 140D; 141D; 142C; 143A–E; 144A–D, F; 145B; 147D; 148E; 149B–D; 150E, F; 151B–D; 154C–D; 155
hatching seal   608
"hatching spines"   688
height preferences   76, 77
Hemimetabola   5, 7
Heremetabola   7
hibernation   567, 677, 713
histamine   250
Holometabola   5
honeydew   229
hormones   204
humidity   61–63, 79, 80, 119, 147, 176, 177, 178, 182, 200, 502, 556, 557, 559, 561, 562, 610. Figs. 190; 191
hydrofuge setae   107
hydropyle canals   553
hydropyle cells   168, 169
hydropyles   110, 163–182, 513, 551, 552, 553, 554, 567, 568. Figs. 187C; 195. Pls. 43C–E; 45D, E; 46F; 47F
hydrostatic pressure   159, 171

incubation   755
incubation period   177, 178, 179, 533, 651, 713, 739
infra-red   263, 264
intruders   289, 290
ipsdienol   65
isotope labelling   211–212, 316

juglone   88

kairomone   74
keel   497, 499. Figs. 152; 153
ketone   69
key to:
    egg-pods of S. Ghana grasshoppers   526–528
    eggs of European Pentatomidae   596–603
    eggs of European Scutelleridae   591–593
    eggs of families Pentatomoidea   590
    eggs of Hydrophiloidea   657–658
    eggs of Muscinae and Stomoxyinae   760–761

key to (*cont.*)
    eggs of Nepidae 634–635
    eggs of N. American Corydalidae 642–643
    eggs of N. American Pentatomidae 595–596
    eggs of some British Nabidae 609
    eggs of Staphylininae 659–663
    eggs of subfamilies Coreidae 587–588
    Fannia 761
    French species of Ephemeroptera 477–479
    N. American grasshopper eggs 523–526
    Tettigoniidae 516–518

larvae 1, 3, 4, 5, 7, 10, 11, 29, 32, 33, 36, 39, 40, 51, 52, 56, 59, 72, 73, 74, 75, 76, 87, 93, 94, 122, 140, 144, 173, 177, 179, 181, 188, 192, 195, 212, 213, 225, 226, 228, 229, 230, 231, 232, 249, 250, 253, 259, 263, 274, 275, 285, 286, 287, 289, 293, 294, 295, 296, 300, 301, 476, 493, 494, 495, 508, 512, 522, 536, 537, 552, 561, 563, 567, 643, 649, 651, 653, 679, 681, 684, 685, 688, 707, 711, 712, 719, 722, 724, 733, 734, 735, 737, 739, 740, 741, 743, 745, 749, 751, 752, 753, 755, 756. Fig. 2
larval ball Fig. 271
larval galleries 476. Fig. 278
larval mines 282. Fig. 112
larval setae 44, 257
lateral edge Pl. 143E
leaf-rolling 278, 688. Figs. 111; 275–277
light 79, 81, 82. Fig. 20
limonene 71
linalool 67
lipid layer 476
lipids 179, 180, 181, 200, 241, 249, 252
longevity 11–31, 33, 42, 43
lupinine 89

macrotype eggs 754
male song 82
malpighian tubules 10, 536
mandibles 651
Manometabola 7
mass provisioning 275–283, 302, 688, 704. Figs. 265–279

maternal care 512, 589, 590, 738
maternal insecticide 647
maternal tissue 495
mating 43, 44, 46, 47, 48
mating song 76
matrone 44, 45, 46, 48
maturation 29, 30, 31, 41, 43, 44, 250, 275
mechanical barrier 258–260
melanins 93
melanization 249
membranous eggs 755–756. Fig. 292C, E, F
metabolism 95, 182, 209
metabolites 30
metamorphosis 39, 251. Figs. 2; 3
microfibrils 209, 715. Fig. 92
micro-organisms 71, 73, 239, 742
micropylar apparatus 509, 642, 645, 652, 687, 763. Fig. 206B–D
micropylar area 147, 167, 571, 582. Figs. 204C; 238B. Pls. 8A, B, D; 10E; 11D, E; 62; 64E; 87A; 88F; 89B, C
micropylar axis 716
micropylar canal 524, 585. Figs. 179A; 223B
micropylar circle 597
micropylar cup 620. Fig. 223B
micropylar knob 642, 645
micropylar plate 539, 542–545, 675. Figs. 178; 182A. Pls. 16B, D–F; 17F
micropylar plug 252
micropylar projections 582, 587, 592, 595, 596, 597, 598, 600, 601, 602, 603, 607. Fig. 213. Pl. 49C, E, F
micropylar ring 586, 590, 591, 592
micropylar stalk 542. Fig. 182A, B. Pl. 15E, F
micropyles 56, 105, 180, 252, 477–479, 493, 499, 506, 507, 509, 510, 512, 513, 515, 516, 522, 523, 525, 538, 542, 543, 545, 547, 551, 553, 554, 564, 567, 568, 572, 573, 574, 575, 579, 581, 582, 583, 584, 585, 586, 588, 589, 590, 591, 592, 593, 594, 595, 596, 597, 598, 599, 600, 601, 602, 609, 610, 611, 613, 617, 618–620, 623, 633, 636, 637, 638, 639, 640, 645–646, 653, 675, 676, 707, 710, 718, 723, 743. Figs. 29D, E; 149; 169A; 187B; 200; 204; 206A; 207; 226; 235; 238B; 240E, F; 241C. Pls. 1D, G; 3D, E; 4E; 8C; 9C, E; 10A, B; 13A, E; 40C; 42C,

micropyles (*cont.*)
    D; 46E; 47E; 60B; 61A, B; 65C, D; 66B, D, F, G; 68D; 72E; 73A–C; 74D–F; 76A, B; 78; 79A, B; 80A; 81A; 82A; 84A; 86E; 92D; 93D; 96B; 97A; 99D; 106B; 107F; 109D; 111F; 112B; 117C; 123D; 132A; 136D; 152A
microtricia 107
microtype eggs 4, 5, 53, 54, 753, 755. Fig. 292B, D, G
"milk" glands 495
mimicry 241, 258, 536, 585. Figs. 94–98
mineral layer in chorion 247
moulds 302
moult 7
mycetangia 304, 305, 308, 309. Fig. 135
myrmecophiles 226, 230
myrtenol 70

natural deception 240–249
nectar 229
nicotine 89
nidification 28, 298
nornicotine 89
nutrition 7, 33, 80
nymphs 2, 3

odour 71, 72, 73, 74, 78, 82
olefin 67
oleoresin 65, 66, 70, 87
olfactory stimuli 56
oöcytes 30, 35, 43, 44, 45, 60, 63, 71, 180, 250, 251, 495, 543, 731
oösorption 29–31
oötheca 44, 83, 185, 188, 189–192, 201, 202, 203, 209, 233, 237, 284, 285, 495–497, 499–505, 506, 513, 546, 591, 607, 676, 677. Figs. 80; 82; 116; 151; 152C; 153; 154; 155–159
oöthecal membranes Fig. 81
oöthecal proteins 201–205
oöthecal tanning 201
opercular strand 555
operculum 106, 171, 512, 536, 539–542, 545, 551, 552, 554, 565, 598, 608, 609, 610, 613, 614, 616, 617, 739. Figs. 181A; 187A, B; 221E; 261A. Pls. 33A; 37E; 39B
osmosis 171, 173, 174, 180

osmotic gradient 172, 180
osmotic pressure 171, 172, 174
ovarian egg 144, 573, 610, 746. Figs. 63; 201B; 238. Pls. 133; 135B–E
ovarioles 571, 607, 634, 647, 679, 681, 755
oviduct 607, 676. Fig. 89
oviduct glands 186
oviparity 284, 286, 495, 549, 570
oviposition 30, 33, 43–47, 51–94, 225, 233, 260, 282, 295, 314, 476, 486–491, 508, 512, 513, 514, 518–520, 533, 546–547, 549–551, 559, 563, 566, 567, 568, 569, 570–571, 572, 575, 576, 577, 579, 580–581, 582, 590, 591, 592, 593, 596, 597, 598, 599, 600, 601, 602, 603, 609, 610, 611, 612, 613, 614, 617, 620, 621–622, 635–636, 637, 638, 639, 640, 642, 643, 649–650, 651, 652, 653, 654, 659–663, 675, 676–679, 683–684, 685, 687, 688, 706, 707, 708, 709, 710, 711, 712–714, 722, 724–725, 728, 733, 734, 735, 736, 737, 738, 739, 740, 741, 742, 743, 745, 747, 748, 749, 750, 751, 752, 753, 756. Figs. 11–14; 17; 110; 115; 141–143; 145; 163; 164; 167; 168
ovipositor 73, 107, 144, 486, 513, 515, 518, 533, 563, 566, 567, 571, 574, 575, 580, 650, 651, 677, 685, 686, 704, 705, 706, 707, 712, 713, 724, 740, 742, 743. Figs. 241A; 242A
ovisac 570
ovoviviparity 123, 284, 286, 303, 485, 495, 497, 508, 549, 563, 676, 711
oxidation 89, 93
oxygen pressure Fig. 67

pallisade scales 258–260. Fig. 102
parasites 51, 53, 54, 55, 56, 59, 194, 211, 238–239, 502
parasitization 295, 297, 708
parental care 269–311, 566, 591, 639, 705. Figs. 123; 127; 128
parthenogenesis 475, 543, 545, 549, 569, 570, 705, 707, 708, 712
paurometabolous insects 7
pedicel 53, 184, 704, 705, 706, 707. Fig. 280
pedicellate eggs 756. Fig. 292A
pedicles 533, 578, 756. Fig. 185

peripheral vision 249
perivaginal pouch 51
pest control 708
pharate adult 7, 31, 34, 35, 66, 142, 475, 679, 734
pharate larva 5, 178, 485, 494, 497, 500, 506, 513, 522, 533, 546, 551, 637, 651, 676, 679, 713, 732, 739
pharate pupa 2
phenols 89, 93
pheremone gland 199
pheremones 46, 56, 63–74, 89, 93, 199, 241, 275, 314, 493
phoretic eggs 57–58
photoperiod 177
physical gills 106, 107, 124, 192, 502, 554, 656, 748
pigmentation 705
pinene 65, 66, 67, 71
pinocarvol 70
planidia 54, 708
plastron 97, 106, 107, 108, 148–162, 188, 508, 509, 521, 571, 578, 581, 607, 612, 617, 621, 624, 625, 626, 628, 632, 633, 634, 635, 636, 639, 640, 651, 681, 715, 723, 726, 735, 740, 741, 744, 745, 748, 749, 751, 752, 757, 758, 759, 761. Figs. 33B; 46A; 66; 161C; 197B, C; 218; 228–231; 232A. Pls. 2A, B, D, F; 15E, F; 34B; 37G; 45C; 47A–C; 48A; 50C, D; 56C; 57A, B; 58; 68B, C; 98A–C; 101; 102; 105C–E, G; 114C; 115D; 125B, D; 129D, E; 131B; 132C, D; 134B, C, E, F; 135E, F; 136C; 137B–F; 138B, C, F; 139; 140A–C, E, F; 141B, D; 142C–F; 144F; 145E; 146D; 147F; 149E, F; 151C, D; 154C, D
  crater Pls. 139C–F; 140
  macro Pls. 7C–E; 133E
  micro Pls. 7C–F; 133A
  resistance to loss of waterproofing 150–151
  resistance to wetting 149–151
  respiration 108–141, 516, 545, 644, 681, 687
plug 526, 527, 528
poison 240, 251, 256
poisonous eggs 250–251
poisonous setae 240, 256–258
polysaccharides 194

posterior
  filament 492
  of egg Pls. 60E, F; 68D; 71A; 74B; 120B; 123B–D; 131A, B; 145F; 150A, B; 154A, B
  pole 477, 478, 479, 507, 510, 551, 555, 592, 608, 637, 645, 720, 726, 739, 742, 744, 760. Figs. 52C–E; 171A, B; 292E. Pls. 3B, C, F; 4A–C; 7B; 8E, F; 9D, F; 10C, D; 13B, F; 48B; 50E; 70C; 105; 132B; 152B, C
  projection 658, 659, 660, 661
precipitin test 212, 213
predators 71, 194, 211–239, 275, 291, 294, 299, 490, 502, 647, 648
protective devices 240–268, 647–648, 654, 712
protozoa 238, 314
pseudoperculum 575, 582, 585, 586, 588, 589, 590, 592, 593, 594, 595, 596, 597, 598, 599, 602, 603, 607, 610, 613, 637, 638, 639, 640, 641
pupae 7, 10, 33, 35, 66, 110, 111, 142, 176, 212, 226, 735. Figs. 268; 273B
pupal cases 713
pupal cocoons 713, 715
pupal cuticle 507, 734

queens 11, 29, 47, 226
quercetin 89
quiescence 485, 684
quinones 89, 93, 179, 201, 205, 261

reabsorption 71, 556
rectal apparatus 678
reflectance 263
repagula 240, 647–648. Fig. 240A, C, D
reproduction 495
reproductive systems 607, 708, 763
resemblance of eggs to
  background 242
  leaf gall 243
  plant seeds 243–247
  plant tendrils 242
  substrate 242
resins 65–66
resistance of eggs to
  desiccation 176–182
  loss of waterproofing 150–151
  wetting 149

1110   SUBJECT INDEX

resistance of plants   86–89
resistance to parasite eggs   89–94
respiration   95–144, 541, 542, 568, 707, 719–720
respiratory
  area   571, 572. Figs. 196–198
  horns   105, 106, 113, 121, 125, 126, 127, 128, 129–142, 144, 159, 162, 612, 613, 621, 622–629, 632, 633, 634, 635, 744, 745, 748, 756, 760. Figs. 25; 26; 27B; 28B; 29A, B; 45; 46; 49B; 51B; 53B, C; 54; 55B, C; 59B; 61B; 62B, D, E; 228–231; 232A, B. Pls. 33E; 34A–D; 37G; 43A, B; 44A–C; 45C; 58; 125A, C; 128C–F; 141A–C; 144E; 145A; 146C, D; 147A, C; 148B, C, F
  papilla   Fig. 62B, E
  pore   542
  systems   10, 95–148, 491, 497, 499, 507, 512, 539, 547, 552, 567, 568, 577, 581, 582, 583, 585, 591, 594, 607, 608, 614, 615, 677, 678, 708, 715, 717, 723, 739, 745, 751, 752, 756
rheotropism   476
rhodanase   251

salicylic acid   89
salt   75, 172, 315
saponin   87
scatoshell   191, 679. Figs. 261; 262
sclerotin   201
sealing bar   616. Pl. 39C
sebaceous glands   199
sequestering poisons   250, 251
serological tests   211–212
serosa   166, 167, 168, 169, 181, 261, 521, 650
serosal
  cells   171, 180
  cuticle   99, 100, 117, 164, 165, 166, 169, 171, 174, 175, 176, 179, 180, 209, 260, 261, 521, 522, 554, 573, 613, 630, 637, 651, 654, 659, 732. Figs. 22; 23A; 247; 251; 252; 255; 256
  cuticle bladder   618, 639
  cuticle hydropyles   164, 169–170, 173, 174, 179, 568, 574
  hydropyles   164, 166–169, 171, 568, 574, 581, 632. Figs. 68; 70
  plug   164, 166–169, 575

seudenol   67
sex attractant   74
sex of eggs   94
shell proteins   205–210
shell section   Figs. 24A, B, D; 29C; 32; 38; 52C; 61A, C; 62C. Pls. 2D, F; 4F; 5C, D; 11A–C; 13C, D; 40D–F; 42E, F; 46C, D; 47D; 54B–D; 81C, D; 82F; 83C, D, F; 84E, F; 87C–F; 88A, C; 89D–F; 91B–D; 92C; 95B–D; 96D, E; 99E, F; 118D; 121B; 123E, F
side of egg   Figs. 28A; 33A; 37; 41; 49A; 51A; 53A; 55A; 62A; 74A. Pls. 1A; 11F; 14C–E; 18F; 25E; 28D; 36B–D; 39A; 47A, C; 49E; 50C, D, F; 55C; 59C; 61C–F; 63; 70D; 84B–D; 85A–C, F; 94; 97B–E; 106A, C; 107A–E; 108; 109B, C, E, F; 110; 111A–C; 112A, C–F; 113; 114A–C; 120F; 143E; 152D–F; 154E, F
sight
  attentive   249
  pre-attentive   249
silk   147, 183, 187, 297, 298, 506, 546, 643, 658, 740
silk balls   740
silken web   546–547, 657
social insects   11, 269, 285, 704
sparteine   89
spermatheca   44, 45, 47, 94
spermatic groove   614, 616. Pls. 38A, B; 41B, C; 42C, D
spermatophragma   48
spiracles   95, 107, 111, 144, 159, 708. Fig. 63C–E
spiracular atrium   95
spiracular gills   110, 111, 142, 149, 171
spumaline   44, 51, 107, 183, 192–195, 197, 200, 240, 242, 257, 258, 259, 264, 265, 294, 486, 489, 491, 493–494, 508, 509, 512, 537, 546, 553, 554, 555, 567, 571, 580, 591, 614, 633–634, 635, 650, 651, 654, 676, 677, 678, 681, 683, 684, 686, 710, 714, 722, 735, 736, 738, 744, 750, 754. Figs. 143; 150; 193; 201A; 264; 281; 287; 291
spumaline membrane   574
stalk   52, 53, 56, 144, 148, 569, 644–645, 647, 704, 706. Figs. 224A; 225
stalked aeropyles   102–103

starch   77
stimuli   275
stridulation   67, 69
stridulatory organs   66
subchoral membrane   96, 117, 119, 140, 165, 171, 174, 176, 177, 179, 181, 513, 733. Fig. 68A
subimaginal cuticle   475
subimaginal instar   475
subimago   475
sub-social insects   269–311, 639
sulcatol   70
suprachorionic membrane   574
surface tension   70, 150, 151, 265
surrogate mothers   293
symbiosis   226
symphiles   226
synoeketes   226

tachygenesis   2
tactile setae   78
tanned membrane   261
tanning   179, 180, 201, 206, 210
tannins   72, 89
techniques   312–316, 727, 764
 cold storage   314
 detection by dogs   314
 egg traps   314
 implanting   313
 marking   315
 obtaining and handling eggs   312
 recording distribution and oviposition of eggs   314
 sampling   313
 sterilization   314
temperature   30–35, 43, 60, 61, 79, 80, 177, 178, 181, 261–265, 314, 502, 556–561, 569, 730, 731. Figs. 4; 188–190
terebra   486, 487, 493. Figs. 143; 144
termitophiles   226
terpenes   65, 66, 70, 71
time   80, 81
tolerance to disturbance   292
triangulins   29, 54, 676

trophic eggs   226, 287
trophobionts   226
tubercles   117, 528, 642, 681, 715, 723. Fig. 201A. Pls. 103; 104B, C
tyrosine   182, 203, 204, 205, 206, 207, 209

ultra-violet   264
uterus   495, 711, 753

ventilation of brood chamber   301
ventral strip   Pl. 16F
verbenol   65, 66, 67, 70
verbenone   66, 67, 69
vibration   75, 76
viruses   238, 239, 314
vitelline membrane   164, 165, 169, 171, 180, 520, 539, 554, 615, 651, 676, 678, 716. Figs. 222; 260B
viviparity   286, 485, 495, 546, 569, 712, 740, 752
volatiles   71, 73

water absorption   163–175, 180, 200. Figs. 75; 84
wax   180, 181, 240
 balls   289
 canal filaments   160
 "eggs"   289
 layer   180
 ovisac   570
waxy
 filaments   240
 material   537, 538
 membrane   119
weight   35, 36. Fig. 7
wetting   561–562
white cuticle   169
wing of mosquito eggs   Pls. 134B, C, E, F; 136B; 137B–F; 138E

yellow cuticle   169

Professor Hinton's Publications, 1930–1977

1930 Observations on two California beetles. *Pan-Pacif. Ent.* **7**, 94–95.

1933 A new *Lycostomus* from Mexico (Coleoptera: Lycidae). *Can. Ent.* **65**, 191–192.

1934 *Xenoheptaulacus*, new genus of Aphodiinae from Panama (Scarabaeidae, Coleoptera). *Ann. ent. Soc. Am.* **27**, 613–615, 1 fig.

Psephenus usingeri, n. sp. from Mexico with notes on the regional *Ps. palpalis* (Coleoptera. Psephenidae). *Ann. ent. Soc. Am.* **27**, 616–618, 2 figs.

Descriptions of new neotropical Histeridae and notes on others (Coleoptera). *Ann. Mag. nat. Hist.* **15** (10) 584–592.

A new melyrid from Mexico (Coleoptera, Melyridae). *Can. Ent.* **66**, 21–22.

Helichus puncticollis Sharp in Arizona (Dryopidae, Coleoptera). *Can. Ent.* **66**, 72.

A new name for *Ataenius consors* Fall (Scarabaeidae, Coleoptera). *Can. Ent.* **66**, 119.

Notes on *Aphodius* in the *cadaverinus* group, with a description of a new species (Coleoptera, Scarabaeidae). *Can. Ent.* **66**, 218–220.

New species of *Terapus* from North America (Histeridae, Coleoptera). *Ent. News* **45**, 270–272.

A new name for *Aphodius smithi* Brown (Coleoptera, Scarabaeidae). *Ent. News* **45**, 277.

Two coleopterous families new to Mexico. *Pan-Pacif. Ent.* **9**, 160–162.

Two genera of Aphodiinae new to Mexico (Scarabaeidae, Coleoptera. *Pan-Pacif. Ent.* **10**, 27–30, 1 fig.

A new species of West Indian *Tytthonyx* (Coleoptera; Cantharidae). *Pan-Pacif. Ent.* **10**, 30–32.

Miscellaneous studies in the Helminae (Dryopidae, Coleoptera). *Rev. Ent. Rio de J.* **4**, 192–201.

A second species of the genus *Termitodius* (Col., Scarabaeidae). *Rev. Ent. Rio de J.* **4**, 340–342, 1 fig.

New species of North American *Aphodius* (Col., Scarabaeidae). *Stylops* **3**, 188–192.

Aphodius colimaensis nom. n. (a new name for *Aphodius nubilus* Hinton). *Stylops* **3**, 200.

(Hinton H. E. and Ancona L.) Fauna de coleopteros en nidos de hormigas (*Atta*), en Mexico y Centro-America. *An. Inst. Biol. Mex.* **5**, 243–248, 2 figs.

1935 Anotaciones acerca de las costumbres micetofagicas de dos especies de *Phanaeus*. *An. Inst. Biol. Mex.* **6**, 129–130, 1 fig.

A short review of the North American species of *Pseudister* (Coleoptera, Histeridae). *Can. Ent.* **67**, 11–15, 4 figs.

Additions to the Histeridae of Lower California. *Can. Ent.* **67**, 78–82.

Notes on the American species of *Colydodes* (Coleoptera, Colydiidae). *Entomologist's mon. Mag.* **71**, 227–232, 8 figs.

New Histeridae from the nests of ants of the genus *Atta* in Mexico (Coleoptera). *Ent. News* **46**, 50–54.

Two new neotropical species of *Murmidius* (Colydiidae, Coleoptera). *Ent. News* **46**, 273–276.

1935 New species of North American *Helichus* (Dryopidae, Coleoptera). *Pan-Pacif. Ent.* **11**,
(*cont.*) 67–71, 5 figs.
New genera and species of neotropical Colydiidae, with notes on others (Col.). *Rev. Ent. Rio de J.* **5**, 202–215, 3 figs.
New American Histeridae (Col.). *Stylops* **4**, 57–65.
Notes on the Dryopoidea (Col.). *Stylops* **4**, 169–179, 2 figs.
(Hinton H. E. and Ancona L.) Fauna de coleopteros en nidos de hormigas (*Atta*), en Mexico y Centro-America. II. *An. Inst. Biol. Mex.* **6**, 307–316, 2 figs.

1936 New species of *Ataenius* allied to *A. cribrithorax* Bates (Coleoptera, Scarabaeidae). *Ann. Mag. nat. Hist.* **17** (10) 413–428, 13 figs.
Lepiceridae—a new name for the Cyathoceridae. *Lepicerinus*—a new name for the Scolytid genus *Lepicerus* Eichh. (Coleoptera). *Ann. Mag. nat. Hist.* **17** (10) 472–473.
Results of the Oxford University expedition to Borneo, 1932. Dryopidae (Coleoptera). Part I. *Ann. Mag. nat. Hist.* **18** (10) 89–109, 3 figs.
Results of the Oxford University expedition to Borneo, 1932. Dryopidae (Coleoptera). Part II. *Ann. Mag. nat. Hist.* **18** (10) 204–224, 12 figs.
New Dryopidae from the Japan Empire (Coleoptera). *Entomologist* **69**, 164–168, 2 figs.
Descriptions and figures of new Brazilian Dryopidae (Coleoptera). *Entomologist* **69**, 283–289, 6 figs.
A new genus and a new species of Elminae (Coleoptera, Dryopidae). *Entomologist's mon. Mag.* **72**, 1–5, 7 figs.
Synonymical and other notes on the Dryopidae (Coleoptera). *Entomologist's mon. Mag.* **72**, 54–58, 2 figs.
Notes on the genus *Lobogestoria* Reitt. (Coleoptera, Colydiidae). *Entomologist's mon. Mag.* **72**, 128–129, 2 figs.
Notes on some American Colydiidae (Coleoptera). *Ent. News* **47**, 185–189.
Miscellaneous studies in the neotropical Colydiidae (Col.). *Rev. Ent. Rio de J.* **6**, 47–97, 10 figs.
A new species of *Ataenius* from Mexico (Col., Scarabaeidae). *Rev. Ent. Rio de J.* **6**, 471–474, 3 figs.
Descriptions of new genera and species of Dryopidae (Coleoptera). *Trans. R. ent. Soc. Lond.* **85**, 415–434, 1 pl., 43 figs.
Notes on the biology of *Dryops luridus* Erichson (Coleoptera, Dryopidae). *Trans. Soc. Br. Ent.* **3**, 67–78, 23 figs.
Studies in the Mexican and Central American Eupariini (Coleoptera, Scarabaeidae). *Univ. Calif. Publ. Ent.* **6**, 273–276, 1 fig.

1937 *Helichus immsi*, sp. n., and notes on other North American species of the genus (Coleoptera, Dryopidae). *Ann. ent. Soc. Am.* **30**, 317–323, 1 pl., 1 fig.
New African Lavinae (Coleoptera, Dryopidae). *Ann. Mag. nat. Hist.* **19** (10) 289–304, 20 figs.
Descriptions of new American *Ataenius*, with notes on others (Coleoptera, Scarabaeidae). *Ann. Mag. nat. Hist.* **20** (10) 177–196, 44 figs.
Additions to the neotropical Dryopidae (Coleoptera). *Arb. morph. taxon. ent. Berlin* **4**, 93–111, 21 figs.
Descriptions and figures of new Peruvian Dryopidae (Coleoptera). *Entomologist* **70**, 131–138, 17 figs.
New species of *Cylloepus* from Brazil (Coleoptera, Dryopidae). *Entomologist* **70**, 279–284, 20 figs.
Descriptions of new Brazilian Dryopidae and distributional records of others. *Entomologist's mon. Mag.* **73**, 6–12, 4 figs.

Notes on some Brazilian Potamophilinae and Elminae (Coleoptera, Dryopidae). *Entomologist's mon. Mag.* **73**, 95–100, 9 figs.

Ataenius chapini, sp. n., from Mexico (Coleoptera, Scarabaeidae). *Proc. ent. Soc. Wash.* **39**, 3–8, 3 figs.

Ceradryops punctatus, new genus and species of Dryopidae from Ceylon (Col.). *Proc. ent. Soc. Wash.* **39**, 79–81, 2 figs.

On the Psephenidae collected by Dr Fritz Plaumann in Brazil (Coleoptera). *Proc. R. ent. Soc. Lond.* (B) **6**, 9–13, 13 figs.

Protoparnus pusillus, new species of Dryopidae from St. Vincent (Col.). *Rev. Ent. Rio de J.* **7**, 302–306, 8 figs.

1938 A key to the genera of the suborder Cyphophthalmi, with a description and figures of *Neogovea immsi,* gen. et sp. n. (Arachnida, Opiliones). *Ann. Mag. nat. Hist.* **2** (11) 331–338, 16 figs.

New species of neotropical Aphodiinae (Col., Scarabaeidae). *Rev. Ent. Rio de J.* **8**, 122–129, 12 figs.

1939 Notes on the Australian Mantidae (Orthoptera). *Ann. Mag. nat. Hist.* **4** (11) 282–289, 13 figs.

Descriptions and figures of new South American Dryopidae (Col.). *Ann. Mag. nat. Hist.* **4** (11) 430–439, 9 figs.

A contribution to the classification of the Limnichidae (Coleoptera). *Entomologist* **72**, 181–186, 11 figs.

Notes on American Elmidae, with descriptions of new species (Coleoptera). *Entomologist's mon. Mag.* **75**, 179–185, 6 figs.

On some new and little known South American *Neoelmis* Musgrave (Coleoptera, Elmidae). *Entomologist's mon. Mag.* **75**, 228–234, 10 figs.

A note on the genus *Austrolimnius* C. and Z., with a description of a new species from French Guiana (Coleoptera, Elmidae). *Proc. R. ent. Soc. Lond.* (B) **8**, 195–199, 8 figs.

On some new genera and species of neotropical Dryopoidea (Coleoptera). *Trans. R. ent. Soc. Lond.* **89**, 23–45, 54 figs.

An inquiry into the natural classification of the Dryopoidea, based partly on a study of their internal anatomy (Col.). *Trans. R. ent. Soc. Lond.* **89**, 133–184, 1 pl., 105 figs.

1940 A synopsis of the Brazilian species of *Neoelmis* Musgrave (Coleoptera, Elmidae). *Ann. Mag. nat. Hist.* **5** (11) 129–153, 17 figs.

On some new Brazilian *Microcylloepus,* with a key to the species (Coleoptera, Elmidae). *Ann. Mag. nat. Hist.* **6** (11) 236–248, 6 figs.

Entomological expedition to Abyssinia, 1926–27. Coleoptera, Psephenidae, Dryopidae, Elmidae. *Ann. Mag. nat. Hist.* **6** (11) 297–306, 12 figs.

A synopsis of the Bolivian species of *Cylloepus* Er. (Coleoptera, Elmidae). *Ann. Mag. nat. Hist.* **6** (11) 393–409, 12 figs.

A synopsis of the genus *Elmoparnus* Sharp (Coleoptera, Dryopidae). *Entomologist* **73**, 183–189, 11 figs.

A synopsis of the Brazilian species of *Microcylloepus* (Coleoptera, Elmidae). *Entomologist's mon. Mag.* **76**, 61–68, 6 figs.

A note on the generic name *Lara* Leconte (Col., Elmidae). *Entomologist's mon. Mag.* **76**, 116.

A monographic revision of the Mexican water beetles of the family Elmidae. *Novit. Zool.* **42**, 217–396, 401 figs.

A revision of the genus *Bufonides* Bolivar (Orthoptera, Tetrigidae). *Proc. R. ent. Soc. Lond.* (B) **9**, 30–38, 34 figs.

1940 A synopsis of the genus *Macronychus* Muller (Coleoptera, Elmidae). *Proc. R. ent. Soc.*
(*cont.*) *Lond.* (B) **9**, 113–119, 15 figs.
The Peruvian and Bolivian species of *Macrelmis* Motsch. (Coleoptera, Elmidae). *Trans. Linn. Soc. Lond.* **1** (3) 117–147, 80 figs.
A monograph of *Gyrelmis* gen. n., with a study of the variation of the internal anatomy (Coleoptera, Elmidae). *Trans. R. ent. Soc. Lond.* **90**, 375–409, 74 figs.
(Hinton H. E. and Yarrow I. H. H.) *Crickets*. Econ. Leaf. Br. Mus. (Nat. Hist.) **5**, 3 pp., 2 figs.

1941 Entomological expedition to Abyssinia, 1926–27. Coleoptera, Colydiidae. *Ann. Mag. nat. Hist.* **7** (11) 145–172, 33 figs.
The Ptinidae of economic importance. *Bull. ent. Res.* **31**, 331–381, 59 figs.
A new *Atomaria* from mushroom-beds in South Africa (Col., Cryptophagidae). *Bull. ent. Res.* **32**, 133–134, 2 figs.
Coleoptera associated with stored Nepal barley in Peru. *Bull. ent. Res.* **32**, 175–183, 7 figs.
The Lathridiidae of economic importance. *Bull. ent. Res.* **32**, 191–247, 67 figs.
A new *Teredolaemus* from New Britain (Coleoptera, Colydiidae). *Entomologist* **74**, 136–137, 2 figs.
The immature stages of *Sericoderus lateralis* (Gyllenhal) (1827) (Coleoptera, Corylophidae). *Entomologist* **74**, 198–202, 11 figs.
A change of name and a key to the British species of *Cartodere* (Coleoptera, Lathridiidae). *Entomologist* **74**, 241–243, 2 figs.
A key to the British species of *Holoparamecus* Curtis (Col., Lathridiidae). *Entomologist's mon. Mag.* **77**, 131–132.
A synopsis of the American species of *Austrolimnius* Carter (Col., Elmidae). *Entomologist's mon. Mag.* **77**, 156–163, 8 figs.
The immature stages of *Acrotrichis fascicularis* (Herbst) (Col., Ptiliidae). *Entomologist's mon. Mag.* **77**, 245–250, 9 figs.
Notes on the internal anatomy and immature stages of *Mycetophagus quadripustulatus* (Linnaeus) (Coleoptera, Mycetophagidae). *Proc. R. ent. Soc. Lond.* (A) **16**, 39–48, 21 figs.
The larva and pupa of *Tachinus subterraneus* (Linnaeus) (Coleoptera, Staphylinidae). *Proc. R. ent. Soc. Lond.* (A) **16**, 93–98, 14 figs.
New genera and species of Elmidae (Coleoptera). *Trans. R. ent. Soc. Lond.* **91**, 65–104, 50 figs.
(Hinton H. E. and Stephens F. L.) Notes on the biology and immature stages of *Cryptophagus acutangulus* Gyll. (Col., Cryptophagidae). *Bull. ent. Res.* **32**, 135–143, 10 figs.
(Hinton H. E. and Stephens F. L.) Notes on the food of *Micropeplus*, with a description of the pupa of *M. fulvus* Erichson (Coleoptera, Micropeplidae). *Proc. R. ent. Soc. Lond.* (A) **16**, 29–32, 3 figs.

1942 A revision of the Cerylonini of Borneo (Coleoptera, Colydiidae). *Ann. Mag. nat. Hist.* **9** (11) 141–173, 26 figs.
Notes on the larvae of the three common injurious species of *Ephestia* (Lepidoptera, Phycitidae). *Bull. ent. Res.* **33**, 21–25, 1 pl., 15 figs.
A new leaf-mining Nitidulid (Coleoptera). *Entomologist* **75**, 126–129, 3 figs.
Review of *Insect Pests in Stored Products*, by H. Hayhurst. *Entomologist* **75**, 207–208.
Secondary sexual characters of *Tribolium*. *Nature, Lond.* **149**, 500–501, 1 fig.
A synopsis of the Old World species of *Murmidius* Leach (Coleoptera, Colydiidae). *Proc. R. ent. Soc. Lond.* (B) **11**, 39–45.
Dermestidae in spider webs. *Proc. R. ent. Soc. Lond.* (C) **7**, 18.

1943 The larvae of the Lepidoptera associated with stored products. *Bull. ent. Res.* **34**, 163–212, 128 figs.

House moths feeding on dead insects in or near spider webs. *Entomologist* **76**, 4–5.

Observations on species of Lepidoptera infesting stored products. III. Characters distinguishing the larvae of the house moths, *Hofmannophila pseudospretella* (Staint.) and *Endrosis sarcitrella* (L.). *Entomologist* **76**, 65–67, 14 figs.

Description and figures of a new *Anthrenus* (Col., Dermestidae). *Entomologist's mon. Mag.* **79**, 14–16, 3 figs.

Notes on two species of *Atta genus* (Col., Dermestidae) recently introduced into Britain. *Entomologist's mon. Mag.* **79**, 224–227, 1 fig.

A key to the species of *Carpophilus* (Col., Nitidulidae) that have been found in Britain, with notes on some species recently introduced with stored food. *Entomologist's mon. Mag.* **79**, 275–277.

Natural reservoirs of some beetles of the family Dermestidae known to infest stored products, with notes on those found in spider webs. *Proc. R. ent. Soc. Lond.* (A) **18**, 33–42.

Stethomezium squamosum gen. et sp. n. infesting stored food in Britain, with notes on a South African Ptinid not previously recorded in stored products (Coleoptera). *Proc. R. ent. Soc. Lond.* (B) **12**, 50–54, 6 figs.

Tinea fuscipunctella Haw. as a wood-boring insect. *Proc. R. ent. Soc. Lond.* (C) **8**, 26.

(Hinton H. E. and Corbet A. S.) *Common Insect Pests of Stored Food Products. A Guide to Their Identification.* Econ. Ser. Br. Mus. (Nat. Hist.) **15**, 44 pp., 87 figs.

(Hinton H. E. and Greenslade R. M.) Observations on species of Lepidoptera infesting stored products. XI. Notes on some moths found in birds' guano. *Entomologist* **76**, 182–184, 2 figs.

1944 A new nitidulid beetle from Burma, *Carpophilus (Urophorus) prodicus* sp. n. *Entomologist* **77**, 172–173, 1 fig.

Some general remarks on sub-social beetles, with notes on the biology of the Staphylinid, *Platystethus arenarius* (Fourcroy). *Proc. R. ent. Soc. Lond.* (A) **19**, 115–128, 15 figs.

Beetles and moths recently introduced into Britain on stored products. *Proc. R. ent. Soc. Lond.* (C) **9**, 2.

The systematic position of *Pyralis glaucinalis* (L.). *Proc. R. ent. Soc. Lond.* (C) **9**, 6.

A remarkable case of parental care of eggs and young larvae amongst beetles. *Proc. R. ent. Soc. Lond.* (C) **9**, 22.

1945 A synopsis of the Brazilian species of *Cylloepus* Er. (Coleoptera, Elmidae). *Ann. Mag. nat. Hist.* **12** (11) 43–67, 16 figs.

The Histeridae associated with stored products. *Bull. ent. Res.* **35**, 309–340, 56 figs.

(In Manton S. M.) The larvae of the Ptinidae associated with stored products. With an introduction by H. E. Hinton. *Bull. ent. Res.* **35**, 341–365, 7 pls., 1 fig.

The species of *Anthrenus* that have been found in Britain, with a description of a recently introduced species (Coleoptera, Dermestidae). *Entomologist* **78**, 6–9, 6 figs.

New and little known species of *Microcylloepus* (Coleoptera, Elmidae). *Entomologist* **78**, 57–61, 1 fig.

Descriptions of two new species of *Elsianus* Sharp, with a key to the *graniger* species-group (Col., Elmidae). *Entomologist's mon. Mag.* **81**, 90–92, 5 figs.

A key to the North American species of *Terapus*, with a description of a new species (Col., Histeridae). *Proc. R. ent. Soc. Lond.* (B) **14**, 38–45, 16 figs.

Stethelmis chilensis, new genus and species of Elmidae from Chile (Coleoptera). *Proc. R. ent. Soc. Lond.* (B) **14**, 73–76, 10 figs.

1945 *A Monograph of the Beetles Associated with Stored Products*, Vol. I, London, Br. Mus. (Nat.
(*cont.*) Hist.), viii + 443 pp., 505 figs.

1946 A synopsis of the Peruvian species of *Cylloepus* Er. (Coleoptera, Elmidae). *Ann. Mag. nat. Hist.* **12** (11) 713–733, 11 figs. (1945).

On some new Indo-Australian *Sosylus*, with a key to the species (Coleoptera, Colydiidae). *Ann. Mag. nat. Hist.* **13** (11) 35–53, 5 figs.

A key to the species of *Xenelmis* Hinton, with descriptions of three new species (Col., Elmidae). *Entomologist's mon. Mag.* **83**, 237–241, 4 figs.

Concealed phases in the metamorphosis of insects. *Nature, Lond.* **157**, 552–553, 1 fig.

The "gin-traps" of some beetle pupae. *Proc. R. ent. Soc. Lond.* (C) **11**, 13.

A new classification of insect pupae. *Proc. zool. Soc. Lond.* **116**, 282–328, 64 figs.

A synopsis of the Brazilian species of *Elsianus* Sharp (Coleoptera, Elmidae). *Trans. R. ent. Soc. Lond.* **96**, 125–149, 47 figs.

On the homology and nomenclature of the setae of lepidopterous larvae, with some notes on the phylogeny of the Lepidoptera. *Trans. R. ent. Soc. Lond.* **97**, 1–37, 24 figs.

The "gin-traps" of some beetle pupae; a protective device which appears to be unknown. *Trans. R. ent. Soc. Lond.* **97**, 473–496, 27 figs.

1947 A new species of Colydiidae associated with stored products, with a key to the species of *Tyrtaeus* Champion (Coleoptera). *Ann. Mag. nat. Hist.* **13** (11) 851–856, 6 figs. (1946).

On some new and little known Indo-Australian Diaperini (Coleoptera, Tenebrionidae). *Ann. Mag. nat. Hist.* **14** (11) 81–98, 11 figs.

Notes on two beetles which may have become established in Britain and are associated with stored products. *Entomologist* **80**, 187.

Some beetles occasionally introduced into the British Isles. *Entomologist's mon. Mag.* **83**, 284–289, 1 pl.

Insect silk-spinners. *Ill. Lond. News* (June), p. 690.

The gills of some aquatic beetle pupae (Coleoptera, Psephenidae). *Proc. R. ent. Soc. Lond.* (A) **22**, 52–60, 1 pl., 12 figs.

The gills of some aquatic beetle pupae. *Proc. R. ent. Soc. Lond.* (C) **12**, 6.

The larva of *Micropteryx* Hübner (Zeugloptera). *Proc. R. ent. Soc. Lond.* (C) **12**, 46.

On the reduction of functional spiracles in the aquatic larvae of the Holometabola, with notes on the moulting process of spiracles. *Trans. R. ent. Soc. Lond.* **98**, 449–473, 5 figs.

(Hinton H. E. and McKenny-Hughes A. W.) A new pest in houses. *Mon. Bull. Min. Health* (Oct.), 173–174, 1 fig.

1948 The dorsal cranial areas of caterpillars. *Ann. Mag. nat. Hist.* **14** (11) 843–852, 6 figs.

Coleoptera: Dryopidae and Elmidae. *Expedition to South-west Arabia 1937–8*. Br. Mus. (Nat. Hist.) **1**, 133–140, 23 figs.

A synopsis of the genus *Tribolium* Macleay, with some remarks on the evolution of its species groups (Coleoptera, Tenebrionidae). *Bull. ent. Res.* **39**, 13–55, 33 figs.

Carpet Beetles. Econ. Leaf. Br. Mus. (Nat. Hist.) **8**, 3 pp., 2 figs. (Anonymous.)

On two species of *Lyphia* introduced with stored products into Britain (Coleoptera, Tenebrionidae). *Entomologist* **81**, 15–19.

Sound production in lepidopterous pupae. *Entomologist* **81**, 254–269, 8 figs.

A new use for a moth-collector's trick. *Ill. Lond. News* (May), p. 584, 2 figs.

A synopsis of the genus *Mecedanum* Erichson (Coleoptera, Colydiidae). *Novit. zool.* **42**, 475–484, 5 figs.

Sound production in the Lepidoptera. *Proc. R. ent. Soc. Lond.* (C) **13**, 22.

On the origin and function of the pupal stage. *Trans. R. ent. Soc. Lond.* **99**, 395–409, 1 fig.

1949 On the function, origin and classification of pupae. *Proc. S. Lond. ent. nat. Hist. Soc.* **1947–48**, 111–154, 39 figs.

Review of *The Songs of Insects with Related Material on the Production, Propagation, Detection, and Measurement of Sonic and Supersonic Vibrations*, by G. W. Pierce. *Entomologist* **82**, 167.

Review of *Handbooks for the Identification of British Insects*. Vol. IX. Part 1. *Diptera I: Introduction and Key to Families*, by H. Oldroyd. *Entomologist* **82**, 239–240.

(Hinton H. E. and Corbet A. S.) *Common Insect Pests of Stored Products. A Guide to Their Identification*. Econ. Ser. Br. Mus. (Nat. Hist.) **15**, 52 pp., 115 figs.

1950 Review of *La Biologie des Lépidoptères*, by P. Portier. *Entomologist* **83**, 142–144.

Aquatic Diptera collected in the River Dove near Dovedale, Derbyshire. *J. Soc. Br. Ent.* **3**, 203–206, 2 figs.

A trichopterous larva with a chelate front leg. *Proc. R. ent. Soc. Lond.* (A) **25**, 62–65, 3 figs.

1951 The Wegener–Du Toit Theory of continental displacement and the distribution of animals. *Advanc. Sci. Lond.* **8**, 74–79.

A new Cylloepus from Argentina (Coleoptera, Elmidae). *Ann. Mag. nat. Hist.* **4** (12) 820–823, 2 figs.

On a little known protective device of some Chrysomelid pupae (Coleoptera). *Proc. R. ent. Soc. Lond.* (A) **26**, 67–73, 2 figs.

New and little known adaptations to environments that are alternately dry and flooded. *Proc. R. ent. Soc. Lond.* (C) **16**, 26.

Insect distribution and the hypothesis of continental drift. *Proc. R. ent. Soc. Lond.* (C) **16**, 62.

Myrmecophilous Lycaenidae and other Lepidoptera—a summary. *Proc. S. Lond. ent. nat. Hist. Soc.* **1949–50**, 111–175, 9 figs.

The structure and function of the endocrine glands of the Lepidoptera. *Proc. S. Lond. ent. nat. Hist. Soc.* **1950–51**, 124–160, 19 figs.

A new chironomid from Africa, the larva of which can be dehydrated without injury. *Proc. zool. Soc. Lond.* **121**, 371–380, 1 pl., 5 figs.

1952 The structure of the larval prolegs of the Lepidoptera and their value in the classification of the major groups. *Lepid. News* **6**, 1–6, 4 figs.

Survival of a chironomid larva after 20 months dehydration. *Trans. 9th Int. Congr. Ent. Amsterdam 1951*, **1**, 478–482.

1953 Further experiments on a chironomid larva that can be dehydrated without injury. *Proc. R. ent. Soc. Lond.* (C) **18**, 10.

Digestion of keratin. *Sci. Progr. Lond.* **41**, 674–682.

Some adaptations of insects to environments that are alternately dry and flooded, with some notes on the habits of the Stratiomyidae. *Trans. Soc. Br. Ent.* **11**, 209–227, 3 figs.

1954 Resistance of the dry eggs of *Artemia salina* L. to high temperatures. *Ann. Mag. nat. Hist.* **7** (12) 158–160.

The initiation, maintenance, and rupture of diapause: a new theory. *Entomologist* **86**, 279–291 (1953).

A pseudoscorpion attached to an adult Tipulid. *Entomologist* **87**, 162.

On the structure and function of the respiratory horns of the pupae of the genus *Pseudolimnophila* (Diptera: Tipulidae). *Proc. R. ent. Soc. Lond.* (A) **29**, 135–140, 9 figs.

Variations in structure and function of the respiratory horns of the pupae of the Tipulidae. *Proc. R. ent. Soc. Lond.* (C) **19**, 22.

Radioactive tracers in entomological research. *Sci. Progr. Lond.* **42**, 292–305.

Insect blood. *Sci. Progr. Lond.* **42**, 684–696.

1955 The structure of the spiracular gill of the genus *Lipsothrix* (Tipulidae), with some observations on the living epithelium isolated in the gill at the pupa–adult moult. *Proc. R. ent. Soc. Lond.* (A) **30**, 1–14, 14 figs.

1955 Sound producing organs in the Lepidoptera. *Proc. R. ent. Soc. Lond.* (C) **20**, 5–6.
(*cont.*) On the respiratory adaptations, biology, and taxonomy of the Psephenidae, with notes on some related families (Coleoptera). *Proc. zool. Soc. Lond.* **125**, 543–568, 30 figs.
Caste determination in bees and termites. *Sci. Progr. Lond.* **43**, 316–326.
Insecticides and the balance of animal populations. *Sci. Progr. Lond.* **43**, 634–647.
On the structure, function, and distribution of the prolegs of the Panorpoidea, with a criticism of the Berlese–Imms theory. *Trans. R. ent. Soc. Lond.* **106**, 455–545, 1 pl., 31 figs.
On the taxonomic position of the Acrolophinae, with a description of the larva of *Acrolophus rupestris* Walsingham (Lepidoptera: Tineidae). *Trans. R. ent. Soc. Lond.* **107**, 227–231, 12 figs.
Protective devices of endopterygote pupae. *Trans. Soc. Br. Ent.* **12**, 49–92, 23 figs.
(Hinton H. E. and Corbet A. S.) *Common Insect Pests of Stored Food Products. A Guide to Their Identification.* Econ. Ser. Br. Mus. (Nat. Hist.) **15**, 61 pp., 126 figs., 3rd edn.

1956 The larvae of the species of Tineidae of economic importance. *Bull. ent. Res.* **47**, 251–346, 216 figs.
A problem in taxonomy. *Proc. R. ent. Soc. Lond.* (C) **21**, 5.
Dietary requirements of insects. Amino acids and vitamins. *Sci. Progr. Lond.* **44**, 292–309.
(Hinton H. E. and Bradley J. D.) Observations on species of Lepidoptera infesting stored products. XVI. Two new genera of clothes moths (Tineidae). *Entomologist* **89**, 42–47, 4 figs.
(Bradley J. D. and Hinton H. E.) A new genus of Tineinae (Lep., Tineidae) from North America. *Entomologist's mon. Mag.* **91**, 307–308, 2 figs.
(Diakonoff A. and Hinton H. E.) Observations on species of Lepidoptera infesting stored products. XV. On a new genus of Nemapogoninae (Tineidae). *Entomologist* **89**, 31–36, 18 figs.

1957 The structure and function of the spiracular gill of the fly *Taphrophila vitripennis*. *Proc. R. Soc.* (B) **147**, 90–120, 12 figs.
The function of the tissue isolated in the pupal gill of a Tipulid at the pupa–adult moult. *Proc. R. ent. Soc. Lond.* (C) **22**, 18.
Biological control of pests. Some considerations. *Sci. Progr. Lond.* **45**, 11–26.
Some aspects of diapause. *Sci. Progr. Lond.* **45**, 307–320.
Some little known respiratory adaptations. *Sci. Progr. Lond.* **45**, 692–700, 1 fig.
The Ptinidae of economic importance. *Bull. ent. Res.* **31**, 33–381 (1941). Translated into Chinese and published in Peking.
The larvae of the Lepidoptera associated with stored products. *Bull. ent. Res.* **34**, 163–212. Translated into Chinese and published in Peking.

1958 The phylogeny of the Panorpoid orders. *A. Rev. Ent.* **3**, 181–206.
The pupa of the fly *Simulium* feeds and spins its own cocoon. *Entomologist's mon. Mag.* **94**, 14–16, 1 fig.
The first household infestation of the Australian carpet beetle *Anthrenocerus australis* (Hope) in Britain. *Entomologist's mon. Mag.* **94**, 192.
On the nature and metamorphosis of the colour pattern of *Thaumalea* (Diptera, Thaumaleidae). *J. Insect Physiol.* **2**, 249–260, 30 figs.
On the pupa of *Spalgis lemolea* Druce (Lepidoptera, Lycaenidae). *J. Soc. Br. Ent.* **6**, 23–25, 3 figs.
The spiracular gills of insects. *Proc. 10th Int. Congr. Ent.* **1**, 543–548, 1 fig.
The pigmented tissue of the Simuliidae. *Proc. R. ent. Soc. Lond.* (C) **23**, 6–7.
Concealed phases in the metamorphosis of insects. *Sci. Progr. Lond.* **46**, 260–275, 5 figs.

1959 Origin of indirect flight muscles in primitive flies. *Nature, Lond.* **183**, 557–558, 1 fig.
Plastron respiration in the eggs of *Drosophila* and other flies. *Nature, Lond.* **184**, 280–281, 4 figs.
The function of chromatocytes in the Simuliidae, with notes on their behaviour at the pupal–adult moult. *Q. Jl microsc. Sci.* **100**, 65–71, 4 figs.
General entomology. Being a review of *A General Textbook of Entomology*, by A. D. Imms, 9th edn. *Sci. Progr. Lond.* **47**, 126–129.
How the indirect flight muscles of insects grow. *Sci. Progr. Lond.* **47**, 321–333, 17 figs.

1960 Plastron respiration in the eggs of blowflies. *J. Insect Physiol.* **4**, 176–183, 9 figs.
Cryptobiosis in the larva of *Polypedilum vanderplanki* Hint. (Chironomidae). *J. Insect Physiol.* **5**, 286–300, 4 figs.
A fly larva that tolerates dehydration and temperatures from $-270°$ to $+102°C$. *Nature, Lond.* **188**, 336–337, 1 fig.
The structure and function of the respiratory horns of the eggs of some flies. *Phil. Trans. R. Soc.* (B) **243**, 45–73, 15 figs.
The chorionic plastron and its role in the eggs of the Muscinae (Diptera). *Q. Jl microsc. Sci.* **101**, 313–332, 9 figs.
The ways in which insects change colour. *Sci. Progr. Lond.* **48**, 341–350, 15 figs.

1961 The structure and function of the egg-shell in the Nepidae (Hemiptera). *J. Insect. Physiol.* **7**, 224–257, 15 figs.
The respiratory systems of insect eggs. *Proc. R. ent. Soc. Lond.* (C) **26**, 9.
How some insects, especially the egg stages, avoid drowning when it rains. *Proc. S. Lond. ent. nat. Hist. Soc.* **1960**, 138–154, 22 figs.
The role of the epidermis in the disposition of tracheae and muscles. *Sci. Progr. Lond.* **49**, 329–339, 5 figs.

1962 A key to the eggs of the Nepidae (Hemiptera). *Proc. R. ent. Soc. Lond.* (A) **37**, 65–68, 5 figs.
The structure and function of the spiracular gills of *Deuterophlebia* (Deuterophlebiidae) in relation to those of other Diptera. *Proc. zool. Soc. Lond.* **138**, 111–122, 2 pls., 4 figs.
The structure of the shell and respiratory system of the eggs of *Helopeltis* and related genera (Hemiptera, Miridae). *Proc. zool. Soc. Lond.* **139**, 483–488, 2 figs.
The fine structure and biology of the egg-shell of the wheat bulb fly *Leptohylemyia coarctata*. *Q. Jl microsc. Sci.* **103**, 243–251, 5 figs.
Respiratory systems of insect egg-shells. *Sci. Progr. Lond.* **50**, 96–113, 23 figs.

1963 The respiratory system of the egg-shell of the blow-fly *Calliphora erythrocephala* Meig. as seen with the electron microscope. *J. Insect Physiol.* **9**, 121–129, 5 figs.
The origin and function of the pupal stage. *Proc. R. ent. Soc. Lond.* (A) **38**, 77–85, 6 figs.
The origin of flight in insects. *Proc. R. ent. Soc. Lond.* (C) **28**, 24–25, 1 fig.
Metamorphosis of the epidermis and hormone mimetic substances. *Sci. Progr. Lond.* **51**, 306–322.
The ventral ecdysial lines of the head of endopterygote larvae. *Trans. R. ent. Soc. Lond.* **115**, 39–61, 22 figs.
A Monograph of the Beetle Associated with Stored Products, Vol. I, pp. 1–443, 505 figs. Reprinted 1945 edition, Johnson Reprint Corp., New York.
(Hinton H. E. and Corbet A. S.) *Common Insect Pests of Stored Food Products. A Guide to Their Identification.* Econ. Ser. Br. Mus. (Nat. Hist.) **15**, 61 pp., 126 figs., 4th edn.

1964 The respiratory efficiency of the spiracular gill of *Simulium*. *J. Insect Physiol.* **10**, 73–80, 5 figs.
Sperm transfer in insects and the evolution of haemocoelic insemination. *Symp. R. ent. Soc. Lond.* **2**, 95–107.

1965 A revision of the Australian species of *Austrolimnius* (Coleoptera: Elmidae). *Aust. J. Zool.* **13**, 97–172, 118 figs.

The spiracular gill of the fly *Orimargula australiensis* and its relation to those of other insects. *Aust. J. Zool.* **13**, 783–800, 1 pl., 4 figs.

Polyphyletic origin of spiracular gills. *Proc. R. ent. Soc. Lond.* (C) **30**, 41.

(Hinton H. E. and Blum M. S.) Suspended animation and the origin of life. *New Scient.* **28**, 270–271, 1 fig.

(Hinton H. E. and Cole S.) The structure of the egg-shell of the cabbage root fly *Erioischia brassicae. Ann. appl. Biol.* **56**, 1–6, 1 pl., 6 figs.

1966 The spiracular gill of the fly *Eutanyderus* (Tanyderidae). *Aust. J. Zool.* **14**, 365–369, 1 pl., 2 figs.

Plastron respiration in marine insects. *Nature, Lond.* **209**, 220–221, 1 fig.

How insects adjust to changes in temperature. *Penguin Science Survey B*, pp. 88–107, 6 figs.

Respiratory adaptations of the pupae of beetles of the family Psephenidae. *Phil. Trans. R. Soc.* (B) **251**, 211–245, 22 figs.

The spiracular gill of the fly *Antocha bifida* as seen with the scanning electron microscope. *Proc. R. ent. Soc. Lond.* (A) **41**, 107–115, 1 pl., 5 figs.

(Jenkin P. M. and Hinton H. E.) Apolysis in arthropod moulting cycles. *Nature, Lond.* **211**, 871, 1 fig.

1967 The structure of the spiracles of the cattle tick *Boophilus microplus. Aust. J. Zool.* **15**, 941–945, 2 pls., 1 fig.

Structure and ecdysial process of the larval spiracles of the Scarabaeoidea, with special reference to those of *Lepidoderma. Aust. J. Zool.* **15**, 947–953, 4 pls., 1 fig.

On the spiracles of the larvae of the suborder Myxophaga (Coleoptera). *Aust. J. Zool.* **15**, 955–959, 1 pl., 1 fig.

The respiratory system of the egg-shell of the common housefly. *J. Insect Physiol.* **13**, 647–651, 18 figs.

Plastron respiration in the marine fly *Canace. J. mar. biol. Ass. UK* **47**, 319–327, 3 pls., 3 figs.

Spiracular gills in the marine fly *Aphrosylus* and their relation to the respiratory horns of other Dolichopodidae. *J. mar. biol. Ass. UK* **47**, 485–497, 5 pls., 3 figs.

Structure of the plastron in *Lipsothrix*, and the polyphyletic origin of plastron respiration in Tipulidae. *Proc. R. ent. Soc. Lond.* (A) **42**, 35–38, 1 pl., 1 fig.

Convergent evolution of respiratory structures of insects and mites. *Proc. R. ent. Soc. Lond.* (C) **32**, 13.

(Hinton H. E. and Dunn A. M. S.) *Mongooses, their Natural History and Behaviour*, Oliver & Boyd, Edinburgh and London, vii + 144, 16 pls., 26 figs. American edition: University of California Press, 1967.

1968 Suspension reversible del metabolismo, con especial referencia a los insectos. *Acta politec. méx.* **8**, 121–140.

Spiracular gills. *Adv. Insect Physiol.* **5**, 65–162, 86 figs.

Observations on the biology and taxonomy of the eggs of *Anopheles* mosquitoes. *Bull. ent. Res.* **57**, 495–508, 6 pls.

Professor J. E. Harris. *Nature, Lond.* **220**, 626–627.

Structure and protective devices of the egg of the mosquito *Culex pipiens. J. Insect Physiol.* **14**, 145–161, 24 figs.

The subgenera of *Austrolimnius* (Coleoptera, Elminthidae). *Proc. R. ent. Soc. Lond.* (B) **37**, 98–102, 1 pl.

Reversible suspension of metabolism and the origin of life. *Proc. R. Soc.* (B) **171**, 43–56,

2 figs.
(Rothschild M. and Hinton H. E.) Holding organs on the antennae of male fleas. *Proc. R. ent. Soc. Lond.* (A) **43**, 105–107, 6 pls.
(Thompson T. E. and Hinton H. E.) Stereoscan electron microscope observations on opisthobranch radulae and shell sculpture. *Bijdr. Dierk.* **38**, 91–92, 4 pls.

1969 Respiratory systems of insect egg shells. *A. Rev. Ent.* **14**, 343–368, 6 pls., 1 fig.
Diffraction gratings in burying beetles (*Nicrophorus*). *Entomologist* **102**, 185–189, 1 pl.
Structure of the plastron of *Hexacylloepus*, with a description of a new species (Coleoptera, Elminthidae). *J. nat. Hist.* **3**, 125–130, 3 pls., 1 fig.
Plastron respiration in adult beetles of the suborder Myxophaga. *J. Zool. Lond.* **159**, 131–137, 3 pls., 4 figs.
Report on a Visit to Bulgaria from 17 April to 1 May 1968 under the Agreement on Exchange Visits of Scientists between the Academy of Sciences of the Socialist Republic of Bulgaria and the Royal Society, 4 pp., Royal Society, London.
(Hinton H. E., Gibbs D. F., and Silberglied R.) Stridulatory files as diffraction gratings in mutillid wasps. *J. Insect Physiol.* **15**, 549–552, 3 pls.
(Hinton H. E. and Gibbs D. F.) An electron microscope study of the diffraction gratings of some carabid beetles. *J. Insect Physiol.* **15**, 959–962, 5 pls.
(Hinton H. E. and Gibbs D. F.) Diffraction gratings in Phalacrid beetles. *Nature, Lond.* **221**, 953–954, 2 figs.

1970 Algunas pequeñas estructuras de insectos observadas con microscopico electronico explorador. *Acta politec. méx.* **10**, 181–201, 6 pls.
The zoological results of Gy. Topal's collectings in South Argentine. 21. A second species of *Stethelmis* (Coleoptera: Elminthidae). *Acta zool. hung.* **16**, 109–113, 2 pls.
Discovery of *Hydroscapha* in Bulgaria (Coleoptera, Myxophaga). *Izv. zool. Inst., Sof..* **30**, 153–157, 2 pls.
Functional morphology of mosquito eggs. *J. Parasit.* **56**, 147.
Some structures of insects as seen with the scanning electron microscope. *Micron* **1**, 84–108, 6 pls.
Insect eggshells. *Scient. Am.* **223** (2) 84–91, 7 pls., 4 figs.
Some little known surface structures. *Symp. R. ent. Soc. Lond.* **5**, 41–58, 4 pls., 1 fig.
Report on a Visit to Mexico City under the Latin American Programme of the Royal Society 6 pp., Royal Society, London.
(Hinton H. E. and Mackerras I. M.) Reproduction and metamorphosis. In *The Insects of Australia*, pp. 83–106, 18 figs., Melbourne University Press.
(Hinton H. E. and Service M. W.) The surface structure of aedine eggs as seen with the scanning electron microscope. *Ann. trop. Med. Parasit.* **63**, 409–411, 2 pls.

1971 Origin and significance of stridulatory behaviour in beetles. *Acta cien. venez.* **22** (Suppl. 2), 112–114, 5 pls.
The Elmidae (Coleoptera) of Trinidad and Tobago. *Bull. Br. Mus. nat. Hist. (Ent.)* **26**, 245–265, 9 pls., 17 figs.
The species of *Dryopomorphus* (Coleoptera, Elmidae). *Entomologist* **104**, 293–297, 8 figs.
Some American *Austrolimnius* (Coleoptera: Elmidae). *J. Ent.* (B) **40**, 93–99, 1 pl., 3 figs.
Plastron respiration in the mite *Platyseius italicus*. *J. Insect Physiol.* **17**, 1185–1199, 14 figs.
Reversible suspension of metabolism. *Proc. 2nd Int. Conf. Theoretical Physics Biol., Versailles*, pp. 69–89.
Some neglected phases in metamorphosis. *Proc. R. ent. Soc. Lond.* (C) **35**, 55–64, 6 figs.
A revision of the genus *Hintonelmis* Spangler (Coleoptera: Elmidae). *Trans. R. ent. Soc. Lond.* **123**, 189–208, 3 pls., 10 figs.
Polyphyletic evolution of respiratory systems of eggshells, with a discussion of structure

1971
(cont.) and density-independent and density-dependent selective pressures. In *Scanning Electron Microscopy* (V. H. Heywood, ed.), pp. 17–36, 5 pls., 2 figs., Academic Press, London.
(Hinton H. E. and Gibbs D. F.) Diffraction gratings in gyrinid beetles. *J. Insect Physiol.* **17**, 1023–1035, 21 figs.
(Hinton H. E. and Wilson R. S.) Stridulatory organs in spiny orb-weaver spiders. *J. Zool. Lond.* **162**, 481–484, 1 pl., 2 figs.

1972 Physiological colour changes in the elytra of beetles. *Abstracts 14th Int. Congr. Ent.*, p. 127.
Sperm transfer in insects of medical and veterinary importance. *Abstracts 14th Int. Congr. Ent.*, pp. 281–282.
Hallazgo de un nuevo *Austrolimnius* en Guerrero, Mexico (Col., Elmidae). *Ciencia, Mex.* **27**, 135–137, 1 pl.
Two new genera of South American Elmidae (Coleoptera). *Coleopts Bull.* **26**, 37–41, 1 pl., 6 figs.
Pilielmis, a new genus of Elmidae (Coleoptera). *Entomologist's mon. Mag.* **107**, 161–166, 2 pls., 6 figs.
Electron microscope studies on the surface pattern of insects and their relation to a precise definition of developmental stages. *Folia ent. Mex.* **23–24**, 104–105.
The Venezuelan species of *Neoelmis* (Coleoptera, Elmidae). *J. Ent.* (B) **41**, 133–144, 1 pl., 5 figs.
New species of *Neoelmis* from South America (Coleoptera, Elmidae). *Papeis Dep. Zool. S. Paulo* **26**, 117–135, 2 pls., 14 figs.
The origin and function of the pupal stage. *Readings in Entomology*, pp. 72–80, 6 figs. (P. Barbosa and T. M. Peters, eds.), W. B. Saunders Co., Philadelphia.
(Hinton H. E. and Corbet A. S.) *Common Insect Pests of Stored Food Products. A Guide to Their Identification.* Econ. Ser. Br. Mus. (Nat. Hist.) **15**, 62 pp., 126 figs., 5th edn.
(Hinton H. E. and Jarman G. M.) Physiological colour change in the hercules beetle. *Nature, Lond.* **238**, 160–161.
(Ghiradella H., Aneshansley D., Eisner T., Silberglied R. E., and Hinton H. E.) Ultraviolet reflection of a male butterfly: interference color caused by thin-layer elaboration of wing scales (*Euremalisa*: Lep. Pieridae). *Science, Wash.* **178**, 1214–1216, 4 figs.

1973 New genera and species of Bolivian Elmidae (Coleoptera). *Coleopts Bull.* **27**, 1–6, 7 figs.
The Venezuelan species of *Hexacylloepus* (Col., Elmidae). *Entomologist's mon. Mag.* **108**, 251–256, 1 pl., 5 figs.
Neglected phases in metamorphosis: a reply to V. B. Wigglesworth. *J. Ent.* (A) **48**, 57–68, 1 fig.
Some recent work on the colours of insects and their likely significance. *Proc. Br. ent. nat. Hist. Soc.* **6**, 43–54, 9 figs.
Reversible suspension of metabolism and the origin of life. In *Anhydrobiosis* (J. H. Crowe and J. S. Clegg, eds.), pp. 65–78, 2 figs., Dowden, Hutchinson, and Ross, Pennsylvania.
Natural deception. In *Illusion in Nature and Art* (R. L. Gregory and E. H. Gombrich, eds.), pp. 96–159, 4 pls., 30 figs., Duckworth, London.
(Hinton H. E. and Jarman G. M.) Physiological colour change in the elytra of the hercules beetle *Dynastes hercules. J. Insect Physiol.* **19**, 533–549, 16 figs.

1974 Lycaenid pupae that mimic anthropoid heads. *J. Ent.* (A) **49**, 65–69, 2 pls.
Accessory functions of seminal fluid. *J. med. Ent.* **11**, 19–25.
Reproduction. In *The Insects of Australia*: Supplement, Chap. 4, 18–21, 2 figs., CSIRO, Melbourne University Press.

(Jarman G. M. and Hinton H. E.) Some defence mechanisms of the hercules beetle *Dynastes hercules. J. Ent.* (A) **49**, 71–80, 9 figs.

1975 (Hinton H. E. and Corbet A. S.) *Common Insect Pests of Stored Food Products. A Guide to Their Identification.* Econ. Ser. Br. Mus. (Nat. Hist.) **15**, 62 pp., 126 figs. Reprint 5th edn.

1976 Notes on neglected phases in metamorphosis, and a reply to J. M. Whitten. *Ann. ent. Soc.. Am.* **69**, 560–566.

Plastron respiration in bugs and beetles. *J. Insect Physiol.* **22**, 1529–1550.

The fine structure of the pupal plastron of simuliid flies. *J. Insect Physiol.* **22**, 1061–1070.

Possible significance of the red patches of the female crab-spider *Misumena vatia. J. Zool. Lond.* **180**, 35–39.

Maternal care in the Membracidae. *Proc. R. ent. Soc. Lond.* (C) **40**, 33.

Hedyselmis, a new genus of Elmidae (Coleoptera) from Malaysia. *Systematic Entomol.* **1**, 259–261.

Colour changes. In *Environmental Physiology of Animals* (J. Bligh, J. L. Cloudsley-Thompson, and A. C. Macdonald, eds.), pp. 389–412, Blackwell Scientific Publications.

Respiratory adaptations of marine insects. In *Marine Insects* (L. Cheng, ed.), pp. 43–78, North-Holland.

Recent work on physical colours of insect cuticle. In *The Insect Integument* (H. R. Hepburn, ed.), pp. 475–496, Elsevier.

(Hinton H. E. and Jarman G. M.) A diffusion equation for tapered plastrons. *J. Insect Physiol.* **22**, 1263–1265.

1977 Subsocial behaviour and biology of some Mexican membracid bugs. *Ecol. Ent.* **2**, 61–79.

Mimicry provides information about the perceptual capacities of predators. *Folio ent. mex.* **37**, 19–29.

Function of shell structures of pig louse and how egg maintains a low equilibrium temperature in direct sunlight. *J. Insect Physiol.* **23**, 785–800.

Enabling mechanisms. *Proc. 15th Int. Congr. Ent. Washington 1976*, 71–83.

(Hinton H. E. and Reichardt H.) On the new world beetles of the family Hydroscaphidae. *Papeis Avulsos de Zoologia* **30**, 1–24.

(Pope R. D. and Hinton H. E.) A preliminary survey of ultraviolet reflectance in beetles. *Biol. J. Linn. Soc.* **9**, 331–348.